Lecture Notes in Biomathematics

Managing Editor: S. Levin

38

Biological Growth and Spread

Mathematical Theories and Applications
Proceedings, Heidelberg 1979

Edited by
Willi Jäger, Hermann Rost, and Petre Tautu

Springer-Verlag
Berlin Heidelberg New York

Lecture Notes in Biomathematics

Vol. 1: P. Waltman, Deterministic Threshold Models in the Theory of Epidemics. V, 101 pages. 1974.

Vol. 2: Mathematical Problems in Biology, Victoria Conference 1973. Edited by P. van den Driessche. VI, 280 pages. 1974.

Vol. 3: D. Ludwig, Stochastic Population Theories. VI, 108 pages. 1974.

Vol. 4: Physics and Mathematics of the Nervous System. Edited by M. Conrad, W. Güttinger, and M. Dal Cin. XI, 584 pages. 1974.

Vol. 5: Mathematical Analysis of Decision Problems in Ecology. Proceedings 1973. Edited by A. Charnes and W. R. Lynn. VIII, 421 pages. 1975.

Vol. 6: H. T. Banks, Modeling and Control in the Biomedical Sciences. V, 114 pages. 1975.

Vol. 7: M. C. Mackey, Ion Transport through Biological Membranes, An Integrated Theoretical Approach. IX, 240 pages. 1975.

Vol. 8: C. DeLisi, Antigen Antibody Interactions. IV, 142 pages. 1976.

Vol. 9: N. Dubin, A Stochastic Model for Immunological Feedback in Carcinogenesis: Analysis and Approximations. XIII, 163 pages. 1976.

Vol. 10: J. J. Tyson, The Belousov-Zhabotinskii Reaktion. IX, 128 pages. 1976.

Vol. 11: Mathematical Models in Medicine. Workshop 1976. Edited by J. Berger, W. Bühler, R. Repges, and P. Tautu. XII, 281 pages. 1976.

Vol. 12: A. V. Holden, Models of the Stochastic Activity of Neurones. VII, 368 pages. 1976.

Vol. 13: Mathematical Models in Biological Discovery. Edited by D. L. Solomon and C. Walter. VI, 240 pages. 1977.

Vol. 14: L. M. Ricciardi, Diffusion Processes and Related Topics in Biology. VI, 200 pages. 1977.

Vol. 15: Th. Nagylaki, Selection in One- and Two-Locus Systems. VIII, 208 pages. 1977.

Vol. 16: G. Sampath, S. K. Srinivasan, Stochastic Models for Spike Trains of Single Neurons. VIII, 188 pages. 1977.

Vol. 17: T. Maruyama, Stochastic Problems in Population Genetics. VIII, 245 pages. 1977.

Vol. 18: Mathematics and the Life Sciences. Proceedings 1975. Edited by D. E. Matthews. VII, 385 pages. 1977.

Vol. 19: Measuring Selection in Natural Populations. Edited by F. B. Christiansen and T. M. Fenchel. XXXI, 564 pages. 1977.

Vol. 20: J. M. Cushing, Integrodifferential Equations and Delay Models in Population Dynamics. VI, 196 pages. 1977.

Vol. 21: Theoretical Approaches to Complex Systems. Proceedings 1977. Edited by R. Heim and G. Palm. VI, 244 pages. 1978.

Vol. 22: F. M. Scudo and J. R. Ziegler, The Golden Age of Theoretical Ecology: 1923–1940. XII, 490 pages. 1978.

Vol. 23: Geometrical Probability and Biological Structures: Buffon's 200th Anniversary. Proceedings 1977. Edited by R. E. Miles and J. Serra. XII, 338 pages. 1978.

Vol. 24: F. L. Bookstein, The Measurement of Biological Shape and Shape Change. VIII, 191 pages. 1978.

Vol. 25: P. Yodzis, Competition for Space and the Structure of Ecological Communities. VI, 191 pages. 1978.

Lecture Notes in Biomathematics

Managing Editor: S. Levin

38

Biological Growth and Spread

Mathematical Theories and Applications
Proceedings of a Conference Held at Heidelberg
July 16 – 21, 1979

Edited by
Willi Jäger, Hermann Rost, and Petre Tautu

Springer-Verlag
Berlin Heidelberg New York 1980

Editors

Willi Jäger
Hermann Rost
Institut für Angewandte Mathematik
Universität Heidelberg
Im Neuenheimer Feld 294
D-6900 Heidelberg 1
Federal Republic of Germany

Petre Tautu
Institut für Dokumentation, Information und Statistik
Deutsches Krebsforschungszentrum
Im Neuenheimer Feld 280
D-6900 Heidelberg 1
Federal Republic of Germany

AMS Subject Classifications (1980): 34 Kxx, 35-xx, 45-xx, 60 Fxx, 60 Gxx, 60 Hxx, 60 Jxx, 60 K35, 60 K99, 92-06

ISBN 3-540-10257-4 Springer-Verlag Berlin Heidelberg New York
ISBN 0-387-10257-4 Springer-Verlag New York Heidelberg Berlin

Printed in Germany

Printing and binding: Beltz Offsetdruck, Hemsbach/Bergstr.
2141/3140-543210

PREFACE

These Proceedings have been assembled from papers presented at the Conference on Models of Biological Growth and Spread, held at the German Cancer Research Centre Heidelberg and at the Institute of Applied Mathematics of the University of Heidelberg, July 16-21, 1979. The main theme of the conference was the mathematical representation of biological populations with an underlying spatial structure. An important feature of such populations is that they and/or their individual components may interact with each other. Such interactions may be due to external disturbances, internal regulatory factors or a combination of both. Many biological phenomena and processes including embryogenesis, cell growth, chemotaxis, cell adhesion, carcinogenesis, and the spread of an epidemic or of an advantageous gene can be studied in this context. Thus, problems of particular importance in medicine (human and veterinary), agriculture, ecology, etc. may be taken into consideration and a deeper insight gained by utilizing (more) realistic mathematical models.

Since the intrinsic biological mechanisms may differ considerably from each other, a great variety of mathematical approaches, theories and techniques is required. The aims of the conference were

(i) To provide an overview of the most important biological aspects.

(ii) To survey and analyse possible stochastic and deterministic approaches.

(iii) To encourage new research by bringing together mathematicians interested in problems of a biological nature and scientists actively engaged in developing mathematical models in biology.

The conference was sponsored in part by the German Cancer Research Centre Heidelberg and chiefly by the Deutsche Forschungsgemeinschaft:

this was one of the meetings organized by the Sonderforschungsbereich 123, 'Stochastic Mathematical Models'.

The Organizing Committee consisted of W.J.Bühler (University of Mainz), K.Dietz and K.P.Hadeler (University of Tübingen), W.Jäger and H.Rost (University of Heidelberg), P.Tautu (German Cancer Research Centre Heidelberg). There were almost 200 registrants from 16 countries.

During five days fifty-one invited papers were presented which included ten survey lectures. Each biological survey lecture was followed by a mathematical one. The necessity to improve communication between mathematicians and biologists imposed this successful experiment. In fact, social intercourse was always necessary for mathematics. It was in a recent article (1978) by A.L.Hammond ("Mathematics - our invisible culture") where we found the remark that, after all, the idea of a proof must have developed from the idea of discourse. To prove something was to convince somebody. The permanent dialogue between mathematicians and biologists will certainly lead to a creative interaction between the experiments, observations and biological theories and the concepts, intuitions and techniques of mathematics.

The order of papers given at the conference was different from that presented here. We have grouped them into six topics as follows:

I. Proliferation, spread and reaction-dispersal processes;

II. Random systems with locally interacting objects;

III. Spatial models in epidemiology and genetics;

IV. Models for cell motility;

V. Stochastic versus deterministic approaches;

VI. Further mathematical approaches and techniques.

We would like to thank all participants and collaborators for their enthusiastic co-operation and support. It is our hope that in particular young researchers will be inspired by these Proceedings to enter the field of mathematically modeling biological phenomena and to help solve some of the many challenging problems remaining. They should read them, as Francis Bacon recommended, "not to contradict and confute, nor to believe and take for granted, nor to find talk and discourse, but to weigh and consider".

Heidelberg, July 1980 The Editors

LIST OF AUTHORS

ALT, W.; Universität Heidelberg, Institut für Angewandte Mathematik, Im Neuenheimer Feld 294, D-6900 Heidelberg

ALTENBURG, H.-P.; Fakultät für klinische Medizin Mannheim, Med. Statistik / Biomathematik, Theodor-Kutzer-Ufer, D-6800 Mannheim 1

BAILEY, N.T.J.; World Health Organisation, Health Statistical Methodology, CH-1211 Geneva

BARBOUR, A.D.; University of Cambridge, Gonville and Caius College, Cambridge CB2 1TA, England

BARTOSZYŃSKI, R.; Polish Academy of Sciences, Institute of Mathematics, Sniadeckich 8, PL-00-950 Warszawa

BASERGA, R.; Temple University School of Medicine, Department of Pathology, 3400 N.Broad Street, Philadelphia, Pennsylvania 19140, USA

BELL, G.I.; University of California, Los Alamos Scientific Laboratory, Los Alamos, New Mexico 87545, USA

BERG, H.C.; University of Colorado, Department of Molecular, Cellular and Developmental Biology, Boulder, Colorado 80309, USA

BIGGINS, J.D.; Sheffield University, Department of Probability and Statistics, Sheffield S3 7RH, England

BÖGEL, K.; World Health Organization, Veterinary Public Health, CH-1211 Geneva

BRAMSON, M.; New York University, Courant Institute of Mathematical Sciences, 251 Mercer Street, New York, N.Y. 10012, USA

BÜHLER, W.J.; Universität Mainz, Fachbereich Mathematik, Saarstr. 21, D-6500 Mainz

CONLEY, C.; University of Wisconsin - Madison, Department of Mathematics, Madison, Wisconsin 53706, USA

DIEKMANN, O.; Mathematisch Centrum, Tweede Boerhaavestraat 49, NL-1091 AL Amsterdam

DOSS, H.; Université de Paris VI, Laboratoire de Calcul des Probabilités, 4 Place Jussieu - Tour 56, F-75230 Paris Cedex 05

FIFE, P.; University of Arizona, Department of Mathematics, Tucson, Arizona 85721, USA

FÖLLMER, H.; Eidgenössische Technische Hochschule Zürich, Mathematikdepartement, ETH Zentrum, CH-8092 Zürich

GRIFFEATH, D.; University of Wisconsin - Madison, Department of Mathematics, Madison, Wisconsin 53706, USA

HAMMERSLEY, J.M.; University of Oxford, Institute of Economics and Statistics, St.Cross Building - Manor Road, Oxford OX1 3UL, England

HOPPENSTEADT, F.C.; University of Utah, Department of Mathematics, Salt Lake City, Utah 84112, USA

JÄGER, W.; Universität Heidelberg, Institut für Angewandte Mathematik, Im Neuenheimer Feld 294, D-6900 Heidelberg

KELLER, E.F.; State University of New York, College at Purchase, Division of Natural Science, Purchase, N.Y. 10577, USA

KESTEN, H.; Cornell University, Department of Mathematics, Ithaca, N.Y. 14853, USA

KURTZ, T.G.; University of Wisconsin - Madison, Department of Mathematics, Madison, Wisconsin 53706, USA

LAPIDUS, I.R.; Stevens Institute of Technology, Department of Physics/Engineering Physics, Hoboken, New Jersey 07030, USA

LAUFFENBURGER, D.; University of Pennsylvania, Department of Chemical & Biochemical Engineering, 220 South 33rd Street - Towne Building, Philadelphia, Pennsylvania 19104, USA

LEVANDOWSKY, M; Pace University, Haskins Laboratories, 41 Park Row, New York, N.Y. 10038, USA

LIGGETT, T.M.; University of California - Los Angeles, Department of Mathematics, Los Angeles, California 90024, USA

LITTLE, W.A.; Stanford University, Department of Physics, Stanford, California 94305, USA

LORD, B.I.; Christie Hospital & Holt Radium Institute, Paterson Laboratories, Manchester M20 9BX, England

NAGASAWA, M.; Universität Zürich, Seminar für Angewandte Mathematik, Freiestrasse 36, CH-8032 Zürich

NGUYEN, X.X.; Universität Heidelberg, Institut für Angewandte Mathematik, Im Neuenheimer Feld 294, D-6900 Heidelberg

NOSSAL, R.; National Institutes of Health, Bldg 12A/Rm 2007, Bethesda, Maryland 20205, USA

PAPANICOLAOU, G.C.; New York University, Courant Institute, 251 Mercer Street, New York, N.Y. 10012, USA

PELETIER, L.A.; Delft University of Technology, Department of Mathematics, NL - Delft

POTTEN, C.S.; Christie Hospital & Holt Radium Institute, Paterson Laboratories, Manchester M20 9BX, England

RITTGEN, W.; Deutsches Krebsforschungszentrum, Institut für Dokumentation, Information und Statistik, Im Neuenheimer Feld 280, D-6900 Heidelberg

ROBERTSON, A.; Biological Research Corporation, Lexington, Georgia 30648, USA

ROTHE, F.; Universität Tübingen, Institut für Biologie 1, Lehrstuhl für Biomathematik, Auf der Morgenstelle 28, D-7400 Tübingen 1

SAWYER, S.; Purdue University, Department of Mathematics, West Lafayette, Indiana 47907, USA

SCHOFIELD, R.; Christie Hospital & Holt Radium Institute, Paterson Laboratories, Manchester M20 9BX, England

SCHUMACHER, K.; Universität Tübingen, Institut für Biologie 2, Lehrstuhl für Biomathematik, Auf der Morgenstelle 28, D-7400 Tübingen 1

SCHÜRGER, K.; Universität Bonn, Fakultät für Gesellschafts- und Wirtschaftswissenschaften, Statistische Abteilung, D-5300 Bonn

SMOLLER, J.; University of Michigan, Department of Mathematics, Ann Arbor, Michigan 48109, USA

SMYTHE, R.T.; University of Oregon, Department of Mathematics, Eugene, Oregon 97403, USA

TAUTU, P.; Deutsches Krebsforschungszentrum, Institut für Dokumentation, Information und Statistik, Im Neuenheimer Feld 280, D-6900 Heidelberg

THIEME, H.R.; Universität Heidelberg, Institut für Angewandte Mathematik, Im Neuenheimer Feld 294, D-6900 Heidelberg

TUCKWELL, H.; University of British Columbia, Department of Mathematics, Vancouver, B.C. Canada V6T 1W5

TURELLI, M.; University of California - Davis, Department of Genetics, Davis, California 95616, USA

WEINBERGER, H.F.; University of Minnesota, School of Mathematics, 206 Church Street S.E., Minneapolis, Minnesota 55455, USA

WOFSY, C. (LIPOW); New Mexico Health and Environment Department, Environmental Evaluation Group, 320 E.Mercy Street, Santa Fe, New Mexico 87503, USA

TABLE OF CONTENTS

IIb. MODELS

TOPIC III. SPATIAL MODELS IN EPIDEMIOLOGY AND GENETICS

TOPIC IV. MODELS FOR CELL MOTILITY

TOPIC V. STOCHASTIC VERSUS DETERMINISTIC APPROACHES

TOPIC VI. FURTHER MATHEMATICAL APPROACHES AND TECHNIQUES

TOPIC I

Proliferation, spread and reaction-dispersal processes

Discussion by the Editors

It was already said by C.H.Waddington (1970) that of all the major classical problems of biology the ones we understand least are those in the area of 'developmental biology'. In fact - he said - from the four main classical concepts in this domain (namely growth, development, differentiation and morphogenesis) none is suitable for today's use in statements which aim for a reasonable degree of precision. Considerable effort must be invested in this field, and it is a hope that the papers presented in these Proceedings will contribute to a better understanding of one or two essential questions in cell biology. Perhaps the main problem in this section is the proliferation depending on size, density or environment. This problem should find a certain equivalent in the biological distinction (H.H.Pattee) between constraints, rules and regulation.

The papers in this section deal with the biological aspects of proliferation of cell populations and the mathematical analysis of proliferation, spread and pattern formation. The biological contribution is composed of three important papers whose content suggests new mathematical models. C.S.Potten presents theoretical models for proliferation in surface epithelia; his work is in a form suitable for mathematical interpretation. In fact, Potten's representation of epithelial proliferative units suggests an analysis in terms of Dirichlet (Voronoi) tessellations. In the paper submitted by B.J.Lord and R.Schofield, the spatial distribution of cell types in the haematopoietic tissue is described in considerable detail. The movement of the stem cells in the bone marrow and particularly the age hierarchy in the stem cell population are new and essential observations. R.Baserga presented evidence that there might be five genes and five gene products that regulate five different steps in the G1 phase; these steps could be essential for the entry of cells into the S phase. His observation can be a new experimental argument against the 'transition probability' model which convinced some scientists by its simplicity.

From the ten remaining mathematical papers, five deal with the dependency hypothesis. The survey lecture given by H.Kesten presents some non-Markovian processes in random environments, where the environment is indexed by time (nonhomogeneous multitype branching processes, selection processes with random fitness coefficients, Lotka-Volterra models with random fluctuations) or by space (random walks and diffusions, random velocities and acceleration). The inter-specific competition with demographic and environmental "stochasticity" is studied by M.Turelli, and the effect of random "disasters" originating in environment is analyzed by H.Tuckwell. These disasters are either constant, size-proportional or density-independent, and the growth processes considered are logistic or Gompertz. The convergence of 'controlled' branching processes to diffusion processes and conditions for positive recurrence are investigated by Carla (Lipow) Wofsy and W.Rittgen, respectively. A.D.Barbour discusses the approximation of density-dependent Markov population processes by utilizing the theory of partial sums of independent random variables, and J.D. Biggins studies the spread of branching random walks. The model constructed by R.Bartoszyński intends to explain two phenomena involving mitotic cells, namely the cell loss and the variability of the inter-mitotic times.

Two papers deal with pattern formation models. Based on experimental data and suggesting new experiments, F.C.Hoppensteadt and W.Jäger examine a reaction-diffusion model for the dispersal of histidine auxotrophic S. Typhimurium; more theoretically, M.Nagasawa investigates a segregation model which is an environment-dependent diffusion process.

Two papers presented at the conference were published elsewhere: the contribution of A.Winfree constitutes a chapter in his book "The Geometry of Biological Time" (Springer, 1979) and that of G.Oster is a research announcement (jointly with G.Odell, B.Burnside and P. Alberch) published in J.Math.Biology (1980).

GENETIC BASIS OF BIOLOGICAL GROWTH

Renato Baserga

Fels Research Institute and
Department of Pathology
Temple University Medical School
Philadelphia, Pa. U.S.A.

ABSTRACT

An increase in the number of cells is one of the fundamental characteristics of biological growth. An increase in cell number depends on cell division and this, in turn, on the ability of the cell to replicate its genetic material. We have therefore studied the $G_0 \rightarrow S$ transition in the cell cycle, which is the phase of the cell cycle that determines whether or not a cell will replicate DNA and its chromosomal proteins. We have assumed that the key to our understanding of cell proliferation lies in the identification of genes and gene products that control it. Using cell cycle-specific temperature sensitive mutants and virally-coded proteins from DNA oncogenic viruses, we have tentatively identified five genes that control the $G_0 \rightarrow S$ transition of mammalian cells in culture. The experiments demonstrating that ribosomal RNA genes are one set of controlling genes are discussed in detail.

It is the general consensus among investigators that the critical period in the cell cycle for the entry of cells into S, and therefore for the control of proliferation in eukaryotic cells, is in the G_1 phase (for reviews see 3, 23). While many biochemical events have been described in G_1 cells and in the prereplicative phase of G_0 cells stimulated to proliferate (see above reviews), our recent approach has been to identify and purify some genes and gene products that regulate the transition of mammalian cells from a resting to a growing stage. To do this, we use two different systems which have been combined into one single model; namely, the use of cell cycle-specific temperature-sensitive mutants and the use of small DNA oncogenic viruses. Cell cycle-specific temperature-sensitive mutants are defined as mutants that arrest at the nonpermissive temperature in a specific phase of the cell cycle (6). The ts mutants of the cell cycle that we have been studying are mutants like tsAF8 cells that arrest at the nonpermissive temperature in the G_1 phase (5, 2). As to the DNA oncogenic viruses, it has been known for a number of years that SV40,

Polyoma and Adenoviruses can induce the $G_0 \rightarrow S$ transition in resting cells (10, 30, 35, 41, 45). Furthermore, at least in the case of SV40, the viral gene product required for the $G_0 \rightarrow S$ transition has been identified (36, 38, 42, and see reviews 1, 16) with the virally-coded protein, the large T antigen that has all the necessary information to induce host DNA synthesis in resting cells. This evidence stems partly from genetic studies with temperature-sensitive and deletion mutants of SV40 (25, 32, 37), and more formally on the demonstration that T antigen manually microinjected into resting mouse kidney cells causes these cells to enter S (38). The present paper describes and summarizes our initial studies using DNA oncogenic viruses to elucidate the mechanisms that regulate the transition of cells from the resting to the growing stage.

rRNA synthesis in cells stimulated to proliferate. When resting cells are stimulated to proliferate, RNA synthesis promply increases within the first 30 to 60 minutes after stimulation, although the cells do not enter the S phase until several hours later. This increase in RNA synthesis has been demonstrated both in vivo and in vitro (for a review see 3) by a variety of methods. Most of the increase is due to an increase in the synthesis of ribosomal RNA (rRNA) as evidenced by a number of studies (18, 19, 27, 29, 31, 39, 44). With this background it seemed natural to us to ask the question whether the SV40 T antigen may not act on nucleolar genes stimulating rRNA synthesis. Previous reports in the literature had already indicated that infection with either SV40 or Polyoma can increase RNA synthesis (4) or RNA accumulation in the cytoplasm of infected cells (21, 42).

However, in vivo studies of $[^3H]$-uridine incorporation into cellular RNA are not a reliable measurement of RNA synthesis because of variations in the pool size of RNA precursors with growth rates (9). In vitro studies have shown that T antigen preparations added to the incubation mixtures stimulate RNA synthesis in isolated nuclei (15) and isolated nucleoli (43). However, although these findings were highly reproducible, in vitro systems are notoriously riddled with artifacts.

Reactivation of silent rRNA genes by SV40 T antigen. For these reason we have tried to demonstrate that SV40 T antigen can reactivate rRNA synthesis in vivo, i.e., in the intact cell. The system we have used has been the one of hybrid cell lines. In 1976 Croce (7) developed a hybrid cell line between human fibrosarcoma HT1080 cells and $Balb_c$ macrophage in which the cells retain all human chromosomes and 18 mouse chromosomes, including chromosomes 12, 15 and 18, where the mouse genes for rRNA are located. These hybrid cells synthesize human rRNA but not mouse rRNA (8), although the mouse rRNA genes are present. Human and mouse rRNA can be distinguished from each other because human 28S rRNA travels in gels somewhat more slowly than 28S rRNA of rodents (12). Perry et al. (24) have demonstrated that it is the synthesis, and not the processing of rRNA, that is affected; so that one can say that in these hybrids, called G5554, the expression of mouse rRNA genes is repressed. When G5554 cells are infected with SV40, mouse 28S rRNA appears in the cytoplasm with a lag period of 16-18 hours (34).

Further studies have clearly shown that the reactivation of silent rRNA genes is due to the A gene of SV40 and specifically to one of the products of the A gene, the large T antigen. These conclusions are based on the following experiments:

1. Silent mouse rRNA genes in G5554 cells are not reactivated when the cells are infected at the nonpermissive temperature with tsA58, which is a ts mutant of SV40 defective in the A gene. On the contrary, wild type SV40 can reactivate silent mouse rRNA genes even at 39°.

2. Adenovirus 2, at variance with SV40 and polyoma, does not stimulate rRNA synthesis (11). Accordingly, infection of G5554 cells with Adenovirus 2 does not reactivate silent mouse rRNA genes. A hybrid virus called D1, which contains an insertion of the SV40 genome, precisely a piece of SV40 genome that maps between .10 and .71 of the SV40 map (14), can reactivate silent rRNA genes in G5554 cells. Since the map coordinates of the SV40 insertion correspond to the SV40 A gene that codes for the T antigen, these experiments clearly demonstrate that the A gene is responsible for the reactivation of silent mouse rRNA genes.

3. The A gene of SV40 codes for two proteins, large T and small t (26). We used a deletion mutant of SV40, dl2005, that does not produce small t, although the large T is normal (33). This mutant reactivated the silent mouse rRNA genes in G5554 cells, thus indicating that this effect is due to the large T and not to the small t.

It therefore seems unequivocally demonstrated that T antigen can stimulate rRNA synthesis in vivo and in vitro. Thus, T antigen can replace the equivalent serum function in stimulating rRNA synthesis.

Effect of serum and polyoma on the induction of DNA synthesis in resting AF8 cells.

AF8 cells were originally isolated from BHK cells by Meiss and Basilico (22) and, as mentioned above, they are ts mutants that arrest at the nonpermissive temperature in G_1. When AF8 cells are made quiescent by serum restriction and subsequently stimulated by 10% serum, about 80% of the cells enter S at 34°, after a lag period; while at 40.6° the percentage of cells labeled by [^{3}H]-Thymidine does not increase above the zero time level (2). Polyoma infection of quiescent AF8 cells has the same effect as serum.

S phase is induced in approximately 75% of the cells at 34°, while at 40.6° there is no increase in the number of DNA-synthesizing cells (27).

Effect of Adenovirus 2 on induction of DNA synthesis in Resting AF8 cells.

At variance with serum or Polyoma, Adenovirus 2 induces an S phase in quiescent AF8 cells at both permissive and nonpermissive temperatures (27). At the nonpermissive temperature for AF8, Adenovirus 2 can induce cellular DNA synthesis whether the cells are incubated postinfection in conditioned medium, or in fresh 10% serum. The cells infected with Adenovirus 2 become T positive by immunofluorescence. In the paper mentioned above Rossini et al., (27) have given evidence that the DNA synthesized by AF8 cells infected with Adenovirus 2 at 40.6° is (a) cellular and not viral, and (b) semiconservatively replicated. Thus, Adenovirus 2 infection can induce the $G_0 \rightarrow S$ transition in AF8 cells even at a temperature nonpermissive for AF8. While Polyoma early proteins, and their role in the stimulation of cellular DNA synthesis are already reasonably well known (20, 40, 42), the situation is somewhat more confused with Adenovirus 2. There are at least six early proteins of Adenovirus 2 (17), if not more (13). And, it is not yet known which of the early proteins coded by Adenovirus 2 is responsible for inducing cellular DNA synthesis in resting cells. Despite these uncertainties, our findings clearly open the possibility of identifying and purifying one or more proteins that regulate two different critical steps in the G_0/G_1 phase of the cell cycle.

The experiments reported above, together with the knowledge that there are at least three well-established G_1 ts mutants of the cell cycle, make it possible to tentatively identify five genes and five gene products that regulate five different steps in G_1, critical for the entry of cells into the S phase. These studies, therefore, clearly indicate that it is now possible to study the regulation of cell proliferation on a firmly established molecular biopsy basis.

REFERENCES

1. Acheson, N.H.(1976): Cell 8:1-12.
2. Ashihara, T., Chang, S.D., and Baserga, R. (1978): J. Cell. Physiol., 96:15-21.
3. Baserga, R. (1976): Multiplication and Division in Mammalian Cells. Dekker, New York.
4. Benjamin, T.L. (1966): J. Mol. Biol. 16:359-373.
5. Burstin, S.J., Meiss, H.K., and Basilico, C. (1974): J. Cell. Physiol. 84:397-408.
6. Chu, E.H.Y. (1978): J. Cell. Physiol. 95:365-366.
7. Croce, C. (1976): Proc. Nat. Acad. Sci. 73:3248-3252.
8. Croce, C.M., Talavera, A., Basilica, C., and Miller, O.J. (1977): Proc. Nat. Acad. Sci. 74:694-697.

9. Cunningham, D.D., and Pardee, A.B. (1969): Proc. Nat. Acad. Sci. 64:1049-1056.
10. Dulbecco, R., Hartwell, L.H., and Vogt, M. (1965): Proc. Nat. Acad. Sci. 53:403-410.
11. Eliceiri, G.L. (1973): Virology 56:604-607.
12. Eliceiri, G.L., and Green, H. (1969): J. Mol. Biol. 41:253-260.
13. Harter, M.L., and Lewis, J.B. (1978): J. Virol. 26:736-749.
14. Hassell, J.A., Lukanidin, E., Fey, G., and Sambrook, J. (1978): J. Mol. Biol. 120:209-247.
15. Ide, T., Whelly, S., and Baserga, R. (1977): Proc. Nat. Acad. Sci. 74:3189-3192.
16. Levine, A.J. (1976): Biochim. Biophys. Acta 458:213-241.
17. Lewis, J.B., Atkins, J.F., Baum, P.R., Solem, R., Gesteland, R.F., and Anderson, C.W. (1976): Cell 7:141-151.
18. Lieberman, I., Abrams, R., and Ove, P. (1963): J. Biol. Chem. 238:2141-2149.
19. Mauck, J.C., and Green, H. (1973): Proc. Nat. Acad. Sci. 70: 2819-2822.
20. May, E., May, P., and Weil, R. (1973): Proc. Nat. Acad. Sci. 70:1654-1658.
21. May, P., May, E., and Bordé, J. (1976): Exp. Cell Res. 100:433-436.
22. Meiss, H.K., and Basilico, C. (1972): Nat. New Biol. 239:66-68.
23. Pardee, A.B., Dubrow, R., Hamlin, J.L., and Kletzein, R.F. (1978): Ann. Rev. Biochem. 47:715-750.
24. Perry, R.P., Kelley, D.E., Schibler, U., Huebner, K., and Croce, C.M. (1979): J. Cell. Physiol. 98:553-560.
25. Postel, E.H., and Levine, A.J. (1976): Virology 73:206-215.
26. Prives, C., Gilboa, E., Revel, M., and Winocour, E. (1977): Proc. Nat. Acad. Sci. 74:457-461.
27. Rossini, M., and Baserga, R. (1978): Biochemistry 17:858-863.
28. Rossini, M., Weinmann, R., and Baserga, R. (in press).
29. Rovera, G., Mehta, S., and Maul, G. (1975). Exp. Cell Res. 89: 295-305.
30. Sauer, G., and Defendi, V. (1966): Proc. Nat. Acad. Sci. 56: 452-457.
31. Schmid, W., and Sekeris, C.E. (1975): Biochim. Biophys. Acta 402:244-252.
32. Scott, W.A., Brockman, W.W., and Nathans, D. (1976): Virology 75:319-334.
33. Sleigh, M.J., Topp, W.C., Hanich, R., and Sambrook, J. (1978): Cell 14:79-88.
34. Soprano, K., Dev, V.G., Croce, C., and Baserga, R. (1979): Proc. Nat. Acad. Sci. (in press)
35. Strohl, W.A. (1969): Virology 39:653-665.
36. Tegtmeyer, P. (1972): J. Virol. 10:591-598.
37. Tegtmeyer, P. (1975): J. Virol. 15:613-618.
38. Tjian, R., Fey, G., and Graessmann, A. (1978): Proc. Nat. Acad. Sci. 75:1279-1283.
39. Tsakada, K., and Lieberman, I. (1964): J. Biol. Chem. 239:2952-2956.

40. Türler, H., and Salomon, C. (1977): INSERM 69:131-144.
41. Weil, R., and Kåra, J. (1970): Proc. Nat. Acad. Sci. 67:1011-1017.
42. Weil, R., Salomon, C., May, E., and May, P. (1975): Cold Spring Harbor Symposium, Quantitative Biology 39:381-395.
43. Whelly, S., Ide, T., and Baserga, R. (1978): Virology 88:82-91.
44. Zardi, L., and Baserga, R. (1974): Exp. Mol. Path. 20:69-77.
45. Zimmerman, J.E., Jr., and Raska, K. (1972): Nat. New Biol. 239:145-147.

SOME OBSERVATIONS ON THE KINETICS OF HAEMOPOIETIC STEM CELLS AND THEIR RELATIONSHIP TO THE SPATIAL CELLULAR ORGANISATION OF THE TISSUE

B.I. Lord and R. Schofield
Paterson Laboratories,
Christie Hospital and Holt
Radium Institute, Manchester
M20 9BX, England

Haemopoietic tissue is widely scattered around the body, residing primarily in the marrow of the skeletal bones. It consists of a large variety of cell types which, in any one site, appear, by the classical techniques of morphology, to be mixed in a totally random manner. There is no obvious geometrical or spatial organisation of these cells. Much is now known about the dynamic behaviour of the cells, their interrelationships and capacities to respond to stress situations. Many reviews on various aspects of the subject have been published and a recent "Clinics in Haematology" (see Lajtha, 1979) represents an excellent, up to date, overview. A number of attempts to model the tissue mathematically has, due to the rapid accumulation of experimental data over the past twenty years, been singularly unsuccessful and though the most recent of them (Wichmann and Loeffler, 1979) has come extremely close to describing most of these data, it, too, has so far been unable to point the way ahead for the experimentalist. In this paper we shall first describe our current understanding of the structure of haemopoiesis, the spatial distribution of the various cell types within the tissue and the regulation of cell proliferation.

The pluripotential stem cell, whose role is the foundation of all haemopoiesis, will then be examined in detail; its distribution in the marrow spaces; its capacity to differentiate and self renew. Finally, a simple mathematical analysis will be made to explore the value of such an approach in extending our understanding of the behaviour and function of the stem cell population.

STRUCTURE OF HAEMOPOIESIS

The haemopoietic cell populations have now been recognised to fit into a three tiered hierarchy of cell stages. The first tier consists of the pluripotential stem cells, few in number ($\sim$1 per 500 cells) and unrecognisable by morphological techniques. They are assayed by the spleen colony technique (Till and McCulloch, 1961) which recognises the cells by their ability to carry out their stem cell function - that of being able to regenerate their own population and at the same time give rise to several types of differentiated functional progeny.

The second tier is derived from the first by one of several differentiation functions which commit the pluripotent stem cell into one or other of the blood cell lines. In this stage, the cells, known as committed precursor cells, enter a one way suicide maturation process and go through a series of amplifying divisions, those at the end of the compartment being "older" than those at the beginning. The cells are still unrecognisable morphologically and are measured by a variety of *in vitro* colony assays for the various cell lines (Pluznik and Sachs, 1965; Bradley and Metcalf, 1966, for granulocyte/macrophage colonies; Stephenson et al, 1971 for erythroid colonies; Metcalf et al, 1975b for lymphoid colonies and Metcalf et al, 1975a for megakaryocytic colonies).

At the end of the second stage, the cells are subjected to a second differentiation step which converts them into the morphologically recognisable third tier where the cells undergo a further series of amplifying divisions at the end of which the cells emerge as mature functional blood cells.

Throughout all these stages, relative cell numbers can be determined and their prolifera tive behaviour observed so that we now have a fairly extensive knowledge of the physiological make up of haemopoietic tissue and its adaptability to stress situations.

Microarchitecture of Haemopoietic Cell Populations

As mentioned above, haemopoietic tissue has no obvious spatial organisation as determine by conventional microscopy, although it was observed as long ago as 1938 (Weinbeck, 1938) that there was a large build up of the relatively mature cell types towards the centre of the marrow spaces. This was recently confirmed by Shackney et al (1975) who studied the distribution of a tritiated thymidine label in the marrow cells. We have found it possible also to measure the distribution of the pluripotent stem cells and some of the committed precursor cells. This was done by taking cylindrical plugs of marrow from the mouse femur and assaying for stem cells (spleen colony forming cells - CFU-S) and committed granulocyte/macrophage precursors (*in vitro* colony forming cells - CFU-C). The technique and the detailed results have been published previously (Lord and Hendry, 1972; Lord, Testa and Hendry, 1975) and figure 1 illustrates diagrammatica the relative distributions of the three stages of haemopoiesis. Stem cells are most concentrated in the vicinity of the bone surface. Subsequent differentiation and maturation take place in the direction of the central axis of the bone giving rise to a peak of committed cells about one quarter of the way from the bone to the axis and a continuously increasing volume of maturing cells closer to the axis.

Regulation of Cell Production

Variation in the output of mature functional cells is achieved by altering the number of amplifying divisions in the committed and/or maturing cell compartments. This is effected by a series of integrated Inhibitor/Stimulator negative feedback loops. For example, in erythropoiesis, (fig. 2) the red blood cells present to the kidney a signal proportional to their mass. Any reduction in this signal (Inhibitor phase) causes increased production of the hormone erythropoietin, EPo (Stimulator phase)

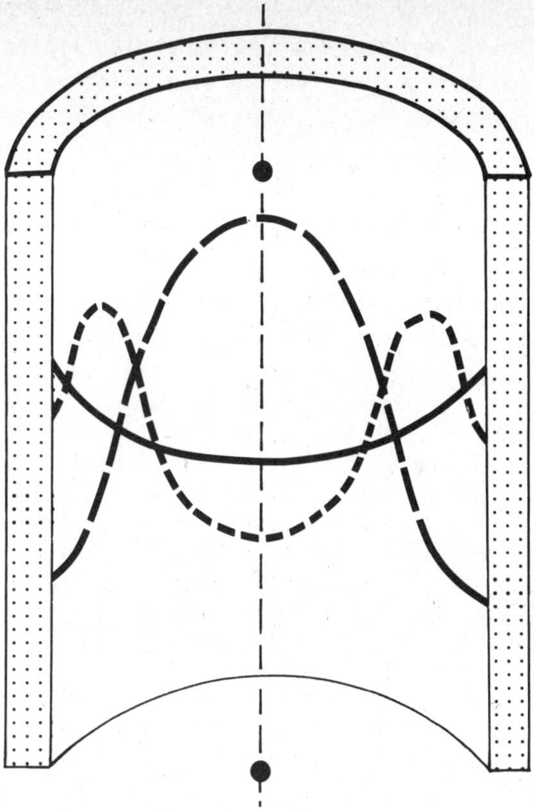

Fig. 1 Diagrammatic representation of the relative concentrations of cells in the three tiers of haemopoiesis. The drawing illustrates half the cylindrical shaft of the mouse femur. ——— Stem cells (CFU-S); ▬ ▬ ▬ Committed precursor cells (CFU-C) ▬▬ ▬▬ Maturing cells.

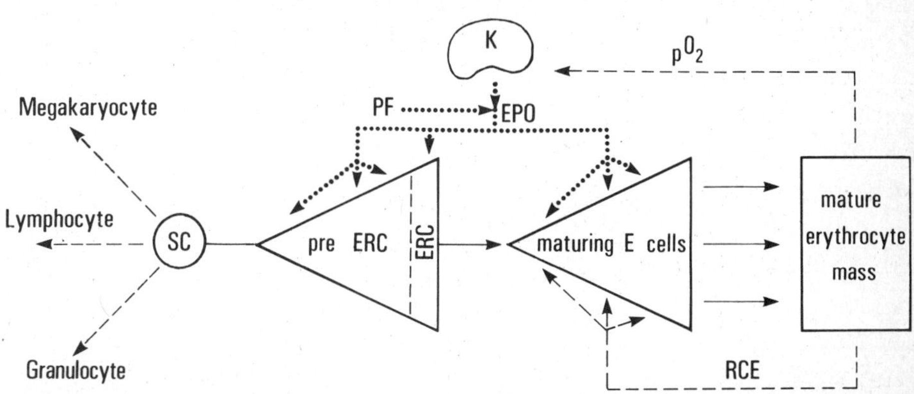

Fig. 2 Regulation of red blood cell production. SC = stem cell; ERC = erythropoietin responsive cells (committed erythroid precursor cells)

which causes the faster output of more erythroid cells. A second inhibitory loop, also directly proportional to this red cell mass, illustrated as RCE (red cell extract, Kivilaakso & Rytömaa, 1971; Lord et al, 1977b) acts as a fine control on the output by modulating the cell cycle of the maturing erythroid cells. Similar systems act for other cell types. Recently, an inhibitor (Lord et al, 1976; Frindel and Guigon, 1977) and a stimulator (Frindel et al, 1976; Lord et al, 1977a) have been recognised for regulating stem cell proliferation. Their source and exact mechanism of action have yet to be identified but it is known that they can act in competition with one another and that excess of one will override the affect of the other (Lord et al, 1977a), i.e. another integrated Inhibitor/Stimulator feedback regulating cell system. It is interesting also to note that the Stimulator has been found to be associated with the bone (Lord 1978) and that this correlates with the relatively high proliferation rate of stem cells near the bone compared with the low overall proliferation rate of the whole stem cell population.

THE STEM CELL POPULATION

The definition of a stem cell requires that it not only produce differentiated cells but that it also be self renewing and in any steady state population, the self renewal probability of the stem cells must be 0.5. When the stem cell population is depleted by any means, if it is to recovery it must exhibit an increased self renewal probabilit The question is whether, under identical conditions, all stem cells can do this to the same extent or equally easily. On the stochastic model of stem cell renewal (Siminovitch et al 1963) the answer is yes. This implies that where recovery from lethal irradiation is effected by transplantation of haemopoietic tissue, then stem cells irrespective of their source can repopulate equally well. This, however, is clearly not the case. When stem cells are transplanted serially into irradiated mice, allowing full recovery between each transfer, increasing numbers are required at each transfer in order to effect that recovery (Lajtha & Schofield, 1971). Finally, the transplant dies out after about 6 or 7 transfers suggesting that the quality of the stem cells decreased with time. We thus attempted to measure the capacity of haemopoietic stem cell populations for self renewal.

The Stem Cell Assay

From the point of view of the following analysis, it is important to understand the relationship of the stem cell to the spleen colony forming cell assay. Haemopoietic tissue containing stem cells, injected into a lethally irradiated mouse, will grow in an attempt to repopulate the haemopoietic tissues of the irradiated animal. This growth is observed as colonies of cells growing in the spleens of the irradiated mice. The number of colonies is proportional to the number of cells injected (Till and McCulloch, 1961) and each colony develops from just one of the injected cells

(Becker et al, 1963). Since colonies contain not only recognisable differentiated cells of all kinds but also more cells capable of producing spleen colonies, the cell initiating the growth of a colony fulfils the definition of a stem cell. Thus, stem cells are generally equated to spleen colony forming cells (CFC-S) i.e. the cells capable of forming colonies. However, only a fraction, f, (~0.1), of the cells capable of producing a colony in the spleen do so (Siminovitch et al, 1963; 1964; Lord & Hendry, 1973). The rest go to bone marrow sites or are trapped in non-haemopoietic sites following injection. Thus, the measured number of colonies is the number of spleen colony forming units (CFU-S) in the grafted tissue

$$\text{i.e. Stem cells} \doteq \text{CFC-S} = \text{CFU-S}/_{f}$$

Measurement of Self Renewal Probability

A normal mouse contains at least 10^6 CFC-S. A countable number of CFU-S (~10) injected into a lethally irradiated mouse is thus under great pressure to reconstitute the whole stem cell population and under these circumstances will show its maximum capacity for self-renewal. The measurement of self-renewal probability therefore is carried out in two stages: (1) establish the growth of about eight colonies per spleen for a period of 11days; (2) excise every spleen colony and re-inject them individually into further irradiated mice to assay the number of CFU-S which has grown in each colony over that first 11 day period. A detailed account of the technique is given by Schofield and Lord (1979). The mean CFU-S per colony and coefficient of variation is determined from the distribution of individual colony counts and the probability of self-renewal of a CFU-S is given by

$$CV^2 = \frac{2 - 2p}{2p - 1} + \frac{1}{M} \qquad (1)$$

where M = mean CFU-S/col and V = coefficient of variation (Vogel et al 1968)

We have carried out experiments on haemopoietic tissue from many sources: different regions of normal marrow, following and during recovery from irradiation (acute or chronic) or cytotoxic drugs etc. Figure 3 shows the results and indicates that CFU-S from different sources do not exhibit the same self-renewal potential. This varies between 0.54 and 0.83 for the cells tested and as might be expected, the bigger the self-renewal probability, the more CFU-S/colony are produced in a standard period of colony growth. Normal bone marrow exhibits a self-renewal probability of 0.683 giving an average of 28.0 CFU-S per colony. This means that 68.3% of CFU-S produce another CFU-S while 31.7% differentiate. Clearly, however, the self-renewal probability, p, is not a constant property of all stem cell populations. A study of the effects of isopropyl-methane sulphonate led Schofield and Lajtha (1973) to suggest an "age" structure within the stem cell population. Schofield later suggested that the CFC-S population might more properly be represented as a true stem cell which may

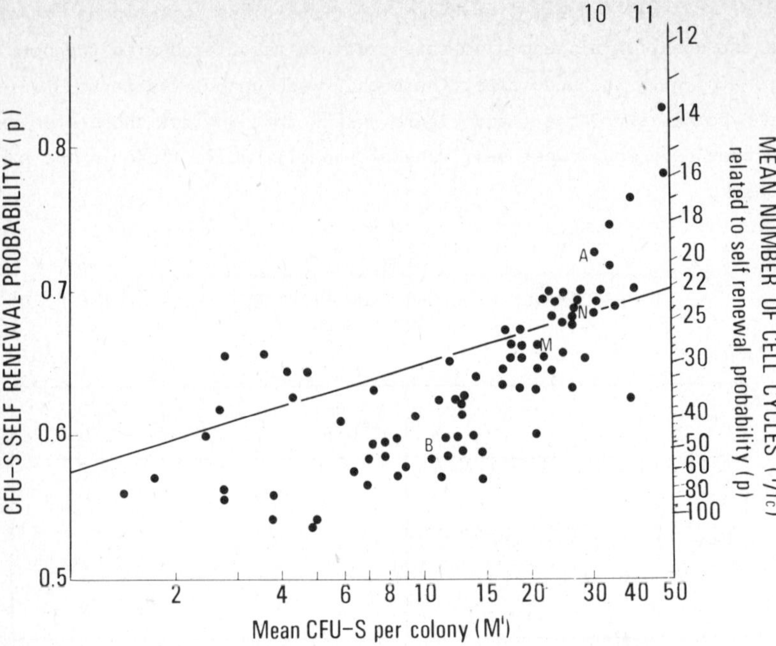

Fig. 3 Self renewal probability of CFU-S under a variety of experimental conditions compared with the mean number of CFU-S produced during a standard interval of colony growth. The right hand ordinate indicates the number of cell cycle required for CFU-S with different self renewal probabilities to complete in order to develop a colony containing 50 CFU-S (see text).
N - normal CFU-S; A - axial CFU-S; M - marginal CFU-S; B-bone derived CFU-S

need to reside in a "niche" followed by an expansion compartment of its own during which, although still pluripotential in its differentiative capacity, gradually loses capacity for self renewal (Schofield 1978). In other words, the stem cell compartment - the CFC-S - is a heterogeneous "ageing" population of cells and in this heterogeneity may lie the variation in self renewal probability.

Spatial Considerations and Self Renewal

Comparisons with the spatial organisation of cells in the epithelial tissues (see Potten this volume) where the stem cells are located in the basal layer, point to the bone surface as a suitable equivalent basal layer for haemopoietic stem cells. The location of a high concentration of stem cells in this region and the apparent movemen of cells from the bone to its central axis encourage this comparison and suggest that the stem cell "niches" may have some affinity with the bone and may house the more primitive, basic stem cell.

Measurements of the self renewal probability of CFU-S close to the bone (marginal), distant from the bone (axial) and also those closely associated with the bone matrix, i.e. those remaining in the bone after thoroughly washing out the marrow were made to test this hypothesis. Table 1 and figure 3 show that in fact those stem cells close to the bone have the lowest self-renewal capacity while those in the axial regions have the highest.

TABLE 1

KINETIC PERFORMANCE OF CFU-S FROM DIFFERENT ZONES OF THE BONE AND MARROW

	Bone derived CFU-S	Marginal CFU-S	Axial CFU-S
Self renewal probability (p)	0.59	0.66	0.73
Mean CFU-S/colony (M)	10.3	21.4	26.5
Mean cell cycle time (T_c)	6.4	8.7	11.4

This is contrary to the idea of the bone located niche and suggests that figure 4 may represent the spatial development of the haemopoietic cell populations. A CFU-S

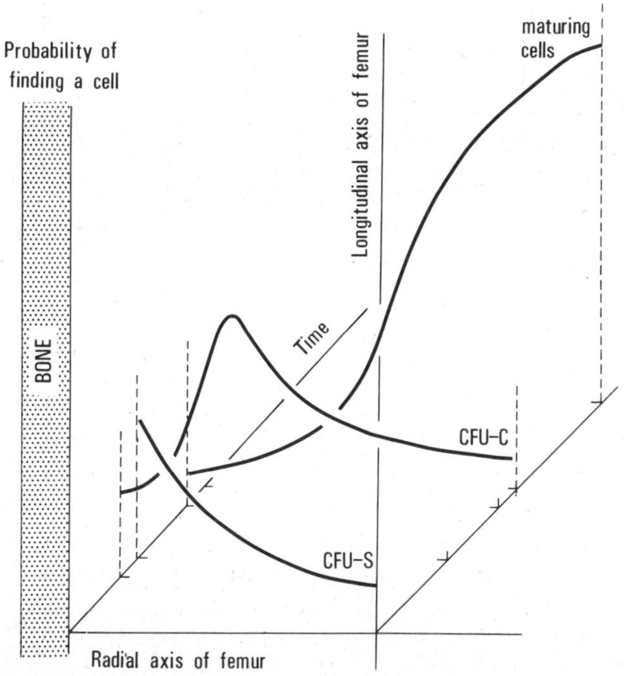

Fig. 4 Diagrammatic representation of the chronological and spatial development of the haemopoietic tissue. CFU-S, stem cells; CFU-C, committed precursor cells.

originating from a "real" stem cell at the central longitudinal axis progresses through an expansion compartment to the bone where the population receives its differentiation stimulus. It then progresses back towards the axis first as a committed precursor and later as a maturing population all the time amplifying its size by successive cell divisions. At least one question remains unanswered in this model, however. That is, what is the mechanism by which the CFU-S population can move against the increasing tidal wave of the more mature cells flowing back from the bone surface? This may be impossible and require us to return to the postulate that bone derived CFU-S contains the primitive 'niche' cells but that they have a less flexible self-renewal capacity; a suggestion somewhat in line with the fixed stem cell hypothesis of Cairns (1975).

Estimation of Cell Cell Cycle Time of CFU-S

Clearly, the stem cell population is extremely complex and with the availability of these data on heterogeneity in the population, one can explore the kinetics of colony growth. The growth of the CFC-S population is limited by the amount of differentiation which is, of course, related to the self renewal probability, p.

∴ growth of the CFC-S population is given by

$$dM = \lambda M (2p - 1) dt \qquad (2)$$

where λ = mean lifetime of the cell $= \frac{0.693}{Tc}$

and Tc is the mean cell cycle time

Equation (2) can be integrated to give the number of CFC-S per colony at time T as

$$M = e^{\lambda (2p - 1) T} \qquad (3)$$

rearranging equation (3) we get

$$p = \frac{3.323 \log M}{2 (T/Tc)} + \frac{1}{2} \qquad (4)$$

Thus, if all CFC-S behave proliferatively similarly then a plot of p versus log M should be linear. Figure 3 shows this not to be the case so one may conclude that not only do CFC-S have different intrinsic self renewal probabilities, but they also exhibit different proliferation characteristics. However, knowing the self renewal probability, one can calculate the number of cell cycles (T/Tc) required to produce a colony containing a given number of CFC-S. Consequently, taking 500 CFC-S (50 CFU-S) as standard, a series of related p and T/Tc values was calculated from equation (4) and plotted on the right hand ordinate of figure 3. Then a line joining p = 0.5, M = 1 with any experimental point projects on to the T/Tc ordinate at the number of cell cycles completed during the growth of the spleen colony. Thus, for normal bone marrow (p = 0.683 and M = 28), the line drawn on figure 3 shows that nearly 21 cell cycles were completed. Since the stem cell population shows little growth over the first three days following injection (Kretchmar and Conover, 1968) then for 11 day old colonies, these 21 cell cycles were completed in 8 days giving an average cell cycle time of 9.1 hrs.

Most of the experimental points (Fig. 3) lie below this line indicating that many experimental CFU-S populations have a reduced mean self renewal probability and undergo more cell cycles during colony growth, i.e. they have a faster proliferation rate and this was exactly what we saw on the CFU-S distribution (Fig. 4). CFU-S near to the bone demonstrate a high proliferation rate and a low p value while those near to the axis have a high p value but low proliferation rate. Calculated from figure 3, the mean cycle time of bone derived CFC-S is 6.4 hrs while that for axial CFC-S is 11.4 hrs (Table 1).

Growth of the CFC-S Population

In order to extend the information we can get from these experimental observations we can now approach the problem from a purely theoretical point of view. ie. indulge in a little modelling. As mentioned above, the growth of the CFC-S population depends on the rate of differentiation into committed precursor cells. Figure 5 sets up a family of curves showing the growth of the CFC-S population assuming a constant cell cycle

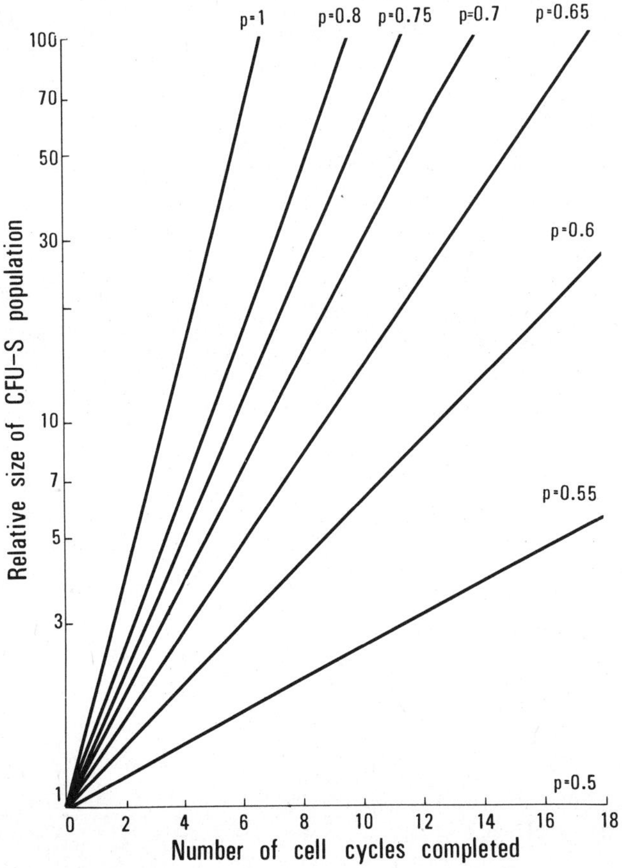

Fig. 5 Growth of the CFU-S population depending on the self renewal probability of the cells - cell cycle time remains constant

time but varying probabilities of self renewal, p. Under those circumstances, reducing p produces an increase in the population doubling time and simultaneously the differentiation rate (1-p) increases. It is a common experimental finding however that in virtually all cases, the growth of the CFC-S population in the spleen exhibits a constant doubling time (Lajtha and Schofield, 1971) In order to maintain a constant population doubling time, it is necessary, therefore, to decrease the cycle time as the differentiation rate increases. This has been done for figure 5 and table 2 shows the relative lengths of the cell cycle for each chosen p value.

TABLE 2

RELATIONSHIP BETWEEN SELF RENEWAL PROBABILITY AND CELL CYCLE TIME

Self Renewal Probability (p)	1	0.8	0.75	0.7	0.65	0.6	0.55	0.5
Relative cell cycle time (Tc^1)	1	0.68	0.58	0.49	0.38	0.27	0.14	0

The line drawn on figure 6 is a plot of table 2 illustrating the theoretical relationship between self renewal probability and cell cycle time. The points drawn on figur 6 are the experimentally-determined cell cycle times calculated from figure 3, the

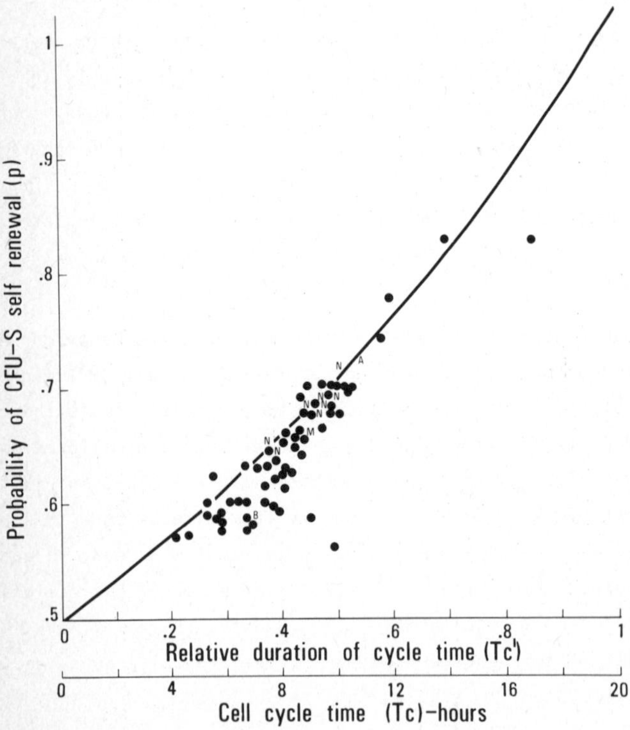

Fig. 6 Variation in cell cy time related to the probabil of self renewal. Population doubling time remains consta N - Normal CFU-S; A - axial CFU-S; M - marginal CFU-S; B - bone derived CFU-S

relationship between the theoretical Tc^1 and the experimental Tc being set by the values for normal bone marrow i.e. for normal bone marrow, $p = 0.683$ and $Tc = 9.1$ hrs. From the theoretical curve (Fig. 6) for $p = 0.683$, $Tc^1 = 0.45 \equiv Tc = 9.1$. It can thus be seen that the experimental data show a remarkably good fit to the theoretical line. In fact the data include eight separate measurements on normal bone marrow and, plotted individually on this graph, show an exact correlation with the predicted line. This approach, therefore, appears to confirm the principles that under conditions of colony growth, the stem cell population has a constant doubling time and that this growth depends not only on the differentiation rate but also on the cycle time of the cells.

Even this argument simplifies the problem because under these conditions all colonies, no matter what the self renewal probability of the seeding CFU-S would, in a fixed time of colony growth, develop the same number of CFC-S. Figure 3 shows this is not true. In calculating cycle times from Figure 3, a constant delay of 3 days has been assumed for the onset of growth of the CFU-S population following transplantation into an irradiated mouse. In experiments where the growth of the CFU-S population has been measured directly, however, it appears that while the population doubling time remains constant, the onset of growth may show a variable delay (Schofield and Lajtha, 1973). Figure 6 in fact demonstrates this delay. It will be noticed that while the experimental points match the slope of the theoretical curve quite satisfactorily, most of them are displaced to the right. This is readily explained if these CFU-S were subjected to a longer delay than 3 days before the population started to grow. For example, if the CFU-S population completed 16 cycles in 8 days (3 days delay), the calculated cycle time would be 12 hrs. If, however, the delay had been 5 days, then the 16 cycles would have taken only 6 days giving a cycle time of 9 hrs. In other words the points would have been plotted to the right of their true position. From this graph, therefore, it can be calculated that an error of 1 hr in the calculated cycle time (ie. displacement from the theoretical curve) is equivalent to an extra proliferation delay of 0.75 day.

As a specific example of the use of this principle, we can return to the behaviour of CFU-S in different zones of the marrow. These are indicated as A (axial), M (marginal) and B (bone derived) CFU-S on Figure 6, and can be seen to be progressively further displaced from the line. A exhibits an error of 0.2 hrs longer than the calibration, while M is 0.8 hrs longer. This predicts that the growth of marginal CFU-S should be delayed 0.45 days behind that of axial CFU-S. We had in fact already done this experiment (Fig. 7) and it is clear that the average delay is about 0.5 days, in good agreement with the predicted value. Similarly, B is 2.2 hrs in error of the calibration line predicting a population delay on transplantation of 1.5 days. A preliminary experiment conducted as a result of this showed a delay not less than 1.3 days, again in good agreement with the prediction.

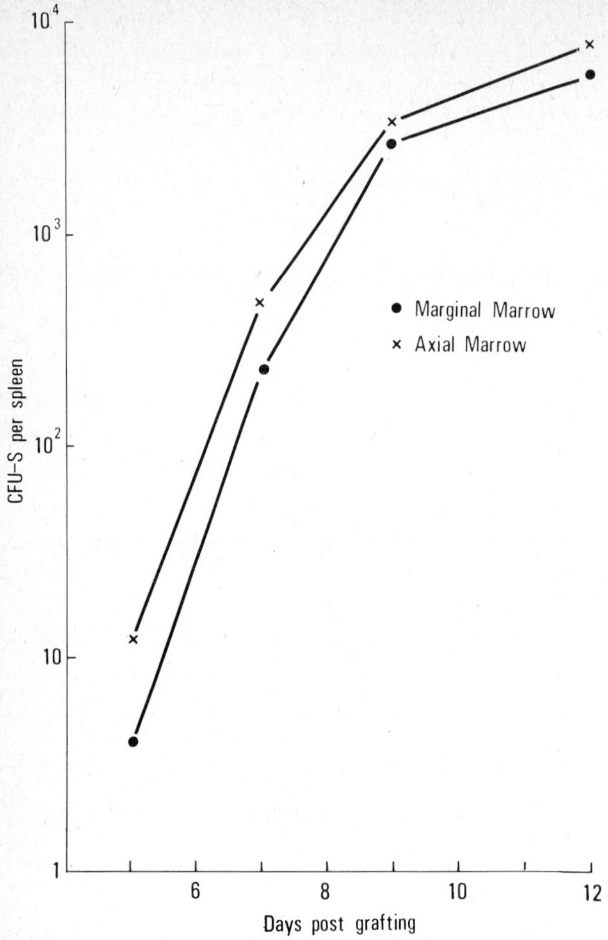

Fig. 7 Growth of the CFU-S population in the spleens of irradiated mice following a graft of axial or marginal bone marrow

SUMMARY

A study of the haemopoietic tissue shows it to have a well organised spatial distribution of the various cell types present. Particular interest is focussed on the stem cell population which is recognised by its ability to reproduce its own kind and at the same time give rise to differentiated functional progeny. This is seen in colonies developing in the spleens of irradiated mice following a graft of haemopoietic tissue. It is shown however that not all of the cells capable of forming colonies contain the same degree of self renewal capacity and as a result it is possible to speculate on the movement of stem cells in the marrow spaces and on the existence of an age structured stem cell compartment.

The growth of the stem cell population is shown to depend both on this self renewal capacity and on the proliferation rate exhibited, the 'younger' CFU-S with a high renewal probability having a long mean cell cycle time while the 'older' ones have a low probability and short cell cycle time.

A simple mathematical analysis carried out alongside the experimental work has made estimates of cell cycle time possible and has clarified our understanding of the heterogeneity and behaviour of the haemopoietic stem cells.

ACKNOWLEDGEMENTS

We would like to thank Mrs. Susan Kyffin and Mrs. Lorna Woolford for expert technical assistance. The work was supported by grants from the Cancer Research Campaign and the Medical Research Council.

REFERENCES

Becker, A.J., McCulloch, E.A., Till, J.E.: Cytological demonstration of the clonal nature of spleen colonies derived from mouse marrow cells. Nature 197, 452-454 (1963)

Bradley, T.R., Metcalf, D.: The growth of mouse bone marrow cells *in vitro*. Aus. J. Exp. Biol. Med. Sci. 44, 287-300 (1966)

Cairns, J.: Mutation selection and the natural history of cancer. Nature, London 255, 197-200 (1975)

Frindel, E., Croizat, H., Vassort, F.: Stimulating factors liberated by treated bone marrow: *in vitro* effect on CFU kinetics. Exp. Hemat. 4, 56-61 (1976)

Frindel, E., Guigon, M.: Inhibition of CFU entry into cycle by a bone marrow extract. Exp. Hemat. 5, 74-76 (1977)

Kivilaakso, E., Rytömaa, T.: Erythrocyte chalone, a tissue specific inhibitor of cell proliferation in the erythron. Cell Tissue Kinet. 4, 1-9 (1971)

Kretchmar, A.L., Conover, W.R.: Early proliferation of transplanted spleen colony-forming cells. Proc. Soc. Exp. Biol. Med. 129, 218-220 (1968)

Lajtha, L.G.: Cellular Dynamics of Haemopoiesis (ed. L.G. Lajtha) in Clinics in Haematology 8 No. 2, 219-528 W.B. Saunders Co. Ltd., London, Philadelphia, Toronto (1979)

Lajtha, L.G., Schofield, R.: Regulation of stem cell renewal and differentiation: possible significance in aging. In: Advances in Gerontological Research (ed. B.L. Strehler) Academic Press N.Y. pp.131-146 (1971)

Lord, B.I.: Cellular and architectural factors influencing the proliferation of hematopoietic stem cells. In: Differentiation of normal and neoplastic hematopoietic cells (Eds. B. Clarkson, P.A. Marks, J.E. Till). Cold Spring Harbor Conferences on Cell Proliferation 5, 775-788 (1978)

Lord, B.I., Hendry, J.H.: The distribution of haemopoietic colony-forming units in the mouse femur, and its modification by X-rays. Br. J. Radiol. 45, 110-115 (1972)

Lord, B.I., Hendry, J.H. Observations on the settling and recoverability of transplanted hemopoietic colony-forming units in the mouse spleen. Blood 41, 409-415 (1973)

Lord, B.I., Mori, K.J., Wright, E.G.: A stimulator of stem cell proliferation in regenerating bone marrow. Biomedicine, 27, 223-226 (1977a)

Lord, B.I., Mori, K.J., Wright, E.G., Lajtha, L.G.: An inhibitor of stem cell proliferation in normal bone marrow. Brit. J. Haematol. 34, 441-445 (1976)

Lord, B.I., Shah, G.P., Lajtha, L.G.: The effects of red blood cell extracts on the proliferation of erythrocyte precursor cells *in vivo*. Cell Tissue Kinet. 10, 215-222 (1977b)

Lord, B.I., Testa, N.G., Hendry, J.H.: The relative spatial distributions of CFU-S and CFU-C in the normal mouse femur. Blood 46, 65-72 (1975)

Metcalf, D., MacDonald, H.R., Odartchenko, N., Sordat, B.: Growth of mouse megakaryocyte colonies *in vitro*. Proc. Nat. Acad. Sci. U.S.A. 72, 1744-1748 (1975a)

Metcalf, D., Warner, N.L., Nossal, G.J.V., Miller, J.F.A.P., Shortman, K., Rabellino, 255, 630-632 (1975b)

Pluznik, D.H., Sachs, L.: The cloning of normal 'mast' cells in tissue culture. J. Cell Comp. Physiology 66, 319-324 (1965)

Schofield, R.: The relationship between the spleen colony-forming cell and the haemopoietic stem cell. Blood Cells 4, 7-25 (1978)

Schofield, R., Lajtha, L.G.: Effects of isopropyl methane sulphonate(IMS) on haemopoietic colony-forming cells. Brit. J. Haematol. 25, 195-202 (1973)

Schofield, R., Lord, B.I., Kyffin, S., Gilbert, C.W.: Self maintenance capacity of CFU-S Submitted for publication (1979)

Shackney, S.E., Ford, S.S., Wittig, A.B.: Kinetic-microarchitectural correlations in the bone marrow of the mouse. Cell Tissue Kinet. 8, 505-516 (1975)

Siminovitch, L., McCulloch, E.A., Till, J.E.: Distribution of colony forming cells among spleen colonies. J. cell comp. Physiol. 62, 327-336 (1963)

Stephenson, J.R., Axelrad, A.A., McCleod, D.L., Shreeve, M.M.: Induction of colonies of hemoglobin-synthesising cells by erythropoietin *in vivo*. Proc. Nat. Acad. Sci. USA 68, 1542-1546 (1971)

Till, J.E., McCulloch, E.A.: A direct measurement of the radiation sensitivity of normal mouse bone marrow cells. Radiat. Res. 14, 213-222 (1961)

Vogel, H., Niewisch, H., Matioli, G.: The self renewal probability of hemopoietic stem cells. J. cell Physiol. 72, 221-228 (1968)

Weinbeck, J.: Die Granulopese des Kindlichen Knockenmarkes und ihre Reaktion auf Infectionen. Beitr. Pathol. Anat. Allg. Pathol. 101, 268-283 (1938)

Wichmann, H.E., Loeffler, M: A comprehensive mathematical model of stem cell proliferation which reproduces most of the experimental results. Submitted for publication.

PROLIFERATIVE CELL POPULATIONS IN SURFACE EPITHELIA: BIOLOGICAL MODELS FOR CELL REPLACEMENT

C.S. Potten
Paterson Laboratories
Christie Hospital
Manchester M20 9BX
England

INTRODUCTION

Cell growth and division can be studied at various levels, two major ones being 1) the characterisation of the complex, interdependent, sequential, biochemical events that occur as any cell passes from one mitosis to the next. This can only be studied effectively on isolated and pure populations of cells under controlled conditions achieved only in culture 2) the characterisation of cellular interactions that are an essential feature of cell replacement in organised tissues. This may involve (a) cells of different types only some of which may be directly involved in cell division, and (b) the maintenance of polarity and architecture within the tissue. In contrast these processes can only be studied effectively *in vivo*.

Models can be devised that attempt to explain these processes and interactions and these may help in understanding the mechanisms involved and the controls that operate. These models also provide a basis for further experimentation. Collaboration between the biologists and mathematicians is helpful in formulating such models. However, for a "marriage" of disciplines to be successful there has to be a relevant and comprehensive dialogue between biologists and mathematicians. For models to be of use they have to be based on 1) a complete appreciation by the mathematicians of the complex biological problem accounting for all the observed data and 2) an understanding by biologists of the mathematical basis of the models and both may pose communication problems. Apart from the danger of any model being based on inaccurate or insufficient information I believe there is a risk that models once presented are accepted as fact and enter the dogma (A theory is a policy not a belief, J.J. Thompson). For example, when a biological process fails to display any apparent order or plan and as a consequence follows apparently random behaviour this can readily be mathematically modelled. Once this occurs there is a tendency to describe the process as being determined by chance almost as if randomness or chance could act as a biological organiser or controller when in fact the whole situation is merely a measure of our ignorance or technical ineptness to study a particular process.

Surface epithelia can have widely differing histological appearances (Fig. 1). All regions are characterised by having a well defined, fixed architecture with a pronounced polarity and some recognisable pattern of cell movement (migration) that is linked to differentiation and maturation. In all cases proliferative activity is restricted

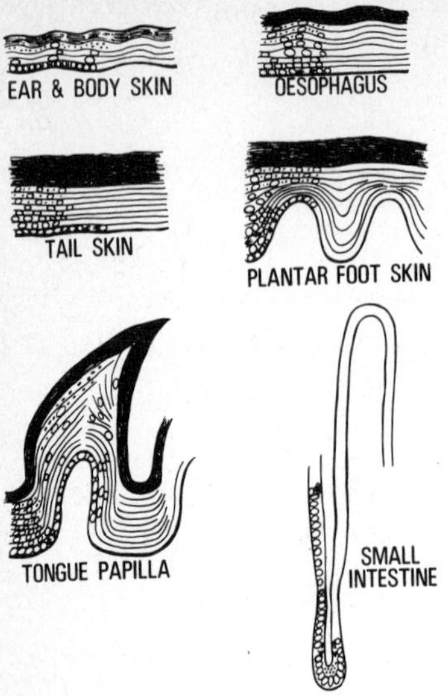

Fig. 1 Schematic representation of the variation in histological appearance in surface epithelia. This is not a comprehensive list but merely provides an indication of the range.

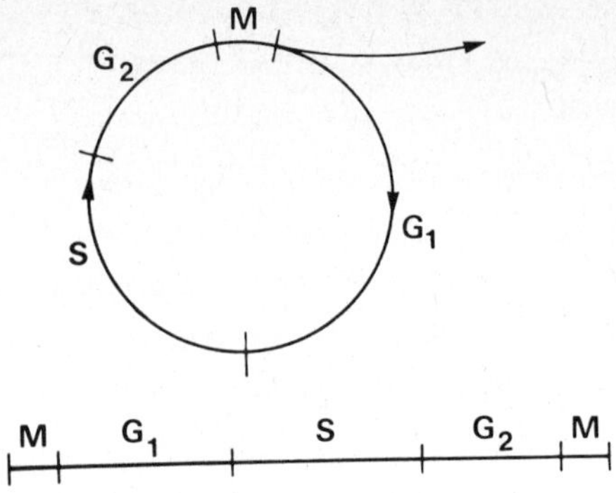

Fig. 2 The cell cycle illustrated as a cycle or as a linear progression.
M = mitosis
S = DNA synthetic phase
G_1 & G_2 = gaps 1 and 2

to a particular area or layer. The complex structure means that lost cells not only have to be replaced but they must be replaced at a particular position in the tissue. Early cell kinetic work was concerned with defining parameters such as the average intermitotic interval (cell cycle or Tc , Fig. 2) for the proliferative basal layer. In simple terms Tc can be expressed as a cyclic or as a linear progression of cells from one mitosis to the next. These two modes of representation may have subtle differences in interpretation. The cyclic presentation implies that the cells at the end of a cycle are identical to those at the beginning; a concept that is perhaps appropriate for permanent self-replacing stem cells. In contrast a linear representation implies that the cells at one end may differ in some slight way from those at the beginning; a concept more appropriate for the progressively'ageing' cell divisions seen in some transit populations. These subtleties of interpretation may have significance for the models to be outlined for surface epithelia. Both schemes outlined in figure 2 are grossly oversimplified since many biochemical transition points can in fact be identified and mapped temporarily in relation to mitosis or DNA synthesis. In some systems cells may 'rest' for a time before proceeding further with the biochemical preparations for mitosis. Commonly these quiescent or rest phases occur in G_1 and the term G_0 has been suggested (Lajtha, 1963).

Measurements of parameters such as T_c alone are quite inadequate in the light of current knowledge since they fail to contribute to our understanding of precisely how cell replacement *in vivo* is achieved on a day to day and cell to cell basis. They do not explain how the tissue architecture is maintained or account for differences in behaviour amongst different classes of cells. They can in some cases be strongly influenced by the most rapidly cycling sub-population within a tissue.

I should like to consider some examples of surface epithelia and review some of the more recent developments in our understanding of the tissue organisation and the models that can be derived. The regions we study are epidermis, an external stratified epithelium, intestinal mucosa, an internal epithelium one cell thick and the dorsal surface of the tongue, an intermediate multi-layered keratinising epithelium. These regions though differing greatly in appearance (Fig. 1) have many similarities which indicate that they can be considered together. We have produced models to explain the proliferative organisation in these tissues (based largely on biological rather than mathematical considerations).

THE EPIDERMIS

Initially we were interested in epidermis and selected as an example one which we believed was as simple, structurally, as possible; namely the ear or dorsal epidermis on small rodents like mice. The other advantage of these systems is that they have been fairly extensively studied histologically, cell kinetically and radiobiologically.

Mouse ear or dorsum epidermis is thin (about 10-15 cell layers; strata) and relatively flat between the hair follicles. The cells on the basal layer normally look much alike except when some specialised techniques are used when melanocytes (2-5%) and Langerhans cells (about 10%) can be seen (Allen & Potten, 1974; Potten, 1976). Experiments 1) where the movement of labelled cells from the basal layer or 2) where the movement of labelled cells through the strata (transit experiments) or 3) where staining for early stages of keratinisation (FITC) are studied suggest that the basal layer contains differentiated cells that have begun maturation (Potten, 1975a; Christophers, 1971a; Iversen *et al*, 1968; Leblond *et al*, 1967). These cells may have finished with cell cycle activity and be post-mitotic cells awaiting their their turn for suprabasal migration. Unfortunately, it is difficult to distinguish, on the basis of these experiments, between truly post-mitotic cells and cells in cycle that have a high probability for migration only at late G1. Experiments with X-rays, neutrons or ultraviolet light suggest that there are two classes of basal cells that respond in different ways to these radiations (Potten & Hendry, 1973; Potten *et al*, 1978a; Potten *et al*, 1976; Potten, 1978a) and that only a fraction (2-7%) of the basal cells are capable of the extensive divisions necessary for regeneration of a depopulated basal layer. These experiments show that although many of the basal cells may be capable of division there are in fact two types of dividing cells, those with a large division potential that are capable of clonal regeneration and those with a limited division potential that lack regenerative capacity.

Thus if we accept clonal regeneration as an exclusive stem cell property this suggests that less than 1 in 10 of the basal cells are clonogenic stem cells. Similar conclusions have recently been drawn for intestinal and tongue epithelium based on some clonal regeneration data but also on cell migration patterns and the life expectancy of cells in the tissue. These studies add weight to the observations in skin and should not be excluded. In these other tissues the stem cell population occupies a strategic fixed position in the tissue. In skin once the tissue organisation is defined similar strategic positions for the stem cells can also be deduced.

Conventional cell kinetic experiments, labelling and mitotic indices, continuous labelling, transit studies, and the frequency of labelled mitoses (FLM) technique all indicate that the average T_c of the basal cells is between 4 and 6 days with the majority of the time being spent in G1/Go (Potten, 1975b). The age distribution of the basal cells is unknown but is likely to be some complex shape with rectangular and exponential portions. The growth fraction is also uncertain but is less than 0.85 and may be as low as 0.5 - 0.6. Continuous labelling experiments invariably reach labelling index values between 0.85 and 1.0 after times between 4 and 6 days (Potten 1980) but because of the uncertainty in the presence or absence of post-mitotic cells this technique has a limited usefulness for growth fraction studies.

The surprising advance in the last 10 years has been the discovery of the high degree of organisation in the cornified and maturing cell layers with these thin, flat, roughly hexagonal cells being arranged into precise columns (Allen & Potten, 1974; Christophers, 1971b; Mackenzie, 1969; Karatschai *et al*, 1971; Potten, 1974). In section the column boundaries can be seen and traced to the basal layer in electron micrographs. These boundaries show a minimal overlap with the 6 neighbouring columns and a regular alternating interdigitation of cell edges (Allen & Potten, 1974; Potten 1976; Potten & Allen, 1975). In surface view (scanning electron microscopy or preparations of full thickness epidermal sheets) the columnar arrangement is evident from the polygonal-hexagonal pattern and minimal overlap with the neighbouring hexagonal cornified cells. The cell kinetic data indicate that each column receives one new cell a day and that there is a strong circadian rhythm. These are the basic facts that have to be accommodated.

Epidermal models:

The columnar structure can be related to the 10-11 basal cells that underlie each column. These must be the proliferative component for each column which then represents the historical record of the proliferative activity of the basal cells. Thus the epidermis must be divided up into a series of discrete proliferative cell groups (proliferative units) (Potten, 1974) (Fig. 3). These must have a long life expec-

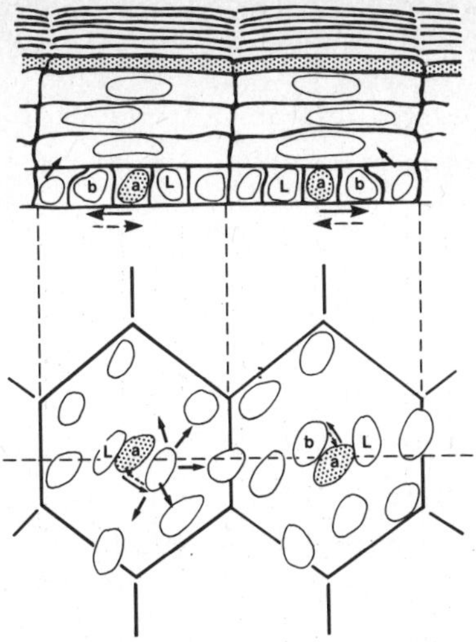

Fig. 3 Schematic representation of Epidermal Proliferative Units in section and surface view (as seen in epidermal sheet preparations). Each unit contains a Langerhans cell (L) and a stem cell (a) which is adjacent to some hypothetical "focal" point in the tissue (see Potten *et al*, 1979) Reproduced by courtesy of Elsevier/North-Holland

tancy since the columnar arrangement is not readily altered once set down (wounding excepted) i.e. the units have a certain autonomy and longevity. The regular alternating interdigitation indicates some inter-unit control of, at least, migrational activity (Potten & Allen 1975). Figure 3 presents the basic epidermal proliferative unit model (EPU) for mouse dorsal epidermis which is based on the following postulates. The epidermis is subdivided into a series of discrete proliferative units each with a basal cell complement (about 10-11 cells). The progeny of such basal cells are aligned into columns immediately above them. Each unit contains one Langerhans cell positioned towards the centre of the basal cell group. This does not contribute cells for the column. One cell in every three units can be expected to be a melanocyte. These are usually amelanotic and do not appear to be distributed in a pattern related to the EPUs. Each unit may contain up to 3 post-mitotic cells peripherally positioned and one centrally positioned clonogenic stem cell. The 6 cycling basal cells produce one cell a day and between them have an average T_c of 4-6 days with a tendency to be in S in the early hours of the morning i.e. show some synchrony. The central position would be the most strategic position for the stem cells. The cells at this position in fact cycle more slowly and respond rapidly and dramatically by moving into DNA synthesis after wounding.

A model to explain the working of the EPU:

The postulates outlined above suggest a proliferative scheme such as that in fig. 4. Figure 5 shows how this could in fact operate on a day to day basis. This is based on an earlier model (Potten, 1976) that has been improved in that the cell numbers and cycle times are more in accord with recent data. The model makes the following predictions: 1) The stem cells cycle once every 8 days 2) about half of the total

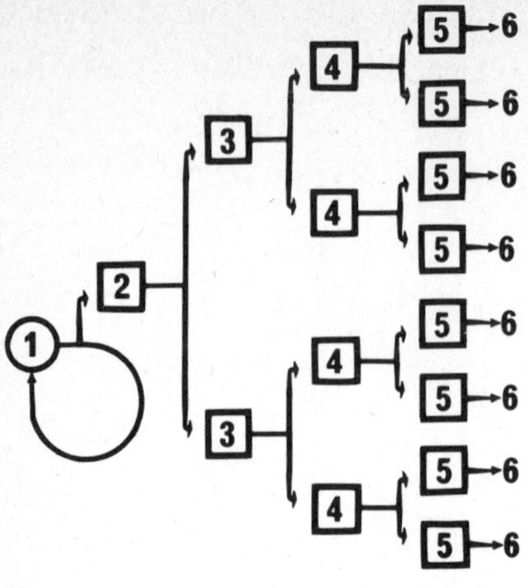

Fig. 4 Proposed proliferative organisation in the epidermis. Stem cells (1) divide asymmetrically to produce a daughter (2) that is displaced from a hypothetical focal point (or is inherently different from the stem cell). These displaced cells then undergo 3 divisions progressively ageing with each division (2-4) until a post-mitotic maturing cell is produced (5) which awaits the signal from migration (6). Reproduced by courtesy of Academic Press (Potten, 1976)

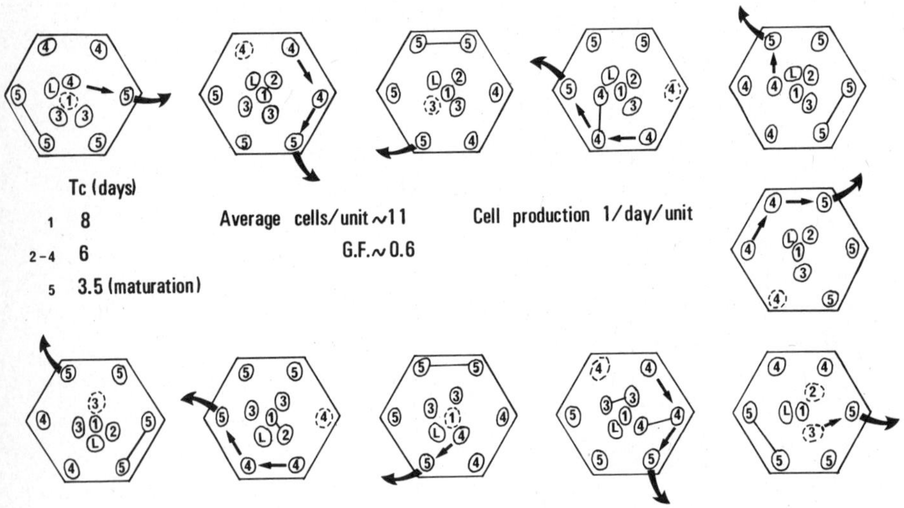

Fig. 5 Model to explain the day to day activity of epidermal proliferative units (EPU The numbering is the same as for fig. 4. L = Langerhans cells. Each arrow leaving a hexagon represents a suprabasal migration of a post-mitotic cell. Each hexagon represents the activity of one day. Dotted lines represent the beginning of mitosis and lines linking two cells represent late telophase. I am indebted to D. Major for his help with this model.

proliferative cells will only divide once i.e. the cells at the end of the division differ from those at the end of the previous division (see earlier) 3) all cells will eventually label under conditions of continuous labelling even though the growth fraction is about 0.6 4) the stem cells either divide in a regular, invariable, asymmetric fashion or the tissue possesses a micromilieu "focal" point at which stem cell function (self-replicative capacity, clonogenicity, selective DNA strand segregation efficiency etc) is at a maximum and differentiation probability is at a minimum (Potten *et al*, 1979) 5) because of the distribution of only one stem cell per unit special mechanisms would be required to ensure that stem cell damage was kept to a minimum (Cairns, 1975; Potten *et al*, 1978b)

Models to explain inter-EPU relationships:

Although each EPU is autonomous it cannot be entirely independent of its neighbours. The regular alternating interdigitation of cornified cell boundaries suggests that the cell migration into the columns is controlled relative to the neighbouring columns. One way in which this might occur is shown in fig. 6 which considers the epidermis as a hexagonal grid. (In fact 74% of the cornified cells are irregular hexagons with the remaining 26% being 5 or 7 sided figures (Menton, 1976b)). This model would predict that a third of the EPUs produce a new cell into the column within an 8 hour period. Then each EPU in the next third produces a cell within the next 8 hours and so on. This pattern of cell migratory activity may or may not be linked directly to mitotic or DNA synthetic activity.

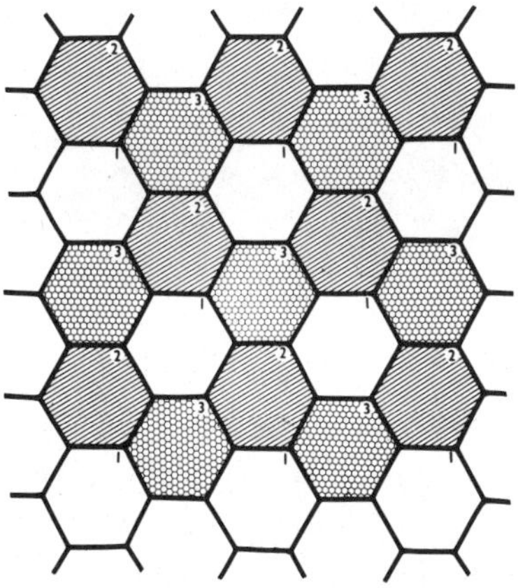

Fig. 6 Model to explain the inter-EPU relationships. Each EPU is surrounded by 6 neighbours that are divided into 2 alternating classes. Thus the hexagonal grid is composed of 3 classes altogether. These represent EPU's synchronised to within 8 hours in terms of cell migration into the columns. Thus every unit adds one cell per day. Reproduced by courtesy of Springer-Verlag (Potten & Allen, 1975)

An alternative model, which reaches the same conclusions, was proposed initially by Menton (Menton, 1976a,b) and was later reviewed by Allen and Potten (Allen and Potten, 1976). This considered the shapes adopted by cells when maximally packed into a given volume. The shape that permits the packing of the maximum number of cells is a tetrakaidecahedron (as described originally by Kelvin(1894) and discussed by D'arcy Thompson, 1961). When packed at maximum density these form columns (Fig. 7) and the columns are arranged in a way that has the same surface pattern as seen in fig. 6. The additional feature of this model is that the packed columns of cells may form a superstructure that aids in the migration and positioning of new cells thus tending to make it a stable self-assembly system.

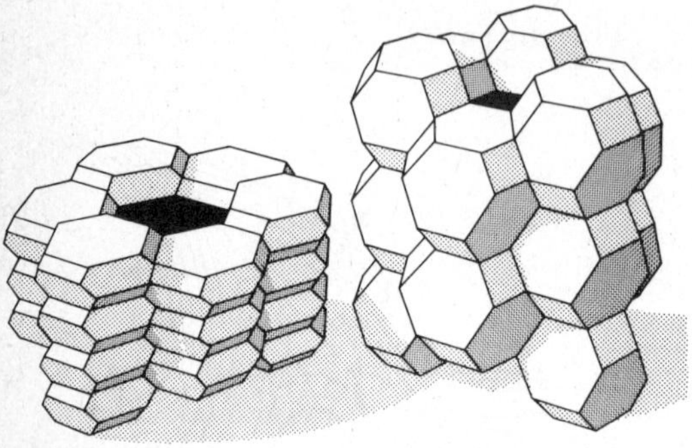

Fig. 7 Tetrakaidecahedra packed at maximum density generate columns Regular tetrakaihedra on the right and flattened (epidermal-like) on the left. This type of arrangement forms a "mould" for the next migrating cell (eg. the central one if the models are inverted)

THE FILIFORM PAPILLA OF MOUSE TONGUE

The dorsal surface of the tongue is a stratified keratinising epithelium like mouse dorsum. It differs in that it is thicker (more strata) and has an undulating basal layer with the undulations related to the surface structure. The surface has many curved, tapering, conical shaped structures, the filiform papilla. Other structures are also present and between the filiform papillae there is a flatter inter-papillary epithelium. The papillae in a given region have the same shape and size and scanning electron microscopy reveals that they have similar surface markings where cell boundaries occur i.e. they are all constructed in the same way according to a common plan. Careful sectioning reveals that these papilla are composed of columns of maturing (keratinising) cells like epidermis, only in this case their shape has been "tailored" to that of the papilla. The columns are tilted and each papilla has two dominant columns that make the anterior and posterior aspects of the papilla. There are 2 or 3 minor lateral buttressing columns of cells. The column boundaries are particularly pronounced and can be traced to the basal layer and thus to the proliferative cells that are associated with the column i.e. Tongue Proliferative Units (TPU) (Hume & Potten, 1976, 1979; Potten 1978b) (Figs. 8, 9). Each column may contain about 20 cells and the basal component consists of 30-40 cells. Tritiated thymidine (^{3}HTdR) labelling shows: 1) marked differences in the number of cells in S with the time of day 2) marked differences in the number of cells in S at different cell positions (cell positions 5 and 6 rarely having cells in S) 3) a movement of cells both along the basal layer and also away from it suprabasally.

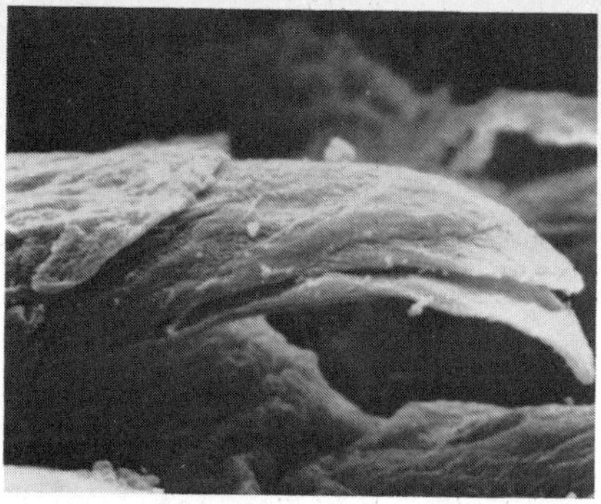

Fig. 8 Scanning electron micrograph of a mouse filiform papilla showing the cell boundaries and the sequence of desquamation for the posterior column
X 3000
I am indebted to T. Allen and W. Hume for this picture

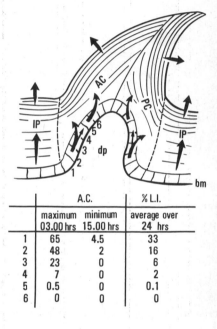

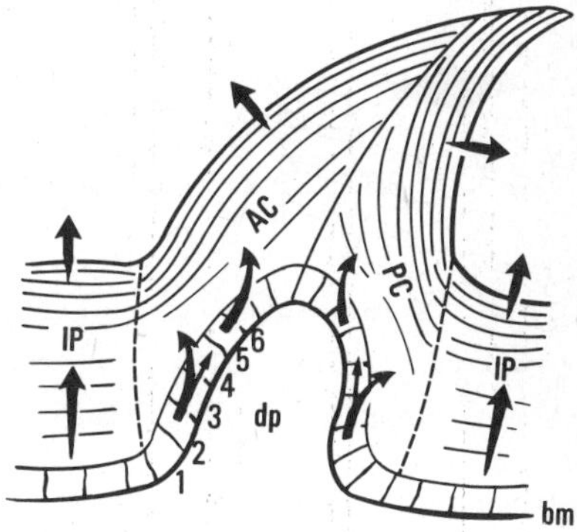

	A.C.		% L.I.
	maximum 03.00 hrs	minimum 15.00 hrs	average over 24 hrs
1	65	4.5	33
2	48	2	16
3	23	0	6
4	7	0	2
5	0.5	0	0.1
6	0	0	0

Fig. 9
Diagrammatic representation of a mouse filiform papilla with the cell migration and cell loss patterns displayed. The basal cells for the anterior tongue proliferative unit (TPU) are numbered with cell position 1 the origin of all the cell flow and therefore the position that is likely to contain the fixed stem cell population. The ^{3}HTdR labelling indices are shown for the anterior TPU. AC - anterior column; PC - posterior column; IP - inter-papillary region, dp dermal papilla and bm basement membrane. Reproduced by courtesy of Cambridge University Press (Potten, 1978b)

The cells at position 1 have a Tc of 24 hrs and are well synchronised since it is easy to obtain 100% labelling at position 1 by three injections of ^{3}HTdR spaced around 03.00 hrs. It is clear that with the tongue our modelling is at an earlier stage than with epidermis. However, a model can be presented to show the structural organisation (Fig. 9). Work is currently underway to determine cell production rates and to formulate a working model that explains the day to day cell replacement.

A model to explain the derivation of the papillary structure:

Figure 10 illustrates a possible means by which structures like the filiform papillae may be derived from a flat columnar epithelium like epidermis (Hume & Potten, 1976). It remains to be seen whether intermediate structures such as those shown in fig. 10

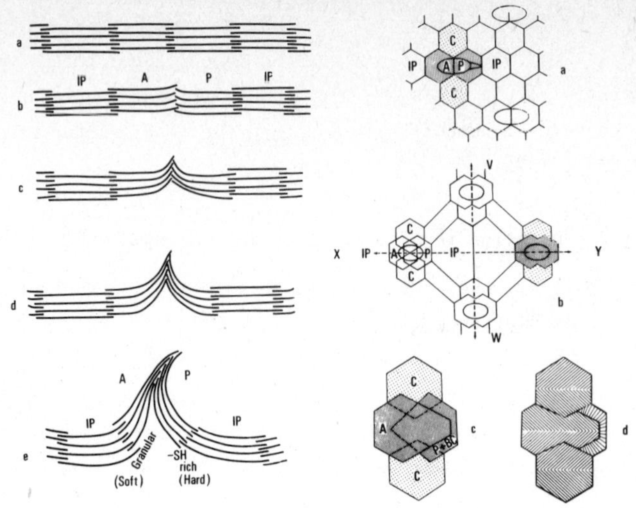

Fig. 10 Model providing one means by which a filiform papilla could be derived from a flat columnar epithelium. For legend see fig. 9 C, lateral columns. Reproduced by courtesy of Cambridge University Press (Hume & Potten 1976)

are seen during development of the tongue or whether they occur in the transitional regions at the tongue edge.

THE SMALL INTESTINAL MUCOSA OF THE MOUSE

The intestinal mucosa is a layer of cells (one cell deep) moulded into a complex shape involving finger-like or ridge projections, the villi (each with several thousand cells), with small bags of about 250 cells, the crypts, around the base of each villus. The proliferative activity is entirely restricted to the crypts. The villus is thus the mature functional component of the mucosa. At the very base of each crypt there are also a few (between 15 and 30) mature cells, the Paneth cells. The proliferative cells (about 150 with Tc values of about 10-14 hours) lie above the Paneth cell region of the crypt. The total cell output from each crypt is thus about 250-300 cells per day. It is clear that with cell production and Tc values like these the life expectancy of a cell near the top of the proliferative compartment is only a few hours while that for the cells lower down might be between a few hours and a day. The upper cells cannot be regarded as permanent, fixed or self-replicating since their probability of being removed to differentiation is high. However, permanent, "anchored" self-replicating cells must exist in the lowest proliferative cell positions i.e. immediately above or in between the Paneth cells. This region contains cells that 1) have longer Tc values (Wright 1978) 2) appear to be the precursors for at least three of the cell lines in the mucosa (Paneth, goblet and columnar) (Cheng & Leblond, 1974) and 3) respond to cell depletion by a dramatic shortening of their cell cycle times (Wright, 1978). Clonal regeneration studies suggest that clonogenic cell numbers are less than the total number of dividing cells (Potten & Hendry, 1975).

Model for cellular organisation in a small intestinal crypt:

Figure 11 shows our current model for the proliferative cell organisation of the crypt. The stem cell population may represent a full ring of about 20 cells at cell position 1, in which case, 3 or 4 amplifying divisions occur in the non-stem cell daughters as they mature and move up the crypt. Alternatively the stem cell number may be less (1-5) with 5-8 amplifying cell divisions and some movement of cells circumferentially at the lowest cell position. Further studies are needed to clarify these points.

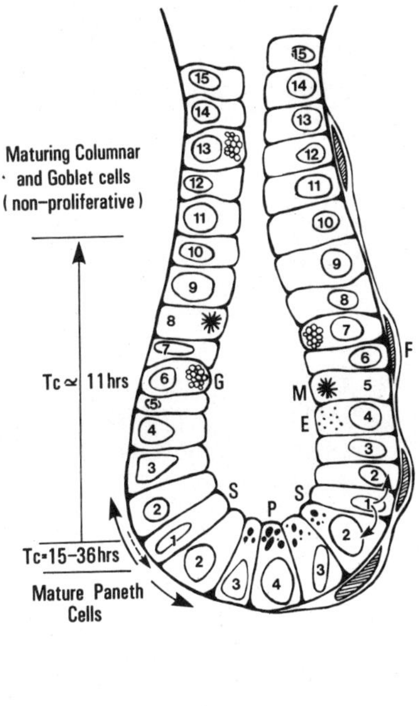

Fig. 11 Schematic representation of a small intestinal crypt. P, Paneth cells; S, stem cells; G, goblet cells; E, enteroendocrine cells; M, mitosis; F, pericryptal fibroblast Reproduced by courtesy of Elsevier North/Holland (Potten *et al*, 1979).

CONCLUSIONS

Biological modelling of surface epithelia can be achieved and these models provide working hypotheses for the proliferative cell organisation of the tissue. Complete confirmation of the models awaits further experimentation. The models outlined have many similarities with those formulated for other systems (eg. bone marrow, see also Gilbert & Lajtha, 1965; Potten *et al*, 1979). Future mathematical modelling could help strengthen the models, provide a mathematical basis for some of the processes and interactions and help in formulating predictions of the response of the tissue to various insults and changing situations including carcinogenic transformation.

ACKNOWLEDGEMENTS

This work was supported by grants from the Medical Research Council and the Cancer Research Campaign. The ideas in this paper have been developed over several years and have stemmed from discussions with many colleagues including Charles Gilbert and Don Major. I am particularly grateful to both Terry Allen and Bill Hume with whose help several ideas were developed. Thanks are also due to Irene Nicholls, Caroline Chadwick, Joan Bulloch, and Dorothy Robinson for their excellent technical assistance. This was a paper presented at the Conference on Models of Biological Growth and Spread, Mathematical Theories and Applications. Heidelberg, F.R.G. July 1979

REFERENCES

Allen, T.D. & Potten, C.S. (1974) Fine structural identification and organization of the epidermal proliferative unit. J. cell Sci. 15 291-319

Allen, T.D. & Potten, C.S. (1976) Significance of cell shape in tissue architecture Nature 264 545-546

Cairns, J. (1975) Mutation selection and the natural history of cancer. Nature 255 197-200

Cheng, H. & Leblond, C.P. (1974) Origin, differentiation and renewal of the four main epithelial cell types in the mouse small intestine. V Unitarian theory of the origin of the four epithelial cell types. Am. J. Anat. 141 537-562

Christophers, E. (1971a) Cellular architecture of the stratum corneum. J. invest. Derm. 56 165-169

Christophers, E. (1971b) Die Epidermale Columnarstruktur. Z. Zellforsch. 114 441-450

Gilbert, C.W. & Lajtha, L.G. (1965) The importance of cell population kinetics in determining response to irradiation of normal and malignant tissue Cellular Radiation Biology. The University of Texas. Williams and Wilkins Co. Baltimore 474-495

Hume, W.J. & Potten, C.S. (1976) The ordered columnar structure of mouse filiform papillae. J. cell Sci. 22 149-160

Hume, W.J. & Potten, C.S. (1979) Advances in epithelial kinetics: an oral view. J. oral Path. 8 3-22

Iversen, O.H., Bjerknes, R. & Devik, F. (1968) Kinetics of cell renewal, cell migration and cell loss in the hairless mouse dorsal epidermis. Cell Tissue Kinet. 1 351-367

Karatschai, M., Kinzel, V., Goerttler, K. & Suss, R. (1971) "Geography" of mitosis and cell divisions in the basal layer of mouse epidermis. Z. Krebsforsch. Klin. Onkol. 76 59-64

Kelvin, W.T. (1894) On the homogeneous division of space. Proc. Roy. Soc. 55 1-16

Lajtha, L.G. (1963) On the concept of the cell cycle. J. cell. comp. Physiol. 62 (suppl.) 143-145

Leblond, C.P., Clermont, Y. & Nadler, N.J. (1967) The pattern of stem cell renewal in three epithelia (Esophagus, intestine and testis). Canadian Cancer Conf.

Pergamon Press Oxford 7 3-30

Mackenzie, I.A. (1969) Ordered structure of the stratum corneum of mammalian skin. Nature 222 881-882

Menton, D.N. (1976a) A minimum-surface mechanism for the organisation of cells into columns in the mammalian epidermis. Am. J. Anat. 145 1-22

Menton, D.N. (1976b) Liquid film model of tetrakaidecahedral packing to account for the establishment of epidermal cell columns. J. invest. Derm. 66 283-291

Potten, C.S. (1974) The epidermal proliferative unit: the possible role of the central basal cell. Cell Tissue Kinet. 7 77-88

Potten, C.S. (1975a) Epidermal transit times. Brit. J. Derm. 93 649-658

Potten, C.S. (1975b) Epidermal cell production rates. J. invest. Derm. 65 488-500

Potten, C.S. (1978a) Cellular and tissue response of skin to single doses of ionising radiation. Current Topics in Rad. Res. 13 1-59

Potten, C.S. (1978b) Epithelial proliferative subpopulations. In Stem Cells and Tissue Homeostasis Ed. Lord, B.I., Potten, C.S. & Cole, R.J. Cambridge Univ. Press 317-334.

Potten, C.S. (1980) Normal epidermopoiesis. In The Epidermis in Disease Springer-Verlag In press

Potten, C.S., Al-Barwari, S.E. & Searle, J. (1978a) Differential radiation response amongst proliferating epithelial cells. Cell Tissue Kinet. 11 149-160

Potten, C.S. & Allen, T.D. (1975) Control of epidermal proliferative units (EPUs). An hypothesis based on the arrangement of neighbouring differentiated cells. Differentiation 3 161-165

Potten, C.S. & Hendry, J.H. (1973) Clonogenic cells and stem cells in epidermis. Int. J. Rad. Biol. 24 537-540

Potten, C.S. & Hendry, J.H. (1975) Differential regeneration of intestinal proliferative cells and cryptogenic cells after irradiation. Int. J. Rad. Biol. 27 413-424

Potten, C.S., Hume, W.J., Reid, P. & Cairns, J. (1978b) The segregation of DNA in epithelial stem cells. Cell 15 899-906

Potten, C.S., Schofield, R. & Lajtha, L.G. (1979) A comparison of cell replacement in bone marrow, testes and three regions of surface epithelium. Biochim. Biophys. Acta Reviews on Cancer 560 281-299

Thompson, D.W. (1961) On growth and form. Cambridge University Press 327pp

Wright, N.A. (1978) The cell population kinetics of repopulating cells in the intestine. In Stem Cells and Tissue Homeostasis. Ed. Lord, B.I., Potten, C.S. & Cole, R.J. Cambridge Univ. Press 335-358

Density dependent Markov population processes.

A.D. Barbour

Gonville and Caius College, Cambridge, U.K.

A sequence $(X_N)_{N\geq 0}$ of Markov processes on the k-dimensional integer lattice $(Z^+)^k$ is called density dependent if X_N has transition rates of the form $X \to X+j$ at rate $Ng_j(N^{-1}X)$, where the g_j are independent of N. Many phenomena of practical interest, such as epidemics, learning processes, chemical reactions and competition between species, can naturally be described in this way, with the components of X_N denoting the counts of the k populations being studied. The mathematical problem is to find approximations to the processes, especially when N is large, which enable one to describe simply the way in which they evolve.

The basic approach to approximation has been to mimic the classical theory of partial sums of independent random variables. This turns out to be successful because, for large N, the process $x_N \equiv N^{-1}X_N$ typically evolves by means of a rapid succession of individually insignificant steps, very much as in the classical setting. The main theorems, analogous to the weak law of large numbers and the central limit theorem, were proved for discrete time chains by Norman (1972) and for continuous time processes by Kurtz (1970)(1971), in each case for a somewhat more general class of processes than that considered here. In the present framework, Kurtz's central limit theorem could be expressed as follows.

Define

$$F(x) = \Sigma_j\, j g_j(x)\,; \qquad \sigma(x) = \Sigma_j\, j j^T g_j(x)\,,$$

and suppose that F has a uniformly continuous matrix of first partial derivatives DF and that σ is bounded and uniformly continuous. Fixing any x_0, let ξ_0 denote the solution of the equation $\dot{x} = F(x)$ which has $\xi_0(0) = x_0$, and write $A_s = DF(\xi_0(s))$.

<u>Theorem 1</u> (Kurtz, 1971) If, in addition,

$$\lim_{m\to\infty} \sup_x \sum_{|j|>m} |j|^2 g_j(x) = 0,$$

and if $\lim_{N\to\infty} \sqrt{N}[x_N(0) - x_0] = z$, then $(\sqrt{N}[x_N(s) - \xi_0(s)])_{0 \le s \le t}$ converges weakly as $N \to \infty$ to the diffusion Z with infinitesimal drift and covariance given by $A_s Z(s)$ and $\sigma(\xi_0(s))$, and with $Z(0) = z$.

In particular, Z(s) has mean $C_s z$ and covariance matrix

$$\Sigma_s = C_s \int_0^s C_u^{-1} \sigma(\xi_0(u)) (C_u^{-1})^T C_s^T \, du,$$

where C_s is the unique matrix solution to the equation $\dot{C}_s = A_s C_s$ which has $C_0 = I$. Σ_s can also be found as the solution of the equation $\dot{\Sigma}_s = A_s \Sigma_s + \Sigma_s A_s^T + \sigma(\xi_0(s))$ which has $\Sigma_0 = 0$. It is, of course, necessary also, for the theorem to hold, that ξ_0 exists throughout [0,t].

The rate of convergence in this central limit theorem has also been well established. For instance, when $x_N(0) = x_0$, Kurtz (1978) has shown, using the theorem of Komlós, Major and Tusnády (1975), that, if all but finitely many of the g_j are always zero, it is possible to construct a version of Z on the same probability space as x_N, in such a way that

$$\sup_{0 \leq s \leq t} |\sqrt{N}[x_N(s) - \xi_o(s)] - Z(s)| = O(N^{-1/2}\log N).$$

Similar results can also be proved under weaker assumptions on the g_j, but with larger error terms. The natural rate of $N^{-1/2}$ for the convergence of the distribution of $\sqrt{N}[x_N(t) - \xi_o(t)]$ to that of $Z(t)$ has also been demonstrated, in considerable generality, by Alm (1978).

In practice, the restriction of s to a fixed finite interval as N increases can be a significant limitation on the usefulness of the results, when x_o is an equilibrium point of the deterministic equations $\dot{x} = F(x)$. To take the simplest case, if x_N is positive recurrent with limiting distribution π_N and x_o is the only point where $F(x_o) = 0$, it is natural to ask whether the limiting distribution of $\sqrt{N}(x_N - x_o)$ is close to that of $Z(\infty)$ - i.e. normal with mean zero and covariance matrix Σ satisfying $A\Sigma + \Sigma A^T + \sigma(x_o) = 0$, where $A \equiv DF(x_o)$. The question is not easily answered by the theorems mentioned above, because the equilibrium distribution depends on the whole process, and not just on its distribution over a finite time interval.

A theorem of this nature was proved by Norman (1974) (for discrete time processes), assuming that x_N was restricted to a compact subset of R^k. When considering unbounded subsets of R^k, however, one encounters the problem that, whereas, for technical reasons, it is important that $G(x) \equiv \Sigma_j\, g_j(x)$ should be bounded as x varies, this is no longer a natural requirement, and indeed would in practice be rather irksome. For a finite time interval [0,t], one is able to ignore the problem because, for large N, few of the sample paths of x_N stray far from the compact set $K = \{\xi_o(s),\ 0 \leq s \leq t\}$; hence one is able to work instead with the process $\bar{x}_N$ obtained by stopping x_N on first leaving $K^{\varepsilon}=\{x\in R^k: |x-K|<\varepsilon\}$, for some $\varepsilon>0$, and $\bar{x}_N$ has bounded transition rates if G is merely continuous, which is a much more natural condition. The equilibrium distribution π_N, however, depends on the behaviour of paths throughout the state space, and so something else is needed.

Fortunately, there is a simple relationship between the equilibrium distributions of continuous time Markov processes which have the same transition probabilities but different residence time parameters. This enables one to analyse the equilibrium distribution for a process with unbounded G by working instead with a process which has bounded G, but the same transition probabilities. The method works well in practice, but has its drawbacks. For instance, if we consider a process x_N in one dimension which, for large x, has transition rates $x \to x+N^{-1}$ at rate x^4-1; $x \to x-N^{-1}$ at rate x^4+1, so that $G(x) = 2x^4$ is unbounded in x, we must compare it with, for example, the process with bounded rates

$$x \to x + N^{-1} \text{ at rate } 1 - x^{-4} \; ; \quad x \to x - N^{-1} \text{ at rate } 1 + x^{-4}.$$

Now this latter process is null recurrent, and so has no equilibrium probability distribution, let alone one localized around x_o. Yet one can compute equilibrium distributions for birth and death processes, and, in particular, for the original process, and show that it is indeed approximately normally distributed, as conjectured.

A rather different situation occurs when x_o is an isolated unstable equilibrium of the deterministic equations and $G(x_o) > 0$, so that the stochastic fluctuation about x_o is in due course enough to send the process x_N drifting away along one of the deterministic trajectories emanating from x_o. The time scale for this to happen is typically of order logN, which again prevents Theorem 1 from being used, but it is still possible to use a diffusion approximation near x_o to describe the eventual behaviour of the process (Barbour, 1976).

There is, however, a closely related problem in which the diffusion approximation is not in itself enough. Suppose, for example, that x_N is one dimensional and confined to the interval [0,1], and that $F(x) = 0$ at exactly three points $0 < a < b < c < 1$, with a and c stable equilibria and b unstable. Then x_N is necessarily positive recurrent, and it is not too difficult to show (or imagine!) that its equilibrium distribution π_N is concentrated around a and c: indeed, conditional on being near a, say, $\sqrt{N}(x_N - a)$ is asymptotically normally distributed with mean zero and variance $-\sigma(a)/2F'(a)$. The asymptotic description of π_N is thus complete, but for the relative weights to be attached to the distributions around a and c. A tempting way of evaluating them is to try replacing x_N everywhere by its diffusion approximation, with drift $F(x)$ and infinitesimal variance $N^{-1}\sigma(x)$, for which the computations can reasonably easily be made. However, the results obtained are not even asymptotically correct: the weights apparently depend in an essential way on the detailed structure of x_N.

There is a much more obvious context in which diffusion approximations cannot be appropriate: that is, when a process is very near a point x_0 on a natural boundary, so that $G(x_0) = 0$. In such a situation, an individual jump may have a marked effect on the whole evolution of the process , since it may either lead to absorption at x_0 or substantially increase the chance of escape from x_0, whereas, if a diffusion is to describe a process satisfactorily, one expects individual jumps to have negligible effect. This makes it necessary to look for a different kind of approximation.

For definiteness, suppose that $x_o^{(i)}$, the ith component of x_o, is zero for each i in some subset $I \subseteq \{1,2,\ldots,k\}$, and that $g_j(x) = x^{(i_j)} f_j(x)$ for each j, where, if $f_j(x)$ is not identically zero, $f_j(x_o) \neq 0$ and $i_j \in I$: assume also that all but finitely many of the f_j are indeed identically zero. This specification is broadly representative of the most common class of problems arising in practice from the presence of boundaries. An example is given by the closed stochastic epidemic, where $k = 2$ and the only non-zero transition rates for x_N are given by

$$x \rightarrow x + N^{-1}(-1,1) \quad \text{at rate} \quad N\alpha x^{(1)} x^{(2)}$$
$$x \rightarrow x + N^{-1}(0,-1) \quad \text{at rate} \quad N\beta x^{(2)}.$$

N here denotes a typical population size, and a physically natural initial condition is $x_N(0) = (1,N^{-1})$, corresponding to the introduction of one infective into a susceptible population of size N. For such an initial condition, it follows that $x_N(0) \rightarrow x_o = (1,0)$ as $N \rightarrow \infty$, and the process can be represented in the above form by taking $I = \{2\}$, $f_{(-1,1)}(x) = \alpha x^{(1)}$ and $f_{(0,-1)}(x) = \beta$. If $\alpha/\beta > 1$, it is easily seen that the ultimate behaviour of the epidemic process depends significantly on whether, for instance, the first transition is of type (-1,1) or (0,-1), rendering a diffusion approximation inappropriate.

The most appealing way of tackling the problem is to approximate the process in its early stages by another density dependent process $\tilde{x}_N$ with the same initial condition, but with $\tilde{g}_j(x) = x^{(i_j)} f_j(x_o)$. Under the further assumptions

$$j^{(i_j)} \geq -1 \ ; \qquad j^{(k)} \geq 0, \quad k \in I \setminus \{i_j\},$$

$N\tilde{x}_N$ can be expressed as a Markov branching process: the extra assumptions entail no real loss of generality because, without them, one or more of the counts X_i, $i \in I$, could become negative, and this would in turn imply the possibility of negative transition rates. Since Markov branching processes are fairly well understood, $\tilde{x}_N$ is a useful candidate for approximating x_N, and the problem becomes to establish how good an approximation it affords.

Let t_r $(\tilde{t}_r)$ denote the time of the rth jump of x_N $(\tilde{x}_N)$; choose any sequence $y_0, y_1, \ldots$ such that $P[\bigcap_{r=0}^{m}\{\tilde{x}_N(\tilde{t}_r) = y_r\}] > 0$, and pick any $0 = u_0 < u_1 < \ldots$. Then the Radon-Nikodym derivative of the measure P_N associated with x_N with respect to the measure $\tilde{P}_N$ associated with $\tilde{x}_N$ evaluated at $\{y_r, u_r\}_{r=0}^{m}$ can be written as

$$L_m \equiv L_m(\{y_r,u_r\}_{r\geq 0}) = \prod_{r=0}^{m-1}\{g_{j_r}(y_r)/\tilde{g}_{j_r}(y_r)\}\exp\{-N[G(y_r)-\tilde{G}(y_r)](u_{r+1}-u_r)\}$$

where $j_r \equiv y_{r+1} - y_r$. Let (Y,U) denote a stochastic process with realizations $\{y_r,u_r\}_{r\geq 0}$ as above. Then $\{L_m(Y,U)\}_{m\geq 0}$ is a martingale under $\tilde{P}_N$ with respect to the σ-fields Σ^m generated by $\{y_r,u_r\}_{r=0}^{m}$, and it follows by direct computation that

$$\tilde{E}_N\{(L_{m+1}-L_m)^2 \mid \{y_r,u_r\}_{r\geq 0}^{m}\} =$$

$$= L_m^2 \cdot \frac{\Sigma_j\{y_m^{(ij)}(f_j(y_m)-f_j(x_o))^2/f_j(x_o)\}}{\Sigma_j\, y_m^{(ij)}(2f_j(y_m)-f_j(x_o))} .$$

Choosing $\varepsilon > 0$ so small that, whenever $|y-x_o| \leq \varepsilon$,

$$|2f_j(y) - f_j(x_o)| > f_j(x_o)/2 \quad \text{for all } j,$$

and assuming that

$$\max_j \sup_{|y-x_o|\leq\varepsilon} |f_j(y) - f_j(x_o)|/|y-x_o| < \infty ,$$

it follows that there exists $K > 0$ such that, whenever $|y_m-x_o| \leq \varepsilon$,

$$\tilde{E}_N\{(L_{m+1} - L_m)^2 \mid \{y_r,u_r\}_{r\geq 0}^{m}\} \leq L_m^2 K |y_m - x_o|^2.$$

Hence, if $\tau(m)$ denotes the first r for which $|y_r - x_0| > m/N$, where $m/N < \varepsilon - \frac{1}{N}$, and if $L_r^* \equiv L_{r\wedge\tau(m)}$, it follows that

$$\begin{aligned}
\tilde{E}_N(L_n^*-1)^2 &= \tilde{E}_N\{\Sigma_{r=1}^n \tilde{E}_N[(L_r^* - L_{r-1}^*)^2|\Sigma^{r-1}]\} \\
&\leq \tilde{E}_N\Big\{\sum_{r=0}^{n-1} KL_r^2|y_r - x_0|^2 I[\tau(m) \geq r]\Big\} \\
&\leq K(m/N)^2\tilde{E}_N\Big\{\sum_{r=0}^{n-1}(L_r^*)^2\Big\} \\
&= K(m/N)^2\Big\{n + \sum_{r=0}^{n-1}\tilde{E}_N(L_r^* - 1)^2\Big\},
\end{aligned}$$

since L_r^* is itself a $\tilde{P}_N$ martingale. Hence, by induction on n,

$$\tilde{E}_N(L_n^* - 1)^2 \leq C_1 nm^2 N^{-2}$$

whenever $m/N < \varepsilon - \frac{1}{N}$ and $nm^2 \leq C_2N^2$.

Now, for any event $A \in \Sigma^n \cap \Sigma^{\tau(m)}$,

$$P_N(A) = \tilde{E}_N\{L_n^* I[A]\} = \tilde{P}_N(A) + \tilde{E}_N\{(L_n^* - 1)I[A]\}.$$

Take any $\eta > 0$, and let B_η denote the event $\{|L_n^* - 1| < \eta\}$. Then

$$\begin{aligned}
&|\tilde{E}_N\{(L_n^* - 1)I[A]\}| \\
\leq\ & \tilde{E}_N\{|L_n^* - 1|I[A\cap B_\eta]\} + \tilde{E}_N\{|L_n^* - 1|I[A\setminus B_\eta]\} \\
\leq\ & \eta\tilde{P}_N[A\cap B_\eta] + \eta^{-1}\tilde{E}_N\{(L_n^* - 1)^2 I[A\setminus B_\eta]\} \\
\leq\ & \eta + \eta^{-1}C_1 n(m/N)^2.
\end{aligned}$$

Thus, choosing $\eta = m\sqrt{n}/N$, it follows that $|P_N(A) - \tilde{P}_N(A)| \leq C_3 m\sqrt{n}/N$ whenever $A \in \Sigma^n \cap \Sigma^{\tau(m)}$, $m/N < \varepsilon$ and $nm^2 \leq C_2N^2$.

So taking, for instance, m and n both of order $N^{2/3-\alpha}$ for some $\alpha > 0$, one can approximate the probability of any event depending only on the first $N^{2/3-\alpha}$ transitions of x_N by its $\tilde{x}_N$ probability, to an accuracy of order $N^{-3\alpha/2}$. This rather detailed form of approximation is more than enough to cope with paths which are eventually absorbed at the boundary: those that are not absorbed are, after $N^{2/3-\alpha}$ transitions, far enough away from the boundary for approximations based on diffusion around deterministic drift to be appropriate for describing their subsequent evolution.

It is interesting to consider the various orders of magnitude of stochastic variation caused by starting X_N near such a boundary. Consider, for the purposes of illustration, X_N to be precisely a super-critical positive regular Markov branching process starting with just one organism, so that the results in Athreya and Ney (1972), Chapter V, Sections 7 and 8, can be applied directly. The huge difference between absorption and non-absorption is naturally the largest effect. Next in importance, in the event of non-absorption, is the random delay of order 1 resulting from the variability in the timing of the first few transitions. This can be described as follows. Let A be the infinitesimal generator of the semigroup of matrices $M(t)$ with $M_{rs}(t) \equiv E\{X_N^{(s)}(t)\,|\,X_N(0)=e_r\}$ (the notation is that of Athreya and Ney: unfortunately, the matrix A_s defined before Theorem 1 is here equal to A^T for all s), and let the eigenvalues of A be, in descending order of real parts, $\lambda_1, \lambda_2, \ldots$. Denote by v and u respectively the strictly positive left and right eigenvectors of A with eigenvalue $\lambda_1 > 0$, normalized by choosing $u.v = 1$, $u.1 = 1$. Then it is known that $u.X_N(t)e^{-\lambda_1 t} \to W$ a.s. as $t \to \infty$, and that, on the set of non-absorption, $X_N(t)/(u.X_N(t)) \to v$ a.s. In other words, the relative magnitudes of the $X_N^{(r)}$ settle down to the values predicted by the deterministic equations, but the time at which their absolute magnitude reaches a given level is random. Thus, in order for $u.X_N$ to attain the magnitude N of interest in the present problem, the time taken is approximately $\lambda_1^{-1}(\log N - \log W)$, which has a dominant deterministic term $\lambda_1^{-1}\log N$, and then a random component

$-\lambda_1^{-1}\log W$ of order 1.

The next question to consider is how large are the variations of the relative magnitudes $X_N^{(r)}(t)/(u.X_N(t))$ about their deterministic values $v^{(r)}$, and how they compare with the $O(N^{-1/2})$ diffusion about the deterministic path which obtains once X_N is of magnitude N. Again the answers are provided, at least in outline, in Athreya and Ney (1972): see also Athreya (1969). Let w be a right eigenvector of A with eigenvalue $\lambda \neq \lambda_1$, and let a denote $\mathrm{Re}\{\lambda\}$. Then there are three cases to be distinguished. If $a > \lambda_1/2$, $\eta_N \equiv (w.X_N(\tau_N))/(u.X_N(\tau_N)) = O(N^{-1+a/\lambda_1})$, where, for some $\varepsilon > 0$, τ_N denotes the time at which $u.X_N$ first reaches the value εN: if $a = \lambda_1/2$, $\eta_N = O(N^{-1/2}\log N)$, and, if $a < \lambda_1/2$, $\eta_N = O(N^{-1/2})$. (Of course, $w.v = 0$.) In the first case, as with W, the significant variability arises from the first O(1) transitions; in the last case, the effects of the first $O(N^{\beta})$ transitions are unimportant for any $\beta < 1$, so that one is essentially able to use Theorem 1 to provide the detailed answer; the intermediate case falls somewhat between the two extremes, in that it is early behaviour which is important, but with the proportion of the variability due to the first m transitions being of order $\sqrt{(\log m/\log N)}$. In all but the last case, the variability arising in the initial stages, even when the delay embodied in W has been conditioned out, is of larger order of magnitude than that arising from the subsequent diffusion described by Theorem 1.

The differences between the three cases can be explained as follows. Away from a zero of F, nearby deterministic paths are almost parallel, and x_N merely superimposes a diffusion of order $N^{-1/2}$ onto what is an almost constant drift. However, near a natural boundary, such as the origin for a positive regular Markov branching process, it is possible for neighbouring deterministic paths to diverge quite rapidly. In such cases, the effects of an initial small amount of diffusion may be greatly magnified by the subsequent divergence of the deterministic paths. Thus, two deterministic paths of the Markov branching process, one starting at v and the other at v + w', where w' is a left eigenvector of A corresponding to the eigenvalue λ, are separated after time t by an amount of modulus $|w'|e^{at}$ ($a = \mathrm{Re}\{\lambda\}$), so that initial discrepancies in the w' direction are magnified by a factor e^{at}. To evaluate the importance of this deterministic magnification, it must be compared with the effects of natural diffusion, which, evolving as the square root of the number of transitions, attains a magnitude $e^{\lambda_1 t/2}$ after time t. Thus, for $a < \lambda_1/2$, the natural diffusion is the more important, and the eventual effect on x_N (net of time delay) is of order $N^{-1/2}$. For $a > \lambda_1/2$, the deterministic separation is the more important, and the initial fluctuations in the first O(1) transitions, which in x_N have magnitude N^{-1}, are magnified over the timespan of $\lambda_1^{-1}\log N$ by a factor of N^{a/λ_1}, and dominate the subsequent diffusion. Note that if $a = \lambda_1$, which can only happen if the process is <u>not</u> positive regular, the lasting effects of the initial fluctuation are of order $N^{-1+a/\lambda_1} = 1$: an example of this kind was examined by Osei and Thompson (1977). In the intermediate case, a more careful balance between the two sources of variation has to be achieved.

It is tempting to look upon the considerable progress that has been made in approximating Markov population processes and conclude that an effective practical means of describing biological processes has been developed. In certain respects, however, this is not the case. For instance, in order to preserve a Markovian model, it is necessary to assume that lifetimes of individuals have negative exponential distributions - or, at the expense of some increase in the number of populations, Erlangian distributions. It turns out that, even at the level of the deterministic approximation, the choice of lifetime distribution is important. The simplest example of this is the age dependent branching process, in which, although the average behaviour is of the form $x_t = x_o e^{\alpha t}$, the constant α is not, in general, as the law of large numbers might suggest, the reciprocal of the mean lifetime. Thus, even when dealing with large population approximations, the dependence upon the underlying distributions is considerably more complex than in the classical theory of sums of independent random variables. Important progress in this direction has been made by Wang (1975,1977a,1977b), but the problems are very much more difficult. Another important feature of real populations is that they are spread out over space, the way in which they are distributed having a significant effect on the progress of contact processes in particular. The recent work of Arnold and Theodosopulu (1979) in extending the theory of density dependent processes to include such models has had considerable success, though once again the problems are much more difficult. At present, it seems likely that Markov population processes give a useful qualitative description of many real life phenomena: but to what extent they actually do, and how their predictions need to be modified to accommodate the various discrepancies between real phenomena and the ideal model, are still relatively little explored.

References.

[1] Alm, S.E. (1978) On the rate of convergence in diffusion approximation of jump Markov processes. Ph.D. Thesis, Uppsala University.

[2] Arnold, L. and Theodosopulu, M. (1979) Deterministic limit of the stochastic model of chemical reactions with diffusion. Private communication.

[3] Athreya, K.B. (1969) Limit theorems for multitype continuous time Markov branching processes, II: the case of an arbitrary linear functional. Z. Wahrscheinlichkeitstheorie 13, 204-214.

[4] Athreya, K.B. and Ney, P.E. (1972) Branching Processes. Springer-Verlag, Berlin.

[5] Barbour, A.D. (1976) Quasi-stationary distributions in Markov population processes. Adv. Appl. Prob. 8,296-314.

[6] Komlós, J., Major, P. and Tusnády, G. (1975) An approximation of partial sums of independent RV'-s and the sample DF. I. Z. Wahrscheinlichkeitstheorie 32, 111-131.

[7] Kurtz, T.G. (1970) Solutions of ordinary differential equations as limits of pure jump Markov processes. J. Appl. Prob. 7, 49-58.

[8] Kurtz, T.G. (1971) Limit theorems for sequences of jump Markov processes approximating ordinary differential processes. J. Appl. Prob. 8, 344-356.

[9] Kurtz, T.G. (1976) Limit theorems and diffusion approximations for density dependent Markov chains. Math. Programming Study 5, 67-78.

[10] Kurtz, T.G. (1978) Strong approximation theorems for density dependent Markov chains. Stoch. Procs Applics 6, 223-240.

[11] Norman, M.F. (1972) Markov processes and learning models. Academic Press, New York.

[12] Norman, M.F. (1974) A central limit theorem for Markov processes that move by small steps. Ann. Prob. 2, 1065-1074.

[13] Osei, G.K. and Thompson, J.W. (1977) The supersession of one rumour by another. J. Appl. Prob. 14, 127-134.

[14] Wang, F.J.S. (1975) Limit theorems for age and density dependent stochastic population models. J. Math. Biol. 2, 373-400.

[15] Wang, F.J.S. (1977a) A central limit theorem for age and density dependent population processes. Stoch. Procs Applics 5, 173-193.

[16] Wang, F.J.S. (1977b) Gaussian approximation of some closed stochastic epidemic models. J. Appl. Prob. 14, 221-231.

A MODEL OF DEVELOPMENT OF A CELL

R. Bartoszyński

Institute of Mathematics
Polish Academy of Sciences
Warszawa

INTRODUCTION

This paper presents some preliminary results of the current research, joint with J.R. Thompson, A. Michałowski and J. Ćwik. The object of the study is to construct a model which would explain the following phenomena: (1) Cell loss, i.e. the fact that some cells die before reaching mitosis, and (2) the variability of the intermitotic times.

The central idea of the model is the assumption that within the cell there occur two stochastic processes. One of them, say $X(t)$, is a birth and death process of some type of organellas. When $X(t)$ reaches the value 0, the cell dies (perhaps not immediately).

The organellas in question are assumed to be responsible for supplying energy to the cell. The energy level, $Y(t)$, is an integer valu stochastic process, whose jump intensity at t is proportional to $X(t)$

The process $Y(t)$ is intended to reflect the "maturity" of the cell for mitosis. This is modelled by the assumption that the intensity of entering the mitosis is proportional to $Y(t)$. When the cell divides, the organellas existing at that time are partitioned randomly among the two daughter cells, which then evolve independently, starting from some initial numbers of organellas, and energy levels 0.

FORMAL ASSUMPTIONS OF THE MODEL

Let us begin with the analysis of life of a single cell until it reaches mitosis. Assume that the cell starts its evolution at $t = 0$, with m organellas present, so that $X(0) = m$, and energy level $Y(0) =$

Since we are interested only in the number of organellas at mitos (which influence the initial states of the daughter cells), we may introduce another process, $Z(t)$, with $Z(t) = k$ signifying that mitosi has occurred at some time $t' < t$, with $X(t') = k$, and $Z(t) = *$ signifying that mitosis has not occurred until t. Observe that we may have $Z(t) = 0$: to achieve a convenient symmetry in the formulas we allow mitosis of a cell with no organellas. To obtain the probability of death of a cell, we shall later collect the appropriate states together.

The state of a cell is therefore described by a vector (x,y,z), where the first two coordinates are nonnegative integers, and the third coordinate is either a nonnegative integer or $*$.

The process of transitions will be assumed to be a Markov process satisfying the following assumptions.

POSTULATE 1. Any state (x,y,z) with $z \neq *$ is absorbing.

POSTULATE 2. Given the state $(x,y,*)$ at t, the probability of transition to $(x+1,y,*)$ and $(x-1,y,*)$ before t+h is independent of y and equal respectively $\lambda xh + o(h)$ and $\mu xh + o(h)$.

This postulate implies that the first coordinate (process X(t)) is a linear birth and death process, with intensities λ and μ independent of the values of other coordinates.

POSTULATE 3. Given the state $(x,y,*)$ at t, the probability of transition to the state $(x,y+1,*)$ before t+h equals $\rho xh + o(h)$.

This assumption means that the second coordinate (process Y(t)) is a pure birth process, with the jump intensity proportional to the number X(t) of organellas.

POSTULATE 4. Given the state $(x,y,*)$ at t, the probability of transition to (x,y,x) before t+h (mitosis with x organellas) equals $\nu yh + o(h)$.

This means that the intensity of mitosis is governed by the energy level Y(t).

Finally, we add

POSTULATE 5. Given the state $(x,y,*)$ at t, the probability of any transition other than those specified in Postulates 2 - 4 is of the order o(h).

ANALYSIS

Let now $P_{k,n}(t) = P(X(t) = k, Y(t) = n, Z(t) = *)$. By standard reasoning we obtain

PROPOSITION 1. The probabilities $P_{k,n}$ satisfy the following system of differential equations:

$$P'_{k,n} = \lambda(k-1)P_{k-1,n} - (\lambda+\mu+\rho)kP_{k,n} + \mu(k+1)P_{k+1,n} + \rho kP_{k,n-1} - \nu nP_{k,n} \qquad (k,n > 0),$$

$$P'_{0,n} = \mu P_{1,n} - \nu n P_{0,n},$$

$$P'_{k,0} = \lambda(k-1)P_{k-1,0} - (\lambda+\mu+\rho)kP_{k,0} + \mu(k+1)P_{k+1,0},$$

$$P'_{0,0} = \mu P_{1,0}.$$

If the cell starts its evolution with m organellas, then the initial conditions are $P_{m,0}(0) = 1$, $P_{k,n}(0) = 0$ for $(k,n) \neq (m,0)$. Let

$$G(z,s,t) = \sum_{k=0}^{\infty}\sum_{n=0}^{\infty} z^k s^n P_{k,n}(t).$$

We obtain then, after some transformations

PROPOSITION 2. <u>The probability generating function</u> $G(z,s,t)$ <u>satisfies the partial differential equation</u>

(1) $$\frac{\partial G}{\partial t} - \left[\lambda z^2 - (\gamma - \rho s)z + \mu\right]\frac{\partial G}{\partial z} + \nu s\frac{\partial G}{\partial s} = 0$$

<u>where</u> $\gamma = \lambda+\mu+\rho$. <u>The initial condition is</u>

(2) $$G(z,s,0) = z^m.$$

Denoting by $Q_k(t)$ the probability of mitosis before t with k organellas (that is, probability of a state (k,y,k) for some y), and putting

(3) $$H(z,t) = \sum_{k=0}^{\infty} z^k Q_k(t),$$

we obtain easily

PROPOSITION 3. <u>The probabilities</u> $Q_k(t)$ <u>and their probability generating function</u> $H(z,t)$ <u>satisfy the equations</u> $Q'_k = \nu\sum_n nP_{k,n}$ <u>and</u>

(4) $$\frac{\partial H}{\partial t} = \nu\left.\frac{\partial G}{\partial s}\right|_{s=1}.$$

The characteristic equations for (1) are

$$\frac{dt}{1} = -\frac{dz}{\lambda z^2 - (\gamma - \rho s)z + \mu} = \frac{ds}{\nu s}.$$

This yields $s = C_1 e^{\nu t}$ and

(5) $$\frac{dz}{dt} = -\lambda z^2 + (\gamma - C_1\rho e^{\nu t})z - \mu,$$

which is a Ricatti equation, in general not solvable explicitly. However, using the results of Iwiński (1962), we get

PROPOSITION 4. <u>An explicit solution for the probability generating function</u> $G(z,s,t)$ <u>exists for</u> $\nu = (\gamma+\sqrt{\Delta})/2$, <u>where</u> $\Delta = \gamma^2 - 4\lambda\mu$. <u>The solution, satisfying the initial condition (2) is given by the formula below, in which</u> $\nu^* = (\gamma - \sqrt{\Delta})/2$:

$$G(z,s,t) = g(z,s,t)^m,$$

where

$$g(z,s,t) = \frac{1}{\lambda}(\nu^* - \rho s e^{-\nu t}) +$$

$$+ \frac{z - \frac{1}{\lambda}(\nu^* - \rho s)}{\exp\left[\frac{\rho s}{\nu}(1-e^{-\nu t}) + t\sqrt{\Delta}\right] - \lambda\left[z - \frac{1}{\lambda}(\nu^* - \rho s)\right] I(t)}$$

and

(6) $$I(t) = \int_0^t \exp\left[\frac{\rho s}{\nu} e^{-\nu t}(e^{\nu u} - 1) + u\sqrt{\Delta}\right] du \, .$$

The proof is by straightforward but tedious calculations. Firstly, we observe that equation (5) has a solution of the form $z = a + be^{\nu t}$ for $\nu = (\gamma \pm \sqrt{\Delta})/2$. This allows determining a and b, and obtaining thus a particular solution of the Ricatti equation (5), say z_1. Classical substitution $z = z_1 + 1/y$ reduces (5) to a linear equation, and consequently, leads to determination of

$$C_2 = \exp\left[\frac{\rho s}{\nu}(1-e^{-\nu t}) + (\gamma - 2\nu^*)t\right]\left[z - \frac{1}{\lambda}(\nu^* - \rho s)\right]^{-1} - \lambda I(t)$$

with I(t) given by (6). Here ν and ν^* are equal $(\gamma \pm \sqrt{\Delta})/2$, chosen with opposite signs. The general solution of (1) is thus $F(C_1, C_2)$ with $C_1 = se^{-\nu t}$, and the initial condition (2) gives $F(s,z) = (1/z + \frac{1}{\lambda}(\nu^* - \rho s))^m$. Finally, analysing the behaviour of the solution for large t we see that for the solution to be probabilistically meaningful, one must take $\nu = (\gamma + \sqrt{\Delta})/2$ and $\nu^* = (\gamma - \sqrt{\Delta})/2$.

To determine the function H(z,t), one may substitute s = 1 in (1), and express $\partial G/\partial s$ by $\partial G/\partial z$ and $\partial G/\partial t$. Substituting to (4), using the fact that H(z,0) = 0, and integrating, we obtain

PROPOSITION 5. In the notations of Proposition 4, we have

(7) $$H(z,t) = z^m - G(z,1,t) + \left[\lambda z^2 - (\lambda + \mu)z + \mu\right] \int_0^t \frac{\partial G(z,1,u)}{\partial z} du.$$

The knowledge of G(z,s,t) and H(z,t) provides useful information about the distributions of various quantities of interest. Firstly,

(8) $$F(t) = H(1,t) - H(0,t) = \sum_{k=1}^{\infty} Q_k(t)$$

is the probability that the cell will go to mitosis before t. Consequently, $F(t)/F(\infty)$ is the conditional distribution of intermitotic time, given that mitosis occurs. Fig. 1 gives the graph of the density of distribution (8) for some selected values of the parameters.

Similarly, $G(0,1,t) = \sum_{n=0}^{\infty} P_{0,n}(t)$ is the probability that death of the cell (or, more precisely, loss of all its organellas) occurs before t.

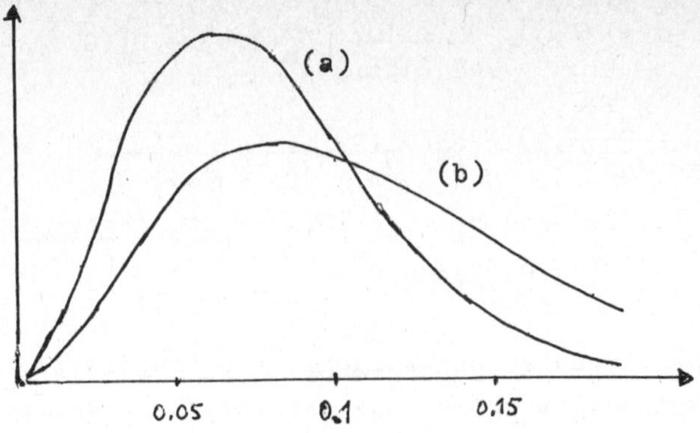

Fig. 1

Densities for $\lambda = 5$, $\mu = 2$, $\rho = 4$, $\nu = 10$, and $m = 10$ (curve (a)), and $m = 5$ (curve (b)).

Since $G(0,1,t) = g(0,1,t)^m$, the probability of death of a cell before mitosis decreases with the initial number m of organellas. Thus m plays the role of an index of "strength" of the cell.

Next, $\frac{\partial}{\partial z}H(z,\infty)\big|_{z=1} = \sum kQ_k(\infty)$ is the expected number of organellas at mitosis.

JOINT DISTRIBUTIONS

Let us now add

POSTULATE 6. <u>When a cell enters the state of mitosis with</u> k <u>organellas</u> (<u>i.e. the state</u> (k,y,k)), <u>it is replaced by two cells in states</u> $(J,0,*)$ <u>and</u> $(k-J,0,*)$, <u>which start evolving independently according to Postulates 1-5. Here</u> J <u>has the symmetric binomial distribution, so that</u>

$$P(J = m) = \binom{k}{m}2^{-k}, \quad m = 0,1,\ldots, k.$$

Thus, it is possible that we have $J = 0$ or $J = k$, which may be interpreted as birth of only one daughter cell. In this way the model accounts for the possibility of "abnormal" family trees (see Powell 1955).

Let now T, T_1 and T_2 denote the intermitotic times of the mothe cell and its two daughter cells, and let $X = \min(T_1,T_2)$, $Y = \max(T_1$ It is possible to obtain the formulas for the joint distribution of (T,T_1,T_2) and (T,X,Y), and their marginals. As an illustration, we give here one such distribution.

PROPOSITION 6. <u>If the mother cell begins its evolution with</u> m <u>organellas, then</u>

$$P(T \leq t, Y \leq x) = \frac{1}{H_m(1,\infty)}\Big[H_m(1,t) - H_m(0,t) - H_m(u_x,t) + $$
$$+ 2H\left(\frac{u_x+v_x}{2},t\right) - H_m(v_x,t)\Big],$$

where $u_x = g(1,1,x)$ *and* $v_x = g(1,0,x)$, *with* g *given by Proposition 4.*

DISCUSSION

The derived formulas are somewhat too complicated to permit the construction of estimators of the parameters. Qualitatively, an "average" cell must go to mitosis with about twice as many organellas as it had at its birth (which implies that we must have $\lambda > \mu$, otherwise the population becomes extinct). Moreover, to allow for an appreciable number of cells dying without mitosis, and appreciable variation in the initial number of organellas, their numbers should not be too large.

The suggested model appears to explain the facts (1) and (2) mentioned in the Introduction, that is, cell loss (as well as abnormal family trees), and the shape of the distribution on Fig. 1 seems to agree remarkably well with the histogram given by Powell (1955; p. 34).

The preliminary computer simulation suggest a low sister-sister correlation, and lack of mother-daughter correlation. The intuitive explanation of the existence of some sister-sister correlation is that when a cell enters mitosis, its organellas tend to split evenly between the two daughter cells, thus leading to sister cells which originate their evolution with numbers of organellas close one to another. This yields a sort of symmetric dependence between sister cells. The effect, however, is rather low, much lower than that reported by Powell (1955). Whether or not this should be regarded as a feature which disqualifies the model is an open question, since it is not clear if Powell's sister-sister correlation is not an artifact.

From the description of his experiment, it would appear that in collecting the data for sister-sister correlation, he used the following technique: upon observing a birth of two cells, one starts two clocks, and one keeps observing the development of the two cells. At some time, one of the cells enters mitosis; the first clock is then stopped, and its time, say x, recorded. After some time, the other cell enters mitosis, whereupon the second clock is stopped, and its time, say y, recorded. One gets then a pair of numbers, (x,y), and after accumulating sufficiently many such pairs, one calculates the sample correlation coefficient.

Such a procedure, however, contains a serious error: unless the order in each pair (x,y) is randomized (or unless sister cells are labelled at birth), we shall always have $x < y$. Thus, instead of calculating correlation between (T_1, T_2), we obtain correlation between $X = \min(T_1, T_2)$ and $Y = \max(T_1, T_2)$, the random variables which are correlated even if T_1 and T_2 are not. The difference between correlation of (X, Y) and correlation of (T_1, T_2) depends on the distribution but it may be as high as 0.5.

Thus, it is possible that the sister-sister correlation between cells reported by Powell is, in fact, lower or even nonexisting.

REFERENCES

Iwiński, T. "On a certain class of linear differential equations". Zastosowania Matematyki, 7 (1963), 59-76.

Powell, E.O. "Some features of the generation times of individual bacteria". Biometrika, 42 (1955), 16-44.

SPATIAL SPREAD IN BRANCHING PROCESSES

J.D. Biggins

University of Sheffield, England.

1. The results to be presented here concern the age-dependent branching process with the extra feature that people have positions in X ($= \mathbb{R}^d$ for some finite d). The process starts with an initial ancestor at the origin. Associated with this initial ancestor is a point process Z on $X \times (0,\infty)$ and a random variable $\ell > 0$. Let $\{(x_r, t_r)\}$ be some enumeration of the points of Z; then t_r is the birth time of the first generation person with the position x_r. The random variable ℓ is the life-length of the initial ancestor; a natural restriction, though we do not impose it, would be that $\ell \geq t_r$ for all r, that is, no child is born after its parent's death. Associated with each child is an independent copy of (Z,ℓ) which gives the positions and birth times of his children relative to his own, and the length of his life, and so it goes on.

Clearly, if we 'forget' about the positions of the people, the process is an age-dependent branching process of the kind described in Jager's book (1975). Let $\{(x_r^{(n)}, t_r^{(n)}, \ell_r^{(n)})\}$ be an enumeration of the positions, birth times and life-lengths of the nth generation people. At time t the positions of those people alive are those $x_r^{(n)}$ with $t_r^{(n)} \leq t < t_r^{(n)} + \ell_r^{(n)}$. It is reasonable to expect that, at least in some cases, the distance to the person furthest from the origin in any particular direction at time t should increase linearly with t. Let

$$C^{(t)} = \{t^{-1} x_r^{(n)} : t_r^{(n)} \leq t < t_r^{(n)} + \ell_r^{(n)}\},$$

so that $C^{(t)}$ is the set of positions of all those people alive at time t scaled by the factor t^{-1}. For any set $\mathcal{D}$ we will write $H\mathcal{D}$ for the convex hull of $\mathcal{D}$. A set H will be called an upper bound for the process if

$$\bigcap_{t>0} \bigcup_{s>t} HC^{(s)} \left(= \limsup_{t\to\infty} HC^{(t)}\right) \subset H \text{ a.s;}$$

roughly, any set larger than H contains $HC^{(t)}$ for large t.

Here some upper bounds for the process will be given. There is some overlap between the results given here and those in Biggins (1978); the argument given there (just after equation (3.4)) to obtain the upper bound is incorrect. An alternative argument is given here and the result extended to the process described above. A new formula for the set H, which covers the majority of cases and is comparatively easy to compute, is given. This formula is illustrated with a few examples.

2. In this section we will introduce more notation and concentrate attention on finding an upper bound when people can live forever, i.e. when $\ell = \infty$ with positive probability. The formula in this case has a natural generalization for the case when ℓ is finite and this is given in the following section.

The inner product of two vectors X and Y, of the same dimension, will be denoted by X.Y. For $\Theta \in \mathcal{X}$ and $\theta \in \mathbb{R}$ let

$$k(\Theta,\theta) = \log\left\{\mathbb{E}\left[\sum_r \exp(-(\Theta,\theta).(X_r,t_r))\right]\right\}.$$

We will assume throughout that $k(\Theta,\theta)$ is finite for some $(\Theta_o,\theta_o) \neq (0,0)$ (and so is finite for (Θ_o,θ) wherever $\theta \geq \theta_o$). When this condition fails the rate of spread in all directions is, in general, faster than linear. More restrictive conditions on the finiteness of k will be imposed in the last section. We will also assume that $k(0,0)>0$; this condition, that the expected number of people in the first generation is greater than one, ensures that the process survives with positive probability.

Let A be the smallest closed convex set in $\mathcal{X}\mathrm{x}\mathbb{R}$ that contains the points $\{(X_r,t_r)\}$ almost surely. For $A \in \mathcal{X}$ and $a \in \mathbb{R}$ let

$$\xi(A,a) = \inf\{k(\Theta,\theta)+(\Theta,\theta).(A,a) : (\Theta,\theta)\}, \tag{2.1}$$

then ξ is a concave function and in fact

$$k(\Theta,\theta) = \sup\{\xi(A,a)-(\Theta,\theta).(A,a) : (A,a)\} \tag{2.2}$$

(see Biggins (1978, eq.(2.5))). This is a result in convex function theory; we will make much use of such results in what follows. The book by Rockafellar (1970) has been used as the source of the necessary theory. For $\varepsilon \geq 0$ let us define the convex set

$$\mathcal{C}(\varepsilon) = \{(A,a) : \xi(A,a) \geq -\varepsilon\}$$

and $\mathcal{C}(0) = \mathcal{C}$, so that

$$\bigcap_{\varepsilon>0} \mathcal{C}(\varepsilon) = \mathcal{C}.$$

For any set $\mathcal{D}$ we will denote its interior by int$\mathcal{D}$, its closure by cl$\mathcal{D}$ and its boundary by $\partial\mathcal{D}$. If $\mathcal{D}$ is convex rint$\mathcal{D}$ denotes its relative interior. A polyhedral set is one that is the intersection of a finite number of half spaces.

Let

$$\mathcal{I}^{(n)} = \{(n^{-1}X_r^{(n)}, n^{-1}t_r^{(n)}) : r\}.$$

For some $\varepsilon>0$ suppose that $C(\varepsilon)\subset P$ where P is a closed polyhedral convex set. We will prove the following lemma.

Lemma 1

$$I^{(n)}\subset P \text{ for all but finitely many } n \quad \text{a.s.}$$

Proof. Let us suppose first of all that P is a half-space in $X\times\mathbb{R}$ given by

$$\{(A,a) : (A,a).(W,w)\geq c\}$$

where (W,w) is a unit vector. By Corollary 6.5.2 of Rockafellar (1970) either $C(\varepsilon)\subset \mathrm{int}P$ or $C(\varepsilon)\subset\partial P$. In the latter case $\{(A,a) : \xi(A,a)>-\infty\}\subset\partial P$ and so, using Lemma 1 of Biggins (1978) $A\subset\partial P$ and the result must hold in this case. In the other case, using Theorem 11.2 of Rockafellar there exists a hyperplane containing the set $\{(A,a,-\varepsilon) : (A,a)\in\partial P\}$ and not cutting the convex set $\{(A,a,y) : y\leq\xi(A,a)\}$. This plane must have the form

$$\left\{(A,a,y) : \phi^{-1}(y+\varepsilon) = (A,a).(W.w)-c\right\}$$

for some ϕ. In fact $\phi>0$ otherwise this plane cuts $\left\{(A,a,y) : y\leq\xi(A,a)\right\}$ for all $(A,a)\in \mathrm{rint}C(\varepsilon)$. If $\phi = \infty$ the plane is vertical and then $\xi(A,a) = -\infty$ for $(A,a)\notin P$; this is true if and only if $A\subset P$, again by Lemma 1 of Biggins (1978), and the stated result follows immediately. This leaves only the case when $0<\phi<\infty$ to consider. We can rewrite the equation of the plane as

$$y = \phi((A,a).(W,w)-c)-\varepsilon$$

and this is above the function ξ, therefore

$$\xi(A,a)\leq\phi((A,a).(W,w)-c)-\varepsilon.$$

Rearranging this gives

$$-\varepsilon\geq\left\{\xi(A,a)-\phi(A,a).(W,w)\right\}+\phi c$$

and so

$$-\varepsilon\geq\sup\left\{\xi(A,a)-\phi(A,a).(W,w) : (A,a)\right\}+\phi c = k(\phi W,\phi w)+\phi c,$$

using (2.2). Consequently

$$-\varepsilon\geq\inf\left\{k(\theta W,\theta w)+\theta c : \theta>0\right\}.$$

Let Ω_n be the event that some point of $I^{(n)}$ is not in P. A straightforward calculation using the branching property, see Biggins (1978, eq.(3.3)), shows that

$$\mathbb{E}\left[\sum_r \exp(-(\Theta,\theta).(X_r^{(n)},t_r^{(n)}))\right] = \exp(nk(\Theta,\theta)).$$

Hence

$$\exp(nk(\theta W,\theta w)) = \mathbb{E}\left[\sum_r \exp(-\theta n(n^{-1}X_r^{(n)},n^{-1}t_r^{(n)}).(W,w))\right]$$

$$\geqslant e^{-\theta nc}\mathbb{P}\left[\Omega_n\right] \text{ for } \theta>0;$$

thus

$$\mathbb{P}\left[\Omega_n\right]\leqslant\exp(\inf\{k(\theta W,\theta w)+\theta c : \theta>0\}n)\leqslant\exp(-\varepsilon n),$$

and so Ω_n occurs for only finitely many n.

Clearly if P is the intersection of a finite number of half-spaces and $C(\varepsilon)\subset P$ then $C(\varepsilon)$ is contained in each half-space and the stated result follows. □

Any closed convex set is the intersection of a countable number of polyhedral sets containing it. Hence for any $\varepsilon>0$

$$\lim_n\sup HI^{(n)}\subset C(\varepsilon) \quad \text{a.s.}$$

and so

$$\lim_n\sup HI^{(n)}\subset C \quad \text{a.s.}$$

Of course this is a result about the branching random walk on $X\times\mathbb{R}$; it is irrelevant up to this stage that $t_r>0$ a.s. for every r.

Let

$$\hat{C} = \mathrm{cl}H\{A : (aA,a)\in C \text{ for some } a>0\}$$

and let

$$\tilde{C} = \mathrm{cl}H\{\hat{C}\cup\{0\}\}.$$

The sets $\hat{C}(\varepsilon)$ and $\tilde{C}(\varepsilon)$ are similarly defined, then (see Biggins (1978, eq.(5.2),(5.4))

$$\bigcap_{\varepsilon>0}\hat{C}(\varepsilon) = \hat{C} \text{ and } \bigcap_{\varepsilon>0}\tilde{C}(\varepsilon) = \tilde{C}. \tag{2.3}$$

We will assume from now on that

$$\#\{(X_r,t_r) : r\}<\infty \quad \text{a.s.},$$

that is there are only a finite number of people in the 1st generation. It is immediate that subsequent generations are also finite.

Theorem 1

$$\lim_{t\to\infty}\sup HC^{(t)}\subset\tilde{C} \quad \text{a.s.}$$

Proof. For some $\varepsilon>0$ suppose that P is a closed polyhedral set in X containing

$\tilde{C}(\varepsilon)$; obviously $\hat{C}(\varepsilon)$ is also contained in P. Let

$$P_1 = \mathrm{cl}\{(cA,c) : A\in P\ c>0\}$$

then, by Theorem 19.7 of Rockafellar (1970), P_1 is a polyhedral set. Furthermore $C(\varepsilon)\subset P_1$ so that for some random N $I^{(n)}\subset P_1$ for $n\geq N$ by Lemma 1. Then $(t_r^{(n)})^{-1}x_r^{(n)}\in P$ for each r when $n\geq N$. As P contains the origin and is convex this implies that $t^{-1}x_r^{(n)}\in P$ for all r whenever $t\geq t_r^{(n)}$ and $n\geq N$. There are only a finite number of people in the first N generations and for each of these $t^{-1}x_r^{(n)}\to 0$ as $t\to\infty$. Therefore

$$\limsup_{t\to\infty} HC^{(t)}\subset P \quad \text{a.s.}$$

As P is any closed polyhedral set containing $\tilde{C}(\varepsilon)$ and $\varepsilon>0$ is arbitrary we see, using (2.3), that

$$\limsup_{t\to\infty} HC^{(t)}\subset \tilde{C} \quad \text{a.s.} \quad \square$$

We will now obtain a slightly different formula for $\tilde{C}$ which fits in well with the results in the next section. Let

$$k^*(\Theta,\theta) = \begin{cases} k(\Theta,\theta) & \text{if } \theta\geq 0 \\ \infty & \text{if } \theta<0 \end{cases}$$

and define ξ^*, C^* etc. in the same way as ξ, C etc. (As k^* is a closed convex function, just as k is, ξ^*, C^* etc. have properties like those of ξ, C etc.) In particular

$$\xi^*(A,a) = \inf\{k(\Theta,\theta)+\Theta.A+\theta a : \Theta,\theta\geq 0\};$$

a calculation like that given in Rockafellar (1970, Ch.16,p.147) shows that

$$\xi^*(A,a) = \sup\{\xi(A,b) : b\leq a\}.$$

Consequently

$$C^* = \{(A,a) : (A,b)\in C \text{ for some } b\leq a\},$$

from which it follows that

$$\hat{C}^* = \tilde{C}$$

and so

$$\limsup_{t\to\infty} HC^{(t)}\subset \hat{C}^* \quad \text{a.s.}$$

3. Let us consider the case when ℓ is a finite random variable with the Laplace transform $\exp(d(\theta))$. Notice that the life-length of an nth generation person is independent of the positions and birth times of the nth generation people and so

$$\log\left\{\mathbb{E}\sum_r \exp(-(\Theta,\theta).(x_r^{(n)},t_r^{(n)}+\ell_r^{(n)}))\right\} = nk(\Theta,\theta)+d(\theta). \tag{3.1}$$

Let

$$\theta^* = \sup\{\theta : \mathbb{E}e^{-\theta\ell} = \infty\};$$

as $\ell \geq 0$ a.s. $\theta^* \leq 0$. Now let

$$k^*(\Theta,\theta) = \begin{cases} k(\Theta,\theta) & \text{if } \theta \geq \theta^* \\ \infty & \text{if } \theta < \theta^*; \end{cases}$$

this definition agrees with that in the previous section since $\theta^* = 0$ in that case. Let

$$k_n^*(\Theta,\theta) = \begin{cases} k(\Theta,\theta) & \text{if } \theta \geq 0 \\ k(\Theta,\theta)+n^{-1}d(\theta) & \text{if } \theta < 0 \end{cases}$$

and define ξ^*, C^* etc. and ξ_n^*, C_n^* etc. in the same way as ξ, C etc. It is easy to check that

$$\xi_n^*(A,a) \downarrow \xi^*(A,a) \quad \text{as } n\to\infty. \tag{3.2}$$

(Here it is possible to show that $\xi^*(A,a) = \sup\{\xi(A,b)-\theta^*(b-a) : b \leq a\}$.)

Let $J^{(n)}$ be the set of points $\{n^{-1}X_r^{(n)}, n^{-1}(t_r^{(n)}+\ell_r^{(n)}) : r\}$. For some $\varepsilon>0$ and some integer m suppose that $C_m^*(\varepsilon) \subset P$ where P is a closed polyhedral set. We then have the following analogue of Lemma 1.

<u>Lemma 2</u>

$$J^{(n)} \subset P \quad \text{for all but finitely many } n.$$

<u>Proof.</u> The proof is much like that of Lemma 1. Suppose first that

$$P = \{(A,a) : (A,a).(W,w) \geq c\}.$$

Let Ω_n' be the event that some point of $J^{(n)}$ is not in P. If $C_m^*(\varepsilon) \subset \partial P$ or the hyperplane through $\{(A,a,-\varepsilon) : (A,a) \in \partial P\}$ not cutting $\{(A,a,y) : y \leq \xi_m^*(A,a)\}$ is vertical then $\xi_m^*(A,a) = -\infty$ for $(A,a) \notin P$. However

$$\xi_m^*(A,a) \geq \inf\{k(\Theta,\theta)+n^{-1}d(\theta)+(\Theta,\theta).(A,a) : (\Theta,\theta)\} \text{ for } n \geq m$$

and so, using Lemma 1 of Biggins (1978),

$$\{X_r^{(n)}, t_r^{(n)}+\ell_r^{(n)} : r\} \subset \{(nX,nx) : (X,x) \in P\} \text{ a.s.}$$

and Ω_n' cannot occur. Otherwise

$$-\varepsilon \geq \inf\{k_m^*(\theta W,\theta w)+\theta c : \theta>0\}$$

$$\geq \inf\{k(\theta W,\theta w)+n^{-1}d(\theta w)+\theta c : \theta>0\} \text{ for all } n \geq m$$

and so, using (3.1)

$$\mathbb{P}\left[\Omega_n'\right] \leq e^{-n\varepsilon} \text{ for } n \geq m.$$

Hence Ω'_n occurs only finitely often. The proof is completed as before. □

Theorem 2

$$\limsup_{t\to\infty} HC^{(t)} \subset \hat{C}^* \quad \text{a.s.}$$

Proof. In the previous section this result was proved when $\ell = \infty$ with positive probability. Hence we assume that $\ell<\infty$ a.s. Suppose that P is a closed polyhedral set containing $\hat{C}^*_m(\varepsilon)$, for some m and $\varepsilon>0$. Notice that, as $C(\varepsilon)\subset C^*_m(\varepsilon)$, this polyhedral set also contains $\hat{C}(\varepsilon)$. Let

$$P_1 = \text{cl}\{(cA,c) : A\in P, c>0\},$$

so that P_1 is a polyhedral set containing both $C^*_m(\varepsilon)$ and $C(\varepsilon)$. Therefore for some finite N both $I^{(n)}$ and $J^{(n)}$ are contained in P_1, when $n\geq N$. As P_1 is convex this implies that $(X_r^{(n)},t)\in P_1$ whenever $n\geq N$ and $t_r^{(n)}\leq t<t_r^{(n)}+\ell_r^{(n)}$. From this it follows that $t^{-1}X_r^{(n)}\in P$ for $n\geq N$ and $t_r^{(n)}\leq t<t_r^{(n)}+\ell_r^{(n)}$. Since there are only a finite number of people in the first N generations and each has a finite lifetime there exists some time T after which they are all dead and so for t>T

$$HC^{(t)}\subset P \quad \text{a.s.,}$$

thus

$$\limsup_{t\to\infty} HC^{(t)}\subset P \quad \text{a.s.}$$

This holds for any closed polyhedral set containing $C^*_m(\varepsilon)$ and so $\limsup_{t\to\infty} HC^{(t)}\subset C^*_m(\varepsilon)$ a.s., but this holds for any m and, using (3.2) and Lemma 5 of Biggins (1978),

$$\bigcap_m \hat{C}^*_m(\varepsilon) = \hat{C}^*(\varepsilon).$$

A second application of Lemma 5 of Biggins (1978) shows that

$$\bigcap_{\varepsilon>0} \hat{C}^*(\varepsilon) = \hat{C}^*.$$

Hence we deduce that

$$\limsup_{t\to\infty} HC^{(t)}\subset \hat{C}^* \quad \text{a.s.}$$

as required. □

4. We will now establish a more explicit formula for $\hat{C}^*$ which holds in some cases.

Theorem 3

If $\{\theta : k(\theta,0)<\infty\}$ contains a neighbourhood of the origin in X then

$$\hat{C}^* = \text{cl}\{A : \inf_\theta\{k^*(\theta,-\theta.A)\}\geq 0\}$$

Proof. For $A\in X$ let

$$\mu(A) = \sup\{\xi^*(aA,a) : a>0\}$$

$$= \sup_{a>0}\{ \inf_{(\Theta,\theta)}\{k^*(\Theta,\theta)+\Theta.Aa+\theta a\}\};$$

notice that if $a\leqslant 0$ then

$$k^*(\Theta_o,\theta)+\Theta_o.Aa+\theta a \to -\infty \text{ as } \theta\to\infty \tag{4.1}$$

and so

$$\mu(A) = \sup_a \inf_{(\Theta,\theta)} \{k^*(\Theta,\theta)+\Theta.Aa+\theta a\}.$$

As

$$\hat{C}^* = \text{cl}\{A : (aA,a)\in C^* \text{ for some } a>0\}$$

$$= \text{cl}\{A : \xi^*(aA,a)\geqslant 0 \text{ for some } a>0\}$$

$$= \text{cl}\{A : \sup\{\xi^*(aA,a) : a>0\}\geqslant 0\}$$

$$= \text{cl}\{A : \mu(A)\geqslant 0\}$$

the result is proved if we show that

$$\mu(A) = \inf_\Theta\{k^*(\Theta,-\Theta.A)\}.$$

For any fixed Θ $k^*(\Theta,\theta)$ is a closed convex function of θ, thus, just as in (2.1) and (2.2) we have

$$\sup_a \inf_\Theta\{k^*(\Theta,\theta)+\theta a-\phi a\} = k^*(\Theta,\phi).$$

Therefore

$$\mu(A) = \sup_a \inf_\Theta \inf_\theta\{k^*(\Theta,\theta)+\Theta.Aa+\theta a\}$$

$$\leqslant \inf_\Theta \sup_a \inf_\theta\{k^*(\Theta,\theta)+\Theta.Aa+\theta a\}$$

$$= \inf_\Theta\{k^*(\Theta,-\Theta.A)\}.$$

We must now consider when this inequality is actually an equality. For any fixed A this will depend on the properties of the function

$$\tau(a,\Theta) = \inf_\theta\{k^*(\Theta,\theta)+\theta a\}+\Theta.Aa.$$

It is fairly easy to see using Theorem 33.3 and Corollary 34.2.2 of Rockafellar (1970) that τ is a closed concave-convex function in his terminology, and so Theorem 37.3(b) gives conditions for

$$\sup_a \inf_\Theta \tau(a,\Theta) = \inf_\Theta \sup_a \tau(a,\Theta) \tag{4.2}$$

to hold. We will now verify that these conditions do hold.

As $\{\Theta: k^*(\Theta,0)<\infty\} = \{\Theta : k(\Theta,0)<\infty\}$ we can choose Θ_1 in the interior of

$\{\theta : k^*(\theta,0) < \infty\}$ and such that, for a fixed A, $\theta_1.A<0$. Clearly

$$\{\theta : k^*(\theta,0) < \infty\} \subset \{\theta : \tau(a,\theta) < \infty \text{ for all } a\}$$

and so θ_1 is also in the interior of the latter set. By (4.1) $\tau(a,\theta_1) = -\infty$ for $a \leq 0$ whilst

$$\tau(a,\theta_1) \leq k(\theta_1,0) + \theta_1.Aa \to -\infty \text{ as } a \to \infty.$$

Hence the concave function $\tau(\cdot,\theta_1)$ has no direction of recession. This suffices for (4.2) to hold and so proves the theorem. □

Notice that whenever $k^* = k$ (which occurs in particular whenever $\theta^* = -\infty$) or $0 \in \hat{C}$ we have

$$\limsup_{t\to\infty} HC^{(t)} \subset \hat{C} \quad \text{a.s.}$$

because $\hat{C}^* = \hat{C}$. These two possibilities probably cover the majority of cases that might arise in practice.

Suppose that C is some point with $\mu(C)>0$ and U is a unit vector. The function $k^*(\theta,-\theta.(C+aU)$ is convex in θ and so its minimum value is easily found numerically for any particular a. A straightforward search procedure will find the value of $a>0$ at which $\mu(C+aU)$ changes sign. Repeating this procedure for different unit vectors U allows $\hat{C}^*$ to be obtained.

I have written a computer program to draw the set $\hat{C}$ when $X = \mathbb{R}^2$, for a given k (though some problems are encountered when $\partial\hat{C}$ and $\partial\hat{A}$ are too close). Figure 1 gives three examples of the output. The three cases are

(1) After a length of time which is exponentially distributed with mean one the initial ancestor has two children. One is at the origin. The other can occupy one of the four positions (1,0), (-1,0), (0,1), (0,-1), choosing each with probability 1/4.

(2) The initial ancestor has 1 child at each of the points (1,0), (-1,0), (0,1), (0,-1). The time to the birth of each child being an exponential random variable with mean four.

(3) The initial ancestor has 12 children. There is one child at each of the points (-2,0), (2,0), (0,2), (0,-2) and two at each of the points (-1,1), (-1,-1), (1,1), (1,-1). The time to the birth of each child is the sum of two independent exponential random variables with mean four.

The first example is essentially the contact birth process as described by Mollison (1978), with nearest-neighbour contact distribution. Mollison (1977, Table 1) gives figures for its rate of spread which agree with those here. In Mollison's terminology (1978) this contact birth process underlies the simple epidemic on the square lattice with nearest-neighbour contacts (i.e. the simple

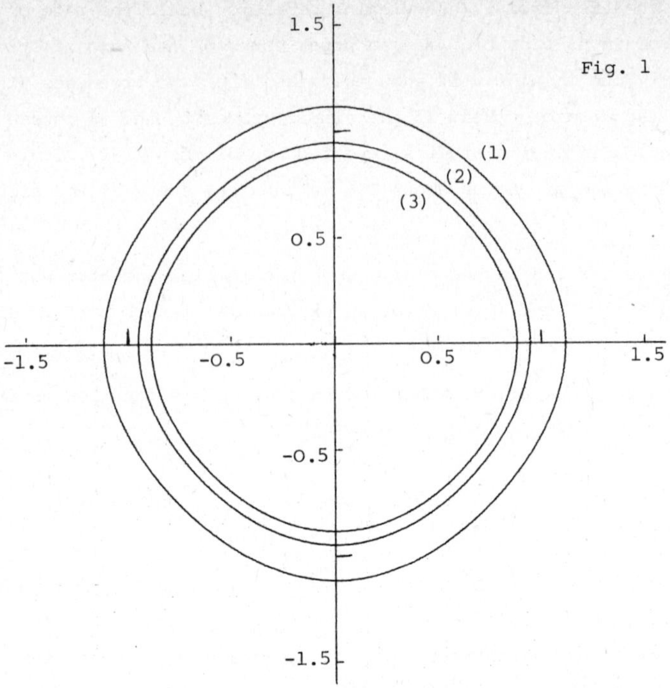

Fig. 1

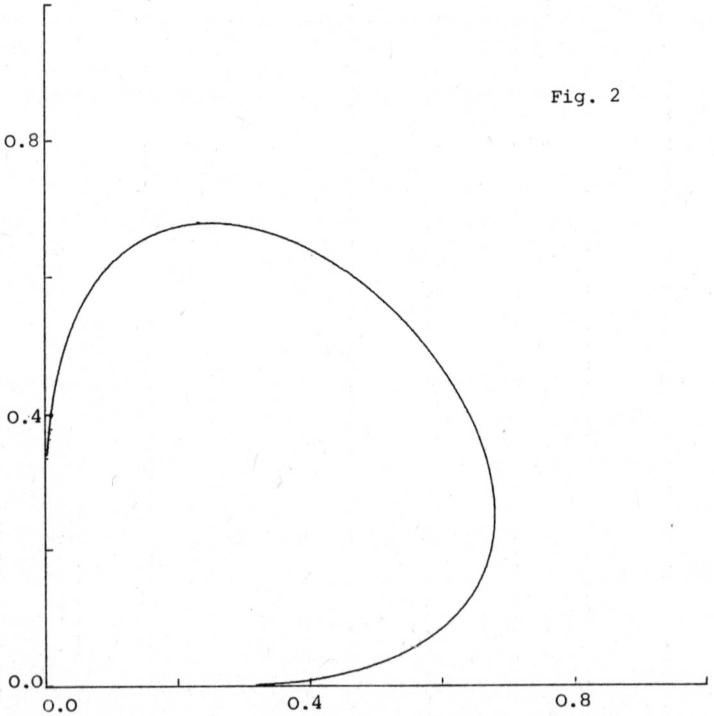

Fig. 2

epidemic can be obtained from it by deleting people). Hence the rate of spread of this contact birth process provides an upper bound on the rate of spread of the simple epidemic. Some thought will show that the other two processes also provide upper bounds on the rate of spread of the simple epidemic, and of course these bounds are better. Simulation studies reported by Mollison (1977, Table 1) suggest that a 'circle' of radius about 0.6 is the true shape of the simple epidemic.

Figure 2 shows the leading edge of $\hat{C}$ when the initial ancestor has 1 child at each of the points (1,0) and (0,1) after an exponential length of time with mean four.

I would like to thank Dr. M. Maher for the use of his function minimization program.

REFERENCES

BIGGINS, J.D. (1978) The asymptotic shape of the branching random walk. Adv.Appl.Prob. 10, 62-84.

JAGERS, P. (1975) Branching Processes with Biological applications. Wiley, London.

MOLLISON, D. (1977) Spatial contact models for ecological and epidemic spread. J.R.Statist.Soc., B 39, 283-326.

MOLLISON, D. (1978) Markovian contact processes. Adv.Appl.Prob. 10, 85-108.

ROCKAFELLAR, R.T. (1970) Convexity Analysis. Princeton University Press, Princeton.

PATTERN FORMATION BY BACTERIA [+]

by

F.C. Hoppensteadt [*]

University of Utah

Salt Lake City, Utah 84112 USA

and

W. Jäger

Universität Heidelberg

D-6900 Heidelberg 1, Germany

---

[*] This work supported in part by NSF Grant No. MCS-7805985

[+] These results were presented at Mathematical Biology Meeting, Universität Heidelberg, July, 1979.

Abstract

A uniform distribution of histidine auxotrophic *Salmonella Typhimurium* on an agar gel exhibits a spatial pattern of growth in response to a diffusing front of histidine. We formulate a mathematical model of this system which describes the histidine concentration ($H(r,t)$) at radius r and time t in a petri dish, the concentration of the growth medium's buffer ($G(r,t)$) and the size of the bacterial population ($B(r,t)$). Histidine diffuses and is taken up by growing cells, the buffer also diffuses but it is neutralized by acids produced as by-products of cell growth, and the bacterial population is fixed, as in a stiff top layer of agar.

The uptake of histidine and the depletion of buffer are modelled by Michaelis-Menten kinetics. The cell growth is modelled as having a hysteretic dependence on the histidine and buffer concentrations: Specifically, cell growth competence depends on combinations of G and H . Cell growth continues until the combination is reduced to a threshold value T_{off} at which cell growth stops. Growth does not begin again until a threshold value T_{on} ($> T_{off}$) is reached.

Typical experiments begin with a uniform distribution of buffer ($G(r,0) = G^o$) and bacteria ($B(r,0) = B^o$), and a drop of histidine solution placed in the center of a petri dish ($H(r,0) = H^o \; \mathcal{H}(R' - r)$*,

---

*) $\mathcal{H}(x)$ is the Heaviside function $\mathcal{H}(x) = 1$ if $x > 0$, but $= 0$ if $x \leqslant 0$.

where R' is the drop's radius). With such initial conditions, the mathematical model is shown to have a central disc of high growth which is surrounded by a ring of even higher growth. This central region is in turn enclosed by concentric rings of heavy growth which are separated by bands of low cell growth. Similar growth patterns are observed in experiments of this kind which have been performed in conjunction with J. Roth and M. Schmid.

1. *Introduction*

Bacteria can grow in spatial patterns in response to diffusion of a needed nutrient. We present here a model of experiments which demonstrate this. The model describes an immobile bacterial population distributed on a petri dish. The diffusing nutrient is taken up by growing cells, and acid is produced as a byproduct of cell growth. We show by a numerical example that the model exhibits behavior similar to that observed in experiments. Experimental data, detailed mathematical analysis and more extensive numerical results will be presented elsewhere.

A novel feature of the model is that the cell's growth has a hysteretic dependence on the amount of nutrient and acid present. It is known that hysteretic kinetics can lead to spatial patterns in chemical systems such as Liesegang rings formed by percipitating colloids. In that case, the hysteresis is described by Ostwald's theory of supersaturation. Other hysteretic systems which lead to patterns arise in electrophysiology and epidemiology.

2. *Experiments*

A lawn of bacteria is fixed on a circular agar gel which contains an acid buffer and minimal growth chemicals, except that these cells need the amino acid histidine to grow (these are called histidine auxotrophs). A drop of histidine solution is placed in the center of the gel, and it diffuses out. The cells grow in response to the histidine in a pattern of concentric circles as shown in Fig. 1. This is a standard procedure, called the crystal test, for determining cell characteristics.

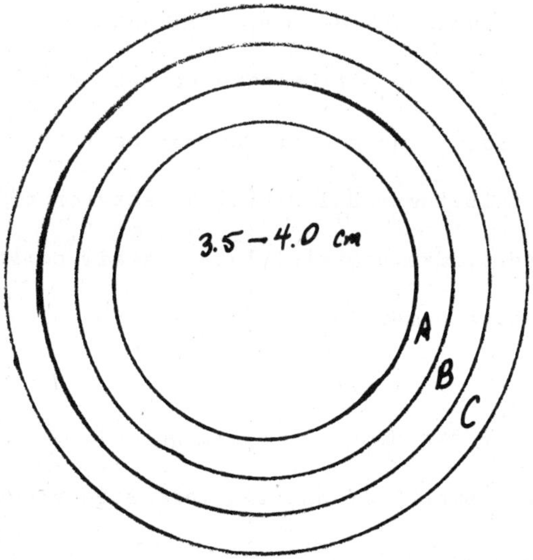

Figure 1: Diagram of a petri plate which has a histidine auxotroph plated in top agar on which 2µℓ of a 1 molar histidine solution was placed in the center. The regions A, B, C are regions of no bacterial growth. The measurements were A = 5 mm, B = 4 mm, C = 2 mm, and the bands of growth which separated A, B and C were 1-2 mm in width.

3. *The Model*

We describe the bacterial population size at radius r $(0 \leqslant r \leqslant R)$ by B(r,t), the concentration of histidine by H(r,t) and the pH by the buffer concentration G(r,t). These functions are determined by the equations

$$\frac{dB}{dt} = \alpha VB \qquad \text{(cell growth)}$$

$$(1) \qquad \frac{\partial H}{\partial t} = D\Delta H - \beta VB \qquad \text{(histidine diffusion and kinetics)}$$

$$\frac{\partial G}{\partial t} = D'\Delta G - \gamma VB \qquad \text{(buffer diffusion and kinetics)}$$

Diffusion is described by the operator Δ in radial coordinates

$$\Delta u = \frac{1}{r}\frac{\partial}{\partial r} r \frac{\partial}{\partial r} u$$

The constants D and D' are the diffusivities of H and G, respectively. The kinetic terms involve the function V which describes the growth competence of sites in the disc. αV is the bacterial growth rate, βV the uptake rate of histidine and γV the acid production by growing cells. The function V depends on the amounts of histidine and buffer. It could be taken to have the form

$$V = \frac{H}{K+H}\,\frac{G}{K'+G}$$

where K and K' are Michaelis-Menten constants. However, it is known from experiments that for adequate histidine concentrations, cells have a hysteresis in growth depending on pH: When growing, cells will continue to grow even as pH decreases until a threshold is reached at which growth stops. As pH is increased from this threshold, growth does not begin again until a higher more favourable pH threshold is reached. This is described in Figure 2.

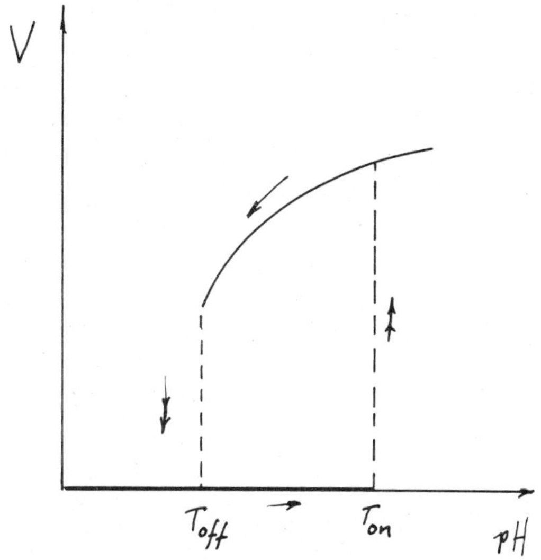

Figure 2: Cells grow when V > 0 . As pH decreases in the presence of adequate histidine, V drops to zero quickly. As pH increases, V remains 0 until the threshold T_{on} is reached at which V quickly moves to the upper branch.

The situation is somewhat more complicated by similar thresholds for histidine. This is illustrated in the GH-phase plane in Figure 3 where ϕ denotes the threshold at which cell growth stops and ψ denotes that

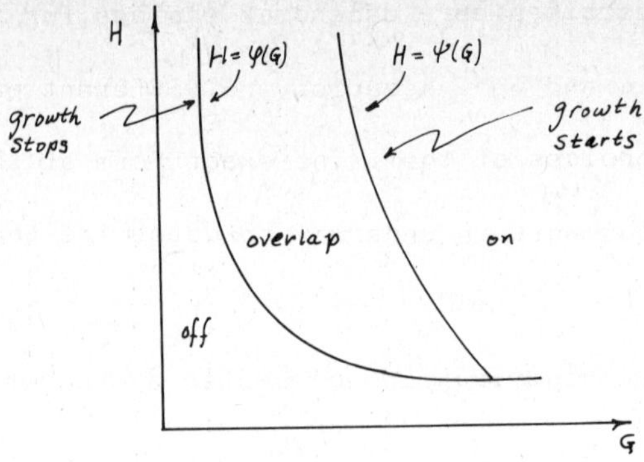

Figure 3: The GH-phase plane provides a useful description of the model dynamics. Here are typical choices for the turnoff threshold (φ) and the turnon threshold (ψ). The overlap region corresponds to the region between ψ_1 and ψ_2 in Figure 3.

at which cells again begin to grow. Figure 4 shows the dependence of histidine uptake on G as G passes into the overlap region.

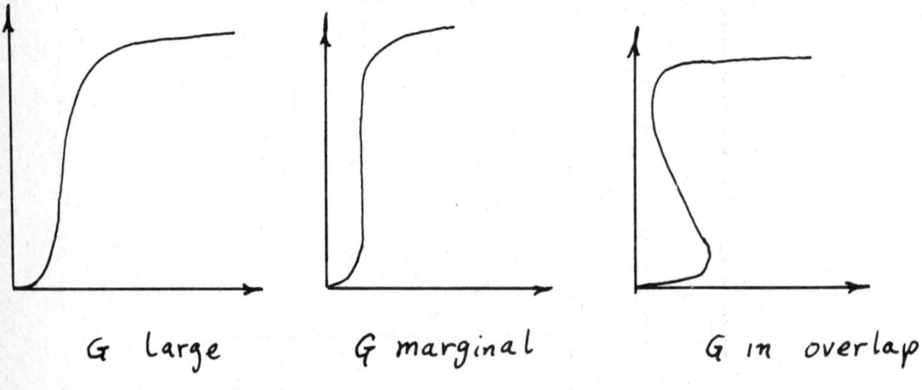

Figure 4: These curves illustrate how histidine uptake varies with

The results described here depend on choices for the initial data and the thresholds ϕ and ψ . A variety of different patterns are possible for various choices of them. The exact form of the functions ϕ and ψ is unknown, but work is underway to determine them experimentally.

Uniform initial distributions of bacteria and buffer are given by

$$B(r,0) = B^0 \quad \text{and} \quad G(r,0) = G^0 ,$$

and the initial histidine distribution is given by

$$H(r,0) = H^0(r) ,$$

as depicted in Figure 5.

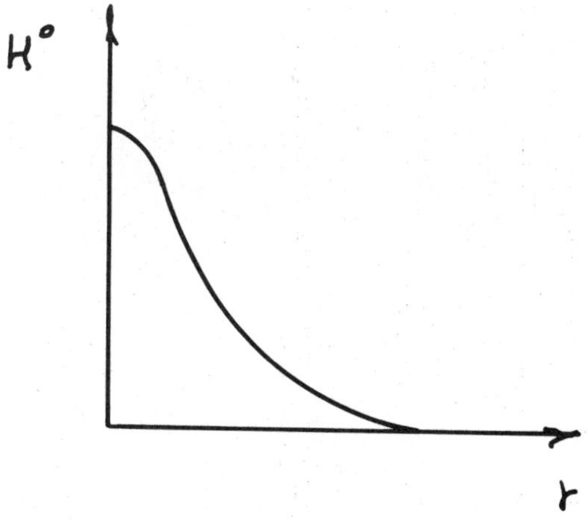

Figure 5: Initial histidine distribution.

4. *Analysis of the Model*

Experimental values of the parameters indicate that diffusion is dominated by kinetics (D, D' << β, γ) when V > 0 . This suggests the following heuristic arguments which we describe in the GH-phase plane.

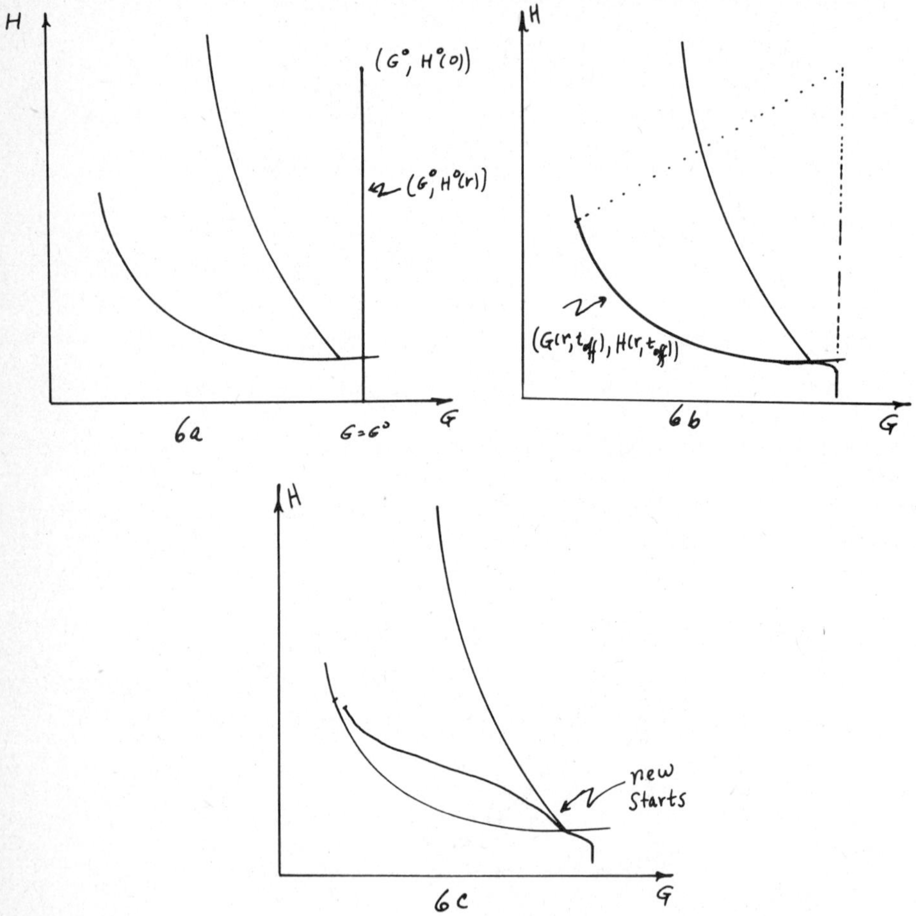

Figure 6: GH-phase plane. 6a. Initial distribution and thresholds.

6b. Approximate distribution of G,H after initial kinetics.

6c. Approximate distribution after diffusion has moved up to the turnon threshold.

Figure 6a shows the initial data and the thresholds. Figure 6b shows the result of kinetics acting on these data. Note that if we ignore diffusion, then $\frac{dH}{dG} = \beta/\gamma$. Thus, the initial data move towards the curve $H = \phi(G)$ along straight lines. Next, diffusion acts on this distribution, slowly raising portions towards the threshold $H = \psi(G)$. Figure 6c shows the results at the point where certain cells are again turned on. These drive the turned on values (G,H) towards the curve $H = \phi(G)$. This process continues to give repeated switching on and off of various sites. It stops when there no longer remains enough histidine or buffer to again reach $H = \psi(G)$. The result is a spatial pattern of growth.

Approximate behaviour of the model can be deduced by iterating the kinetic and diffusion phases. The kinetics were described in the preceding paragraph: Motion is along straight lines having slope β/γ in the GH-plane. Once all growth stops, the model reduces to two diffusion equations for G and H . These can be solved in terms of Fourier-Bessel expansions, and these, in turn, used to approximate where the next competent sites will be. This procedure and the construction of appropriate boundary layers for matching these solutions together is lengthy, and will not be presented here.

. *Numerical Example*

We present in this section the results of a numerical simulation f (1) which is based on a method of lines solver. The parameters chosen for the system are based on rescaling, nondimensionalizing and substituting realistic values into the result. The choice of the threshold curves ϕ and ψ is difficult because there is little data on this phenomenon. However, we define

$$\text{Growth stops} \quad \{(G,H): \min(G,H) = 1.\}$$

$$\text{Growth starts} \quad \{(G,H): H \geq 1,\ G = \frac{a}{H} + b\} \quad .$$

These are shown in Figure 7.

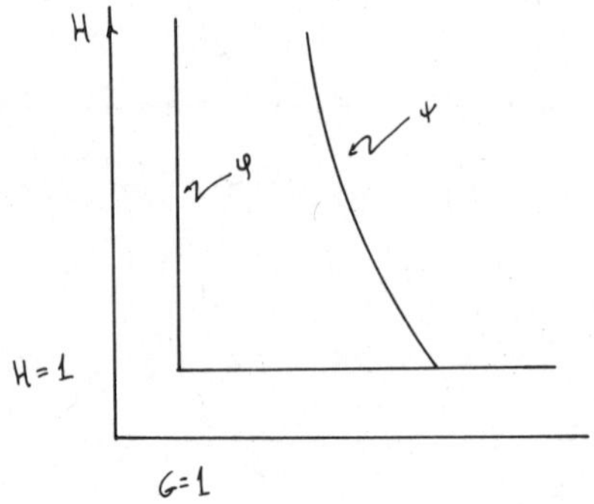

Figure 7: GH phase plane depicting threshold curves used in the numerical computation.

The function V is taken to be 1 if a site is competent and zero otherwise. Figure 8 shows the result of the calculations for

$$\alpha = 1,\ \beta = 5,\ \gamma = 5,\ D = 1E - 3,\ D' = .5E - 3\ ,$$

$$G^o = 500,\ a = 100,\ b = 100\ ,$$

$$H^o = 2000 \quad \text{if } r \leq R/10, = 0 \quad \text{for} \quad r > R/10\ .$$

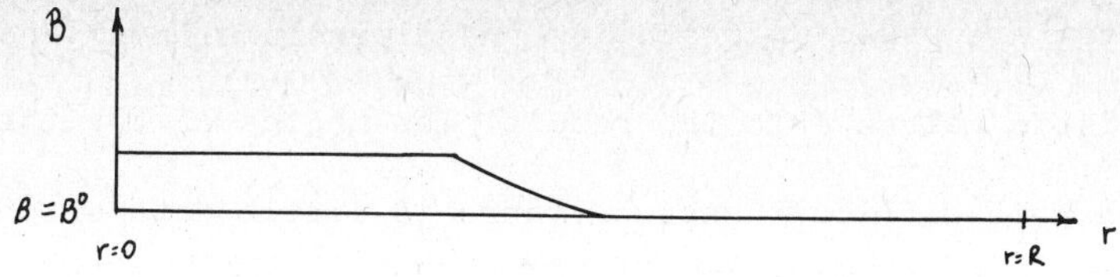
B
B = B°
r=0
r=R
r

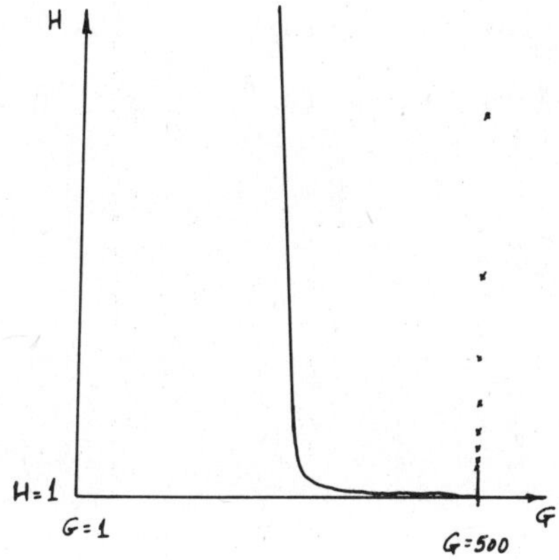
H
H=1
G=1
G=500
G

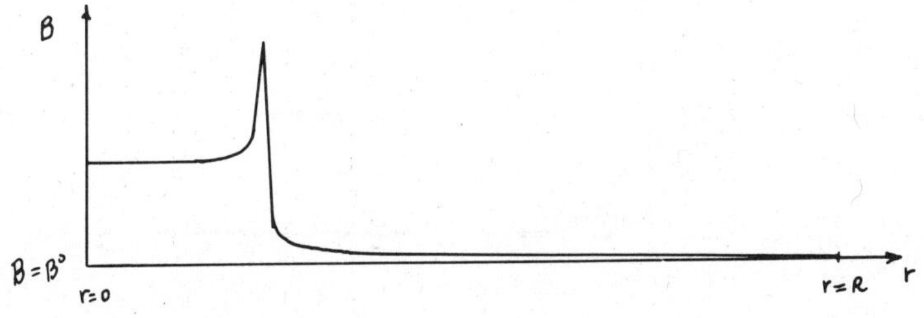
B
B=B°
r=0
r=R
r

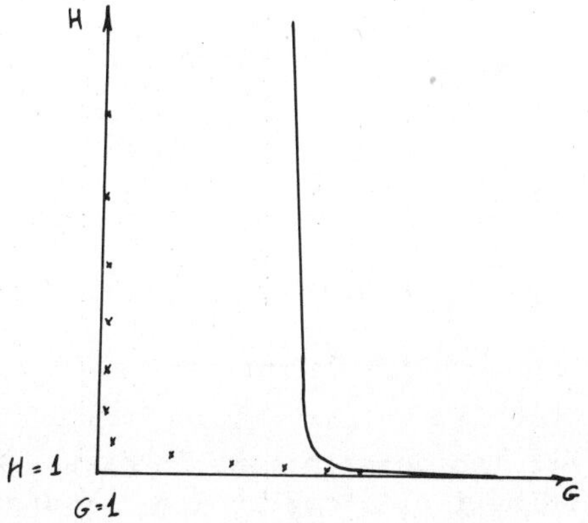
H
H = 1
G=1
G

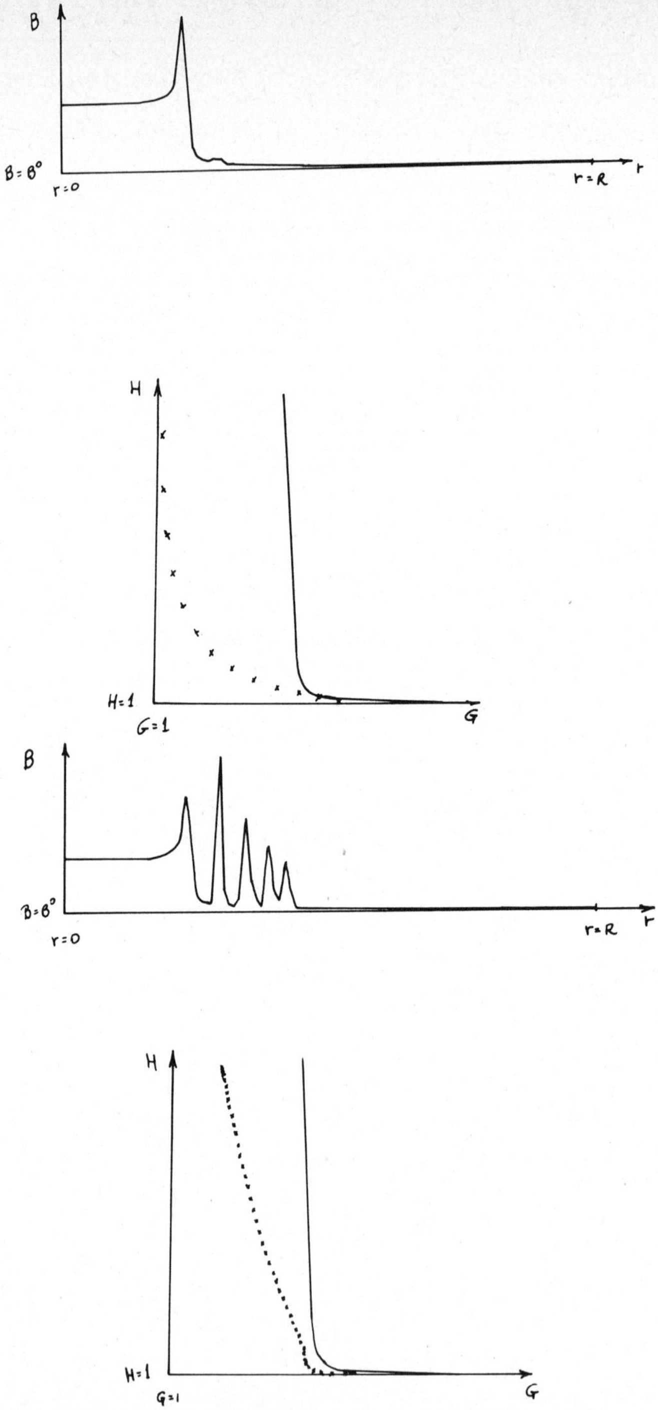

Figure 8: 8a-8c illustrate the descriptions in Figure 6a-6c. 8d shows the distributions after a time comparable to that at which Figure 1 was drawn

The numerical result described in Figure 8d corresponds to the passage of approximately 3 days for growing bacteria as observed in experiments. It compares favourably with the experimental observations in Figure 1.

Random processes in random environments

by Harry Kesten

1. Introduction.

We shall give a brief survey of some random processes in random environments. Roughly speaking, our examples arise from models in which a phenomenon is described by a Markov process or chain. The Markov process in question is specified by a (usually infinite) number of parameters. Instead of assuming the parameters to be constant, we assume that the parameters themselves are subject to random fluctuations. In such situations it is conceptually helpful to describe the model in two stages. In the first stage the values of the parameters are chosen. Each realization of the parameter values constitutes a choice of the "random environment". Once the environment has been fixed, it stays fixed forever. The second stage describes the evolution of the random process given the environment. In all our examples, the conditional behavior of the process, given the environment, is that of an inhomogeneous Markov process. In sections 3, 4 and 6 this Markov process actually has no random element left in it, but is deterministic.

Since we only observe the values of the process, but not the environment, the observable process is non-Markovian in general. This is so, because as one gathers more observations, more and more information about the environment becomes available. Our examples fall into two classes. In the first class (sect. 2-4) the environment is indexed by time, and in the second class (sect. 5 and 6) by space. Since an observer can return to the same space point repeatedly, but not to the same time point, the non-Markovianness in the second class models is usually more difficult to handle.

We only describe the nature of some results. The reader has to look up precise theorems and conditions in the references. No attempt has been made to give complete references; only some more recent references have been listed and it is hoped that the reader can work backwards from them.

2. Branching processes in random environments.

Assume a population contains d types of individuals and has non-overlapping generations. Denote by $Z_n(i)$ the number of individuals of type i in the n-th generation. The $(n+1)$-st generation consists of the offspring of the n-th generation; it is assumed that the n-th generation itself does not survive and does not become part of the $(n+1)^{th}$ generation. The fundamental hypothesis for a branching process is that all individuals produce offspring independently of each other. The parameters needed to describe the process are the offspring distributions:

$$p_n(\vec{k}|i) = P\ \{\text{an individual of type } i \text{ in the n-th generation has } k_\ell \text{ children of type } \ell,\ 1 \leq \ell \leq d\}$$
$$(\vec{k} = (k_1, k_2, \ldots, k_d)).$$

In the classical Bienaymé - Galton - Watson process the $p_n(\vec{k} \mid i)$ do not depend on n and are non-random. In the random environment model ([1], [2]) one takes the sequence $\Pi = \{p_n(\vec{k} \mid i)\}_{n \geq 0}$ as random environment. Given the environment, i.e., given a realization of Π, $Z_0, Z_1, \ldots$ is a non-homogeneous branching process. Classical Bienaymé - Galton - Watson processes are classified as subcritical, critical or supercritical according as $\rho < 1$, $\rho = 1$ or $\rho > 1$, where ρ is the largest eigenvalue of the expectation matrix

$$(2.1)\quad M_n = (m_n(i,j))_{1 \leq i,j \leq d} = \sum_{\vec{k}} k_j\, p_n(\vec{k} \mid i).$$

Note that M_n is independent of n in the classical case. ρ plays the role of a Malthusian parameter. If $\rho \leq 1$ then (with only trivial exceptions) the population becomes extinct, i.e. $Z_n = 0$ eventually, with probability 1 (w.p.1). In the supercritical case ($\rho > 1$) there exists a random variable $W \geq 0$ such that w.p.1

$$(2.2)\quad \lim_{n\to\infty} \rho^{-n} Z_n = Wv\ ,$$

where v is the left eigenvector corresponding to ρ of M_n. Moreover, Z_n becomes extinct almost everywhere on the set $\{W = 0\}$. Thus, the process either dies out or grows exponentially. In the random environment case one has partial analogues of these results. One now assumes that Π is a stationary ergodic sequence. Unless one takes the special case of [1], where the p_n are independent identically distributed, Z_n is no longer Markovian. Still one obtains under suitable irreducibility and regularity conditions that

$$(2.3)\quad P\{\ Z_n \text{ becomes extinct, or } X = \lim_{n\to\infty} \frac{1}{n} \log |Z_n| \text{ exists and } X > 0\ \} = 1\ ,$$

i.e., one again has either extinction or exponential growth. (See [3] - [6])). Beyond the existence of the expectation matrices M_n of (2.1) this result requires no moment conditions for the offspring distributions. X is a constant on the set of non-extinction and equals

$$(2.4) \qquad \lim \frac{1}{n} \log \|M_1 M_2 \cdots M_n\| ,$$

where $\|M\|$ denotes $\max_j \sum_i |m_{ij}|$ for a matrix $M = (m_{ij})$. The limit in (2.4) is known to exist under mild moment conditions. In the one-type case (d = 1) [2,part II] has a much sharper result, analogous to (2.2). Denote by m_n the single element of M_n in this case. Then there exists a random variable W such that w.p.1

$$\lim_{n\to\infty} (m_1 \cdots m_n)^{-1} Z_n = W ,$$

and $W > 0$ on the set of non-extinction. It is still not known whether (2.2) has a decent analogue in the multi-type random environment case. Thus we have the following

Problem: Do there exist a sequence of vectors $V_n = (V_n(1), \ldots, V_n(d))$ which are functions of $M_1, \ldots, M_n$ only, and a random variable W such that w.p.1 $\{V_n(i)\}^{-1} Z_n(i) \to W$?

Other limit theorems are proven in [2, part II].

3. Selection with random fitness coefficients.

We only discuss the simplest case where the deterministic model is that of Haldane, Fisher and Wright for selection at one locus with two possible alleles, A and a. Again we consider non overlapping generations and concentrate on the gene frequencies in the gametes.

$$(3.1) \qquad p_n = \frac{\text{Number of gametes of type A in n-th generation}}{\text{Total number of gametes in n-th generation}} .$$

The selection acts on the zygotes, and the relative fitnesses in the nth generation of the types AA, Aa and aa are σ_n, 1 and τ_n. Set $q_n = (1 - p_n)$. The p_n are governed by the well known recursion (see [7], Ch. 1.6)

$$(3.2) \qquad p_{n+1} = \frac{p_n^2 \sigma_n + p_n q_n}{p_n^2 \sigma_n + 2p_n q_n + q_n^2 \tau_n}$$

The environment in this model consists of the sequence $\{\sigma_n, \tau_n\}_{n \geq 0}$. In the random environment model it is again assumed that this is a stationary ergodic sequence. Once the environment is given the p_n develop deterministically according to (3.2). The points $p = 0$ and $p = 1$ are fixed points for the recurrence relation (3.2), and correspond to fixation in a, respectively A. Among other results, [8] - [10] discuss the stability of these fixed points. In the random environment case $\bar{p}$ is called a stochastically (locally) stable fixed point if for every $\epsilon > 0$ there exists a $\delta > 0$ such that for all $|\bar{p} - p| <$

(3.3) $$P\{\lim_{n\to\infty} p_n = p \mid p_o = \overline{p}\} \geq 1 - \varepsilon .$$

Since for small p_n ,

$$p_{n+k} \approx \frac{1}{\tau_{n+k-1}} \cdot \frac{1}{\tau_{n+k-2}} \cdots \frac{1}{\tau_n} p_n$$

we can expect $\overline{p} = 0$ to be stable if

(3.4) $$\frac{1}{\tau_n} \cdot \frac{1}{\tau_{n-1}} \cdots \frac{1}{\tau_1} \to 0 \quad \text{w.p. } 1.$$

If $E\{\log \tau\}$ exists, (3.4) occurs if and only if

(3.5) $$E\{\log \tau_1\} > 0 .$$

Indeed, as shown in [8] - [10] $\overline{p} = 0$ is stable if (3.5) holds; moreover

$$P\{\lim_{n\to\infty} p_n = 0\} = 0$$

if $E\{\log \tau_1\} < 0$. Similarly $p = 1$ is stable if $E\{\log \sigma_1\} > 0$. If $E\{\log \tau_1\} < 0$ and $E\{\log \sigma_1\} < 0$, then p_n cannot converge to zero, nor to one, and hence will move around somewhere between these points. We can expect in these cases that p_n will have a limit distribution which is concentrated on the open interval (0,1). This was proved by Norman [11] in case the (σ_n, τ_n), $n = 0, 1, \ldots$ are independent and identically distributed. (see also [10], sect. 5).

There is no easily interpretable continuous time analogue of this model. However, when the fitness coefficients are close to one (i.e. the selection effects per generation are small, then one can use diffusion approximations (sec [12] and [13]).

. Lotka-Volterra equations with random fluctuations.

This is our first continuous time example. Consider a population consisting of d species, the amount of the i-th species present at time t being measured by $X_i(t)$. The usual deterministic Lotka-Volterra equations are

(4.1) $$dX_i(t) = X_i(t)\{k_i - \sum_{j=1}^{d} b_{ij} X_j(t)\} dt,$$

$$i = 1,\ldots,d, \qquad X_i(t) \geq 0 ,$$

for some constants k_i , b_{ij} with $b_{ii} > 0$. The signs of the off-diagonal b_{ij} depend on whether one deals with a predator-prey, competing species or symbiosis model. $k_i > 0$ ($k_i < 0$) indicates that the i-th species on its own increases (dies out) approximately exponentially, as long as only small quantities of this species are present.

If one allows the k_i to fluctuate randomly in time several replacements of (4.1) can be defended. Here we choose the version

$$(4.2)\quad d\,X_i(t) = X_i(t)\,\{k_i - \sum_{j=1}^{d} b_{ij}X_j(t)\}dt$$

$$+ X_i(t)\sum_{j=1}^{d} \sigma_{ij}dW_j(t)\ ,\ i = 1,\ldots,d,\ X_i(t) \geq 0,$$

where σ is a non-singular (constant) matrix and $(W_1,\ldots W_d)$ a Brownian motion. (4.2) is interpreted as an Ito equation: see [14] for a justification of this equation. Here each specification of the sample path of the Brownian motion W corresponds to a specification of the environment, and the solution of (4.2) is a deterministic function of the sample path of the Brownian motion.

One question here is: "When is coexistence of all species possible?" This is translated into the mathematical question, under what conditions on the k_i, b_{ij} and t_{ij} is $X(t)$ positive recurrent, or the essentially equivalent question, under what conditions does the distribution of $X(t)$ converge to a probability distribution concentrated on the <u>open</u> quadrant $\{x_1 > 0,\ldots, x_d > 0\}$? $X_i(t) \to 0$ in distribution is interpreted as the eventual disappearance of the ith species. For $d = 2$, and $k_i, b_{ij} > 0$ for all i, j, Turelli and Gillespie [15] conjectured necessary and sufficient conditions for coexistence and proved sufficiency of their conditions. Sufficient conditions in some special cases are given in [16] and [17]. Necessary and sufficient conditions for coexistence in pratically all two-dimensional cases can be found in [18].

5. Random walk and diffusion in random environment.

Here the environment is indexed by the space variable. In the one-dimensional situation the environment consists of a doubly infinite sequence $\{\alpha_i\}_{-\infty < i < \infty}$ of independent identically distributed random variables α_i, which take values in $(0,1)$. For a given environment, the observed process, $\{X_n\}_{n \geq 0}$, is a Markov chain on Z with transition probabilities

$$(5.1)\quad P\,\{X_{n+1} = X_n + 1 \mid \{\alpha_i\}\} = \alpha_j \quad \text{on} \quad \{X_n = j\}\ ,$$

$$P\,\{X_{n+1} = X_n - 1 \mid \{\alpha_i\}\} = 1 - \alpha_j \quad \text{on} \quad \{X_n = j\}\ .$$

Thus X moves one step to the right (left) with probability equal to the α $(1-\alpha)$ assigned to its present position. This model in a slightly more complicated form was introduced by Chernov [19] as a model for replication of DNA. The environment represents an (infinitely long) DNA strand which acts as a template. In this model the synthesis of the complementary DNA strand proceeds from one end to the other, and $X_n = j$ indicates that at time n the synthesis has proceeded from the position $-\infty$ to the position j. Sometimes X goes backwards, because a nucleotide disassociates itself. Temkin (see[20]) used Chernov's model for various physical phenomena.

The first question now is: "When does $X_n \to \infty$ w.p.1?" (I.e. when does DNA replication succeed?) This happens if and only if

$$(5.2) \qquad E \log \frac{\beta_0}{\alpha_0} < 0 \qquad (\beta_i = 1 - \alpha_i)$$

(see [21]). Next, if (5.2) holds, so that $X_n \to \infty$, can we find a limit theorem for X_n? $n^{-1}X_n$ converges w.p.1 to a constant, but as discovered independently by Kolmogorov and Spitzer, this constant may be zero. Thus it is possible that $X_n \to \infty$ w.p.1, but slower than linearly. This surprising phenomenon comes about because the process is slowed down greatly in stretches with relatively small α_i. The full limit theorem for X_n is given in [22]; part of these results had been conjectured by Kolmogorov and Spitzer. In all cases of [22], X_n (after suitable normalization) converges in distribution to the inverse of a stable law. Quite different is the situation when

$$(5.4) \qquad E \log \frac{\beta_0}{\alpha_0} = 0 .$$

In this case

$$(5.5) \qquad \liminf X_n = -\infty \text{ and } \limsup X_n = +\infty \text{ w.p.1} .$$

However, $|X_n|$ grows very slowly. It is proved in [23] that in this case $\max_{k \leq n} |X_k|$ is only of the order of $(\log n)^2$. Most likely $(\log n)^{-2} X_n$ has a limit distribution in this case.

It is also possible to prove a limit theorem for the environment as seen from the random position X_n, i.e. for the vector $\{\alpha_{X_n+i}\}_{-\infty < i < \infty}$ (see [24]). In the language of DNA replication such a theorem states that the types of nucleotides which the replicating strand sees around its front end on the template, have a steady state limit distribution.

We have almost no results when the $\{\alpha_i\}$ are not independent identically distributed, or when $X_{n+1} - X_n$ can take other values than ± 1. Such results, however, seem within reach. Much harder are the multidimensional questions, and most of them are still unanswered. For illustration consider the two-dimensional case. The environment now is an array $\{\alpha_x\}_{x \in Z^2}$ of independent identically distributed 4-vectors:

$$\alpha_x = (\alpha_x(1), \alpha_x(2), \alpha_x(3), \alpha_x(4)) , \quad \alpha_x(i) \geq 0 ; \sum_{i=1}^{4} \alpha_x(i) = 1.$$

Given the $\{\alpha_x\}$, X_n has the transition probabilities

$$P\{X_{n+1} = X_n + e_i \mid \{\alpha_x\}\} = \alpha_x(i) \quad \text{on} \quad \{X_n = x\} ,$$

where $e_1,\ldots,e_4$ are the four unit vectors $(\pm 1,0)$, $(0,\pm 1)$. Again we would like to know conditions for $|X_n| \to \infty$, and limit theorems for X_n. In a very ingeneous paper [25], Kalikow gives sufficient conditions for $X_n(1) \to \infty$ w.p.1 ($X_n(1)$ = first component of X_n). Most likely these conditions are for from necessary; in fact $\lim\inf n^{-1} X_n(1) > 0$ w.p.1 under Kalikow's conditions. Here are some unsolved Problems. i) Find better conditions for $P\{|X_n| \to \infty\} = 1$, and for $P\{X_n(1) \to \infty\}$. In particular, does every random walk in a random environment in dimension $d \geq 3$ go to ∞ ? ii) Do there exist zero-one laws, i.e. is $P\{|X_n| \to \infty\}$ necessarily equal to 0 or 1? iii) Find limit laws for X_n.

It is possible to formulate analogous questions for continuous media. One is then interested in the asymptotic behavior of a diffusion X_t , governed by a generator

$$\mathcal{L} = \frac{1}{2} \sum_{i,j} a_{ij}(x) \frac{\partial^2}{\partial x_i \partial x_j} + \sum_i b_i(x) \frac{\partial}{\partial x_i} ,$$

whose coefficients themselves are random. It is remarkable that Papanicolaou and Vavadhan ([26] and a forthcoming paper) proved a limit theorem for $t^{-1/2} X(t)$ in this set up, when $\mathcal{L}$ is self adjoint or $b_i \equiv 0$.

6. Random velocity and acceleration.

These are continuous time models in which the random environment is a random field $F : \mathbb{R}^d \times \Omega \to \mathbb{R}^d$ or $F : \mathbb{R}^d \times \mathbb{R}^d\ \Omega \to \mathbb{R}^d$, indexed by the points in $\mathbb{R}^d$, respectively $\mathbb{R}^d \times \mathbb{R}^d$ (Ω is a probability space). Note that F does not depend on the time variable. Given the environment, the observed process is a deterministic function of the environment, in fact, the solution of a differential equation.

The random velocity situation is supposed to describe the trajectory of a particle in a slightly turbulent atmosphere. The wind velocity is a constant nonzero vector $v \in \mathbb{R}^d$ plus a small random perturbation, ϵF . The observed process X is the solution of

$$(6.1) \quad \frac{dX^\epsilon(t,\omega)}{dt} = v + \epsilon\, F\,(X^\epsilon(t,\omega),\omega),\ X^\epsilon(0,\omega) = x_0 .$$

The random acceleration equation models the motion of a particle in a weak force field, e.g.an electrically charged particle in an electromagnetic field. The equations are

$$(6.2) \quad \frac{dX^\epsilon(t,\omega)}{dt} = V^\epsilon(t,\omega) \quad , \quad X^\epsilon(0,\omega) = X_0 ,$$

$$\frac{d^2 X^\epsilon(t,\omega)}{dt^2} = \frac{d\, V^\epsilon(t,\omega)}{dt} = \epsilon\, F(X^\epsilon(t,\omega),\ V^\epsilon(t,\omega),\ \omega)\ ,\ V^\epsilon(0,\omega) = v_0\ .$$

[27] and [28] prove limit theorems as $\epsilon \downarrow 0$. One assumes that F is stationary in x, has mean zero, and satisfies some smoothness and moment conditions, and most importantly, satisfies a mixing condition. The mixing condition states (in a quantitative way) that $F(x_1)$ and $F(x_2)$ (respectively $F(x_1, v_1)$ and $F(x_2, v_2)$ are almost independent when $|x_1 - x_2|$ is large. Under this type of conditions, in the case of (6.1), as $\epsilon \downarrow 0$, $X^\epsilon(\frac{t}{\epsilon^2}) - \frac{t}{\epsilon^2} v$ converges weakly in $C = C([0,\infty); \mathbb{R}^d)$ to a diffusion. In case (6.2), $V^\epsilon(\frac{t}{\epsilon^2})$ converges weakly in C to a diffusion, $V(t)$ say, and $\epsilon^2 X^\epsilon(\frac{t}{\epsilon^2})$ converges weakly to the integral of $V(\cdot)$. The generators of the limiting diffusions are calculated in [27], [28].

References

Section 2.

[1] Smith, W. L. and Wilkinson, W. E. (1969) On branching processes in random environments. Ann. Math. Statist. 40, 814-827.

[2] Athreya, K. B. and Karlin, S. (1971) On branching processes in random environments I and II. Ann. Math. Statist. 42, 1499-1520 and 1843-1858.

[3] Kaplan, N. (1974) Some results about multidimensional branching processes with random environments, Ann. Prob. 2, 441-455.

[4] Weissner, E. W. (1971) Multi-type branching processes in random environments, J. Appl. Prob. 8, 17-31

[5] Tanny, D. (1977) Limit theorems for branching processes in a random environment, Ann. Prob. 5, 100-116.

[6] Tanny, D. (1980) On multi-type branching processes in a random environment, I and II, to appear.

Section 3.

[7] Ewens, W. J. (1969) Population genetics, Methuen's monographs on applied probability and statistics.

[8] Gillespie, J. (1973) Polymorphism in random environments, Theor. Pop. Biol. 4, 193-195.

[9] Karlin, S. and Lieberman, U. (1974) Random temporal variation in selection intensities: case of large population size, Theor. Pop. Biol. 6, 355-382.

[10] Karlin, S. and Liberman, U. (1975) Random temporal variation in selection intensities: one locus two-allele model, J. Math. Biol 2, 1-17.

[11] Norman, F. (1975) An ergodic theorem for evolution in a random environment, J. Appl. Prob. 12, 661-672.

[12] Karlin, S. and Levikson, B(1974) Temporal fluctuations in selection intensities: case of small population size, Theor. Pop. Biol. 6, 383-412.

[13] Leviksn, B. and Karlin, S.(1975) Random temporal variation in selection intensities acting on infinite deploid populations: Diffusion method analysis, Theor. Pop. Biol. 8, 292-300.

Selection 4.

[14] Turelli, M. (1977) Random environments and stochastic calculus, Theor. Pop. Biol. 12, 140-178.

[15] Turelli, M. and Gillespie, J. H. (1979), Conditions for the existence of asymptotic densities in two-dimensional diffusion processes with applications in population biology, preprint.

[16] Polansky, P. (1978) Invariant distributions for multi-population models in random environments, preprint.

[17] Barra, M., Del Grosso, G., Gerardi, A., Koch, G. and Marchetti, F. Some basic properties of stochastic population models, Working conference on system theory in immunology, Rome, to appear in Lecture notes in biomathematics, Springer-Verlag.

[18] Kesten, H. and Ogura, Y. (1979) Recurrence properties of Lotka-Volterra models with random fluctuations, preprint.

Section 5.

[19] Chernov, A.A. (1967) Replication of a multicomponent chain by the lightning mechanism, Biophysics 12, 336-341.

[20] Temkin, D. E. (1972) One-dimensional random walks in a two-component chain, Soviet Math. 13, 1172-1176.

[21] Solomon, F. (1975) Random walks in a random environment, Ann. Prob. 3, 1-31.

[22] Kesten, H., Kozlov, M.V. and Spitzer, F. (1975) A limit law for random walk in a random environment, Comp. Math. 30, 145-168.

[23] Ritter, G. A. (1976) Random walk in a random environment, Ph.D. thesis, Cornell Univ.

[24] Kesten, H (1977) A renewal theorem for random walk in a random environment, Proc. Symp. Pure Math. 31, 67-77, Amer. Math. Soc.

[25] Kalikow, S. A. (1980) Generalized random walk in a random environment, Ann. Prob. 8.

[26] Papanicolaou, G. C. and Varadhan, S. R. S. (1979) Boundary value problems with rapidly oscillating random coefficients, preprint.

Section 6.

[27] Kesten, H. and Papanicolaou, G. C. (1979) A limit theorem for turbulent diffusion, Comm. Math. Phys. 65, 97-128.

[28] Kesten, H and Papanicolaou, G. C. (1979) Stochastic acceleration, submitted to Comm. Math. Phys.

Cornell University
Ithaca, N. Y. 14853
U.S.A.

SEGREGATION MODEL AND ITS APPLICATIONS

Masao Nagasawa
Seminar für Angewandte Mathematik
Universität Zürich
Freiestrasse 36, 8032 Zürich

When we investigate distributions of a population (of people, animals, insects, bacteria, molecules and so on), we find typical interesting spatial patterns. Among them we see distributions like doughnut, like hamburger, and of segregated groups. We assume that individuals of a population move at random under the influence of the environment. We can also interprete, for example, the septation of bacteria as the segregation of molecules in a bacterium into two groups.

The purpose of this note is to present a mathematical model which can explain those spatial patterns and the segregation of a population. The model is based on the theory of diffusion processes under the influence of external forces. Thus we assume that individuals of a population move as sample paths of a diffusion process whose transition probability is determined by

$$(1) \qquad \frac{\partial u}{\partial t} = \frac{1}{2}\sum_{i,j=1}^{d} a^{ij}(x,N)\frac{\partial^2 u}{\partial x^i \partial x^j} + \sum_{i=1}^{d} b^i(x)\frac{\partial u}{\partial x^i}$$

in d-dimensional Euclidean space ($d \geq 1$), where $a^{ij}(x,N)$, which may depend on the space variable x and the population size N, is assumed to be positive definite. We regard the diffusion coefficient as intrinsic for a population under consideration. The drift coefficient $b^i(x)$, however, is determined by the external force (the environment potential) which can affect the individuals in the population. For simplicity we assume, in this note, that the diffusion coefficient a^{ij} does not depend on the space variable x (for general case cf. [3]).

Suppose the diffusion process has reached an equilibrium state with the distribution density $\varphi(x)$. Then, in the neighbourhood of the peaks of the density (i.e. areas where the intensity is relatively higher), the population has higher <u>population pressure</u> and the average movement of individuals in these high intensity areas is restricted, i.e., the population has lower <u>kinetic energy</u>. In the areas populated rarely (i.e. near the foot of the distribution density $\varphi(x)$), the population pressure is lower and individuals can move freely and rapidly, i.e. the population has higher kinetic energy there. Thus it is plausible to define population pressure (or intensity energy) $Q(x)$ and kinetic energy $K(x)$ of a population at a point x by the following formulas defined in terms of the distribution density $\varphi(x)$.

$$(2)\qquad Q(x) = \frac{1}{2}\frac{1}{\varphi(x)}\Big(-\sum_{i,j=1}^{d} a^{ij}\frac{\partial^2\varphi(x)}{\partial x^i\partial x^j}\Big),\ \text{and}$$

$$(3)\qquad K(x) = \frac{1}{2}\sum_{i,j=1}^{d} a^{ij}\Big(\frac{1}{2}\frac{1}{\varphi(x)}\frac{\partial\varphi(x)}{\partial x^i}\Big)\Big(\frac{1}{2}\frac{1}{\varphi(x)}\frac{\partial\varphi(x)}{\partial x^j}\Big).$$

Fig.1.

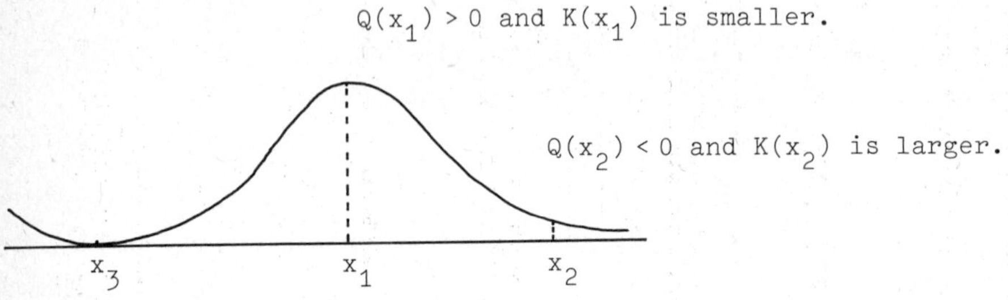

The external force is described in terms of an "*environment potential*" $V(x)$, which is chosen suitable for the problems under consideration. Then, we define the *total energy* $H(x)$ of the population *at* x by

$$(4)\qquad H(x) = K(x) + Q(x) + V(x).$$

When the population reaches an equilibrium state, the $H(x)$ should be equal to a constant λ everywhere, i.e. if

$$(5)\qquad H(x) = \lambda \quad \text{for all } x,$$

the population is in an *equilibrium state of energy* λ.

The relation (5) is of "macroscopic" nature (in other words "thermodynamical"), and gives possible equilibrium densities of diffusion processes under external influence. The equilibrium distribution density, as we will see, determines a diffusion process (in general a pair of diffusion processes in duality). Then we can analyse the detailed behaviour of sample paths of the diffusion process (i.e. that of individuals of the population).

The relation between the distribution density $\varphi(x)$ and the drift coefficient is given by

$$(6)\qquad b^i(x) = \frac{1}{2}\frac{1}{\varphi(x)}\sum_{j=1}^{d} a^{ij}\frac{\partial\varphi(x)}{\partial x^j} + B^i(x).$$

Remark. In [2] the relation is stated as follows

$$b^i(x) + \hat{b}^i(x) = \frac{1}{\varphi(x)} \sum_{j=1}^{d} a^{ij} \frac{\partial\varphi(x)}{\partial x^j}.$$

Here $\hat{b}^i(x)$ is the drift coefficient of the dual diffusion process with respect to the equilibrium density $\varphi(x)$. Therefore in order to determine the diffusion processes in duality we must give another vector function $B^i(x)$ and put

$$b^i(x) - \hat{b}^i(x) = 2B^i(x).$$

Actually we will assume that

$$B^i(x) = \sum_{j=1}^{d} a^{ij} \frac{\partial\beta}{\partial x^j}. \tag{7}$$

Remark. Kinetic energy defined in (3) is for the case $\beta\equiv 0$. If β is not identically equal to zero, we must add an additional term

$$\frac{1}{2} \sum_{i,j=1}^{d} a^{ij} \frac{\partial\beta}{\partial x^i} \frac{\partial\beta}{\partial x^j} \tag{8}$$

to the right hand side of (3).

Now, let us assume that the equilibrium density $\varphi(x)$ is the square of some function $\psi(x)$, i.e.,

$$\varphi(x) = |\psi(x)|^2 \tag{9}$$

and $\psi = |\psi|e^{i\beta}$ (cf. (7)). Thus ψ is in general complex valued, and if $\beta\equiv 0$, it is real valued. Then, by simple calculations, we can show

$$K(x) + Q(x) = -\frac{1}{\psi(x)}\left(\frac{1}{2} \sum_{i,j=1}^{d} a^{ij} \frac{\partial^2\psi(x)}{\partial x^i \partial x^j}\right). \tag{10}$$

Therefore, combining (5) and (10), we can reduce the problem of finding out the equilibrium density $\varphi(x)$ to the eigenvalue problem

$$\frac{1}{2} \sum_{i,j=1}^{d} a^{ij} \frac{\partial^2\psi}{\partial x^i \partial x^j} + (\lambda - V(x))\psi = 0. \tag{11}$$

When the dimension $d=1$, the eigenvalues are not degenerated and we can take ψ to be real valued (this corresponds to the fact that one-dimensional diffusion processes are selfadjoint with respect to the equilibrium density $\varphi(x)$). However, since the eigenvalues (except the lowest one) may be degenerated for $d \geq 2$, the eigenfunction ψ is, in general, a linear combination (with complex coefficients) of orthogonal eigenfunctions of (11). Thus $\beta \not\equiv 0$ in general (cf. [3]).

Since the eigenfunctions have zeros (or nodal surfaces), except the ground state, we must analyse the behaviour of sample paths of the diffusion process near the nodal surfaces. We can prove that the diffusion process with the distribution density $\varphi = |\psi|^2$ cannot approach the nodal surfaces of ψ, i.e., the population is segregated by the nodal surfaces (at x_3 in Fig. 1, cf. [3]). This fact has important bearings in the following examples.

Example 1. Let us take an Escherichia coli, for example. Since we are interested in the length of E. coli, we take a one-dimensional diffusion process to describe molecules in an E. coli and assume an environment potential $V(x) = |x|$ in the E. coli. Then (11) becomes

$$\frac{d^2\psi}{dx^2} + (\lambda - |x|)\psi = 0. \tag{12}$$

We interprete the distribution $\varphi_1 = |\psi_1|^2$ of the ground state, corresponding to the smallest eigenvalue λ_1, as an E. coli of normal size (this means that there is a minimal size for E. coli). Since the second eigenvalue λ_2 is greater than $2\lambda_1$, the E. coli will not grow up to the second equilibrium distribution $\varphi_2 = |\psi_2|^2$, but it splits into two normal sized E. coli.

Besides this normal septation of E. coli, we can apply the model to interesting septation phenomena of a mutant E. coli. For this, cf. [4], [6].

Example 2. The spatial distribution patterns depend on population size (cf. [1], also [5]). Assume that there are favourable areas and unfavourable areas for insects (when the insects are kept in a box, we must treat diffusion processes with reflecting boundary condition cf. [2], [3]). If the population size is small enough, then insects inhabit the favourable areas. If the population size grows up, they start to inhabit unfavourable areas, and then finally the distribution becomes almost uniform. We can treat this problem allowing the diffusion coefficient to depend on the population size N, say,

$$a^{ij}(N) = (\alpha N + \beta)\, D^{ij}. \tag{13}$$

For the environment potential of this problem we take, for example, a set function

$$V(x) = \begin{cases} c > 0, & \text{in unfavourable areas,} \\ 0, & \text{in favourable areas.} \end{cases}$$

The smallest eigenvalue increases as the population size N, and we obtain the equilibrium distribution depending on the population size N as described above (cf. [3]).

Example 3. If a population in two dimension is attracted to a centre, the equilibrium distributions show interesting spatial patterns. For the lowest eigenvalue, we obtain a distribution like a hamburger. For the second, a doughnut distribution appears. Besides this, the population can be segregated into two groups like a broken doughnut. If the population is excited to the third eigenvalue, it distributes like hamburger + doughnut (there are several other types, cf. [3]). These are observed in a population of animals, insects, or bacteria. These patterns of segregation of population are observed also in residence distributions of cities with large population size (it's called "doughnut phenomena"). If we speak of the segregation of politicians, who are attracted by political fruits, into parties in terms of this segregation theory, perhaps one might wonder whether we went too far away.

References

[1] Morisita, M. (1971) Measuring of habitat value by the "environmental density" method. Statistical ecology vol. 1., pp. 379-401, Ed. Patil, G.P., Pielou, E.C., Waters, W.E. The Pennsylvania State Univ. Press, University Park and London.

[2] Nagasawa, W. (1961) The adjoint process of a diffusion with reflecting barrier. Kodai Math. Sem. Rep. 13, 235-248.

[3] Nagasawa, M. (to appear) Segregation of a population in an environment.

[4] Nagasawa, M. (to appear) An application of the segregation model for septation of Escherichia coli.

[5] Shigesada, N., Teramoto, E. (1977) A consideration on the theory of environmental density (Japanese with English summary). Japanese J. Ecology.

[6] Yamada, M., Maruyama, T., Hirota, Y. (1979) A model for distribution of septation sites in Escherichia coli. U.S.-Japan Intersociety Microbiology Congress, Los Angeles-Honolulu, May 1979.

POSITIVE RECURRENCE OF MULTI-DIMENSIONAL POPULATION-DEPENDENT BRANCHING PROCESSES

W. Rittgen

Institut für Dokumentation,
Information und Statistik

Deutsches Krebsforschungszentrum
Heidelberg

0. INTRODUCTION

Continuous time branching processes are appropriate stochastic models for the temporal description of cell populations as long as the cells existing simultaneously can be assumed to develop independently. A one-dimensional process is sufficient for the examination of the number of cells, but if one wants to consider internal states (for example cell cycle phases, see Baserga, 1976, for the biological aspects), one has to pass to the multi-dimensional processes. As up to now the information about the distributions of the phase durations is limited (see Prescott, 1976 especially Chapter 3), it is surely sufficient to confine oneself to Markov branching processes, which are easier to handle than the age-dependent processes. This restriction is not so severe as it seems because the sojourn times in the different biological phases can be represented as sums of independent exponentially distributed random variables without losing the Markov property of the whole process (see for example Kendall, 1948; Rittgen & Tautu, 1976; Rittgen, 1978).

The fundamental assumption of non-interaction between cells limits the applicability of such models. For example, at a low density cells have a high division activity and low death rate whereas at a high density cells have a low division activity and high death rate (see Smith & Martin, 1974). Therefore a desirable generalization of the branching cell models is that the reproduction behaviour at time t might depend on the state of the population at time t.

These population-dependent branching processes can also be interesting for other problems if we do not restrict ourselves to the offspring structure given by the cell development, that means death of an object, transition to another state or division into two new objects. For example the development of an animal population or the development of an epidemic can be described with the aid of such processes.

Branching models in which the reproduction behaviour depends on the state of the population have proved to be very difficult to handle because the usual branching process tools such as martingales or iterates of generating functions are no longer available for these generalized processes. But one can expect that some analogues of "classical" results can be proved (see Lipow, 1977; Wofsy(Lipow), 1980). Our results, conditions for for positive recurrence, however, can have no "classical" counterpart.

For the discrete time case the reader is referred to Levina et al. (1968), Vasil'ev et al.(1968), Labkovskii (1972), Sevast'yanov & Zubkov (1974), Zubkov (1974), Yanev (1975) and Fujimagari (1976).

1. DEFINITION OF A POPULATION-DEPENDENT MARKOV BRANCHING PROCESS

For the examination of the generalized processes we need the following definitions:

d := number of different types of objects (dimension)

$T := \mathbb{N}_0^d$ (state space)

$T(N) := \{ x = (x_1,\dots,x_d)' \in T : \sum_{i=1}^{d} x_i \geq N \}$

$T^* := \{ x = (x_1,\dots,x_d)' : x_i \in N_0 \cup \{-1\}$, only one x_i can be negative $\}$

where x' denotes the transpose of x (vector or matrix).

A multi-dimensional population-dependent Markov branching process can be described verbally as follows:

The process starts at time 0 with $Z(0) \in T(1)$ objects, where $Z(t) := (Z_1(t),\dots,Z_d(t))'$ and $Z_i(t)$ is the number of objects of type i alive at time $t \geq 0$.

If at a certain time an object of type i is alive, the waiting time until its death is exponentially distributed with expectation $1/\lambda_i$ ($0<\lambda_i<\infty$ for $i=1,\dots,d$).

At its death it is replaced by $y = (y_1,\dots,y_d)' \in T$ objects with probability $p_y^i(x)$ if x T(1) is the population vector immediately before the death of the object.

The expectation vector and the matrix of the second central moments respectively of the offspring distribution of an object of type i , if before its death there are $x=(x_1,\dots,x_d)' \in T(1)$ objects alive, are

$(m_{i1}(x),\dots,m_{id}(x))'$ with $m_{ij}(x) := \sum_{y \in T} y_j \, p_y^i(x)$ and

$$(\sigma^i_{jk}(x))_{jk} \quad \text{with} \quad \sigma^i_{jk}(x) := \sum_{y \in T} y_j \, y_k \, p^i_y(x) \quad \text{for } i,j,k=1,\dots,d.$$

For the simplification of the notation later on let us define:

$p^i_y(x) := 0 \quad$ for $x \in T(1)$, $y \notin T$, $i=1,\dots,d$,

$\sum_i := \sum_{i=1}^{d}$ and $\sum_y := \sum_{y \in T^*}$,

$e_i := (0\dots0,1,0\dots0)'$ with the 1 in the i-th component,

$M(x) := (m_{ij}(x))_{ij}$,

$I := (\delta_{ij})_{ij}$ with $\delta_{ij} := \begin{cases} 1 & \text{if } i=j \\ 0 & \text{if } i \neq j \end{cases}$,

$\Lambda := (\lambda_i \, \delta_{ij})_{ij}$,

$\lambda := (\lambda_1, .., \lambda_d)'$

$(x,y) := \sum_i x_i \, y_i$ for $x,y \in \mathbb{R}^d$ (scalar product),

$\| x \| := (x,x)^{1/2}$ for $x \in \mathbb{R}^d$,

$\| M \|$:= operator norm of the matrix M with respect to the vector norm defined above,

$\rho(M)$:= eigenvalue with the largest real part of the matrix M which is non-negative outside the diagonal. Note that $\rho(M)$ is real.

For the transition probabilities of the population-dependent process

$$P_{xy}(t) := P\left(Z(t) = x+y \mid Z(0) = x \right) \text{ with } t \geq 0 \text{ , } x \in T(1) \text{ , } x+y \in T$$

we have

$$P_{xy}(h) = h \sum_i x_i \, \lambda_i \, p^i_{y+e_i}(x) + o(h) \quad \text{for } y \in T^* \text{ , } y \neq 0$$

$$P_{x0}(h) = 1 - h \sum_i x_i \, \lambda_i \, p^i_{e_i}(x) + o(h) \quad \text{and}$$

$$P_{xy}(h) = o(h) \quad \text{for } y \notin T^*$$

with the boundary condition $x+y \in T$.

For the further considerations we assume that the following conditions hold:

(I) $\quad p^i_0(e_i) = 0$ for $i=1,\dots,d$

(II) $\quad p^i_{e_i}(x) \leq c_0 < 1$ for $i=1,\dots,d$, $x \in T(1)$

(III) $\sigma^i_{jk}(x) \leq c_2 < \infty$ for $i,j,k=1,\dots,d$, $x \in T(1)$ including

$m_{ij}(x) \leq c_1 < \infty$ for $i,j=1,\dots,d$, $x \in T(1)$

(IV) $\{Z(t)\}$ is irreducible.

The problem is now what conditions guarantee that the total number of objects of such a population-dependent process cannot escape to infinity.

For this investigation we separate the defined process into two processes,

one the process of split times, i.e. the times at which particles split,

the other the process of jumps, i.e. the discrete time process representing the sequence of states into which the original process goes at the split times.

(See Chung, 1967,p.259 for continuous time Markov chains and Athreya & Karlin, 1967, or Athreya & Ney, 1972, Section III.9 for Markov branching processes.)

Let us denote by τ_n the time of the n-th split and by W_n the state into which the process jumps at that split.

Then we have

$$Z(0) = W_0 \ , \ \tau_0 = 0$$

$$Z(t) = W_n \quad \text{for } \tau_n \leq t < \tau_{n+1} \ , \ n=0,1,2,\dots$$

where $\{W_n\}$ is a discrete time Markov chain with stationary transition probabilities.

The general assumption (I) ensures that $\{Z(t)\}$ cannot die out whereas the following two conditions ensure that $\{Z(t)\}$ can only have a finite number of jumps in a finite time interval and has no absorbing state.

Then the process $\{Z(t)\}$ is positive recurrent or transient if the same holds for the process $\{W_n\}$ and the transition probabilities of the latter can easily be written down:

$$Q_{xy} := P\left(W_{n+1} = x+y \mid W_n = x \right) =$$

$$= (x,\lambda)^{-1} \sum_i x_i \lambda_i p^i_{y+e_i}(x) \quad \text{for } x \in T(1) \ , \ y \in T^* \ , \ n=0,1,2,\dots$$

To prove the positive recurrence we use the following result due to Foster (1953).

THEOREM 1

$\{W_n\}$ is positive recurrent if and only if there exists a positive function $\Phi(x)$, $x \in T^*$ and two constants $N \geq 1$ and $\varepsilon > 0$ such that the following two conditions hold:

(1) $\sum_y Q_{xy} \ \Phi(x+y) < \infty$ for $x \in T(1)$

(2) $\sum_y Q_{xy} \ \Phi(x+y) \leq \Phi(x) - \varepsilon$ for $x \in T(N)$

2. THE ONE-DIMENSIONAL CASE

THEOREM 2

If $m(x) := m_{11}(x) \leq 1 - \varepsilon < 1$ for $x \geq N > 1$ then $\{W_n\}$ is positive recurrent and the limiting distribution has finite mean.

Proof:

Carla Wofsy (under the name Lipow, 1975) proved similar results for a process with a supercritical offspring distribution below a certain population size and a subcritical one otherwise.

1) positive recurrence

$$Q_{xy} = p_{y+1}(x) := p^1_{y+1}(x)$$

$\Phi(x) := x$

Condition (1) of Theorem 1 is trivially satisfied and we only have to look at Condition (2).

$$\sum_y Q_{xy} \ \Phi(x+y) = \sum_y p_{y+1}(x) \ (x+y) = x + m(x) - 1 \leq$$

$$\Phi(x) - \varepsilon \quad \text{for } x \geq N$$

2) finite mean

Let $\{\nu_x\}$ be the limiting distribution (obviously $\nu_0 = 0$) with

$$\sum_{y-x \geq -1} \nu_x \ Q_{x,y-x} = \nu_y \quad \text{or}$$

$$\sum_{x=1}^{y+1} \nu_x \ p_{y-x+1}(x) = \nu_y \quad \text{for } y=0,1,2,\ldots \ .$$

Substituting z for y-x+1 , multiplying by s^{y+1} ($|s| \leq 1$) and summing with respect to y we get

$$\sum_{y=0}^{\infty} \sum_{z=0}^{y} \nu_{y-z+1} \ s^{y-z+1} \ p_z(y-z+1) \ s^z = s \sum_{y=0}^{\infty} \nu_y \ s^y$$

and by interchanging the summation

$$\sum_{z=0}^{\infty}\sum_{y=z}^{\infty} \nu_{y-z+1}\, s^{y-z+1}\, p_z(y-z+1)\, s^z = s\sum_{y=0}^{\infty} \nu_y\, s^y \quad .$$

Replacing y-z+1 by x and interchanging the summation again we obtain

$$\sum_{x=1}^{\infty} \nu_x\, s^x \sum_{z=0}^{\infty} p_z(x)\, s^z = s\sum_{y=1}^{\infty} \nu_y\, s^y \quad .$$

With $f(s;x) := \sum_{z=0}^{\infty} p_z(x)\, s^z$ for $|s| \leqq 1$ we have

$$\sum_{x=1}^{\infty} \nu_x\, s^x\, (\, f(s;x) - s\,) = 0$$ and when dividing by $1-s$ for $s \neq 1$

$$\sum_{x=1}^{\infty} \nu_x\, s^x\, \left(1 - \frac{1-f(s;x)}{1-s}\right) = 0 \quad .$$

Differentiation with respect to s and $s \to 1$ from the left implies

$$\sum_{x=1}^{\infty} x\, \nu_x\, (\, 1 - m(x)\,) = \frac{1}{2}\sum_{x=1}^{\infty} \nu_x\, (\, \sigma(x) - m(x)\,) \leqq \frac{1}{2}\sum_{x=1}^{\infty} \nu_x\, c_2 = \frac{1}{2}\, c_2 \, .$$

Combining this result with $1 - m(x) \geqq \varepsilon > 0$ for $x > N$ completes the proof.

When I had proved this theorem, I expected that the same would hold with the analogous condition for the multi-dimensional processes, that means under the condition $\rho(M(x)) \leqq 1 - \varepsilon < 1$ for $x \in T(N)$ with $N > 1$ (Rittgen & Tautu, 1976). But, in general, this result does not hold as I will show later on in a counterexample. Positive recurrence can only be proved under an additional condition on the structure of the expectation matrices M(x).

3. THE MULTI-DIMENSIONAL CASE

THEOREM 3

If $\lim_{\|x\| \to \infty} M(x) = M$ and $\rho(M) < 1$ then $\{W_n\}$ is positive recurrent.

Proof:

$\Phi(x) := (x,Bx)$ where B is a symmetric, positive definite matrix which will be specified later in the proof (see also Kingman, 1961).

$\Phi(x+y) = (x+y,B(x+y)) = (Bx,x) + 2(x,By) + (y,By)$

Again Condition (1) of Theorem 1 is trivially satisfied and we only have to look at Condition (2).

$$\sum_y Q_{xy}\,\Phi(x+y) = (x,\lambda)^{-1}\sum_y\sum_i x_i\,\lambda_i\,p^i_{y+e_i}(x)\,[\Phi(x) + 2(Bx,y) + (y,By)]$$

$$= \Phi(x) + 2(x,\lambda)^{-1}\,[B\,x\,,\sum_i x_i\,\lambda_i \sum_y y\,p^i_{y+e_i}(x)] +$$

$$(x,\lambda)^{-1}\sum_i x_i\,\lambda_i\sum_y (y,By)\,p^i_{y+e_i}(x)$$

The last term is uniformly bounded for all $x \in T(1)$ by $c_2\,c_B$ where c_B is a constant depending only on B .

In the second term we have

$$(x,\lambda)^{-1}\sum_i x_i\,\lambda_i\sum_y y\,p^i_{y+e_i}(x) = (x,\lambda)^{-1}\,[M(x)' - I]\,\Lambda\,x\,,$$

this is the vector of the mean drift of the jump process $\{W_n\}$ (see e.g. Kesten, 1976; Kingman, 1961).

Then we get

$$\sum_y Q_{xy}\,\Phi(x+y) \leq \Phi(x) + c_2\,c_B + 2(x,\lambda)^{-1}\,[B\,x\,,\,(M(x)' - I)\,\Lambda\,x]\,,$$

and we only have to specify the matrix B .

$\rho(M) < 1$ is equivalent to $\rho(\,\Lambda(M-I)) < 0$ and therefore by Ljapunow's theorem (see Gantmacher, 1966,pp.165-167) there exists a symmetric positive definite matrix B which solves the equation

$$\Lambda\,(\,M - I\,)\,B + B\,(\,M' - I\,)\,\Lambda = -\,I\,.$$

Therefore by the convergence of M(x)

$$2\,(\,B\,x\,,\,(M(x)' - I)\,\Lambda\,x\,) =$$

$$2\,(\Lambda\,(M(x) - M)\,B\,x\,,\,x) + [(\Lambda\,(M - I)\,B + B\,(M' - I)\,\Lambda)x\,,\,x] \leq$$

$$[-1 + \|\Lambda\,(M(x) - M)\,B\|\,]\,\|x\|^2 \leq -\gamma\,\|x\|^2$$

for $\gamma > 0$ and $x \in T(N)$ if N is sufficiently large.

Combining the results above we obtain

$$\sum_y Q_{xy}\,\Phi(x+y) \leq \Phi(x) + c_2\,c_B - \gamma\,(x,\lambda)^{-1}\,\|x\|^2 \leq \Phi(x) - \gamma$$

for $x \in T(N^*)$ for some $N^* \geq N$.

COROLLARY 1

If there exists a matrix M, non-negative outside the diagonal, and two constants $N > 1$ and $\varepsilon > 0$ with

$\rho(M) < 1$ and

$$(\Lambda\ (M(x) - M)\ B\ x\ ,\ x\ ) \leqq \frac{1}{2}\ (1 - \varepsilon)\ \|x\|^2 \qquad \text{for } x \in T(N)$$

where B is defined by $\Lambda(M - I)\ B + B\ (M' - I)\ \Lambda = -I$

then $\{W_n\}$ is positive recurrent.

This corollary follows directly from the proof of Theorem 3.

COROLLARY 2

If one of the following conditions

(1) $M(x)$ is normal (i.e. $M(x)'M(x) = M(x)\ M(x)'$) and

$\rho(M(x)-I) \leqq -\varepsilon < 0$ for $x \in T(N)$ for some $N > 1$,

(2) $\Lambda M(x)$ is normal and

$\rho(\ \Lambda(M(x)-I)) \leqq -\varepsilon < 0$ for $x \in T(N)$ for some $N > 1$

holds then $\{W_n\}$ is positive recurrent.

Proof:

By a spectral radius argument for commuting operators

$\rho(M+M') \leqq 2\ \rho(M)$.

Combining this result with

$B = \Lambda^{-1}$ for Condition (1) and

$B = I$ for Condition (2)

completes the proof.

COUNTEREXAMPLE

The condition $\rho(M(x)) \leqq 1 - \varepsilon < 1$ for $x \in T(N)$ for some $N > 1$ is not sufficent for the positive recurrence of $\{W_n\}$.

Proof:

We assume $d = 2$, $\lambda_1 = \lambda_2 = 1$ and choose a constant $c \geqq 4$ and even. Then we have $x = (x_1,x_2)$ and define $|x| := x_1+x_2$.
We define the offspring distribution in a way that an object of either type can only have 0 or c offspring and that the according probabilities depend on the parity of the population size only.

(1) $\left.\begin{array}{l} p^1_{(0,c)}(x) := 1 \\ p^2_{(c,0)}(x) := 1 \end{array}\right\}$ if $|x| = 2$

(2) $\left.\begin{array}{l} p^1_{(0,c)}(x) := 1 \\ p^2_{(0,0)}(x) := 1 \end{array}\right\}$ if $|x| > 2$ and even

(3) $\left.\begin{array}{l} p^1_{(0,0)}(x) := 1 \\ p^2_{(c,0)}(x) := 1 \end{array}\right\}$ if $|x| > 2$ and odd

The expectation matrices are

$$M(x) = \begin{pmatrix} 0 & c \\ c & 0 \end{pmatrix} \quad \text{if } |x| = 2 ,$$

$$M(x) = \begin{pmatrix} 0 & c \\ 0 & 0 \end{pmatrix} \quad \text{if } |x| > 2 \text{ and even} ,$$

$$M(x) = \begin{pmatrix} 0 & 0 \\ c & 0 \end{pmatrix} \quad \text{if } |x| > 2 \text{ and odd} .$$

Therefore we have $\rho(M(x)) = 0$ for $|x| > 2$,but because of (1) the process cannot die out.

If $|x| > 2$ the process can have the following jumps:

$\lvert x \rvert$ even	$\lvert x \rvert$ odd	probability
(-1, c)	(-1, 0)	$\frac{x_1}{x_1+x_2}$
(0,-1)	(c,-1)	$\frac{x_2}{x_1+x_2}$

For the two-step process $\{W_{2n}\}$ we have the symmetric jump possibilities (-1,-1) , (-2,c) , (c,-2) and (c-1,c-1) but with ugly non-symmetric probabilities. Nevertheless this process is easier to handle than the one-step process.

By direct but laborous calculations we get

$$|E\,(\, W_2 - W_0 \mid W_0 = x \,,\ |x| > 2 \ \text{ and even })| =$$

$$c - 2 + \frac{c^2 x_1(x_1-1) + c(c-2)x_1x_2}{(x_1+x_2)(x_1+x_2-1)(x_1+x_2-1+c)}$$

and an expression of similar form for $|x| > 2$ and odd.

Combining these results and (1) we have

$$|E\,(\, W_2 - W_0 \mid |W_0| > 2 \,)| \geqq c - 2 \geqq 2$$

and by the Markov property of $\{W_{2n}\}$

$$|E\,(\, W_{2n} \mid |W_0| > 2 \,)| \to \infty .$$

It is obvious that the process $\{W_{2n}\}$ cannot be positive recurrent for each c sufficiently large, but moreover it has a positive mean drift bounded below in the direction of the unit vector $(1/\sqrt{2}, 1/\sqrt{2})$ all over its state space and therefore the process cannot be recurrent but must be transient. For a detailed proof Corollary 2 of Kesten (1976) can be used; its conditions trivially hold.

The transience of the two-step process means that the total number of the objects observed at each second step tends to infinity with positive probability. The increase and decrease respectively of the total population by one jump are uniformly bounded and therefore the one-step process $\{W_n\}$ is transient too.

4. CONCLUSIONS

The population-dependent branching process of the counterexample was only constructed with a view to showing that the eigenvalue condition is not sufficient for the positive recurrence and there is probably no comparable natural population process. However, this counterexample shows that the reproduction behaviour must not be too irregular but that a certain regularity is essential for an efficient growth control. Such a regularity should exist in most of the biological populations by the fact that with increasing population density the mean number of offspring of an object decreases monotonically. Thus the additional conditions on the expectation matrices M(x) in Theorem 3 , Corollary 1 and Corollary 2 are not really restrictions from the biological view point.

REFERENCES

1. Athreya,K.B., Karlin,S.(1967): Limit theorems for the split times of branching processes. J.Math.Mech. 17, 257-277
2. Athreya,K.B., Ney,P.(1972): Branching Processes. Berlin-Heidelberg-New York: Springer
3. Baserga,R.(1976): Multiplication and Division in Mammalian Cells. New York and Basel: Marcel Dekker Inc.
4. Chung,K.L.(1967): Markov Chains with Stationary Transition Probabilities. 2nd edition, Berlin-Heidelberg-New York: Springer
5. Foster,F.G.(1953): On stochastic matrices associated with certain queuing processes. Ann.Math.Stat. 26, 355-360
6. Fujimagari,T.(1976): Controlled Galton-Watson process and its asymptotic behaviour. Kōdai Math.Sem.Rep. 27, 11-18
7. Gantmacher,F.R.(1966): Matrizenrechnung. Teil II, Berlin: VEB Deutscher Verlag der Wissenschaften

8. Kendall,D.G.(1948): On the role of variable generation times in the development of a stochastic birth process. Biometrika 35, 316-330

9. Kesten,H.(1976): Recurrence criteria for multi-dimensional Markov chains and multi-dimensional linear birth and death processes. Adv.Appl.Prob. 8, 58-87

10. Kingman,J.F.C.(1961): The ergodic behaviour of random walks. Biometrika 48, 391-396

11. Labkovskii,V.A.(1972): A limit theorem for generalized random branching processes depending on the size of the population. Theor. Probability Appl. 17, 72-85

12. Levina,L.V., Leontovich,A.M., Pyatetskii-Shapiro,I.I.(1968): A controllable branching process. Problems of Information Transmission 4, 72-82

13. Lipow,C.(1975): A branching model with population size dependence. Adv.Appl.Prob. 7, 495-510

14. Lipow,C.(1977): Limiting diffusions for population size dependent branching processes. J.Appl.Prob. 14, 14-24

15. Prescott,D.M.(1976): Reproduction of Eukariotic Cells. New York-San Francisco-London: Academic Press

16. Rittgen,W.(1978): Zellerneuerungssysteme. Medizinische Informatik und Statistik 8, 310-333

17. Rittgen,W., Tautu,P.(1976): Branching models for the cell cycle. Lecture Notes in Biomathematics 11, 109-126

18. Sevast'yanov,B.A., Zubkov,A.M.(1974): Controlled branching processes. Theor.Probability Appl. 19, 14-24

19. Smith,J.A., Martin,L.(1974): Regulation of cell proliferation. In Cell Cycle Controls (G.M.Padilla, I.C.Cameron, A.Zimmerman eds.) pp.43-60. New York-San Francisco-London: Academic Press

20. Vasil'ev,N.B., Levina,L.M., Leontovich,A.M., Pyatetskii-Shapiro, I.I.(1968): Regulation of numbers in a population of dividing cells. Problems of Information Transmission 4, 84-91

21. Wofsy(Lipow),C.(1980): Behavior of limiting diffusions for density-dependent branching processes. this volume

22. Yanev,N.M.(1975): Conditions for degeneracy of φ-branching processes with random . Theor.Probability Appl. 20, 421-428

23. Zubkov,A.M.(1974): Analogies between Galton-Watson processes and φ-branching processes. Theor.Probability Appl. 19, 309-331

POPULATION GROWTH WITH LARGE RANDOM FLUCTUATIONS

HENRY C. TUCKWELL

Department of Mathematics
University of British Columbia
Vancouver
B.C. Canada V6T 1W5

ABSTRACT

Real population trajectories often have relatively large changes in short time intervals. Though deterministic models can display such changes it is an advantage to have models in which large fluctuations may occur randomly. Previous random differential equations for population growth have had solutions which are diffusion processes and whose sample paths are continuous. A class of Markov models can be obtained in which random discontinuities are superimposed on otherwise continuous trajectories. Some simple examples are considered including logistic growth with random disasters which are either of constant magnitude or proportional to population size and Gompertzian growth with density independent disasters. The quantity of interest is the first passage time of the population size to some low level at which extinction occurs. The equations for calculating the moments of this random variable have been solved in certain cases.

1. INTRODUCTION

It has long been realized that models for the growth of populations should be random rather than deterministic. Even in the classic demonstration of logistic growth by Gause (1934) with an experimental paramecium population the number density fluctuated about the logistic curve. Results such as these can possibly be explained by deterministic models. Equations of the kind

$$(1)\qquad N_{t+1} = F(N_t) \ ,$$

where $F(\cdot)$ is a suitable nonlinear function and N_t is the population size at time t , can produce large fluctuations which may be periodic or have the appearance of randomness (see May, 1976 for a succinct summary).

In field populations extremely large changes often occur in very short time intervals. Classic examples are provided by crashes in ungulate populations (Scheffer, 1951; Klein, 1968). The U.S. government introduced 25 reindeer to St. Paul Island (off Alaska) in 1911. The animals were free of predators and their numbers grew steadily to about 2000 in 1938. By 1950 there were only 8 reindeer on the island. More sudden was the crash on St. Matthew Island. Twenty-nine animals were released there in 1944 and their numbers grew steadily to 6000 by 1963, but by 1966 there were only 42 reindeer on the island.

The explanation offered for these crashes was that the food supply had diminished by overbrowsing. Furthermore the reindeer depend on lichen during the winter months which they detect beneath up to four feet of snow. At the time of the 1964 crash on St. Matthew, the winter was one of the most severe on record. This random event solidified the snow making it nearly impossible for the animals to obtain food. Examination of the carcasses showed that the animals died from starvation.

Deterministic models have been advanced to explain these crashes. In particular, these populations have been construed as the herbivores in plant-herbivore systems. Letting H be the number of herbivores and letting V be the vegetation, these models take the form

$$\text{(2A)} \quad \frac{dH}{dt} = f(H,V) \, ,$$

$$\text{(2B)} \quad \frac{dV}{dt} = g(H,V) \, ,$$

and with the appropriate choice of functions $f(\cdot)$ and $g(\cdot)$ the data on the reindeer crashes can be fairly well duplicated (see Caughley, 1976, for a summary).

It seems therefore that deterministic models can incorporate large and fairly rapid fluctuations in population sizes. Nevertheless, it is apparent that many large fluctuations are essentially random. Usually these large fluctuations take the form of disasters caused by the following: floods, fire, severe weather leading to death directly or by depletion of food supply, devastation by pests, excessive hunting, earthquakes, disease, etc. Sometimes a large random increase may occur due to a sudden influx of immigrants or suddenly favorable conditions for food supply and reproduction. It should be mentioned in this context that the eruptions in the reindeer populations which preceded the crashes mentioned above were essentially random (colonizing) events.

2. RANDOM PROCESSES WITH JUMP DISCONTINUITIES

We are interested in processes whose sample paths admit the possibility of large changes in short times. Discontinuous Markov processes were considered a long time ago by Kolmogorov (1931) and Feller (1940). The theory of processes with deterministic, diffusion and jump components has been subsequently developed by Ito (1951), Skorohod (1965) and Gihman & Skorohod (1972).

The general stochastic differential equation for such a temporally homogeneous process, which is to be interpreted in terms of the relevant stochastic integral, can be written

$$(3) \quad dX_t = f(X_t)dt + g(X_t)dW_t + \int_R h(X_t,u)n(dt,du) ,$$

where $f(\cdot)$, $g(\cdot)$, $h(\cdot,\cdot)$ are 'deterministic' functions and W_t is a standard (zero mean, variance t) Wiener process. Integration is over the whole real line and $n(\cdot,\cdot)$ is a Poisson random measure. This means that if A is some Borel set then $n(t,A)$ is a Poisson process with some (positive) rate parameter $\lambda(A)$. That is

$$(4) \quad P\{ n(t,A) = k \} = (t\lambda(A))^k e^{-t\lambda(A)}/k! ,$$

and

$$(5) \quad n(0,A) = 0 .$$

If X_t has a transition probability density $p(y,t|x,s)$, where (x,s) are backward variables, then this density satisfies the Kolmogorov equation (see Gihman & Skorohod, 1972):

$$(6) \quad -\frac{\partial p}{\partial s} = -\Lambda p + f(x)\frac{\partial p}{\partial x} + \frac{1}{2}g^2(x)\frac{\partial^2 p}{\partial x^2} + \int_R p(y,t \mid x + h(x,u),s)\lambda(du) ,$$

where

$$(7) \quad \Lambda = \int_R \lambda(du) < \infty$$

is the total jump rate.

Darling and Siegert (1953) had obtained a recursion system of differential equations for the moments of the first exit time from an interval for a diffusion process. These equations have been generalized (Tuckwell, 1976) to cover processes defined by the stochastic equation (3). Let (a,b) be some interval on the real line and suppose that initially the process X_t takes a value $x \in (a,b)$. Let $T(x)$ be the time of first escape of X_t from (a,b), the idea being sketched in Figure 1. Technically we have

$$(8) \quad T(x) = \inf\{ t \mid X_t \notin (a,b) \mid X_0 = x \in (a,b) \} .$$

The moments of $T(x)$,

$$(9) \quad \mu_n(x) = E[T^n(x)] ,$$

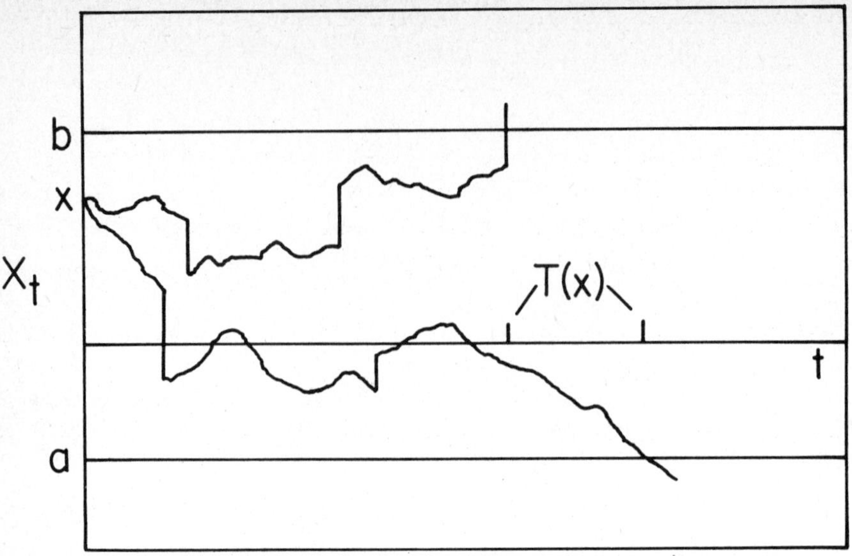

Figure 1. Schematic representation of sample paths of the random process X_t commencing at the value x and first escaping from the interval (a,b) at the (random) time $T(x)$.

assuming that

(10) $$P\{T(x) < \infty\} = 1 ,$$

satisfy the recursion system of integro-differential equations

(11) $$-\Lambda\mu_n(x) + f(x)\frac{d\mu_n}{dx} + \frac{1}{2}g^2(x)\frac{d^2\mu_n}{dx^2} + \int_R \mu_n(x + h(x,u))\lambda(du)$$

$$= -n\mu_{n-1}(x), \quad n = 1,2,\ldots, \quad x \in (a,b)$$

with $\mu_o(x) = 1$. This system of equations is solved with the boundary condition

(12) $$\mu_n(x) = 0 , \quad x \notin (a,b) ,$$

and the constraint that $\mu_n(x) < \infty$ for $x \in (a,b)$. In most cases, except for purely discontinuous processes, $\mu_n(x)$ is also continuous on (a,b).

3. SIMPLE MODELS FOR POPULATION GROWTH WITH LARGE RANDOM FLUCTUATIONS

3.1 LOGISTIC GROWTH WITH RANDOM DISASTERS

We have studied in detail two simple examples of processes satisfying stochastic equations such as (3) (Hanson & Tuckwell, 1978; 1979). In both cases the growth has been according to the logistic equation in the absence of random discontinuities, so that the deterministic part of (3) is

(13) $$f(X_t) = rX_t(1 - X_t / K) ,$$

where r is the "intrinsic growth rate" and K is the carrying capacity. We have set the diffusion contribution at zero,

(14) $$g(X_t) = 0 ,$$

and inserted two kinds of random disaster.

Disasters of Constant Magnitude

In the first case an event in a Poisson process removes a certain fixed number ε of individuals. To translate this into the appropriate stochastic differential equation (3) we let A be an arbitrary set with $\lambda(A) = \lambda > 0$, with the function $h(\cdot,\cdot)$ defined thus

(15) $$h(X_t,u) = \begin{cases} -\varepsilon, & u \in A , \\ 0, & \text{otherwise} \end{cases}$$

the total jump rate being $\Lambda = \lambda$. The stochastic differential equation can also be written in a very simple form

(16) $$dX_t = rX_t(1 - X_t / K)dt - \varepsilon d\Pi_t^\lambda ,$$

where Π_t^λ is a Poisson process of rate parameter λ . This equation can also be interpreted as

(17) $$\frac{dX_t}{dt} = rX_t(1 - X_t/K) - \varepsilon \sum_{i=1}^{\infty} \delta(t - t_i) ,$$

where t_i is the time of occurrence of the i-th event in a Poisson process of rate λ . The process defined by either (16) or (17) will be called logistic growth with random disasters of constant magnitude.

Let $T(x)$ be the first exit time of X_t from the interval (0,K) given that $X_0 = x \in (0,K)$. Since X_t can never go above K in this model, $T(x)$ is also the time at which X_t first goes below zero. Hence $T(x)$ is the "extinction time" or "persistence time" of the population with initial value x. That is,

$$(18) \quad T(x) = \inf\{\, t \mid X_t \notin (0,K) \mid X_o = x \in (0,K) \,\}$$

$$= \inf\{\, t \mid X_t < 0 \mid X_o = x \in (0,K) \,\} .$$

Inserting the appropriate functions f, g and h and the appropriate measure $\lambda(\cdot)$, the set of recurrence equations (11) become

$$(19) \quad rx(1 - \frac{x}{K}) \frac{d\mu_n}{dx} + \lambda[\mu_n(x-\varepsilon) - \mu_n(x)] = -n\mu_{n-1}(x) , \quad x \in (0,K) .$$

We can get rid of two parameters by scaling. Letting $y = x/\varepsilon$ and $\lambda\mu_n(x) = m_n(y)$ and $k = K/\varepsilon$ we have

$$(20) \quad \frac{y}{\gamma}(1 - \frac{y}{k}) \frac{dm_n}{dy} + m_n(y-1) - m_n(y) = -nm_n , \quad y \in (0,k) .$$

where $\gamma = \lambda/r$. The scaling implies that population numbers are measured in units of ε and times are measured in units of $1/\lambda$. The system of equations (20) is solved with the constraints $m_n = 0$ for $y \notin (0,k)$ and with m_n continuous and bounded on $(0,k)$. To illustrate the method of solution consider the equation for $n = 1$ and set $m_1(y) = F(y)$, so that $F(y)$ satisfies

$$(21) \quad \frac{y}{\gamma}(1 - \frac{y}{k}) \frac{dF}{dy} + F(y-1) - F(y) = -1 .$$

Define

$$(22) \quad F_j(y) = F(y) , \quad j-1 < y < j ; \quad j = 1,2,\ldots,k .$$

On $(0,1)$, because $F(y) = 0$ for $y < 0$, we have

$$(23) \quad \frac{y}{\gamma}(1 - \frac{y}{k}) \frac{dF_1}{dy} - F_1(y) = -1 ,$$

which is easily solved to give

$$(24) \quad F_1(y) = 1 + c_1((k-y)/y)^{-\gamma} ,$$

which involves one constant of integration, c_1. The boundary condition

$$(25) \quad \lim_{y \downarrow 0} F(y) = 1 ,$$

is satisfied. Physically this means that as the initial value of the process gets closer to zero, the mean extinction time approaches the mean time of arrival of the first (disaster) event in the Poisson process (recall that in the scaled variables this has a unit rate). Note that the integration constant c_1 is not yet determined

We now use the expression for F_1 on (0,1) to obtain an ordinary differential equation for F_2 :

$$(26)\quad \frac{y}{\gamma}(1-\frac{y}{k})\frac{dF_2}{dy} + F_1(y-1) - F_2(y) = -1 , \quad 1 \le y \le 2 ,$$

which is now (in principle) integrated to obtain F_2 . There is no new constant of integration involved because continuity requires that $F_1(1) = F_2(1)$. We continue in this fashion until the k-th subinterval:

$$(27)\quad \frac{y}{\gamma}(1-\frac{y}{k})\frac{dF_k}{dy} + F_{k-1}(y-1) - F_k(y) = -1 , \quad k-1 \le y \le k .$$

On this last subinterval we have in principle, by continuity, still the one unknown constant of integration c_1 which was introduced on the first subinterval. The requirement that $F(y)$ is bounded on (0,k) implies that

$$(28)\quad F(k^-) = F(k^- - 1) + 1 .$$

Physically this means that the expected extinction time from carrying capacity consists of the expected waiting time for one disaster (which takes X_t to within the interval (0,K)) plus the expected extinction time after that first disaster occurred. The relation (28) determines the initial constant c_1 that has been carried along with the stepwise integration.

In practice results for the mean persistence time $F(y)$ are difficult to obtain. When $0 < k \le 2$ one can obtain closed form expressions for certain values of γ . For higher k values multiple integrals arise as one integrates solutions obtained on previous intervals. Numerical methods must then be employed to find $F(y)$ and these are described briefly elsewhere (Hanson & Tuckwell, 1978), a more detailed version being presented in Hanson (1979). Numerically computed solutions for $k = 2$, 6 and 10 and various values of γ reveal an interesting dependence of the mean survival time on the initial population size. In particular when $k = 10$ the value of $F(y)$ for $\gamma = .5$ changes from 1 to more than 10^8 as y increases from 0 to 1 , but increases very little from $y = 1$ to $y = 10$. Thus a population living under these conditions of fairly low frequency disasters and of fairly small magnitude gains little advantage in maintaining a high population level. Results such as these and those pertaining to the survival of colonizing species in hazardous environments are presented in detail in Hanson & Tuckwell (1978).

Disasters Proportional to Population Size

When a disaster strikes a sparsely populated region it has a minor effect on the population size but if it occurs in a region of high population density it will remove a large number of individuals. It seems that the above model can be replaced by a more realistic one if we set

$$(29) \quad dX_t = rX_t(1 - X_t/K)dt - \varepsilon X_t d\Pi_t^{\lambda} ,$$

If $\varepsilon = 1$ then the first event in the Poisson process will annihilate the population. If $0 < \varepsilon < 1$, however, we find a new feature in that the value 0 can never be attained by the population size in a finite time. The reason of course is that the smaller X_t becomes the smaller the amount of the population removed by a disaster. Under these conditions we cannot define extinction as the time to reach zero. Instead we choose some small number Δ and define the extinction time as

$$(30) \quad T(x) = \inf\{ t | X_t \notin (\Delta,K) | X_o = x \in (0,K)\}$$

$$= \inf\{ t | X_t < \Delta | X_o = x \in (0,K)\} .$$

The first moment, $\mu_1(x)$, of this random variable now satisfies

$$(31) \quad rx(1 - \frac{x}{K})\frac{d\mu_1}{dx} + [\mu_1((1-\varepsilon)x) - \mu_1(x)] = -1, \quad x \in (\Delta,K) ,$$

with $\mu_1(x) = 0$ for $x \notin (\Delta,K)$. If one proceeds to integrate this differential-difference equation by the 'method of steps' one now finds that the subintervals grow geometrically.

In Hanson & Tuckwell (1979) equation (31) has also been integrated by numerical methods. Since interest focussed on a comparison of results for the mean persistence times for the models described by the two stochastic differential equations (16) and (29), a 'normalizing' definition of 'extinction' had to be made. The criterion chosen was that for a given K and Δ, extinction occurred if the population size became less than or equal to Δ and that in each model the same number of consecutive disasters (in the absence of any logistic recovery) would take a population from an initial value K to extinction.

For the cases examined in detail the mean extinction times were not very different in the two models. As might be expected the model with density dependent disasters (equation (29)) usually gave rise to larger persistence times. It is suspected however that values of Δ smaller than the one employed (unity) would lead to a much greater advantage for the population in which disasters were proportional to population size. This would be expected to manifest itself in a long tail in the density of $T(x)$, a feature that was just discernible in the computer simulations performed for the two models.

3.2 GOMPERTZIAN GROWTH WITH DENSITY INDEPENDENT DISASTERS

Growth is not always adequately described by the logistic equation. Cell populations grow in a way which can often be described by the Gompertzian law (Lala, 1971) and when we superimpose Poisson disasters proportional to the population size we obtain the stochastic differential equation

$$(32) \qquad dX_t = rX_t \ln(K/X_t)dt - \varepsilon X_t d\Pi_t^{\lambda} , \qquad \varepsilon > 0 ,$$

where in the absence of the random term the solution approaches, for an initial value $X_o \in (0,K)$, the upper limit K according to

$$(33) \qquad X_t = K \exp[\ln(X_o/K)e^{-rt}] ,$$

which shows the slow approach to K as $t \to \infty$.

It is an advantage to study the model described by equation (32) because it transforms to a stochastic equation related to a problem in neurobiology for which many numerical results have already been obtained. Assume that $0 < \varepsilon < 1$. If we set

$$(34) \qquad k = \ln K , \quad \alpha = -\ln(1-\varepsilon), \quad Y_t = k - \ln X_t ,$$

we obtain the stochastic equation for Y_t ,

$$(35) \qquad dY_t = -rY_t dt + \alpha d\Pi_t^{\lambda} , \quad Y_o \in (0,k) .$$

Here Y_t is a process with random jumps up of magnitude α which in the absence of jumps decays exponentially to zero with a time constant r^{-1} . This is the model for the depolarization of a nerve cell proposed by Stein (1965) in which random excitation arrives until the depolarization reaches a certain threshold (k in this case) for generation of an action potential. If we define extinction for the model described by equation (32) as

$$(34) \qquad T(x) = \inf\{ t | X_t < 1 | X_o = x \in (0,K)\} ,$$

then this is the same as the time between firings of the nerve cell in Stein's model,

$$(35) \qquad T(y) = \inf\{ t | Y_t > k | Y_o = y \in (0,k)\} ,$$

where $y = k - \ln x$. The differential-difference equation for $G(y) = E[T(y)]$ satisfies

$$(36) \qquad -ry \frac{dG}{dy} + \lambda[G(y+\alpha) - G(y)] = -1 , \qquad y \in (0,k) ,$$

and solutions of this equation have been obtained for many values of the parameters of the model (Tuckwell, 1975; Tuckwell & Richter, 1978). The results of Tsurui & Osaki (1976) in which jumps of a single magnitude in (35) are replaced by an exponential distribution of jump amplitudes can also be directly related to

Gompertzian growth with random removal of a given fraction of the population.

Acknowledgements. The majority of the work reported here has been joint work with Floyd Hanson of University of Illinois at Chicago Circle. I thank Michael Levandowsky for his interest and discussions and Carol Aarssen for her technical assistance. Supported in part by NRC of Canada grants A4559 and A9259. The author is now at the Department of Biomathematics, School of Medicine, U.C.L.A., Ca. 90024, U.S.A.

REFERENCES

Caughley, G. (1976). In, Theoretical Ecology, R.M. May, Ed. Blackwell, Oxford.

Darling, D.A. and Siegert, A.J.F. (1953). Ann. Math. Statist. 24, 624-639.

Feller, W. (1940). Trans. Amer. Math. Soc. 48, 488-515.

Gause, G.F. (1934). The Struggle for Existence. Williams and Wilkins, Baltimore.

Gihman, I.J. and Skorohod, A.V. (1972). Stochastic Differential Equations. Springer-Verlag. New York.

Hanson, F.B. (1979), in preparation.

Hanson, F.B. and Tuckwell, H.C. (1978). Theoret. Pop. Biol. 14, 46-61.

Hanson, F.B. and Tuckwell, H.C. (1979). Theoret. Pop. Biol. (submitted).

Ito, K. (1951). Mem. Amer. Math. Soc. 4., pp. 1-51.

Klein, D.R. (1968). J. Wildl. Manage. 32, 350-367.

Kolmogorov, A.N. (1931). Math. Ann. 104, 415-458.

Lala, P.K. (1971). In, Methods in Cancer Research, H. Busch, Ed. Academic, New York.

May, R.M. (1976). In Theoretical Ecology, R.M. May, Ed., Blackwell, Oxford.

Scheffer, V.B. (1951). Scientific Monthly 73, 356-362.

Skorohod, A.V. (1965). Studies in the Theory of Random Processes, Addison-Wesley, Reading.

Stein, R.B. (1965). Biophys. J. 5, 173-194.

Tuckwell, H.C. (1975). Biol. Cybernetics 17, 225-237.

Tuckwell, H.C. (1976). J. Appl. Prob. 13, 39-48.

Tuckwell, H.C. and Richter, W. (1978). J. Theor. Biol. 71, 167-183.

NICHE OVERLAP AND INVASION OF COMPETITORS IN RANDOM ENVIRONMENTS
II. THE EFFECTS OF DEMOGRAPHIC STOCHASTICITY

Michael Turelli
Department of Genetics
University of California, Davis 95616

1. INTRODUCTION

This paper treats a class of models and approximations that are useful for describing the dynamics of a rare species attempting to invade a community consisting of established resident competitors. Such dynamics are a central feature of faunal buildup via immigration and thus bear on the theory of island biogeography as well as the theory of the limiting similarity of competitors. The purpose of the models is to study the joint effects of interspecific competition, demographic stochasticity, and environmental stochasticity. Here "demographic stochasticity" refers to random individual to individual variation in offspring production; whereas "environmental stochasticity" refers to random variation in the mean per capita growth rate for the population as a whole.

Competition, environmental stochasticity, and demographic stochasticity have been previously treated, at least briefly, in all possible pairs. The joint effects of competition and environmental stochasticity have been extensively studied (e.g. May and MacArthur, 1972; May, 1974; Abrams, 1975; Turelli, 1978a, 1979) as have the joint effects of demographic and environmental stochasticity (e.g. Athreya and Karlin, 1971; Leigh, 1975; Keiding, 1975, 1976; Ludwig, 1976; Hanson and Tier, 1979). In contrast, although there is a considerable literature on the consequences of demographic stochasticity in determining the outcome of unstable competition and in establishing quasi-stationary distributions about stable equilibria (e.g. Bartlett, 1960; Mangel and Ludwig, 1977), I know of only one brief treatment of the joint effects of demographic fluctuations and moderate levels of competition on a founding propagule (MacArthur, 1972, pp. 121-126). To my knowledge, this is the first attempt to study all three factors simultaneously. As in Turelli (1978, 1979), the quantity investigated is the probability that a rare invader will successfully establish itself. In particular, a general approximation for this probability, applicable to a large class of models, is provided then applied to two closely related models to assess the relative importance of i) the presence of demographic fluctuations, ii) the level of competition faced, and iii) the presence of small to moderate levels of environmental stochasticity.

To put the results reported below in perspective, it is useful to review briefly the earlier results on these three factors when taken two at a time. The techniques and results presented below directly extend those of Turelli (1979) to include demographic stochasticity. There it was shown that small to moderate levels of environmental stochasticity have only a trivial effect on the probability that a rare invader wil establish itself. In effect, whether this probability is zero or one can be determined by ignoring the random fluctuations and performing a deterministic analysis analogous to the original "limiting similarity" calculations of MacArthur and Levins (1967). In the presence of demographic fluctuations, the probability of successful invasion can vary between zero and one. Recalling my results on environmental stochasticity and competition and those of Keiding (1975) on the joint effects of environmental and demographic fluctuations on undamped, single-species growth, it is natural to expect that the effects of the demographic fluctuations will dominate those of the environmental fluctuations in determining the invasion probabilities. This expectation is confirmed below. It is much less clear how these probabilities will vary with the intensity of competition. MacArthur's (1972) work on the joint effects of competition and demographic fluctuations centered on the mean persistence time of a small founding propagule. As suggested by the rough approximation of MacArthur (1972) for single-species dynamics and the more precise analysis by Hanson and Tier (1979), once an invasion attempt is successful, the mean persistence time of the established population is likely to be an exponential function of its deterministic equilibrium level. Hence as long as the probability of initial establishment is not trivially small, the mean persistence time of a founding population will tend to be quite large, obscuring the fact that perhaps very few invasions are successful. Thus a complete picture of the consequences of competition and demographic fluctuations requires knowledge of both mean persistence times and invasion probabilities. By using an extremely small equilibrium population size of 20, MacArthur (1972) showed that the mean persistence time could be drastically reduced by the presence of even a moderate level of competition. With larger equilibrium values, more intense competition is required to produce a similar effect. The results presented below on the reduction of invasion probabilities as competition intensifies are analogous and complementary to MacArthur's findings.

2. MODELS AND ANALYSIS

To illustrate the approximation techniques in a fairly general setting, I'll assume as in Turelli (1979) that in the absence of demographic fluctuations the dynamics of the competitors can be described by stochastic difference equations of the form

$$X_{i,t+1}=f_i(X_{0,t},\underline{X}_t,z_{i,t};\alpha) \qquad i=0,1\ldots,n. \tag{1}$$

Here $X_{0,t}$ denotes the abundance of the invader, $\underline{X}_t=(X_{1,t},\ldots,X_{n,t})$ denotes the abundances of the n resident species, $\{z_{i,t}\}$ is a stationary, zero mean stochastic process, independent of the state variables $X_{j,t}$, that models the environmental fluctuations experienced by species i, and α is a measure of the intensity of interspecific competition. I assume that when the invader is absent and there is no environmental noise, i.e. $X_{0,t}=z_{i,t}\equiv 0$, (2.1) possess a unique, globally stable, feasible equilibrium, denoted $\hat{\underline{X}}$, for the residents. More assumptions will be imposed on the deterministic equilibrium, $\hat{\underline{X}}$, and the stochastic perturbations, $z_{i,t}$, as needed below.

To incorporate the effects of demographic stochasticity, I follow Ludwig (1976) and replace (1) by the assumption that conditional on the environmental process, $z_{i,t}$, and the abundances, $(X_{0,t},\underline{X}_t)$, of all n+1 species, $X_{i,t+1}$ is a random variable with a Poisson distribution and mean

$$f_i(X_{0,t},\underline{X}_t,z_{i,t};\alpha).$$

This will be denoted

$$(X_{i,t+1}|X_{0,t},\underline{X}_t,z_{i,t}) \overset{D}{=} P(f_i(X_{0,t},\underline{X}_t,z_{i,t};\alpha)). \tag{2}$$

Thus the population sizes remain integer valued; and because (2) assigns a non-zero probability to $X_{i,t}$ reaching zero each generation, all species are assured of local extinction in a finite (but possibly extremely long) time.

To understand the dynamics of the invader, it is useful to decompose f_0 as

$$f_0(X_{0,t},\underline{X}_t,z_{0,t};\alpha)=X_{0,t}g(X_{0,t},\underline{X}_t,z_{0,t};\alpha).$$

In keeping with (2), I assume that each invader produces an independent, Poisson distributed number of offspring with mean g. As in Turelli (1978a, 1979), the key to analyzing the behavior of $X_{0,t}$ is to assume that $X_{0,0}$ is small relative to its potential single-species equilibrium and that the resident competitors are fluctuating near their deterministic equilibria, which are assumed to be much larger than $X_{0,0}$. Then during the initial stages of the invasion, the invader's population size will have little effect on its own per capita growth rate or the dynamics of its competitors, i.e.

$$g(X_{0,t},\underline{X}_t,z_{0,t};\alpha)\simeq g(0,\underline{X}_t,z_{0,t};\alpha) \text{ and}$$

$$f_i(X_{0,t},\underline{X}_t,z_{i,t};\alpha)\simeq f_i(0,\underline{X}_t,z_{i,t};\alpha) \text{ for } i=1,\ldots,n.$$

Hence while $X_{0,t}$ is small, its dynamics can be approximated well by

$$(X_{0,t}|\underline{X}_t,z_{0,t}) \overset{D}{=} P(X_{0,t}g(0,\underline{X}_t,z_{0,t};\alpha)) \tag{3}$$

in which the random vector $\underline{X}_t$ can be regarded as a stationary process, independent of $X_{0,t}$. Equation (3) can be rewritten as

$$X_{0,t+1}=\sum_{i=1}^{X_{0,t}} Y_i(\underline{X}_t,z_{0,t}) \tag{4}$$

where, conditional on $(\underline{X}_t,z_{0,t})$, the random variables $Y_i(\underline{X}_t,z_{0,t})$ are independent $P(g(0,\underline{X}_t,z_{0,t};\alpha))$. For convenience, I will abbreviate $g(0,\underline{X}_t,z_{0,t};\alpha)$ as $G_t(\alpha)$. Equation (4) shows that the behavior of $X_{0,t}$, at least when it is small, can be described by a branching process in a random environment (denoted BPRE, cf. Athreya and Karlin, 1971). Note that the environmental randomness experienced by $X_{0,t}$ comes from three sources: the direct environmental variation, $z_{0,t}$, the environmental variance present in $\underline{X}_t$, and the demographic variance present in $\underline{X}_t$. Even if the environmental process $\{z_{0,t}\}$ is assumed to possess no autocorrelation, the random process $G_t(\alpha)$ will be autocorrelated because of the autocorrelation of $\underline{X}_t$. To isolate the separate effects of environmental and demographic variation on $X_{0,t}$ and to check the diffusion approximation applied below, it is useful to assume that the $\hat{X}_i$'s are so large that the demographic fluctuations of the $X_{i,t}$, whose coefficient of variation is proportional to $(\hat{X}_i)^{-1/2}$, are negligible. Then in the absence of environmental variance, (4) becomes a standard Galton-Watson process in which the Y_i's have mean $g(0,\hat{\underline{X}},0;\alpha)$, the approximate deterministic growth rate of the rare invader.

From the theory of BPRE (see Karlin and Athreya, 1971, or Keiding, 1975), (4) implies that

$$P(X_{0,t}\to 0 \text{ as } t\to\infty) \begin{cases} =1 & \text{if } E\ln G_t \le 0 \\ <1 & \text{if } E\ln G_t > 0. \end{cases} \tag{5}$$

If extinction does not occur, the theory asserts that $X_{0,t}\to\infty$ as $t\to\infty$. Clearly this would not occur for density-dependent growth models. Because the BPRE approximation is valid only while the invader is rare, I will interpret $1-P(X_{0,t}\to 0 \text{ as } t\to\infty)$ as the probability that the rare invader successfully establishes itself in the community. The diagnostic quantity, $E\ln G_t$, in (5) is the same one that appears when only environmental stochasticity is considered (Turelli, 1979); but in that context, $E\ln G_t>0$ insures that $P(X_{0,t}\to 0 \text{ as } t\to\infty)=0$. Here, when $E\ln G_t>0$, there is no formula available for computing $P(X_{0,t}\to 0 \text{ as } t\to\infty)$. However, as shown by Kurtz (1978), the discrete time process (4) can be approximated by a diffusion process. This approximation will be most accurate when $EG_t\simeq 1$ and $\mathrm{Var}G_t\simeq 0$, i.e. when the BPRE is nearly critical and the level of environmental variation is small. Under these conditions, the critical inequality

$$E\ln G_t>0 \tag{6}$$

from (5) is approximately equivalent to

$$EG_t-(1/2)\mathrm{Var}G_t>1. \tag{7}$$

Using Corollary (2.18) of Kurtz (1978) as motivation, I will approximate the process (4) by a diffusion process with infinitesimal mean and variance

$$M(x)=mx \text{ and } V(x)=x+vx^2, \tag{8a}$$

respectively, where

$$m=EG_t-1+\mathrm{Var}G_t S,\quad v=\mathrm{Var}G_t(1+2S), \tag{8b}$$

$$S=\sum_{k=1}^{\infty} r_k, \text{ and } r_k=\mathrm{Cov}(G_t,G_{t+k}). \tag{8c}$$

Applying the Feller (1952) boundary classification (cf. Keiding, 1975, pp. 58-60), one obtains the following approximations:

$$P(X_{0,t}\to 0 \text{ as } t\to\infty | X_{0,0}=x)\simeq \begin{cases} 1 & \text{if } m-v/2<1 \\ \exp(-2mx) & \text{if } v=0 \text{ and } m>1 \\ \exp[(1-2m/v)\ln(1+vx)] & \text{if } v>0 \text{ and } m-v/2>1. \end{cases} \tag{9}$$

As expected, the critical inequality, $m-v/2>1$, in (9) reduces to (7). When $\sigma=0$, (4) is a Galton-Watson process under the assumptions described above. In this case,

$P(X_{0,t}\to 0 \text{ as } t\to\infty | X_{0,0}=x)$ can be approximated more directly as p^x where p is the unique solution less than one of

$$p=\exp[g(0,\hat{\underline{X}},0;\alpha)(p-1)] \tag{10}$$

(see Feller, 1968, Sec. XII.4). This provides a direct check on the accuracy of the diffusion approximation.

Applying (9) requires approximating only the first two moments of $\{G_t(\alpha)\}$. My approximation procedure is straightforward and is presented in detail in Turelli (1979). Only an outline will be provided here. The first step is to introduce the scaled variables

$$\chi_{i,t}=(X_{i,t}-\hat{X}_i)/\hat{X}_i \qquad i=1,\dots,n$$

and expand $G_t(\alpha)$ in a Taylor series about $(\hat{\underline{X}},0)$. I assume that the environmental variation terms, $z_{i,t}$, are sufficiently small and the deterministic equilibria, $\hat{X}_i$, are sufficiently large that terms of the form

$$\chi^a_{i,t}z^b_{j,t} \qquad \text{for } a+b\geq 3$$

are negligible and that the variation in $\chi_{i,t}$ can be attributed solely to environmental stochasticity. Given these assumptions,

$$G_t(\alpha)=g(0,\hat{\underline{X}},z_{0,t};\alpha)\simeq g+\sum_{i=1}^{n}(\partial g/\partial X_i)\hat{X}_i\chi_{i,t}+(\partial g/\partial z)z_{0,t} \tag{11}$$
$$+(1/2)\sum_{i=1}^{n}\sum_{j=1}^{n}[\partial^2 g/\partial X_i\partial X_j]\hat{X}_i\hat{X}_j\chi_{i,t}\chi_{j,t}+(1/2)(\partial^2 g/\partial z^2)z^2_{0,t}$$

with all of the functions on the right evaluated at $(0,\hat{\underline{X}},0)$. Hence approximations for the first two moments of G_t can be obtained by approximating

$E(\chi_{i,t})$ and $Cov(\chi_{i,t},\chi_{j,t+k})$ for $i,j=1,\dots,n$; $k=0,1\dots$.

The covariances can be approximated to order $E(z^2)$ by linearizing the dynamics of the residents about $(X_{0,t},\underline{X}_t,z_{i,t})=(0,\hat{\underline{X}},0)$. A general result is available for n=2, but it is quite cumbersome. Instead I'll present a simple n-species case in which all the residents are assumed to have identical deterministic dynamics, compete in a completely symmetric way, and experience the same level of nonautocorrelated, but symmetrically cross-correlated, environmental fluctuations. That is, for each $i,j,\underline{x}$, z and α, I assume that

$$f_i(0,\underline{x},z;\alpha)=f_j(0,\underline{x}^{ij},z;\alpha),$$

where $\underline{x}^{ij}$ is obtained from $\underline{x}$ by interchanging the ith and j^{th} coordinates, and

$$Cov(z_{i,s},z_{j,t})=\delta_{st}\theta^{1-\delta_{ij}}\sigma^2,$$

where $\delta_{ij}=1$ if i=j and 0 otherwise.

After considerable algebra, one obtains the approximations

$$\chi_{i,t}=(d/n)\sum_{j=1}^{n}\sum_{s=1}^{\infty}[\lambda_1^{s-1}+(n\delta_{ij}-1)\lambda_2^{s-1}]z_{j,t-s} \quad \text{and} \tag{12}$$

$$\mathrm{Cov}(\chi_{i,t},\chi_{j,t+k})=[(\sigma d)^2/n]\{[1+(n-1)\theta]\lambda_1^k/(1-\lambda_1^2)+(n\delta_{ij}-1)(1-\theta)\lambda_2^k/(1-\lambda_2^2)\} \tag{13}$$

where $d=\hat{X}^{-1}\partial f_1(0,\hat{\underline{X}},0)/\partial z$, $\hat{X}$ denotes the common deterministic equilibrium for the residents, and λ_1 and λ_2 are the eigenvalues of the linearized system describing the stochastic fluctuations of the residents about their joint deterministic equilibrium. These eigenvalues have multiplicity one and n-1, respectively, and are expressable as $\lambda_1=a+(n-1)b$ and $\lambda_2=a-b$ with $a=\partial f_1(0,\hat{\underline{X}},0)/\partial X_1$ and $b=\partial f_1(0,\hat{\underline{X}},0)/\partial X_2$. According to the linear approximation (12), $E\chi_{i,t}\equiv 0$. However a significant effect of environmental variation in density-dependent growth models is to reduce the mean population sizes below the corresponding deterministic equilibria (see Turelli and Gillespie, 1980). A refined approximation for $E\chi_{i,t}$, accurate to order $E(z^2)$, can be obtained from a second-order expansion of f_i and application of (13) (cf. Bartlett, 1960, Sec. 4.3 and Turelli, 1979, Sec. 2E). The resulting expression is quite complicated even for the symmetrical class of models under consideration. The result appropriate to the models treated in the next section will be quoted there.

Imposing the additional symmetry assumptions that the invader interacts identically with each resident and experiences the same level of environmental variation yields the approximations, accurate to order σ^2,

$$EG_t \simeq g+n\hat{X}(\partial g/\partial X_1)E\chi_1+(n/2)(\hat{X})^2[(\partial^2 g/\partial X_1^2)\mathrm{Var}\chi_1+(n-1)(\partial^2 g/\partial X_1\partial X_2)\mathrm{Cov}(\chi_1,\chi_2)] \tag{14a}$$

$$+(\sigma^2/2)(\partial^2 g/\partial z^2),$$

$$\mathrm{Var}G_t \simeq \sigma^2\{n[\hat{X}d(\partial g/\partial X_1)^2[1+(n-1)\theta]/(1-\lambda_1^2)+(\partial g/\partial z)^2\}, \text{ and} \tag{14b}$$

$$S \simeq n[\sigma\hat{X}d(\partial g/\partial X_1)]^2[1+(n-1)\theta]\lambda_1/(1-\lambda_1^2)^2 \tag{14c}$$

with all of the functions evaluated at $(0,\hat{\underline{X}},0)$. These expressions can be plugged into (8) and (9) to approximate $P(X_{0,t}\to 0 \text{ as } t\to\infty|X_{0,0}=x)$. In the next section, I will apply these approximations to specific competition models, compare their predictions to estimates obtained via computer simulation, and comment on the biological significance of the results.

3. APPLICATIONS AND DISCUSSION

The following two closely related, discrete time, stochastic analogs of the standard Lotka-Volterra competition equations will be considered:

$$(X_{i,t+1}|X_{0,t},\underline{X}_t,z_{i,t})\overset{D}{=}P\{X_{i,t}[1+r_i(1-\sum_{j=0}^{n}\alpha_{ij}X_{j,t}/K_i)](1+z_{i,t})\} \quad \text{and} \tag{15a}$$

$$(X_{i,t+1}|X_{0,t},\underline{X}_t,z_{i,t})\overset{D}{=}P\{X_{i,t}[1+r_i(1-\sum_{j=0}^{n}\alpha_{ij}X_{j,t}/K_i(1+z_{i,t}))]\} \tag{15b}$$

for $i=0,1,\ldots,n$. These models extend Models III and II, respectively, of Turelli (1979) to include demographic stochasticity. Models (15a) and (15b) differ only in that the stochastic environmental perturbation terms, $z_{i,t}$, enter the mean per capita growth rates multiplicatively in (15a) but additively in (15b). To satisfy the symmetry assumptions applied in Section 2, one must assume that $r_i \equiv r$ and $K_i \equiv K$ for $i=1,\ldots,n$ with $0<r<2$ and $K>0$ and that $\alpha_{ij}=\beta^{1-\delta_{ij}}$ and $\alpha_{0i}=\alpha^{1-\delta_{ij}}$ for $i,j=1,\ldots,n$ with $0\leq\beta<1$. For simplicity I will further restrict attention to totally symmetric cases without cross-correlation, i.e. I will assume that $r_0=r$, $K_0=K$, $\beta=\alpha$, and $\theta=0$. Under these conditions, it is shown in Turelli (1979) that for n=1,2, and 3 the behavior of analogs of (15a) and (15b) without demographic stochasticity is essentially identical for $.5<r<1.5$ and $\sigma\leq.1$ but that the analog of (15a) behaves in a more biologically reasonable manner for more extreme values of r and σ. In that paper, it is also shown that for both models

$$E\chi_{1,t}\simeq-(\sigma^2/nr)\{(2-r)^{-1}+(n-1)[2-r(1-\alpha)/(1+(n-1)\alpha)]^{-1}\}. \tag{16}$$

Using this, it is easy to apply formulas (8), (9), (10), and (14) to predict analytically the probability of successful invasion once values of n, α, σ, r, and $X_{0,0}$ are specified. The predictions obtained with $\sigma=0$ from the Galton-Watson approximation (10) will be denoted by P', those obtained from the diffusion approximation (9) will be denoted by P.

Even with all of the symmetry assumptions imposed, there remains a five-dimensional parameter space to be explored. For models (15a) and (15b), the qualitative dependence of P on each of these parameters is easily described. Holding the other four parameters fixed, P decreases as α, the level of interspecific competition, n, the number of residents, and σ, the level of environmental variation, increase. However as shown below, the effect of σ, at least for $\sigma\leq.2$, is extremely slight. On the other hand, P increases as $X_{0,0}$, the initial population size of the invader, and r, the growth rate parameter (which governs both the deterministic rate of growth for the rare invader and the rate of convergence toward the internal deterministic equilibrium for the residents), increase. I expect that all of these qualitative results will be robust within the class of "Lotka-Volterra-like" competition models but that the quantitative effects on r will depend critically on whether or not it multiplies the stochastic environmental perturbation terms (see the discussions of Models I-III in Turelli, 1979, and the general discussion of stochastic growth models in Turelli, 1978b). Documenting the numerical dependence of P on all of these parameters would be quite tedious. Instead I will concentrate on illustrating the effects of α, σ, and n with r fixed at 1 and $X_{0,0}$ fixed at 2 and comparing the analytical predictions with Monte Carlo results. The r value was set in the middle of the range of values yielding deterministic stability to maximize the likelihood of agreement between the predictions and Monte Carlo results. The initial population size was set at two to emphasize the effect of demographic stochasticity on small propagules.

Tables 1 and 2 below display analytically predicted values of the probability of successful invasion (denoted P or P' depending on whether (9) or (10) was employed) for various values of α and σ as well as empirical estimates, denoted $\hat{P}$, obtained from extensive computer simulations. The simulations were conducted as follows. At the beginning of each replicate, the resident species were set at the deterministic equilibrium values they would achieve in the absence of the invader and stochasticity, i.e. I set $X_{i,0}=K/[1+(n-1)\alpha]$ for $i=1,\ldots,n$. (In all of the simulations reported below, I used $K=1000$.) The invader was initialized at $X_{0,0}=2$. The recursions (15a) or (15b) were then iterated until the invader or one of the residents hit zero or the invader reached $K/(1+n\alpha)$, the equilibrium value it would share with the resi-dents if it coexisted with them in a deterministic environment. The latter event was tallied as a successful invasion. This simulation procedure was repeated 1000 times for each set of parameters and the proportion of successful invasions was denoted by $\hat{P}$. The standard error, denoted SE, of this estimate was approximated by $\sqrt{\hat{P}(1-\hat{P})/1000}$; and approximate 95% confidence intervals were obtained via $\hat{P}\pm 2(SE)$.

In simulating (15a), psuedo-random lognormal deviates with mean one and variance σ^2 were used for the stochastic inputs $(1+z_{i,t})$; for (15b), pseudo-random normal deviates with mean zero and variance σ^2 were used for $z_{i,t}$. For both models, if the Poisson mean was above 25, integerized normal deviates with the appropriate mean and variance were used in place of pseudo-random Poisson deviates. Before describing the results, I should briefly justify my initialization convention. Ideally the initial values for the residents would be random variables drawn from a multivariate distribution describing their quasi-stationary fluctuations. This could be approximated by iterating the recursions (15) many times before introducing the invader. The procedure used requires much less computing, making large numbers of replications possible. It provides slightly conservative values of $\hat{P}$ because, as shown by (16), the stationary mean values of the residents lie below the deterministic equilibrium. The general agreement of the empirical results with the analytical predictions suggests that the bias introduced by this procedure is small.

Table 1 reports values of P', P, and $\hat{P}\pm 2(SE)$ for model (15a) with one resident, $\sigma=0$, .1, and .2, and various levels of competition. There are several notable results. To begin with, P clearly provides an adequate approximation for P'. Although the diffusion approximation displays a slight upward bias that increases with the mean per capita growth rate, even with $\alpha=.5$, which corresponds to a deterministic mean per capita growth rate of 1.5, P is only 5% above the true value. As σ is increased from 0 to .1, P falls only very slightly. Even with $\sigma=.2$, which would allow fluctuations in the mean per capita growth rate on the order of 50%, the P's are at most 10% below the $\sigma=0$ values. For $\alpha<.98$, the proportional decreases in the $\hat{P}$ values as σ increases from 0 to .2 agree quite well with the small predicted decreases. Moreover, the predicted values "fit" the observed values fairly well. For the narrow band $.99<\alpha<1$, there is an unexpected, but I expect biologically insignificant, increase in the small observed invasion rates as σ increases.

In contrast to the small effects of σ, α has a large impact on P and $\hat{P}$ whose values are determined primarily by the effects of the demographic fluctuations. According to both the analytical predictions and the simulation results for $\sigma\leq .2$,

Table 1. Analytical and Monte Carlo estimates of the probability of successful invasion for model (15a) with r=1, K=1000, n=1, $X_{0,0}$=2 and $X_{1,0}$=1000.

	σ=0			σ=.1		σ=.2	
α	P'	P	P$\pm$2(SE)	P	P$\pm$2(SE)	P	P$\pm$2(SE)
.999	.004	.004	.003$\pm$.004	.004	.013$\pm$.008	.004	.041$\pm$.012
.99	.039	.039	.036$\pm$.012	.038	.045$\pm$.014	.035	.060$\pm$.016
.98	.076	.077	.078$\pm$.016	.075	.078$\pm$.016	.069	.088$\pm$.018
.95	.179	.181	.169$\pm$.024	.176	.165$\pm$.024	.163	.139$\pm$.022
.9	.321	.330	.318$\pm$.030	.321	.310$\pm$.030	.298	.286$\pm$.028
.8	.529	.551	.538$\pm$.032	.539	.495$\pm$.032	.506	.477$\pm$.032
.5	.826	.865	.835$\pm$.024	.854	.815$\pm$.024	.820	.795$\pm$.026

when α>.95 less than 20% of all invasion attempts would be successful according to this model. If r is increased to 1.5, P is increased to .24 for α=.95 and σ=.2, with r=1 and $X_{0,0}$ increased to 5, the corresponding P is .35; whereas if both r and $X_{0,0}$ are increased, P increases to .45. Hence, as one would expect, the invasion probabilities are sensitive to r and the size of the invading propagule. This rules out the possibility of general quantitative predictions concerning limits on similarity imposed by demographic fluctuations.

Table 2 presents analytical and simulation results for model (15b) with three residents and σ restricted to 0 and .1. The qualitative conclusions drawn from Table 1 concerning the small effects of σ and the general agreement between the predictions and Monte Carlo data hold here as well. There is however a significant quantitative difference; for a given α, the invasion probabilities are considerably smaller, especially for $\alpha\geq.8$. A better indication than α of the level of competition faced by a rare invader is

$$\gamma=\sum_{i=1}^{n} \alpha_{0i}\hat{X}_i/K_0$$

which measures the fraction of the invader's "niche space" that is already occupied by the resident competitors. For the symmetrical models under consideration, $\gamma=n\alpha/[1+(n-1)\alpha]$ which, for a fixed α, increases as n increases. To see that γ is the critical variable in determining the effects of competition, observe that for both (15a) and (15b) the deterministic mean per capita growth rate is $1+r(1-\gamma)$. This is empirically reflected in the similarity of the results for α=.99 in Table 1 and α=.97 in Table 2, the corresponding γ value is .99 in both cases. Thus as n increases, demographic stochasticity exerts more pressure in limiting the similarity of the competitors, as measured by α.

As shown above, probabilities of successful invasion are strongly dependent on $X_{0,0}$, r, n, and α but only weakly influenced by the presence of small to moderate

Table 2. Analytical and Monte Carlo estimates of the probability of successful invasion for model (15b) with r=1, K=1000, n=3, $X_{0,0}=2$, and $X_{i,0}=1000/(1+2\alpha)$ for i=1,2,3.

	$\sigma=0$			$\sigma=.1$	
α	P'	P	$\hat{P}\pm2(SE)$	P	$\hat{P}\pm2(SE)$
.999	.001	.001	.000	.001	.002±.002
.99	.013	.013	.007±.006	.013	.013±.008
.98	.027	.027	.020±.008	.026	.015±.008
.97	.040	.040	.042±.012	.039	.032±.012
.96	.053	.053	.049±.014	.053	.049±.014
.95	.066	.067	.090±.018	.066	.076±.016
.9	.132	.133	.151±.022	.131	.143±.022
.8	.259	.265	.263±.014	.262	.261±.028
.7	.382	.393	.375±.030	.389	.372±.030
.5	.605	.632	.607±.030	.627	.607±.030

levels of environmental noise. It is hard to say which combinations of these parameters are most likely to characterize natural systems; and even for a specific system, which would undoubtedly be quite asymmetric, the relevant parameters are likely to be extremely hard to estimate. Moreover, the problem of deducing the ecological implications of demographically induced extinctions of invading species is further complicated by the fact that the probability of successful invasion must be weighed against the frequency of arrival of propagules. This will be a function of the biogeography of the habitat and species in question; in particular it will depend on the spatial distribution and dispersal patterns of the potential residents. In spite of these considerable qualifications, it seems reasonable to assert on the basis of the results presented above that although demographic fluctuations may only play a small role in limiting the similarity of competitors, its importance is at least commensurate with and probably greater than that of small levels of environmental fluctuations. In fact in multispecies competition communities, rare invaders may face at least low levels of competition from many resident species, leading to values of γ near 1. The small invasion probabilities in such circumstances could lead to considerable departures of observed levels of species packing from those predicted to be tolerable by deterministic analyses.

Besides reducing invasion probabilities, there is another mechanism whereby demographic fluctuations can limit similarity. In multispecies guilds, asymmetries will surely be the rule. As deterministic limits to similarity are approached, the deterministic equilibrium of one or more species is likely to approach zero (cf. Roberts, 1974). Species whose niches lie near these deterministic thresholds would be vulnerable to fairly rapid extinction from a combination of demographic and environmental fluctuations even if they successfully invade. This is the sort of situation emphasized in MacArthur's (1972) discussion of the joint effects of competition and demographic stochasticity. A more complete study of its consequences would require an approximation of mean persistence times for multispecies competition systems subject to both demographic and environmental fluctuations.

Acknowledgements.

I thank Joe Felsenstein for suggesting the possible importance of demographic fluctuations and John Gillespie for several useful mathematical suggestions. This research was supported by U. C. Davis Faculty Research Grant No. D-1264 and U. C. California Agricultural Experiment Station Hatch Grant No. CA-D*-GEN-3580-H.

REFERENCES

Abrams, P. 1976. Niche overlap and environmental variability. Math. Biosciences 28:357-372.

Athreya, K. B., and Karlin, S. 1971. On branching processes with random environments I, II. Ann. Math. Statist. 42:1499-1520, 1843-1858.

Bartlett, M. S. 1960. "Stochastic Population Models in Ecology and Epidemiology". Methuen, London.

Feller, W. 1952. The parabolic differential equations and the associated semigroups of transformations. Ann. Math. 55:468-519.

Feller, W. 1968. "An Introduction to Probability Theory and Its Applications, Third Ed." Wiley, New York.

Hanson, F. B., and Tier, C. 1979. An asymptotic solution of the first passage problem for singular diffusion in population biology. Unpublished.

Keiding, N. 1975. Extinction and exponential growth in random environments. Theor. Pop. Biol. 8:49-63.

Keiding, N. 1976. "Population Growth and Branching Processes in Random Environments." Preprint No. 9, Institute of Mathematical Statistics, University of Copenhagen.

Kurtz, T. G. 1978. Diffusion approximations for branching processes. In "Branching Processes", ed. Joffe, A., and Ney, P., pp. 269-292. Academic Press, New York.

Leigh, E. G. 1975. Population fluctuations, community stability, and environmental variability. In "Ecology and Evolution of Communities", ed. Cody, M. L., and Diamond, J. M., pp. 51-73. Belknap. Cambridge, Massachusetts.

Ludwig, D. 1976. A singular perturbation problem in the theory of population extinction. SIAM-AMS Proc. 10:87-104.

MacArthur, R. H. 1972. "Geographical Ecology." Harper and Row, New York.

MacArthur, R. H., and Levins, R. 1967. The limiting similarity, convergence and divergence of competing species. Amer. Natur. 101:377-385.

Mangel, M., and Ludwig, D. 1977. Probability of extinction in a stochastic competition. SIAM J. Appl. Math. 33:256-266.

May, R. M. 1974. On the theory of niche overlap. Theor. Pop. Biol. 5:297-332.

May, R. M., and MacArthur, R. H. 1972. Niche overlap as a function of environmental variability. Proc. Nat. Acad. Sci. USA 69:1109-1113.

Roberts, A. 1974. The stability of a feasible random ecosystem. Nature 251: 607-608.

Turelli, M. 1978a. Does environmental variability limit niche overlap? Proc. Nat. Acad. Sci. USA 75:5085-5089.

Turelli, M. 1978b. A reexamination of stability in randomly varying versus deterministic environments with comments on the stochastic theory of limiting similarity. Theor. Pop. Biol. 13:244-267.

Turelli, M. 1979. Niche overlap and invasion of competitors in random environments. I. Symmetric models without correlation and demographic stochasticity. Theor. Pop. Biol.: in press.

Turelli, M., and Gillespie, J. H. 1980. Conditions for the existence of asymptotic densities for some two-dimensional diffusion processes with applications in population biology. Theor. Pop. Biol. 17: in press.

BEHAVIOR OF LIMITING DIFFUSIONS FOR
DENSITY-DEPENDENT BRANCHING PROCESSES

by Carla (Lipow) Wofsy

Abstract

A sequence of density-dependent analogues of continuous time branching processes is shown to converge weakly to a diffusion. The diffusion has infinitesimal mean displacement (drift) $\lambda x\alpha(x)$ and infinitesimal variance $2\lambda x\beta$, where λ and β are positive constants and α is a bounded, continuous, real-valued function on $[0,\infty)$. Boundary behavior of the diffusion is studied, in the case where $\alpha(x)$ is negative for large enough x values. This corresponds to a "controlled" population in which growth is retarded when the population is too large.

Introduction

In most animal and cell populations, growth patterns are observed to change with population size. Deterministic models for population growth generally describe size-dependent or density-dependent population growth, and some work has been done on density-dependent analogues of the most popular stochastic growth models: branching processes. (See, for example, Levina et al., 1968; Fujimagari, 1972; Labkovskii, 1972; Sevast'yanov and Zubkov, 1974; Lipow, 1975; Rittgen and Tautu, 1977; Rittgen, 1980). I will discuss a diffusion approximation for density-dependent or "controlled" branching processes.

Branching Process Background

In a classical branching process, an individual lives a unit time or a random length of time, governed by some probability distribution; then the individual dies, leaving a random number of offspring. The number of offspring produced is governed by another probability distribution. Individuals die and reproduce independently of each other. In particular, the number of offspring an individual produces does not depend on population size.

Since individuals of a given type all follow the same lifetime and offspring number distribution throughout the evolution of a classical branching process, patterns arise which are useful in studying the long term behavior of the process. The mathematical tools which reflect these patterns include transforms of nth generation population size distributions, renewal equations and appropriate martingales.

If the formulation of a branching process is changed so that generation times or numbers of offspring produced depend on population size, the standard branching process tools break down. In fact, the process one looks at becomes a quite general birth-death process, with no special iterative tools.

An advantage of retaining the branching process framework for thinking of such processes, in spite of the fact that the branching process tools are not available, is that this framework may suggest results which can be proved in the more general setting. One such result is a limit theorem, first considered by Feller for classical branching processes, in which a suitably scaled sequence of branching processes is shown to converge to a diffusion. I will show that an analogue of the theorem holds in the case where the branching processes are modified so that the probability distribution of the number of offspring an individual produces depends on the size of the population at the time of reproduction.

Limit Theorem Background

Feller proposed the limit theorem in 1951 and indicated a proof. In 1969, Jirina stated the assumptions of the theorem more precisely and gave a rigorous proof.

Feller and Jirina considered a sequence $\{Z_n\}_{n=1}^{\infty}$ of discrete time (Galton-Watson) branching processes. Let M_n and V_n denote the mean and variance of the number of offspring produced by an individual in the nth process. The principal assumptions are that for some real numbers α and β ($\beta>0$):

$$\lim_{n\to\infty} n[M_n-1] = \alpha$$

and

$$\lim_{n\to\infty} V_n = 2\beta \quad .$$

A third moment assumption is also needed. Then if $Z_n(0) = n$ and $X_n(t) = Z_n([tn])/n$, the process $X_n(t)$ converges to a diffusion $X(t)$ with infinitesimal mean displacement αx and infinitesimal variance $2\beta x$. Equivalently, the infinitesimal operator A of the limiting diffusion has the form:

$Af(x) = \alpha x f'(x) + \beta x f''(x)$.

In 1971, P. Jagers extended the result to the case of continuous time, age-dependent branching processes. In the continuous time case, when the mean lifetime is $1/\lambda$, the limiting diffusion has the infinitesimal operator $Af(x) = \lambda[\alpha x f'(x)+\beta x f''(x)]$.

Feller, Jirina and Jagers proved convergence in the sense of finite-dimensional distributions. Kurtz (1975) gave an alternative proof of Jagers' theorem in which he proved weak convergence of the processes in question. Tools developed by Kurtz also yield weak convergence in the Feller-Jirina case and in the density-dependent case I shall discuss.

Density-Dependent Case

In the nth branching process Z_n considered by Feller, Jirina and Jagers, the mean per capita change in population size in a generation is α/n. If α is positive, the population tends to increase, if α is negative, it decreases.

If one wants to model a population which tends to grow when it is small and to level off or decrease when it is larger, a constant α is inappropriate. In this and other density-dependent situations, one wants the mean per capita change in population size to be a function of population size. The density-dependent analogue to the classical limit theorem is the following. (For a more precise definition of the density-dependent branching processes involved, in terms of jump processes, see Lipow, 1977.)

Theorem 1. Let α be a bounded, continuous, real-valued function on $[0,\infty)$ and let β and λ be positive numbers. For each positive integer n, let $Z_n(t)$ be the size of a population which grows according to the following rules:

(i) The initial population size, $Z_n(0)$, is n.

(ii) Individuals live a random length of time, exponentially distributed with parameter λ.

(iii) When an individual dies, it produces a random number of offspring; probabilities governing the number of offspring produced depend on the population size at the time of reproduction. Let $p_j^{n,i}$ denote the probability that an individual in a population of size i has j offspring.

(iv) The mean number M_n of offspring produced by an individual in a population of size i satifies:

$$M_n = 1 + \alpha(i/n)/n + \varepsilon(i,n)$$

where $\lim_{n\to\infty} n\varepsilon(i,n)=0$ uniformly in i.

(v) The variance V_n and the third moment, $\sum_{j=0}^{\infty} j^3 p_j^{n,i}$, of the number of offspring produced by an individual in a population of size i satisfy:

$$\lim_{n\to\infty} V_n = 2\beta$$

and

$$\lim_{n\to\infty} \sum_{j=0}^{\infty} j^3 p_j^{n,i}/n = 0$$

uniformly in i.

Let $X_n(t) = Z_n(tn)/n$.

Then X_n converges weakly to a diffusion with infinitesimal operator A given by:

$$Af(x) = \lambda x\{\alpha(x)f'(x)+\beta f''(x)\} \quad .$$

Proof.

Details of the proof of Theorem 1 appear in Lipow (1977).

The proof depends on two semigroup convergence theorems due to Kurtz. The first (Kurtz, 1969, Theorem 2.1) gives necessary and sufficient conditions under which convergence of infinitesimal operators A_n of strongly continuous contraction semigroups implies convergence of the semigroups. If the limiting operator A is known to be the infinitesimal operator of a strongly continuous contraction semigroup, as in the present case, the conditions are satisfied. The second theorem (Kurtz, 1975, Theorem (4.29)) shows that if the semigroups $T_n(t)$ correspond to Markov processes $X_n(t)$ (i.e. $T_n(t)f(x) = E(f(X_n(t))|X_n(0)=x)$), and converge uniformly in x for f continuous with compact support, then the processes converge weakly.

In the present setting, the Markov processes $X_n(t)$ induce semigroups $T_n(t)$ defined by:

$$T_n(t)f(\tfrac{i}{n}) = E(f(X_n(t))|X_n(0)=\tfrac{i}{n})$$

on the Banach space of functions on $\{i/n\}_{i=0}^{\infty}$, vanishing at 0 and ∞. The infinitesimal operator for T_n is defined by:

$$A_n f(\tfrac{i}{n}) = \lim_{t\to 0} \frac{E(f(Z_n(tn)/n)|Z_n(0)=n)-f(i/n)}{t} \quad .$$

Hence,

$$A_n f(\tfrac{i}{n}) = \sum_{j=i-1}^{\infty} (f(\tfrac{j}{n})-f(\tfrac{i}{n}))\ \lambda in\ p_{j-i+1}^{n,i} \tag{1}$$

The operator A defined by $Af(x) = \lambda x\{\alpha(x)\ f'(x) + \beta f''(x)\}$ is the infinitesimal operator of a strongly continuous contraction semigroup T(t) on the space of continuous functions on $[0,\infty)$, vanishing at 0 and ∞. The semigroup T(t) corresponds to a diffusion process X(t). (See Feller, 1952 and 1954.) The norms of the Banach spaces in question are sup norms. The operator convergence involved in Kurtz' theorems is defined by:

$$A = \operatorname{ex-lim}_{n\to\infty} A_n$$

if for every $f\varepsilon D(A)$, there is a sequence of functions $\{h_n\}$, $h_n \varepsilon D(A_n)$, such that:

$$\lim_{n\to\infty} \sup_i |h_n(\tfrac{i}{n}) - f(\tfrac{i}{n})|=0 \tag{2}$$

and

$$\lim_{n\to\infty} \sup_i |A_n h_n(\tfrac{i}{n}) - Af(\tfrac{i}{n})|=0 \quad . \tag{3}$$

Thus, it must be shown that for $f\varepsilon D(A)$ (i.e. for f satisfying: f, xf' and xf'' are continuous functions on $[0,\infty)$ vanishing at 0 and ∞), there is a sequence of functions h_n on $\{i/n\}_{i=0}^{\infty}$, vanishing at 0 and ∞, such that equations (2) and (3) hold.

Special case: Suppose $f \varepsilon D(A)$ has compact support and three continuous derivatives. Let h_n be the restriction of f to the set $\{i/n\}_{i=0}^{\infty}$. Then equation (2) is satisfied trivially.

By Taylor's theorem and equation (1) for $A_n f$, there are numbers $\eta_{i,j,n}$ between i/n and j/n such that:

$$A_n f(\tfrac{i}{n}) = \lambda i n \sum_{j=i-1}^{\infty} \{f'(\tfrac{i}{n})(\tfrac{j}{n} - \tfrac{i}{n}) + \tfrac{1}{2}f''(\tfrac{i}{n})(\tfrac{j}{n} - \tfrac{i}{n})^2$$

$$+ \tfrac{1}{6}f'''(\eta_{i,j,n})(\tfrac{j}{n} - \tfrac{i}{n})^3\} \, p_{j-i+1}^{n,i} \quad .$$

Then:

$$A_n f(\tfrac{i}{n}) = \lambda(\tfrac{i}{n}) \quad \{f'(\tfrac{i}{n})n(\sum_{j=i-1}^{\infty} (j-i+1)p_{j-i+1}^{n,i} - 1)$$

$$+ \tfrac{1}{2}f''(\tfrac{i}{n}) \sum_{j=i-1}^{\infty}(j-i)^2 p_{j-i+1}^{n,i}$$

$$+ \frac{1}{6n} f'''(\eta_{i,j,n}) \sum_{j=i-1}^{\infty}(j-i)^3 p_{j-i+1}^{n,i}\} \quad .$$

By the moment assumptions (iv) and (v) of the theorem,

$$A_n f(\tfrac{i}{n}) - \lambda(\tfrac{i}{n}) \quad \{f'(\tfrac{i}{n})\, \alpha(\tfrac{i}{n}) + f''(\tfrac{i}{n})\beta\}$$

tends to 0 uniformly in i as n tends to ∞. Thus equation (3) is satisfied.

General case: $f \varepsilon D(A)$. The construction of $h_n \varepsilon D(A_n)$ such that equations (2) and (3) are satisfied depends on constructing a sequence of functions $f_n \varepsilon D(A)$, with compact support and three continuous derivatives, such that f_n tends to f and Af_n tends to Af, uniformly on $[0,\infty)$. A sequence of functions $h_n \varepsilon D(A_n)$ satisfying equations (2) and (3) can be constructed from restrictions of a subsequence of the f_n.

Behavior of Limiting Diffusion

Two important questions regarding any real population or stochastic population model are:

1. What is the probability that the population will become extinct?
2. What is the probability that the population will ever reach a given level?

In the case of the diffusion X(t) which was obtained as a limit of rescaled density-dependent branching processes, the following theorem holds.

Theorem 2. Let X(t) be a diffusion on $[0,\infty)$ with infinitesimal operator A given by:

$$Af(x) = \lambda x \, \{\alpha(x)f'(x) + \beta f''(x)\}$$

where α, β and λ are as in Theorem 1.

a) Let $u_M(x) = P(\max_t X(t) < M | X(0)=x)$. Then

$$u_M(x) = \frac{\int_x^M e^{-\int_0^z (\alpha(s)/\beta)ds} dz}{\int_0^M e^{-\int_0^z (\alpha(s)/\beta)ds} dz} . \tag{4}$$

b) If $\alpha(x) \leqq 0$ for x large enough, then X(t) hits 0 (i.e. extinction occurs) with probability 1.

c) In general (i.e. for any bounded, continuous α), 0 is an exit boundary and ∞ is a natural boundary for X(t).

Remark. By c), there is always a positive probability that X(t) will be absorbed at 0 in finite time, but the probability that X(t) tends to ∞ in finite time is 0. Also by c), the infinitesimal operator A determines the diffusion X(t) uniquely, up to the initial distribution.

Proof.

a) The function $u_M(x)$ can be expressed as the probability that X(t) hits 0 before M, given that X(0)=x. Thus $u_M(x)$ satisfies:

$$Au_M(x)=0; \qquad u_M(0)=1; \qquad u_M(M)=0 . \tag{5}$$

The solution to (5) is the expression given in (4).

b) Equation (4) can be rewritten as:

$$u_M(x) = 1 - \frac{\int_0^x e^{-\int_0^z (\alpha(s)/\beta)ds} dz}{\int_0^M e^{-\int_0^z (\alpha(s)/\beta)ds} dz} .$$

The denominator of the second term can be shown to approach ∞ as M approaches ∞. Thus, the probability that X(t) is absorbed at 0 before hitting M tends to 1 as M tends to ∞. It follows that X(t) hits 0 with probability 1.

c) Feller's (1952) boundary classification criteria, applied to the infinitesimal operator A, yield the results. The criteria are well stated in Iosifescu and Tautu (1973, Vol. 1, p. 290). I will consider the boundary at 0, following their notation. Let:

$$h(x) = \exp\{-\int_1^x (\alpha(s)/\beta)ds\}$$

$$g(x) = \frac{1}{\lambda x \beta\ h(x)}$$

$$\sigma = \iint_{0<y<x<1} g(x)h(y)dxdy$$

$$\mu = \iint_{0<y<x<1} h(x)g(y)dxdy .$$

To show that 0 is an exit boundary, one must show that $\sigma < \infty$ and $\mu = \infty$. In the present case:

$$\sigma = \frac{1}{\lambda\beta} \iint_{0<y<x<1} \frac{1}{x} e^{\int_y^x (\alpha(s)/\beta) ds} dxdy$$

$$\mu = \frac{1}{\lambda\beta} \iint_{0<y<x<1} \frac{1}{y} e^{-\int_y^x (\alpha(s)/\beta) ds} dxdy \quad .$$

The exponential terms can be bounded above and below by positive numbers. Thus, $\sigma < \infty$ because:

$$\iint_{0<y<x<1} \frac{1}{x} \, dxdy = -\int_0^1 \ln y \; dy < \infty$$

and $\mu = \infty$ because:

$$\iint_{0<y<x<1} \frac{1}{y} \, dx \, dy = \int_0^1 \left(\frac{1}{y} - 1\right) dy = \infty \quad .$$

References

1. Feller, W. (1951) Diffusion processes in genetics. Proc. 2nd Berkeley Symp. Math. Statist. Prob. 227-246.

2. Feller, W. (1952) The parabolic differential equations and the associated semigroups of transformations. Ann. Math. 55, 468-519.

3. Feller, W. (1954) Diffusion processes in one dimension. Trans. Amer. Math. Soc. 27, 1-31.

4. Fujimagari, T. (1972) Controlled Galton-Watson process and its asymptotic behaviour. 2nd Japan-USSR Symp. on Probability Theory, Vol.2, pp.252-262.

5. Iosifescu, M., and P. Tautu (1973) Stochastic Processes and Applications in Biology and Medicine, Vol. I. Berlin-Heidelberg-New York:Springer-Verlag.

6. Jagers, P. (1971) Diffusion approximations to branching processes. Ann. Math. Statist. 42, 2074-2078.

7. Jirina, M. (1969) On Feller's branching diffusion processes. Casopis. Pest. Mat. 94, 84-90.

8. Kurtz, T.G. (1969) Extensions of Trotter's operator semigroup approximation theorems. J. Funct. Anal. 3, 111-132.

9. Kurtz, T.G. (1975) Semigroups of conditioned shifts and approximation of Markov processes. Ann. Prob. 3, 618-642.

10. Labkovskii, V.A. (1972) A limit theorem for generalized random branching processes depending on the size of the population. Theor. Probabilities Appl. 17, 72-85.

11. Lavina, L.V., A.M. Leontovich, and J.J. Pyatetskii-Shapiro (1968) On a regulative branching process. Problems of Information Transmission 4, 72-82.

12. Lipow, C. (1975) A branching model with population size dependence. Adv. Appl. Prob. 7, 495-510.

13. Lipow, C. (1977) Limiting diffusions for population-size dependent branching processes. J. Appl. Prob. 14, 14-24.

14. Rittgen, W. and P. Tautu (1977) Branching models for the cell cycle. Mathematical Models in Medicine: Workshop, Mainz, March 1976, Lecture Notes in Biomathematics 11, 109-126.

15. Rittgen, W. (1980) Positive recurrence of multi-dimensional population-dependent Markov branching processes. Proceedings, Conference on Models of Biological Growth and Spread, Heidelberg, this volume.

16. Sevast'yanov, B.A., and A.M. Zubkov (1974) Controlled branching processes. Theor. Probability Appl. 19, 14-24.

TOPIC II

Random systems with locally interacting objects

Discussion by the Editors

The papers in this section are divided into two groups: those dealing with the theory of infinite particle systems, introduced in 1970 by F.Spitzer, and those dealing with the construction of the corresponding models in biology. In some sense, this distinction is not (and cannot be) a sharp one. For example, in his paper, H.Föllmer starts with a sociological motivation: the behaviour of a 'voter' depends not only on the opinions of his neighbours but also on the overall conditions of the system (e.g. publication of a poll). Also the study by K. Schürger is partly stimulated by the process of viral infection, particularly the plaque formation. However, the above classification gives us the possibility to discuss some biological applications, some deduced results and their limitations. Although neurobiological models were presented in the book "Locally Interacting Systems and Their Application in Biology" (Lecture Notes in Math., Vol.653, 1978), the question of their biological significance was not taken into consideration. Such a meditation is necessary for at least three reasons. The biological applications of the theory of infinite particle systems - motivated by the rigorous results in statistical mechanics - bring forth a new discussion about the reductionist explanation in biology. In our conference, for instance, A.Robertson expressed clearly this tendency: "We know that mathematics is required and we know that in one way biology will become a branch of physics but it is hard to say how." The second motive may be the confrontation of some random biological systems with the argument by E.P.Wigner that self-reproduction is virtually impossible in a quantum mechanical system. The third question is the appropriateness of the theory of infinite particle systems as a whole in biology. As A.Martin-Löf (Lecture Notes in Physics, Vol. 101, 1979) defined statistical mechanics, it investigates and tries to relate macroscopic properties of systems of many interacting microscopic subsystems to what is known about the laws of the interactions of these microscopic parts. Do biological assemblies behave under the same conditions of invariance, ergodicity, entropy,fluctuation as physical systems? Such questions did not yet find a definite answer.

In his appreciated survey lecture, T.Liggett introduces two birth-death systems (the contact process and the voter model) as well as the exclusion process. It is perhaps suggestive to mention that approximately at the same time, at the 9th Conference on Stochastic Processes (Evanston, USA), two invited reviews (by R.Durrett and D.Griffeath) have dealt with the same topic. The (unbiased) voter model (and the 'biased' one) and the contact processes are the most applied, given their tractability and intuitive representation. The reader shall find some other applications of these models in the papers by J.Biggins (Topic I) and S.Sawyer (Topic III). In this topic we classified applications in neurobiology (W.A.Little) and carcinogenesis (M.Bramson and D.Griffeath; P.Tautu). Here, the following fact is noteworthy: although the Williams-Bjerknes model is, strictly speaking, not a model for carcinogenesis, and the authors never used the ideas of the theory of infinite particle systems, this model is most frequently analyzed and discussed (see e.g. this volume). Its simplicity and the fact that the problems it raises can be studied and even partly solved mathematically, can explain the interest showed by some biomathematicians.

In the theoretical subsection, H.Föllmer introduced a new interaction model with global signal, K.Schürger a general branching model with interactions, and F.Spitzer new countable systems with a random clock at each site. The contribution by F.Spitzer will be published in the Annals of Probability (as Wald Lectures).

LOCAL INTERACTIONS WITH A GLOBAL SIGNAL: A VOTER MODEL

H. Föllmer
Mathematikdepartement ETH-Zentrum
CH-8092 Zürich

The purpose of this note is to summarize some results concerning lattice interactions, where the behavior of a single site (voter) is influenced not only by its local environment (opinions of the neighbors) but also by the overall condition of the system (publication of a poll). The main question is to what extent the introduction of such a global signal changes the long run behavior of the system, in comparison to a purely local interaction. This was motivated by the interpretation as a voter model, but the structure of the interaction may be relevant for other applications as well.

Let $I = Z^d$ be the d-dimensional lattice. We are going to study a Markov chain with transition probability $P(x,dy)$ on the space $E = \{0,1\}^I$ of all configurations $x: I \to \{0,1\}$. For $x \in E$ and $i \in I$ we define

$$r(x) = \limsup_n \frac{1}{|V_n|} \sum_{j \in V_n} x(j) \quad , \quad r_i(x) = \frac{1}{|N_i|} \sum_{j \in N_i} x(j)$$

where V_n denotes the box $\{j \in I \mid \|j\| \leq n\}$ and $N_i = V_m + i$ for some fixed $m \geq 0$. We assume that $P(x,.)$ is a product measure

(1) $$P(x,.) = \prod_{i \in I} P_i(x,.) \quad ,$$

and we first consider the specific example

(2) $$P_i(x,\{1\}) = \alpha p_i + \beta r_i(x) + \gamma r(x)$$

with $\alpha, \beta, \gamma \geq 0$ and $\alpha + \beta + \gamma = 1$. Thus, the probability that site i assumes state 1 is a convex combination of three components: an intrinsic propensity $p_i \in [0,1]$, the proportion of ones in the neighborhood of i, and the proportion of ones throughout the population.

Let us first recall what happens in the case $\gamma = 0$ where the interaction is purely local. For $\alpha \neq 0$, the convergence theorem of Wasserstein [7] shows that there is a unique equilibrium distribution ν , and that

(3) $$\lim_n \mu P^n = \nu$$

(in the weak topology) for each initial distribution μ on E. For $\alpha = 0$ we have a voter model in the sense of Holley and Liggett [5]: In dimension $d=1,2$ the extremal equilibrium distributions are the Dirac measures concentrated on $x \equiv 1$ and $x \equiv 0$, and in dimension $d \geq 3$ the extremal equilibrium distributions are of the form

(4) $$\nu_c = \lim_n \beta_c P^n \qquad (0 \leq c \leq 1)$$

where β_c is the Bernoulli measure on E with $\beta_c[x(i)=1] = c$. Bramson and Griffeath [1] have shown that for $d \geq 3$ the measures ν_c exhibi[t] non-classical fluctuations: For block spins

(5) $$|V|^{-\frac{\delta}{2}} \sum_{j \in V} (x(j) - \int x(j)\,d\nu)$$

one has to take $\delta = 1+2d^{-1}$ instead of $\delta = 1$ in order to get convergence to a normal distribution resp. renormalization to a Gaussian random field.

Let us now consider the case $\gamma \neq 0$ where the global signal $r(x)$ comes in. For $c \in [0,1]$ let $\bar{\nu}_c$ denote the unique equilibrium for the purely local interaction $P_c(x,dy)$ which arises if we replace $r(x)$ b[y] c in (2). Let $(\Omega, (X_n)_{n \geq 0}, (P_x)_{x \in E})$ denote the canonical model for the Markov chain with transition probability $P(x,dy)$. The macroscopic process $r_n = r(X_n)$ performs, P_x-almost surely for each $x \in E$, the deterministic motion

(6) $$r_n = p + (1-\alpha)^n (r_0 - p), \quad p \equiv \limsup_n \frac{1}{|V_n|} \sum_{j \in V_n} p_j .$$

For $\alpha \neq 0$ we obtain $r_n \rightarrow p$ almost surely, and this implies

(7) $$\lim_n \mu P^n = \bar{\nu}_p$$

for each initial distribution μ; see [2] or [3]. Thus, in the case $p_i \equiv p$, the equilibrium distribution is the same as if the global signa[l] were cut off and all the individual sites had added the weight γ to their intrinsic propensity p. The only effect of the global signal i[s] to slow down the speed of convergence in (6).

For $\alpha = 0$ the effect of the global signal becomes more significant. For each $c \in [0,1]$ the set $E_c = \{x \in E \mid r(x)=c\}$ is now absorbing, and on E_c the process behaves like the purely local interaction $P_c(x,dy)$. In any dimension $d \geq 1$ the measures $\bar{\nu}_c$ $(0 \leq c \leq 1)$ are the extremal equilibrium distributions for $P(x,dy)$, and we have

$$(8) \qquad \lim_n \mu P^n = \int \mu(dx)\, \bar{\nu}_{r(x)}$$

for each initial μ. Contrary to the case of a purely local interaction, the fluctuations are now classical: $\bar{\nu}_c$, being the equilibrium distribution of $P_c(x,dy)$, has exponentially decaying correlations by the arguments in [1] or [4], and by Malysev's theorem [6] we get the central limit theorem resp. renormalization to a constant multiple of white noise with $\delta = 1$ in (5).

Let us finally see to what extent the behavior of the voter model (2) is typical. Replace $\{0,1\}$ by a general state space S, let A,B,C be transition probabilities on S, and replace (2) by

$$(9) \qquad P_i(x,.) = \alpha A(x(i),.) + \beta \int \rho^o_{i,x}(ds) B(s,.) + \gamma \int \rho^o_x(ds) C(s,.)$$

where $\rho^o_{i,x}$ is the empirical distribution on S induced by the local configuration $x: N_i \to S$, and where ρ^o_x is the empirical distribution induced by $x: I \to S$. The set E^o of configurations for which ρ^o_x exists (as a weak limit in the space of probability measures on S) is absorbing, and on E^o the macroscopic process $\rho^o_n \equiv \rho^o_{X_n}$ performs almost surely the deterministic motion

$$(10) \qquad \rho^o_{n+1} = \rho^o_n(\alpha A + \beta B + \gamma C) .$$

In this sense, the macroscopic evolution of the infinite particle system can be described by a one-particle model, i.e., by the Markov chain on S with transition probability $Q = \alpha A + \beta B + \gamma C$. This is a special effect of the linearities in (9) and does not hold in general. For general homogeneous interactions of type (1), one has to pass from the measure ρ^o_x on S to the underlying ergodic measure ρ_x on E in order to get the right analogue of (10) or (6). Under suitable contraction assumptions of Dobrushin-Wasserstein type, the convergence in (7) resp. (8) can then be extended to these general models. We refer to [3] for the details.

References

[1] Bramson,M.and Griffeath,D.: Renormalizing the 3-dimensional voter model. To appear.

[2] Föllmer,H.: Zur Dynamik interdependenter Präferenzen. In: Quantitative Wirtschaftsforschung (Ed. H.Albach et al.), Mohr, Tübingen (1977)

[3] Föllmer,H.: Macroscopic convergence of Markov chains on infinite product spaces. Proceedings of the Colloquium on Random Fields, Esztergom 1979 (to appear)

[4] Gross,L.: Decay of correlations in classical lattice models. To appear.

[5] Holley,R. and Liggett,T.: Ergodic theorems for weakly interacting systems and the voter model. Ann. Probability 3, 643-663 (1975)

[6] Malycev,V.A.: The central limit theorem for Gibbsian random fields. Soviet Math.Dokl. 16, 1141-1145 (1975)

[7] Vasserstein,L.N.: Markov processes over denumerable products of spac describing large systems of automata. Problems of Information Transmission 5,no.3,47-52 (1969)

INTERACTING MARKOV PROCESSES

Thomas M. Liggett
University of California, Los Angeles
Los Angeles, CA 90024/USA

1. Introduction. Interacting Markov processes are obtained by superimposing some type of interaction on many otherwise independent Markovian subsystems. As a result of the interaction, the subsystems fail to have the Markov property; the system as a whole remains Markovian, however. This subject has grown rapidly during the past decade. It is a branch of modern probability theory, but it draws much of its inspiration and motivation from various areas of science, including physics and biology.

Two factors enter into the choice of models to be studied: The desire to make the models scientifically reasonable and useful, and the requirement that the resulting problems be mathematically tractable. Of course these two goals often conflict with each other, and in fact there is an art to choosing models which satisfy both criteria to a substantial extent. It is often easy in this area to take a model which is nontrivial, but which is nevertheless accessible to mathematical treatment, and to change it into a model which is mathematically either trivial or impossible via a seemingly minor and scientifically irrelevant modification. Thus it is quite reasonable to let the mathematics play a substantial role in the determination of the models to be considered.

This lecture is intended as a brief glimpse into some of the ideas and results in this area. Essentially no proofs will be given, although there will be some discussion of the roles of coupling and duality in the proofs, since these are two of the most important tools in the area. Questions involving the technical matter of constructing the processes will not be considered. In general, we will concentrate on the simpler aspects of the problems to be presented, and will often state only special cases of more general results. For a more extensive and complete survey, [16] or [6] should be consulted. The remainder of this introduction will be devoted to some of the notation and definitions which will be needed later.

S will denote a countable set, which in most cases will be Z^d, the d-dimensional integer lattice. The state space of the process to be studied will be $\{0,1\}^S$, which is the set of all configurations of zeros and ones on S. Configurations will be denoted by η or ζ. Among the possible interpretations of having a zero or a one at site x are: x is vacant or occupied, x is a normal or abnormal cell or individual, x is an inactive or active neuron, and x is occupied by an individual of type zero or of type one. Different interpretations lead to different choices of the basic subsystem and of the type of interaction. The process to be studied will be denoted by $\{\eta_t\}$, and will be a continuous time Markov process on $\{0,1\}^S$. A verbal description of the Markov property is the following: if one wishes to predict the future $\{\eta_t : t \geq t_0\}$ on the basis of the past $\{\eta_t : t \leq t_0\}$, the only relevant information for that prediction is the present η_{t_0} itself.

If μ is a probability measure on $\{0,1\}^S$, let $\mu S(t)$ be the distribution of the process at time t. The set of invariant probability measures for $\{\eta_t\}$ is defined by

$$\mathcal{I} = \{\mu : \mu S(t) = \mu \quad \text{for all} \quad t > 0\}.$$

Of course convex combinations of invariant measures are again invariant, so we are interested primarily in $\mathcal{I}_e$, the set of extreme points of $\mathcal{I}$. As a result of the compactness of $\{0,1\}^S$, $\mathcal{I}$ is usually the closed convex hull of $\mathcal{I}_e$.

With this notation, we can now state the basic problems which will be considered for each model: (a) find an explicit description of $\mathcal{I}$, or equivalently of $\mathcal{I}_e$, and (b) find the (weak) limit of $\mu S(t)$ as $t \to \infty$ for "nice" initial distributions μ. The solution to (a) is the first step in solving (b), since the limit of $\mu S(t)$, if it exists, will be invariant in most of the cases to be described here.

The next section is devoted to general birth-death systems, in which each individual has a characteristic which changes from time to time. Sections 3 and 4 are devoted to two particular birth-death systems: contact processes and voter models. Section 5 deals with some particle motion systems known as exclusion processes.

2. Birth-death systems. In a birth-death system, which is often called a spin-flip system by analogy with the physics context, only one coordinate $\eta_t(x)$ changes at a time. In general, however, infinitely many coordinates change in any interval of time. The basic motion of each coordinate is that of a two state Markov chain. The form of the superimposed interaction is that the transition rate of each coordinate is allowed to depend on the rest of the configuration. The transition rate of coordinate $\eta(x)$ when the system is in configuration η is a prescribed nonnegative function $c(x,\eta)$ which is assumed to satisfy certain technical assumptions [16]. The evolution of the process can be described by saying that as $t \downarrow 0$,

$$P^\eta[\eta_t(x) \neq \eta(x)] = c(x,\eta)t + o(t),$$

and

$$P^\eta[\eta_t(x) \neq \eta(x),\ \eta_t(y) \neq \eta(y)] = o(t)$$

for $x \neq y$, where $o(t)$ represents a quantity which tends to zero more rapidly than t. The superscript η gives the initial configuration of the process.

The following definition describes the simplest possible solution to the problems posed at the end of the introduction.

Definition 2.1. The process $\{\eta_t\}$ is ergodic if

(a) $\mathcal{I} = \{\nu\}$ is a singleton, and

(b) $\lim_{t\to\infty} \mu S(t) = \nu$ for all probability measures μ on $\{0,1\}^S$.

A simple but not very interesting example of an ergodic birth-death system is that in which $c(x,\eta) \equiv 1$. In this case, the coordinates $\eta_t(x)$ are independent two-state Markov chains, and therefore the unique invariant measure is the product measure ν with marginals given by $\nu\{\eta : \eta(x) = 1\} = \frac{1}{2}$ for each x. As will be seen in Theorem 2.2 below, the process is in fact ergodic provided that $c(x,\eta)$ is sufficiently close to being constant.

If S is finite and $c(x,\eta) > 0$ for all x and η, then η_t is an irreducible finite state Markov chain, and it is therefore ergodic. As will be seen, this is not in general the case when S is infinite. Since the situation is so simple when S is finite one might ask whether there is any point in considering the case of infinite S with its attendant mathematical difficulties. After all, one might argue, all real systems are finite. The fact of the matter is that if the real system is very large, and one is interested in large but finite times, the limiting behavior of the corresponding infinite system gives much more insight into the large finite system than one obtains from the simple results for finite systems. Nonergodicity of the infinite system is reflected in the fact that a large finite system will approach equilibrium very slowly. Ergodicity of the infinite system, on the other hand, means that the corresponding finite system will approach equilibrium at a rate which is roughly independent of the size of the system.

An example of a nonergodic process with positive rates is provided by the stochastic Ising model ([10],[13]) on $S = Z^d$, in which the rates take the form

$$c(x,\eta) = \exp\{-\beta[2\eta(x) - 1] \sum_{|y-x|=1} [2\eta(y) - 1]\}.$$

This is a model for magnetism, and was one of the first examples studied in this field. Note that if $\eta(x)$ takes the values 0 and 1, then $2\eta(x) - 1$ takes the values ± 1, so that $2\eta(x) - 1$ gives the spin of the iron atom at site x. The parameter β is positive and represents the reciprocal of the temperature. If $d \geq 2$, this process is nonergodic for sufficiently large values of β. This is a consequence of the fact that phase transition (in the sense of statistical mechanics) occurs for the Ising model for those values of the parameters. In one dimension, the process is ergodic for all $\beta \geq 0$. In two or more dimensions, the process is ergodic for sufficiently small β. The latter fact can be seen from the following theorem, which is due in various forms to Dobrushin [3] and Sullivan [22]. In this form, it appears in [13] and [16]. If $\eta \in \{0,1\}^S$ and $u \in S$, $\eta_u \in \{0,1\}^S$ is defined by $\eta_u(v) = \eta(v)$ for $v \neq u$ and $\eta_u(u) = 1 - \eta(u)$.

Theorem 2.2. Suppose

$$\sup_x \sum_{u \neq x} \sup_\eta |c(x,\eta_u) - c(x,\eta)| < \inf_{x,\eta} [c(x,\eta_x) + c(x,\eta)]. \tag{2.3}$$

Then $\{\eta_t\}$ is ergodic.

Note that the summand in (2.3) can be interpreted as a measure of the amount to which the rates at x depend on the u'th coordinate of η, so that this result gives a precise meaning to the statement that weakly interacting processes are ergodic. One interesting consequence of Theorem 2.2 is that if one adds a sufficiently large constant to virtually any collection of rates, then one obtains an ergodic system.

As seen above, the stochastic Ising model does not provide a counterexample to ergodicity in one dimension. Thus one of the important open problems in the field is to determine whether η_t is always ergodic whenever $S = Z^1$, $c(x,\eta) > 0$, $c(x,\eta)$ has finite range, and $c(x,\eta)$ is translation invariant.

There is nothing substantially stronger than Theorem 2.2 which can be said about completely general birth-death systems. More interesting problems occur when one specializes to more concrete models. Two of these will be discussed in Sections 3 and 4. Note that in both cases, $c(x,\eta) = 0$ for some η.

3. <u>Contact processes</u>. The basic contact process was introduced by Harris in [7]. It is a birth-death system in which $S = Z^d$ and

$$c(x,\eta) = \begin{cases} \lambda \sum_{|y-x|=1} \eta(y) & \text{if } \eta(x) = 0 \\ 1 & \text{if } \eta(x) = 1, \end{cases}$$

where λ is a positive parameter. It can be thought of as a model for the spread of infection: A healthy individual becomes infected at a rate which is proportional to the number of its neighbors which are infected, while an infected individual recovers at a constant rate 1 independently of the configuration. Let δ_η be the pointmass at the configuration η. For the contact process, $\delta_0 \in \mathcal{I}$ since there is no provision in the model for spontaneous infection. (0 and 1 are sometimes used to denote the configurations $\eta \equiv 0$ or $\eta \equiv 1$.) The main question is then whether δ_0 is the only invariant measure. The answer turns out to depend on λ, as might be expected.

In order to discuss this question, we need to introduce some coupling ideas ([9]), beginning with the definition of a partial order on the set of all probability measures on $\{0,1\}^S$. Let $\mathfrak{M}$ be the set of all continuous functions f on $\{0,1\}^S$ which satisfy $f(\eta) \leq f(\zeta)$ whenever $\eta \leq \zeta$. The latter inequality is understood in the coordinatewise sense.

<u>Definition 3.1</u>. If μ_1 and μ_2 are probability measures on $\{0,1\}^S$, then $\mu_1 \leq \mu_2$ if $\int f\, d\mu_1 \leq \int f\, d\mu_2$ for all $f \in \mathfrak{M}$.

An equivalent formulation of this definition is that $\mu_1 \leq \mu_2$ if and only if there exists a probability measure γ on $\{0,1\}^S \times \{0,1\}^S$ with respective marginals μ_1 and μ_2 such that $\gamma\{(\eta,\zeta) : \eta \leq \zeta\} = 1$. One direction of this equivalence is easy, but the other is fairly hard.

Since $c(x,\eta)$ is increasing in η for $\eta(x) = 0$ and decreasing in η for $\eta(x) = 1$, it is not hard to see that if $\eta \leq \zeta$, then one can construct two copies η_t and ζ_t of the contact process on the same probability space so that $\eta_0 = \eta$, $\zeta_0 = \zeta$, and $\eta_t \leq \zeta_t$ for all t. This joint construction is called a coupling. The fact that a construction with these properties is possible implies that $E^{\eta}f(\eta_t) \in \mathfrak{M}$ for each $f \in \mathfrak{M}$ and $t > 0$, and hence that $\mu_1 S(t) \leq \mu_2 S(t)$ for all $t > 0$ whenever $\mu_1 \leq \mu_2$. Since $\mu \leq \delta_1$ for all μ, this gives $\mu S(t) \leq \delta_1 S(t)$. Applying this to $\mu = \delta_1 S(s)$ and using the Markov property implies that $\delta_1 S(t)$ is decreasing in t in the sense of the partial order of Definition 3.1. Therefore $\nu = \lim_{t\uparrow\infty} \delta_1 S(t)$ exists and is invariant. It further follows from these monotonicity considerations that for any μ, all weak limits of $\mu S(t)$ as $t \to \infty$ lie between δ_0 and ν, so that the process is ergodic if and only if $\nu = \delta_0$.

One can also couple together two copies of the process with different values of λ. This gives $\mu S_{\lambda_1}(t) \leq \mu S_{\lambda_2}(t)$ whenever $\lambda_1 < \lambda_2$, where $\mu S_\lambda(t)$ is the distribution of the process with parameter λ at time t with initial distribution μ. Therefore $\nu_\lambda = \lim_{t\uparrow\infty} \delta_1 S_\lambda(t)$ is monotone in λ. As a consequence of this, there exists a $\lambda_c \in [0,\infty]$ so that the process is ergodic for $\lambda < \lambda_c$ and is not ergodic for $\lambda > \lambda_c$. It is not known whether or not the process is ergodic for $\lambda = \lambda_c$. The above monotonicity arguments do not rule out the possibilities $\lambda_c = 0$ or ∞. These do not occur, and substantially more information is provided by the following result.

<u>Theorem 3.2.</u> $\frac{1}{2d-1} \leq \lambda_c \leq \frac{2}{d}$.

The lower bound, and the fact that $\lambda_c < \infty$ are due to Harris [7], while the explicit upper bound was obtained by Holley and Liggett [12]. One interesting aspect of Theorem 3.2 is that the ratio of the upper to the lower bound remains bounded as $d \to \infty$.

Duality is a technique which in general relates an infinite system to a finite system which is in principle more tractable than the original infinite system. The basic contact process is self dual, which means that one can relate the system with an infinite number of infected individuals to the same system with only finitely many infected individuals. A special case of the duality relation for the basic contact process is

$$\delta_1 S(t)\{\eta(0) = 0\} = P^{\{0\}}[\eta_t \equiv 0],$$

where on the left, the process begins with an infected individual at every site, while on the right it begins with a single infected individual at the origin. One implication of this is that the process is ergodic if and only if the infection dies out eventually with certainty when initially there is only one infected individual.

The following results come close to giving a complete description of the limiting behavior of the process in the nonergodic case in one dimension. The corresponding problems in higher dimensions are open. The first result was proved by Harris [8] who showed that δ_0 and ν were the only extremal invariant measures which are also translation invariant, and Liggett [17] who removed the translation invariance restriction. The second was proved by Griffeath [5] for large λ and by Durrett [4] for $\lambda > \lambda_c$. Note that Theorem 3.3 is not quite implied by Theorem 3.4, since it is not known whether the process is ergodic for $\lambda = \lambda_c$.

<u>Theorem 3.3</u>. If $d = 1$, then $\mathcal{J}_e = \{\delta_0, \nu\}$.

<u>Theorem 3.4</u>. If $d = 1$ and $\lambda > \lambda_c$, then

$$\lim_{t\uparrow\infty} \delta_\eta S(t) = \alpha\,\delta_0 + (1-\alpha)\nu,$$

where $\alpha = P^\eta[\eta_t \equiv 0 \text{ for some } t]$. In particular,

$$\lim_{t\uparrow\infty} \delta_\eta S(t) = \nu$$

for all η with $\sum_x \eta(x) = \infty$.

4. <u>Voter models</u>. The voter model was introduced by Clifford and Sudbury [2] as a model for invasion and independently by Holley and Liggett [11] as a mathematically tractable special case of a more general class of processes. It is the birth-death system with $S = Z^d$ and

$$c(x,\eta) = \frac{1}{2d} \sum_{\substack{|y-x|=1 \\ \eta(y)\neq\eta(x)}} 1. \tag{4.1}$$

The following is perhaps a more intuitive description of the voter model which is mathematically equivalent to the above: (a) an individual is located at each site in S, (b) there is a social or political issue on which each individual can have one of two positions which are denoted by 0 and 1, (c) at independent exponential times with parameter one, individuals decide to re-evaluate their position on the issue, and (d) when a given individual is to do so, he simply adopts the position of a randomly chosen neighbor. In Clifford and Sudbury's context, the interpretation is that there are two populations which are denoted by 0 and 1. A configuration η describes the territory controlled by each population, and a shift at x from 0 to 1, for example, is interpreted as having population 1 invade x, which was previously controlled by population 0.

Note that δ_0 and δ_1 are both invariant for the voter model, so the question of interest is whether or not these are the only members of $\mathcal{J}_e$. The answer, which is given in [11] turns out to depend on the dimension d. It can be described intuitively by saying that a consensus on the issue is approached as $t \to \infty$ if $d \leq 2$, but not if $d \geq 3$.

<u>Theorem 4.2.</u> (a) Suppose $d \leq 2$. Then $\mathcal{I}_e = \{\delta_0, \delta_1\}$. If μ is translation invariant, then

$$\lim_{t\to\infty} \mu S(t) = \alpha\,\delta_1 + (1-\alpha)\delta_0,$$

where $\alpha = \mu\{\eta : \eta(x) = 1\}$ is the initial proportion of ones. (b) Suppose $d \geq 3$. Then $\mathcal{I}_e = \{\gamma_\rho,\ 0 \leq \rho \leq 1\}$ where for each ρ, γ_ρ is translation invariant and ergodic and $\gamma_\rho\{\eta : \eta(x) = 1\} = \rho$. If μ is translation invariant and ergodic, then

$$\lim_{t\to\infty} \mu S(t) = \gamma_\rho$$

where $\rho = \mu\{\eta : \eta(x) = 1\}$.

There is also a duality theory for the voter model, which plays a crucial role in the proofs of the above results. A special case of the duality relation is given for $x \neq y$ by

$$\nu_\rho S(t)\{\eta : \eta(x) = 1, \eta(y) = 1\} = \rho P^{y-x}[\tau \leq 2t] + \rho^2 P^{y-x}[\tau > 2t], \tag{4.3}$$

where ν_ρ is the product measure with $\nu_\rho\{\eta : \eta(x) = 1\} = \rho$ for each x, and τ is the hitting time of the origin of a simple random walk on Z^d (starting at $y-x$). One can begin to see from this identity that the dependence on the dimension of the results for the voter model is intimately related to the corresponding dimension dependence of the recurrence properties of simple random walks on Z^d. Take the limit as $t \to \infty$ in (4.3) to obtain

$$\lim_{t\to\infty} \nu_\rho S(t)\{\eta : \eta(x) = 1,\ \eta(y) = 1\} = \rho P^{y-x}[\tau < \infty] + \rho^2 P^{y-x}[\tau = \infty].$$

If $d \leq 2$, the simple random walk on Z^d is recurrent, so

$$\lim_{t\to\infty} \nu_\rho S(t)\{\eta : \eta(x) = 1,\ \eta(y) = 1\} = \rho$$

independently of x and y. On the other hand, if $d \geq 3$,

$$\lim_{y-x\to\infty} \lim_{t\to\infty} \nu_\rho S(t)\{\eta : \eta(x) = 1,\ \eta(y) = 1\} = \rho^2$$

Thus relative to the limit of $\nu_\rho S(t)$ as $t \to \infty$, the random variables $\eta(x)$ and $\eta(y)$ are equal if $d \leq 2$, but are approximately independent if $y-x$ is large if $d \geq 3$. When $d \geq 3$, γ_ρ is this limit of $\nu_\rho S(t)$ as $t \to \infty$.

The voter model provides another illustration of the way in which limiting results for an infinite system give information about the behavior of a large finite system at large but finite times. When S is finite, voter models always reach a consensus eventually, since the process is a finite Markov chain with only two absorbing states $\eta \equiv 0$ and $\eta \equiv 1$, each of which can be reached from any other state. The implications of Theorem 4.2, on the other hand, are that for large finite "two-dimensional" S, the consensus is reached at a rate which is relatively independent of the size of S, while that for large finite "three-dimensional" S, the consensus is reached at a rate which becomes substantially slower as the size of S increases.

The basic voter model can be modified in various interesting ways. First consider the process in which $\varepsilon > 0$ is added to all the rates, so that the new rates are

$$c_{\varepsilon}(x,\eta) = \varepsilon + c(x,\eta),$$

where $c(x,\eta)$ is given in (4.1). Thus in addition to the previous voting mechanism, voters can also change their views spontaneously at rate ε. Of course it follows from Theorem 2.2 that this process is ergodic for sufficiently large ε. However, it is not hard to see directly that it is ergodic for all $\varepsilon > 0$. Suppose that μ is translation invariant, and let

$$f(t) = \mu S(t)\{\eta : \eta(x) = 1\}.$$

A simple computation of the derivative implies that $f(t)$ satisfies the differential equation

$$f'(t) = \varepsilon[1 - 2f(t)].$$

Since $\varepsilon > 0$, it follows that

$$\lim_{t\to\infty} f(t) = \frac{1}{2}. \tag{4.4}$$

Just as in the case of contact processes in Section 3, one can see by coupling that $\delta_1 S(t)$ and $\delta_0 S(t)$ are monotone in t, and therefore converge to limits μ_1 and μ_0 as $t \to \infty$ which satisfy $\mu_0 \leq \mu_1$. By (4.4),

$$\mu_0\{\eta : \eta(x) = 1\} = \mu_1\{\eta : \eta(x) = 1\} = \frac{1}{2},$$

so that $\mu_0 = \mu_1$. A further coupling argument shows that $\mu S(t)$ converges as $t \to \infty$ to the common value of μ_0 and μ_1 for any probability measure μ on $\{0,1\}^S$. Thus the process is ergodic.

A second modification of the basic voter model was introduced by Schwartz [19], and is known as the biased voter model. It has rates given by

$$c_{\kappa}(x,\eta) = \begin{cases} c(x,\eta) & \text{if } \eta(x) = 1 \\ \kappa c(x,\eta) & \text{if } \eta(x) = 0 \end{cases}$$

where $\kappa > 1$ and $c(x,\eta)$ is as in (4.1). Thus changes from 0 to 1 are given an advantage. Schwartz proved that $\mathcal{I}_e = \{\delta_0, \delta_1\}$ for all d. In fact, if μ is translation invariant and puts no mass on $\eta \equiv 0$,

$$\lim_{t\to\infty} \mu S(t) = \delta_1.$$

For $d \geq 3$, this convergence statement can be obtained by coupling from part (b) of Theorem 4.2. For $d = 1$, the result is easy to prove directly, while for $d = 2$, it follows from much stronger theorems of Bramson and Griffeath [1].

If the initial configuration of the biased voter model has only finitely many ones, then this process is exactly the Williams-Bjerknes tumor growth model (see [1]). In that model, ones are interpreted as cancerous cells, zeros as normal cells, and κ is the carcinogenic advantage.

A final voter type model is known as the majority vote process. It is described as follows: Each site $x \in S = Z^1$ waits an exponential time with parameter one; it then surveys the neighborhood $\{y : |y - x| \leq k\}$ to determine which of the symbols 0 or 1 is in the majority in this neighborhood, and it flips to the majority symbol with probability $1 - \varepsilon$ and to the minority symbol with probability ε. Here $0 < \varepsilon \leq \frac{1}{2}$. This process is ergodic for ε sufficiently close to $\frac{1}{2}$ by Theorem 2.2. When $k = 1$, it can be shown using the stochastic Ising model that the process is ergodic for all $\varepsilon \in (0,\frac{1}{2}]$. For $k > 1$, it is not known whether the process is ergodic for all ε in this range. This is in fact a good example of a situation in which the analysis of the process can be vastly complicated by small modifications in the definition of the process. This can be seen by comparing the majority vote process for $k > 1$ with it for $k = 1$, or the majority vote process with the voter model, with or without the additional ε's in the voter model.

5. <u>Exclusion processes</u>. The processes to be described in this section are particle motion processes, in which the basic motion of each particle is that of a continuous time Markov chain on S. Superimposed upon this motion is an interaction which excludes multiple occupancy of points in S. We will consider two such interactions, both of which were proposed by Spitzer in [20]. In contrast to birth-death processes, exclusion processes have the property that two coordinates of η_t change at a time - one from a zero to a one, and the other from a one to a zero. These two coordinate changes represent the motion of a particle from one site to the other. If S is finite, the number of particles remains unchanged in time, so that the Markov chain η_t has many closed classes, and hence many invariant measures. Therefore, if S is infinite, it should not be surprising that exclusion processes normally have infinitely many extremal invariant measures. This again is in contrast to birth-death systems.

In each case, we will consider only the situation in which the process is translation invariant. Therefore it will be described in terms of the transition probabilities $p(x,y)$ for an irreducible random walk on $S = Z^d$. These should satisfy $p(x,y) \geq 0$, $p(x,y) = p(0,y-x)$, and $\sum_y p(x,y) = 1$.

In the simple exclusion process, a particle at x waits an exponential time with parameter one, and then it attempts to move to a y chosen according to the probabilities $p(x,y)$. If y is vacant at that time, it goes to y, while otherwise, it remains at x and begins a new exponential waiting time. It is easy to check that ν_ρ is invariant for the process for each $\rho \in [0,1]$, where ν_ρ is again the product measure with density ρ. The primary question of interest is whether there exist other extremal invariant measures.

The simple exclusion process is self-dual if $p(x,y) = p(y,x)$. A special case of the duality relation in that case is given by

$$P^\eta[\eta_t(x) = 1] = \sum_y P^x[X_t = y]\eta(y),$$

where on the right, X_t is the one-particle random walk which moves according to $p(x,y)$ at exponential times with parameter one. This of course makes it possible to compute probabilities of certain events for the infinite system in terms of the simpler one particle motion. If one integrates the above relation with respect to a probability measure $\mu \in \mathcal{I}$, one obtains the identity

$$\mu\{\eta : \eta(x) = 1\} = \sum_y P^x[X_t = y]\mu\{\eta : \eta(y) = 1\}.$$

Since an irreducible random walk has no nonconstant bounded harmonic functions, it follows that $\mu\{\eta : \eta(x) = 1\}$ is independent of x for $\mu \in \mathcal{I}$. Applying a similar, but more involved, analysis to the general duality relation, one can prove the following result ([14],[21]).

Theorem 5.1. Suppose $p(x,y) = p(y,x)$. Then
(a) $\mathcal{I}_e = \{\nu_\rho, 0 \le \rho \le 1\}$, and
(b) if μ is translation invariant and ergodic, then

$$\lim_{t\to\infty} \mu S(t) = \nu_\rho$$

where $\rho = \mu\{\eta : \eta(x) = 1\}$.

Duality techniques fail in the asymmetric case, and in fact, statement (a) in Theorem 5.1 does not hold, and statement (b) has not been proved, without the symmetry assumption. Using coupling techniques, one can prove ([15]) that among invariant measures which are also translation invariant, the extremal elements are exactly $\{\nu_\rho, 0 \le \rho \le 1\}$. Even in the case

$$(5.1) \qquad d = 1,\ p(x,x+1) = p,\ p(x,x-1) = q, \qquad p + q = 1,\ p \ne q,$$

there are invariant measures which are not translation invariant. In spite of this, one can find $\mathcal{I}_e$ explicitly in this case ([15]), and one can prove some limit theorems for very special classes of initial distributions.

In the long range exclusion process, particles move as above, except that when a particle at x tries to move to an occupied site y, instead of remaining at x, it immediately chooses a z with probability $p(y,z)$ to try to go to. It continues to search in this way until it finds an unoccupied site, at which it remains for another exponential time. Since no time is spent at intervening occupied sites, this is a form of exclusion mechanism. The fact that particles can move great distances in short times in this model makes it technically much more difficult to work with.

In spite of the technical difficulties, this process has two attractive features by comparison with the simple exclusion process. The first is that even with the interaction, the motion of an individual particle is a time change on the motion of an ordinary random walk with transition probabilities $p(x,y)$. The sequence of sites visited by the particle is a subsequence of those it would have visited if there had been no interaction. To describe the second attractive feature, if $\pi(x)$ is a positive function on S, let ν_π be the product measure on $\{0,1\}^S$ with marginals given by

$$\nu_\pi\{\eta : \eta(x) = 1\} = \frac{\pi(x)}{1+\pi(x)} .$$

Formal computations indicate that ν_π should be invariant whenever

$$\sum_x \pi(x)p(x,y) = \pi(y). \tag{5.3}$$

In fact ν_π is invariant under this assumption in the technically much simpler case in which Z^d is replaced by a finite set S and the translation invariance assumption on $p(x,y)$ is removed. The corresponding fact is not correct for the simple exclusion process. In order that ν_π be invariant for the simple exclusion process, one needs in addition to (5.3), the assumption that either $\pi(x)$ is constant or that $\pi(x)p(x,y) = \pi(y)p(y,x)$.

In the translation invariant case, the general solution to (5.3) is a mixture of exponential functions. For example, in Case (5.2), the general solution is given by

$$\pi(x) = c + d\left(\frac{p}{q}\right)^x$$

for nonnegative constants c and d. The formal computations mentioned above are misleading, as is shown by the following result [18], which of course includes the Case (5.2).

<u>Theorem 5.4</u>. Suppose $p(0,x) > 0$ for only finitely many x. If $\pi P = \pi$, then $\nu_\pi \in \mathcal{I}$ if and only if π is constant.

References

1. M. Bramson and D. Griffeath (1979). On the Williams-Bjerknes tumour growth model, to appear.

2. P. Clifford and A. Sudbury (1973). A model for spatial conflict, Biometrika 60, 581-588.

3. R. L. Dobrushin (1971). Markov processes with a large number of locally interacting components, Problems of Information Transmission 7, 149-164.

4. R. Durrett (1979). On the growth of one dimensional contact processes, to appear.

5. D. Griffeath (1978). Limit theorems for nonergodic set-valued Markov processes, Annals of Probability 6, 379-387.

6. D. Griffeath (1979). Additive and Cancellative Interacting Particle Systems, Springer Lecture Notes in Mathematics, 724.

7. T. E. Harris (1974). Contact interactions on a lattice, Annals of Probability 2, 969-988.

8. T. E. Harris (1976). On a class of set-valued Markov processes, Annals of Probability 4, 175-194.

9. R. Holley (1972). An ergodic theorem for interacting systems with attractive interactions, Z. Wahr. verw. Geb. 24, 325-334.

10. R. Holley (1974). Recent results on the stochastic Ising model, Rocky Mountain Journal of Mathematics 4, 479-496.

11. R. Holley and T. M. Liggett (1975). Ergodic theorems for weakly interacting infinite systems and the voter model, Annals of Probability 3, 643-663.

12. R. Holley and T. M. Liggett (1978). The survival of contact processes, Annals of Probability 6, 198-206.

13. R. Holley and D. Stroock (1976). Applications of the stochastic Ising model to the Gibbs states, Communications in Mathematical Physics 48, 249-265.

14. T. M. Liggett (1973). A characterization of the invariant measures for an infinite particle system with interactions, Transactions of the A.M.S. 179, 433-453.

15. T. M. Liggett (1976). Coupling the simple exclusion process, Annals of Probability 4, 339-356.

16. T. M. Liggett (1977). The stochastic evolution of infinite systems of interacting particles, Springer Lecture Notes in Mathematics 598, 188-248.

17. T. M. Liggett (1978). Attractive nearest neighbor spin systems on the integers, Annals of Probability 6, 629-636.

18. T. M. Liggett (1979). Long range exclusion processes, Annals of Probability, to appear.

19. D. Schwartz (1977). Applications of duality to a class of Markov processes, Annals of Probability 5, 522-532.

20. F. Spitzer (1970). Interaction of Markov processes, Advances in Mathematics 5, 246-290.

21. F. Spitzer (1974). Recurrent random walk of an infinite particle system, Transactions of the A.M.S. 198, 191-199.

22. W. G. Sullivan (1974). A unified existence and ergodic theorem for Markov evolution of random fields, Z. Wahr. verw. Geb. 31, 47-56.

ON A CLASS OF BRANCHING PROCESSES ON A LATTICE WITH INTERACTIONS

Klaus Schürger
German Cancer Research Center
Heidelberg

1. Introduction

This paper summarizes some of the results we have recently obtained for a class of branching processes on a lattice with interactions (an extended version of this, containing additional results and complete proofs, will be published elsewhere). A branching process with interactions (BPI) is a spread process in the d-dimensional square lattice Z^d ($d\geq 1$) consisting of points ("sites") $x=(x^1,\dots,x^d)$ where each component x^i is an integer. Each site may or may not be occupied by a particle (all particles being of the same type). Hence the state in which such a process is at time $t\geq 0$ is uniquely determined by the set ξ_t of sites which are occupied at time t. Let Ξ denote the family of all subsets of Z^d, which are called configurations. (Interpretation: Site x is occupied ("infected") in $\xi\in\Xi$ if $x\in\xi$, and vacant (not infected) otherwise.) Let us put $\Xi_0=\{\xi\,|\,\xi\in\Xi,\ 0<|\xi|<\infty\}$ and $\Xi_1=\{\xi\,|\,\xi\in\Xi,\ \xi\neq\emptyset\}$ (if B is any set, we denote its cardinality by $|B|$).

The random mechanism of a BPI can now intuitively be described as follows. Suppose that at time t a BPI is in state $\xi\in\Xi_0$, and consider any site x which may or may not belong to ξ. Then the probability that during a (short) time interval $(t,t+h)$ an event called "splitting" occurs at x, is given by $c(x,\xi)h+o(h)$, where the intensity $c(x,\xi)$ depends on x and ξ. Given a splitting at x occurs, a set $\eta\in\Xi_0$ such that $x\in\eta$, is chosen with probability $p(x,\xi,\eta)$ which depends on x,ξ and η. If η has been chosen, the BPI undergoes the transition $\xi\to\xi\cup\eta$. (Interpretation: In the case $x\in\xi$, we can imagine that the particle at x splits into a finite (random) number of offspring. The latter is placed exactly at the sites of a (random) set η. At sites in $\xi\cap\eta$, only one particle rather than two, remains. If, however, $x\notin\xi$, we may imagine that at x an "auxiliary" unobservable particle (not being of ordinary type) is situated which is able to split, thereby producing a finite (random) number of ordinary (!) particles.) The processes just described are, in a way, more general than those introduced in Holley and Liggett (1975) under the name of branching processes with interference.On the other hand, a BPI is more general than a nearest-neighbour interaction (NNI) considered by Harris (1974), provided an NNI has the additional property that once a site is occupied it remains so forever (a special case of this, in turn, is given by the Williams-Bjerknes process (see Williams and Bjerknes (1972)) with a "carcinogenic advantage" equal to infinity). We would like to mention that our study of branching processes with interactions is partly motivated by a simulation model of the formation of virus plaques; see Schwöbel, Geidel and Lorenz (1966).

Our main interest is in the asymptotic geometrical behaviour of the set ξ_t of sites which are occupied in a BPI at time t. This problem is stated more precisely in Section 3 which also contains the main result. The latter is applied in Section 4 to a class of so-called generalized Eden processes. In Section 2, we formulate a general theorem concerning the possibility to construct a BPI with given intensities $c(x,\xi)$ and probabilities $p(x,\xi,\eta)$ as a Hunt process.

2. Existence theorem

In this section, we give conditions which are sufficient for the existence of a Hunt process with state space $(\Xi,\mathcal{E})$ ($\mathcal{E}$ denoting the family of Borelian subsets of Ξ with respect to the usual product topology of Ξ) which corresponds to the intuitive description of a BPI, given in Section 1.

Let $\mathcal{C}$ denote the family of all continuous functions $f:\Xi \rightarrow R$. With the supremum norm, C is a (real) Banach space. Let $C_o \subset C$ denote the set of all functions depending only on finitely many coordinates (observe that C_o is dense in C). Put $\Xi_{ox}=\{\xi \mid \xi\in\Xi_o,\ x\in\xi\}$, $x\in Z^d$, and

(2.1) $\quad H=\{(x,\eta) \mid x\in Z^d,\ \eta\in\Xi_{ox}\}$.

We will always assume that, for all $x\in Z^d$ and $(x,\eta)\in H$, the functions

$$\xi \mapsto c(x,\xi),\ \xi\in\Xi,$$

and

$$\xi \mapsto p(x,\xi,\eta),\ \xi\in\Xi,$$

are continuous. Additionally, the following will be assumed throughout:

(2.2) $$\sum_{\eta\in\Xi_{ox}} p(x,\xi,\eta)\le 1,\ x\in Z^d,\ \xi\in\Xi,$$

and, for some constant $\hat{c}>0$,

(2.3) $\quad c(x,\xi)\le\hat{c},\ x\in Z^d,\ \xi\in\Xi.$

The idea is now to start with a linear operator of the form

(2.4) $$Af(\xi)=\sum_{x\in Z^d}\sum_{\eta\in\Xi_{ox}} c(x,\xi)p(x,\xi,\eta)[f(\xi\cup\eta)-f(\xi)],\ \xi\in\Xi,$$

which is defined for all functions f in a sufficiently large class $\mathcal{D}$ such that $C_o\subset\mathcal{D}\subset C$ (the sum in (2.4) being absolutely convergent in C for all $f\in\mathcal{D}$). Using a general existence theorem in Liggett (1972), we can derive the result below which gives sufficient conditions for the closure (in $C\times C$) of the operator A to be the infinitesimal generator of a (unique) strongly continuous semigroup of positive contraction operators T_t ($t\ge 0$) on C (once such a semigroup has been constructed, a well-known existence theorem in Blumenthal and Getoor (1968) (p.46) yields the desired Hunt process). In the sequel we denote by $p(x,\eta)$ ($(x,\eta)\in H$) any numbers for which

(2.5) $\quad p(x,\xi,\eta)\le p(x,\eta),\ (x,\eta)\in H,\ \xi\in\Xi.$

We then have

(2.6) *Theorem. Assume that the numbers* $p(x,\eta)$ $((x,\eta)\in H)$ *can be chosen in such a way that, for some constant* $L_1>0$,

$$\sum_{\substack{(x,\eta)\in H\\ y\in\eta}} p(x,\eta)\le L_1,\quad y\in Z^d, \tag{2.7}$$

and furthermore (writing $\xi\cup z$ *instead of* $\xi\cup\{z\}$)

$$\sum_{z\in Z^d}\ \sup_{\xi\in\Xi} |p(y,\xi\cup z,\zeta)-p(y,\xi,\zeta)|\le p(y,\zeta),\quad (y,\zeta)\in H. \tag{2.8}$$

Finally let, for some constant $L_2>0$,

$$\sum_{z\in Z^d}\ \sup_{\xi\in\Xi} |c(y,\xi\cup z)-c(y,\xi)|\le L_2,\quad y\in Z^d. \tag{2.9}$$

Then the desired BPI *with intensities* $c(x,\xi)$ *and probabilities* $p(x,\xi,\eta)$ *can be constructed as a Hunt process with state space* (Ξ,E).

Following the notation of Blumenthal and Getoor (1968) as closely as possible we designate such Hunt processes by $(\Omega,M,M_t,\xi_t,\theta_t,P_\xi)$ or, more briefly, by $\{\xi_t\}$.

3. *The main result*

For the rest of the paper it will be assumed that the conditions of Theorem (2.6) are satisfied.

Our main interest is in the asymptotic geometrical behaviour of the set ξ_t of sites which are occupied in a BPI at time t. In order to make this more precise put

$$\tau(x)=\tau(\omega,x)=\inf\{t\,|\,x\in\xi_t\},\quad x\in Z^d,\quad \omega\in\Omega. \tag{3.1}$$

It is useful to extend $\tau(\cdot)$ to all of R^d as follows. Introduce the cubes

$$Q_x=\{y\,|\,y\in R^d,\ x^i-\tfrac{1}{2}<y^i\le x^i+\tfrac{1}{2},\ i=1,\dots,d\},\quad x\in Z^d, \tag{3.2}$$

and put

$$\tau(y)=\tau(x),\quad y\in Q_x,\quad x\in Z^d. \tag{3.3}$$

We will study the asymptotic behaviour of the set $\{x\,|\,\tau(x)\le t\}$ (as $t\to\infty$). Let us put

$$\tau_\Delta(x)=\Delta\tau(\tfrac{1}{\Delta}x),\quad \Delta>0,\quad x\in R^d. \tag{3.4}$$

We will need the following

(3.5) Condition. For each $t>0$ there exists a number $0<p(t)<1$ such that, for all $x\in Z^d$ and $t>0$,

(3.6) $$P_\xi(\tau(x)\le t)\ge p(t) \text{ for all } \xi\in\Xi_1 \text{ such that } x\notin\xi \text{ and } \xi\cap N_x\ne\emptyset.$$

Here, we put $N_x=\{y|y\in Z^d, \|x-y\|=1\}$, $x\in Z^d$, where

(3.7) $$\|z\|=\sum_{i=1}^{d}|z^i|, \quad z\in R^d.$$

We have (see Schürger (1979))

(3.8) Lemma. Under Condition (3.5) there exists a constant $r>0$ such that, for all $\xi\in\Xi_1$, $\varepsilon>0$ and $z\in R^d$,

(3.9) $$\limsup_{\Delta\downarrow 0}\ \sup_{x:\|x-z\|\le r\varepsilon}|\tau_\Delta(x)-\tau_\Delta(z)|\le\varepsilon \quad \text{a.s. } (P_\xi).$$

A special case of the following monotonicity condition occurred already in Schürger (1979).

Condition (M). For all $(x,\eta)\in H$ and $\xi,\zeta\in\Xi$ such that $\xi\subset\zeta$ and $\eta\not\subset\zeta$, we have that

(3.10) $$c(x,\xi)p(x,\xi,\eta)\le c(x,\zeta)p(x,\zeta,\eta).$$

Using a simple idea of Sullivan (1975), one can show

(3.11) Lemma. Assume that Condition (M) is satisfied. Let $J\ne\emptyset$ be any finite or countable index set. Then, for all $\xi,\zeta\in\Xi$ such that $\xi\subset\zeta$, we have

(3.12) $$P_\xi(\tau(x_j)\ge t_j,\ j\in J)\ge P_\zeta(\tau(x_j)\ge t_j,\ j\in J),\quad x_j\in Z^d,\ t_j>0,\ j\in J,$$

and

(3.13) $$P_\xi(\tau(x_j)\le t_j,\ j\in J)\le P_\zeta(\tau(x_j)\le t_j,\ j\in J),\quad x_j\in Z^d,\ t_j>0,\ j\in J.$$

A BPI is called translation-invariant if

(3.14) $$c(x+y,t_y(\xi))=c(x,\xi),\quad x,y\in Z^d,\ \xi\in\Xi,$$

and

(3.15) $$p(x+y,t_y(\xi),t_y(\eta))=p(x,\xi,\eta),\quad (x,\eta)\in H,\ y\in Z^d,\ \xi\in\Xi,$$

where the translations $t_y(\cdot)$ $(y\in Z^d)$ are given by

(3.16) $$t_y(\xi)(x)=\xi(x-y),\quad x,y\in Z^d,\ \xi\in\Xi.$$

A BPI is called point-symmetric if

(3.17) $c(x,\xi)=c(-x,\tilde{\xi})$, $x\in Z^d$, $\xi\in\Xi$,

and

(3.18) $p(x,\xi,\eta)=p(-x,\tilde{\xi},\tilde{\eta})$, $(x,\eta)\in H$, $\xi\in\Xi$,

where the configurations $\tilde{\xi}\in\Xi$ and $\tilde{\eta}\in\Xi_o$ are given by, respectively, $\tilde{\xi}(y)=\xi(-y)$, $\tilde{\eta}(y)=\eta(-y)$, $y\in Z^d$.

Proceeding as in Schürger (1979) and using Lemma (3.11), one can show

(3.19) Lemma. Let a BPI be translation-invariant and point-symmetric. Then, if Conditions (3.5) and (M) are satisfied, we have, for all $\xi\in\Xi_o$, $x\in R^d$ and $m\geq 1$, that the limit

(3.20) $$\lim_{\Delta\downarrow 0} (E_\xi \tau_\Delta^m(x))^{1/m}=N(x)$$

(E_ξ denoting the expectation formed with respect to P_ξ) exists and is independent of $\xi\in\Xi_o$ and $m\geq 1$. $N(\cdot)$ is a seminorm.

It is easy to formulate a condition which (under the hypotheses of Lemma (3.19)) is sufficient for $N(\cdot)$ to be a norm. For a given BPI, let

$$\beta_1=\min\{k \mid \forall (x,\eta)\in H \ \forall \xi\in\Xi_1: \max_{y,z\in\eta} \|y-z\|>k \Longrightarrow p(x,\xi,\eta)=0\}$$

and

$$\beta_2=\min\{k \mid \forall (x,\eta)\in H \ \forall \xi\in\Xi_1: \min_{y\in\xi} \|x-y\|>k \Longrightarrow p(x,\xi,\eta)=0\}$$

(here, we put $\min \emptyset = \infty$). The magnitude $\max(\beta_1,\beta_2)$ (which may be infinite) will be called the span of the BPI. Using estimates similar to those in the proof of Theorem 3 of Richardson (1973), and a result analogous to Lemma (3.6) of Harris (1974), one can show that, under the hypotheses of Lemma (3.19), the finiteness of the span implies that the limit $N(\cdot)$ in (3.20) defines a norm on R^d. Now we can formulate our main result (generalizing Theorem (3.48) in Schürger (1979)).

(3.21) Theorem. Let the hypotheses of Lemma (3.19) be satisfied. Let the BPI have a finite span $\beta\geq 1$. Then there exists a norm $N(\cdot)$ on R^d with the following properties: For all $\xi\in\Xi_o$ and $0<\varepsilon<1$ we have that

(3.22) $$\{x \mid N(x)\leq(1-\varepsilon)t\}\subset\xi_t\subset\{x \mid N(x)\leq(1+\varepsilon)t\} \text{ a.s. } (P_\xi)$$

for all sufficiently large t.

For obvious reasons, this result might be called a <u>strong law of large configurations</u>. In Richardson (1973) it has been shown for a certain class of spread processes in R^d, satisfying certain axioms, that the probability that (3.22) holds tends to one as $t\to\infty$. Theorem (3.21) can be proved by using some techniques of Richardson (1973) and Schürger (1979), and by utilizing Lemmas (3.19) and (3.8) as well as a convergence theorem due to Kesten (1973).

(3.23) <u>Corollary. Under the hypotheses of Theorem</u> (3.21), <u>we have</u>

(3.24) $$\lim_{t\to\infty} \frac{1}{t^d}|\xi_t| = L_d\{x|N(x)\le 1\} \quad \underline{\text{a.s.}}\ (P_\xi),\ \xi\in\Xi_o,$$

<u>and</u>

(3.25) $$\lim_{t\to\infty} \frac{1}{t^d}|\xi_t| = L_d\{x|N(x)\le 1\} \quad \underline{\text{in}}\ L^m(P_\xi),\ m\ge 1,\ \xi\in\Xi_o,$$

L_d <u>denoting the</u> d-<u>dimensional Lebesgue measure</u>.

4. <u>Generalized Eden processes</u>

In this section, we first introduce a class of simple (discrete time) spread processes $\{\eta_n^{(k)}\}_{n\ge 0}$ $(k\ge 1)$ on Z^d, generalizing a process introduced by Eden (1961). Theorem (3.21) and Corollary (3.23) can then be applied to get results about the asymptotic geometrical behaviour of these processes. Put

(4.1) $$\partial_k(\zeta)=\{z|z\notin\zeta,\ \|z-\tilde{z}\|\le k \text{ for some } \tilde{z}\in\zeta\},\ k\ge 1,\ \zeta\in\Xi$$

(the norm $\|\cdot\|$ given by (3.7)).

Now fix any $k\ge 1$. The process $\{\eta_n^{(k)}\}_{n\ge 0}$ starts at a finite nonvoid configuration $\eta_0^{(k)}\subset Z^d$. Given that the process is in state $\eta_n^{(k)}\in\Xi_o$ $(\eta_n^{(k)}\subset Z^d)$, one of the sites of $\partial_k(\eta_n^{(k)})\subset Z^d$ (say x_o) is chosen at random, all possible choices being equiprobable. Then the state of the process undergoes the transition $\eta_n^{(k)}\to\eta_{n+1}^{(k)}\cup\{x_o\}$. We can prove (utilizing Theorem (3.21) and Corollary (3.23))

(4.2) *Theorem. Let* $k\geq 1$. *Then there exists a norm* $N^*(\cdot)$ *on* R^d *such that, for all* $\xi\in\Xi_0$ *and* $0<\varepsilon<1$, *we have for the generalized Eden process* $\{\eta_n^{(k)}\}$, *for which* $\eta_0^{(k)}=\xi$, *that*

(4.3) $$\{x|x\in Z^d,\ N^*(x)\leq(1-\varepsilon)n^{1/d}\}\subset\eta_n^{(k)}\subset\{x|x\in Z^d,\ N^*(x)\leq(1+\varepsilon)n^{1/d}\}$$

almost surely for all sufficiently large n.

References

BLUMENTHAL,R.M. and GETOOR,R.K.(1968) *Markov Processes and Potential Theory*. Academic Press, New York.

EDEN,M.(1961) A two-dimensional growth process. *Proc.4th Berkeley Symp. Math.Statist.Prob.* *4*, 223-239.

HARRIS,T.E.(1974) Contact interactions on a lattice. *Ann.Prob.* *2*, 969-988.

HOLLEY,R.A. and LIGGETT,T.M.(1975) Ergodic theorems for weakly interacting infinite systems and the voter model. *Ann.Prob.* *3*, 643-663.

KESTEN,H.(1973) Contribution to the discussion in Kingman (1973), p.903.

KINGMAN,J.F.C.(1973) Subadditive ergodic theory. *Ann.Prob.* *1*, 883-909.

LIGGETT,T.M.(1972) Existence theorems for infinite particle systems. *Trans.Amer.Math.Soc.* *165*, 471-481.

RICHARDSON,D.(1973) Random growth in a tessellation. *Proc.Camb.Phil. Soc.* *74*, 515-528.

SCHÜRGER,K.(1979) On the asymptotic geometrical behaviour of a class of contact interaction processes with a monotone infection rate. *Z.Wahrscheinlichkeitsth.* *48*, 35-48.

SCHWÖBEL,W.,GEIDEL,H. and LORENZ,R.J.(1966) Ein Modell der Plaquebildung. *Z.Naturforsch.* *21*, 953-959.

SULLIVAN,W.G.(1975) *Markov Processes for Random Fields*. Comm.Dublin Inst.Adv.Studies Ser. A, No.23.

WILLIAMS,T. and BJERKNES,R.(1972) Stochastic model for abnormal clone spread through epithelial basal layer. *Nature* *236*, 19-21.

The Asymptotic Behavior of a Probabilistic Model for Tumor Growth

by

Maury Bramson and David Griffeath
University of Minnesota University of Wisconsin

We discuss here the behavior for large time of a cancer model introduced by Williams and Bjerknes [5]. Cells are of two types, normal and abnormal, and are located on the planar lattice, one at each site. With each cellular division, one daughter cell stays put; the other usurps the position of a neighbor, who disappears from the population. Abnormal cells are assumed to reproduce at a faster rate than normal cells; the cancer therefore has a tendency to spread. It is shown that, conditioned on its nonextinction, a tumor commencing from a single abnormal cell will have an asymptotically linear rate of radial growth; asymptotically, the tumor assumes a fixed shape.

There has recently been considerable interest in the construction of mathematical models for the spread of cancer cells within a medium. The actual phenomena are more complex than can be treated analytically, and the correspodning mathematical models have, as a matter of necessity, been severe simplifications. One of the few mathematically tractable models for the long-term rate of spread of cancer cells was first solved by Richardson [3]. Richardson's model is a special case of a model proposed by Williams and Bjerknes [5]. Based on biological considerations, they restricted attention to the basal layer of an epithelium, thereby obtaining a two dimensional setting. In their model, cells are assumed to be of two types, normal and abnormal (cancerous), and are to be located on the planar lattice, one at each site. With each cellular division, one daughter cell stays put, while the other usurps the position of a neighbor, who disappears from the population. Abnormal cells are assumed to reproduce at a faster rate than normal cells; it seems reasonable that the cancer should therefore have a tendency to spread. Richardson showed that this is indeed what occurs in the special case where it is assumed that abnormal cells, but not normal cells, reproduce: the tumor tends to grow in each direction at a linear rate. He showed moreover that, under the appropriate scaling, a tumor commencing from a single abnormal cell asymptotically assumes a fixed shape. We are concerned here with the asymptotic behavior of the more general Williams-Bjerknes model. We show why the same basic asymptotic behavior as in the Richardson model still occurs. For a complete account, the reader is referred to [1] and [2].

To be concise, we denote the presence of a normal cell at a site by a 0, and the presence of an abnormal cell by a 1. As time evolves, normal cells will occasionally be displaced by abnormal cells, and abnormal cells by normal cells. The mechanism is as follows: with each cellular division, one of the two daughter cells will usurp the position of one of the four immediate neighbors of that site. In the accompanying diagram, we see that an abnormal cell has just displaced a normal neighbor to its immediate right; (i) exhibits the region just before division, (ii) exhibits the region immediately after division.

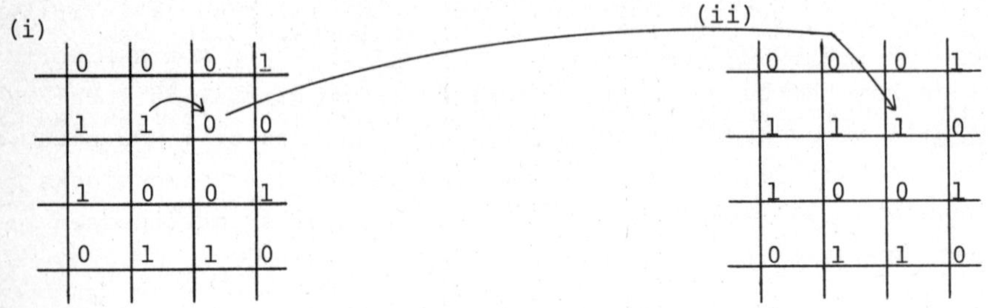

The model is assumed to be in continuous time. We wish for the stochastic process associated with this model to possess the Markov property. It is therefore assumed that cells split independently of one another after exponential holding times. Normal cells split at exponential rate 1, whereas due to "carcinogenic advantage", abnormal cells split at rate $\kappa > 1$. We denote by ξ_t^A the state of this stochastic process at time t given that its initial state is A. We think of ξ_t^A as the set of abnormal sites at time t, i.e. the tumor.

Independently, the Williams-Bjerknes model has been formulated in the field of interacting particle systems as the biased voter model. (See Schwartz [4].) In that context, each site of the two (or more generally, d) dimensional integers is thought of as being occupied by an individual, who is voting on some issue. Each individual holds an opinion: 1 indicates a "for" vote on the issue, and 0 indicates an "against" vote. As in the Williams-Bjerknes model, the behavior at a site is influenced by the behavior of its neighbors. Each individual occasionally reassesses his opinion. When he does so, he observes the opinion of one of his nearest neighbors, which he then proceeds to copy. Built into the system is an assumption of bias towards the "for" opinion: individuals voting "for" tend to reassess their opinions less frequently than those individuals voting "against". ξ_t^A will denote the set of sites voting "for" at time t given that the initial configuration is A.

In both models, one may inquire about the asymptotic behavior of ξ_t^A as $t \to \infty$ for initial configurations A of finite cardinality. Due to the bias towards the abnormal state in the Williams-Bjerknes Model (the "for" state in the biased voter model), it seems reasonable to expect the tumor to eventually spread, as long as it does not first die out. Indeed, the associated magnitude process

$$X_t = |\xi_t^A| ,$$

defines a rescaled (in time) simple random walk with absorption at 0. Since $\kappa > 1$, X_t has a positive drift, and will tend to infinity conditioned on its nonabsorption at 0. However, it is not a priori at all clear <u>how</u> the tumor grows. Does the tumor persist in local regions, or will it eventually completely leave any box of fixed size, as illustrated in the diagram below? Does the tumor form solid blocks or does it have a large surface area, with healthy cells interspersed throughout? How quickly does the tumor grow?

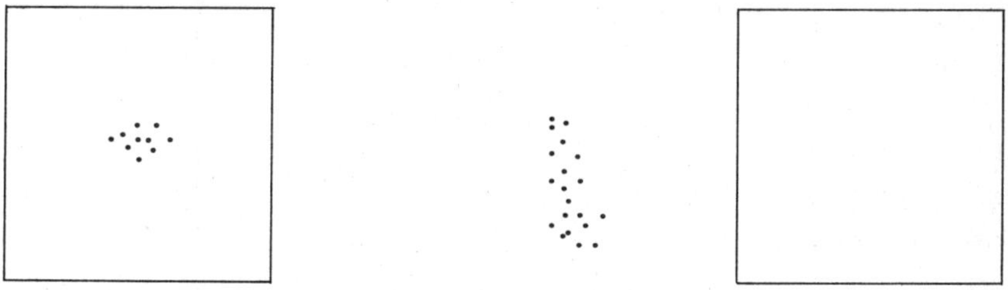

tumor at earlier time situated within large box

tumor at later time having drifted from box

The questions are answered by the following result.

<u>Theorem</u>. There exists a norm $\| \|'$ on R^2 with the following property: if D'_R denotes the set $\{x \in Z^2: \|x\|' \le R\}$ then, for all $\varepsilon > 0$,

(1) $$P(\exists t^* : D'_{(1-\varepsilon)t} \subset \xi_t^{\{0\}} \subset D'_{(1+\varepsilon)t} \text{ for } t \geqq t^* \mid \xi_s^{\{0\}} \neq \emptyset \ \forall s) = 1.$$

In other words, conditioned on its nonextinction, the tumor starting from a single cell at 0 will eventually have a linear rate of radial growth, and the tumor will approach a fixed shape given by some norm. (Note: the Williams-Bjerknes model and the biased voter model may just as easily be formulated for dimensions other than 2; the Theorem remains valid. Also, the initial set $\{0\}$ may be replaced by arbitrary A, with $|A| < \infty$.)

The knowledgeable reader will observe that the statement of the Theorem is essentially the same as that given in [3] for the more specific Richardson model. (In the case of Richardson's model, one need not condition on nonextinction.) We mention in passing that the precise nature of the norm is not known in either case, although the problem has attracted a considerable amount of attention. The demonstration for the present model is, however, considerably longer, and is therefore broken into two parts. We first show that: I. for some $c > 0$,

(2) $P(\exists t^* : D_{ct} \subset \xi_t^{\{0\}} \text{ for } t \geqq t^* \mid \xi_s^{\{0\}} \neq \emptyset \; \forall s) = 1.$

($D_R = \{x : \|x\| \leqq R\}$, where $\|\cdot\|$ denotes the Euclidean norm.) Statement (2) is a weaker version of (1), and says that a tumor which starts at 0, and does not become extinct, will eventually contain a ball of linearly increasing radius. To demonstrate (2), techniques from the field of interacting particle systems are applied. We then show that: II. (1) does indeed follow from (2). The method of proof applies a generalization of Richardson's arguments, and again involves techniques from interacting particle systems. The remainder of the article is devoted to outlining Steps I and II.

Step I can be broken into two claims, which together yield (2).

<u>Claim 1</u>. Let $R > 0$. Then,

$$P(\xi_t^{\{0\}} \supset D_{x,R} \text{ for some } t>0,\ x \in Z^2 \mid \xi_s^{\{0\}} \neq \emptyset \;\forall s \geq 0) = 1,$$

where $D_{x,R} = D_R + x$. Claim 1 states that, conditioned on the tumor not becoming extinct, the tumor will at some random time contain a ball of radius R. No assertion is made regarding the location of the ball. We introduce the notation

$$E_c^\Lambda = \{\exists t^* < \infty : D_{ct} \subset \xi_t^\Lambda \;\forall t \geqq t^*\},$$

where $c > 0$ and $\Lambda \subset R^2$, which we use to state

<u>Claim 2</u>. For some $c > 0$ and each $\varepsilon > 0$,

$$\exists R < \infty : P(E_c^{D_{x,R}}) \geqq 1 - \varepsilon .$$

Claim 2 states that if the tumor starts from a large enough ball, then the tumor will with overwhelming probability not become extinct, and will grow at (at least) a radially linear rate.

It is not difficult to see that the two claims together yield (2). By the first claim, the tumor will eventually contain a ball of radius R; then, applying the second claim, we see that the tumor will grow as desired in (2). The proofs of the two claims are of differing degree of difficulty. Claim 1 involves an essentially trivial argument, which says that for fixed T, a tumor starting from at least one cell will have a positive probability ε of containing a ball of radius R at time T. Since cellular divisions are independent over disjoint time intervals, one obtains by a repetition of this reasoning that, conditioned on nonextinction of the tumor, the probability of the tumor never containing a ball of radius R by time nT is at most $(1-\varepsilon)^n \downarrow 0$ as $n \to \infty$. The demonstration of Claim 2 is considerably more lengthy, and involves the concept of duality for interacting particle systems and the selection of certain embedded processes. See [1] for more details.

For Step II, we follow the outline presented in [3]. Richardson used subadditivity as his basic tool to establish his version of (1). The approach he used was to show that (1) follows from five assumptions about growth models, which we will list and discuss in a moment. The problem then reduces to demonstrating that the model at hand satisfies these properties.

We first introduce the following notation:

$$\tau(x) = \inf\{t : x \in \xi_t^{\{0\}}\}$$

and

$$\tau_\Delta(x) = \Delta\tau(x/\Delta) \qquad \text{for} \quad \Delta > 0.$$

$\tau(x)$ is the first infection time of the site x. We conceive of Δ as being small; therefore $\tau_\Delta(x)$ is the infection time on a reduced time and spatial scale. (Nonlattice x/Δ is to be identified with the nearest lattice point.) We also introduce a family of random variables $V_\Delta(x)$ and constants $\varepsilon_i(\Delta,x)$ which satisfy

$$E[V_\Delta^2(x)] = 0_x(\Delta)$$

and

$$\varepsilon_i(\Delta,x) = 0_x(\Delta).$$

Richardson's assumptions (in slightly modified form) are:

(A1) $\forall x,y \in \mathbb{R}^2$,

$$\tau_\Delta(x+y) \leqq \tau_\Delta(x) + \sigma_\Delta(y) + V_\Delta(x),$$

where $\sigma_\Delta(y)$ is a copy of $\tau_\Delta(y)$ and is independent of $\tau_\Delta(x)$.

(A2) $\exists L > 0$:

$$E[\tau_\Delta(x)] \geqq \|x\|/L + \varepsilon_2(\Delta,x).$$

(A3) For some $r' > 0$,

$$P(\{y : \|x-y\| \leqq r'\delta\} \subset \{y : |\tau_\Delta(x) - \tau_\Delta(y)| \leqq \delta\}) \geqq 1 - \varepsilon_3(\Delta,\delta).$$

(A4) For some $r > 0$,

$$E[\tau_\Delta^2(x)] \leqq \|x/r\|^2 + \varepsilon_4(\Delta).$$

(A5) $E[\tau_\Delta(x)] \leqq E[\tau_\Delta(-x)] + \varepsilon_5(\Delta,x)$.

Properties (A1) and (A3) are the primary assumptions. (A1) states that the first infection time of sites is essentially subadditive for pairs of sites which are far enough apart. (A3) should be thought of as a regularity condition on the infection time for sites which are nearby on the reduced space scale; all such nearby sites will be first infected within a comparatively small margin of time. (A4) is clearly a moment condition on the infection time $\tau(x)$ for large x, and is to be used in conjunction with (A1). Assumption (A2) states that the process grows asymptotically at at most a linear rate. Finally, (A5) is just a simple symmetry statement, so that $\| \ \|'$ in (1) will be a norm, and is otherwise inessential. To show that his analogue of (1) could be derived from properties (A1) through (A5), Richardson used subadditivity arguments which are fairly standard by now. For more details, see [2] or [3].

For the Williams-Bjerknes model, as for Richardson's model, we wish to verify these five properties. However, whereas in the Richardson model (A1) through (A5) are simple to show, verification is more intricate for the general Williams-Bjerknes model. We stop short of any details, but mention the crucial difference for properties (A1) and (A3).

For the Richardson model, not only does (A1) hold, but $\tau(x)$ is strictly subadditive; the term $V_\Delta(x)$ may be dropped. This is because one possible route of infection from 0 to $x + y$ is through x, and once at x, the infection can spread from there to $x + y$; a tumor starting at x behaves in the same manner as does a tumor starting at 0. This strictly subadditive property is no longer true for the Williams-Bjerknes model conditioned on nonextinction of the tumor starting at 0. (We must condition, otherwise the expected infection time would be infinite.) For, a tumor starting at x does not behave in the same manner as does a tumor starting at 0; the tumor starting at 0 cannot become extinct because of the conditioning, whereas the tumor starting at x may.

For property (A3), differences in the models also become apparent. We wish to show that for two points x and y which are relatively close to one another, $\tau(x)$ and $\tau(y)$ will be close. To demonstrate this for the Richardson model, one need only construct a path from x to y along the lattice. The difference in hitting times will be at most the time it takes an infection to travel along this path; this time is easily computed. However, if one attempts the same stratagem for the Williams-Bjerknes model, one runs into difficulties. Since recovery of cells is allowed, the tumor may temporarily recede from an entire region, including the path. Nevertheless, a variant of

(2) may be applied. This will ensure sufficiently rapid growth of the process so that (A3) may be checked.

Figure 1 Moderate κ

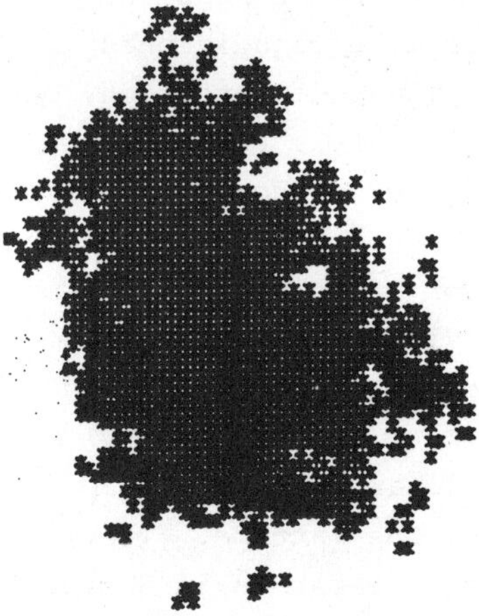

Figure 2 Small κ

References

[1] M. Bramson and D. Griffeath (1979). On the Williams-Bjerknes tumour growth model I. To appear in Annals of Probability.

[2] M. Bramson and D. Griffeath (1979). On the Williams-Bjerknes tumour growth model II. To appear.

[3] D. Richardson (1973). Random growth in a tessellation, Proc. Cambridge Phil. Soc. 74, 515-528.

[4] D. Schwartz (1977). Applications of duality to a class of Markov processes, Annals of Probability 5, 522-532.

[5] T. Williams and R. Bjerknes (1972). Stochastic model for abnormal clone spread through epithelial basal layer, Nature 236, 19-21.

An Ising Model of a Neural Network

W. A. Little
Physics Department
Stanford University
Stanford, Ca. 94305

We show that the behavior of a neural network can be mapped onto a generalized spin ½ Ising model. The conditions for the existence of short term memory are related to the existence of long range order in the corresponding Ising model. Even in the presence of noise and fluctuation in the properties of the network, precisely defined states can exist nevertheless. Long term memory appears to result from the modification of the synaptic junctions as a result of signals propagating through the network. The essentially non-linear problem can be linearized and leads to a holographic-like storage of information and means for recall.

The processing of information in the central nervous system is believed to occur through the, so called, "spike activity" of the neurons. If a neuron is stimulated beyond a certain threshold, the neuron "fires" and an action potential propagates down the axon of this neuron to a synaptic junction on the surface or dendrites of another neuron. The action potential causes the release of a transmitter substance across the synaptic cleft to the next neuron. The resulting changes of ionic concentrations within the neuron cause a change of its electrochemical potential. These changes may tend to excite or inhibit the neuron and the resulting electrical changes are referred to as excitatory postsynaptic potentials, or inhibitory postsynaptic potentials, respectively. If the sum of all the postsynaptic potentials of a given neuron exceeds the threshold, the neuron "fires" and the above process continues. The potential of the neuron is determined by the integrated effect of all the postsynaptic potentials delivered to it within a period of a few milliseconds, a period known as the period of latent summation. This potential decays in a few milliseconds if the neuron does not fire while if it does fire, a sharp positive pulse or spike occurs which propagates as an action potential down the axon. It is followed by a negative going excursion which returns to the resting potential again within a few milliseconds. During this latter refractory period the neuron is recovering and cannot fire. From the above we see that the natural cycle time, within the central nervous system, is of the order of a few milliseconds. It can be shown that this is appreciably longer than the time of flight (0.1 millisecs) of a signal from one neuron to another.

In addition to axonal connections between neurons within the network, similar connections exist between external sensory receptors and synaptic junction on the neurons within the central nervous system. Likewise, outgoing axons carry signals from the network to outlying motor effectors. All together there are some 10^9 - 10^{10} neurons in the human brain and each of these has, on the average, about 10^4 synaptic junctions bringing information from other neurons. Understanding the behaviour of the brain is thus a many body problem of strongly interacting units.

In recent years much progress has been made in the understanding of other strongly interacting many body problems in physics such as superconductivity, superfluidity, the liquid-gas phase transition and magnetic systems. Progress in these has resulted from the recognition of the existence of an order-parameter which characterizes the ordered phases. By applying similar methods of analysis to the behaviour of the brain one can hope to gain an understanding of this more complex system. Indeed, we have been able to show that the neural problem can be mapped onto a generalization of the Ising spin 1/2 model. [1]. By so doing one can gain insight into the behaviour of the brain, to the nature of short- and long-term memory and of the storage capacity of the network([2] [3]). We present here an outline of our model and highlight two or three points which have been eluded to previously but have not been discussed in detail, hitherto. For the full mathematical treatment of the model the reader is referred to the above three papers.

The essence of the model can be understood as follows: suppose one could take a snapshot of the brain. At that instant each neuron has either just fired or has not. We associate a spin 1/2 particle with each of the N neurons of the network. If a particular neuron has fired we take the corresponding spin as "up" and if it has not, as "down." A configuration of all the spins constitutes a description of the instantaneous state of the brain. It is convenient to imagine these N spins arranged in a column. A signal from a fired neuron propagates down the axon to synaptic junctions on other neurons, thereby influencing the probability of these neurons firing. As a result a new state of the brain will evolve which again can be described by a column of N spins. This configuration is influenced by the configuration of spins in the previous column and hence there is an effective spin - spin interaction between adjacent columns. As this firing process repeats itself the corresponding time evolution of the neural network can be mapped onto a succession of columns of spins, generating thereby a two dimensional Ising spin system. It should be noted that while the spins can be arranged on a plane to give a 2D representation, the large number of connections to "near neighbors" ($\sim 10^4$) gives an effective dimensionality to the problem which is much larger (10^3 - 10^4). A function, analogous to the partition function of the spin system describes the behaviour of the neural network.

An important element of the model is that it is not a deterministic model. Because of the small number of transmitter molecules released at the synapses, statistical fluctuations in these numbers cause fluctuations in the postsynaptic potentials. This indeterminacy gives rise to noise in the network which is directly analogous to thermal noise in the spin system [4]. Because of this an effective temperature T* can be defined for the network. In spite of this noise well defined states of the system can occur which are analogous to well defined phases, like the ferromagnetic or antiferromagnetic phases of material systems which, of course, exist at finite temperatures in the presence of thermal noise.

Certain assumptions have been made in formulating the model. It has been assumed that the various parameters of the network such as the strength of the post synaptic potentials and the threshold are fixed and independent of time. If this were strictly so the network would be incapable of learning. However, learning involves changes which occur over a long period of time so that over shorter periods, of the order of a second, the parameters may be treated as fixed and the learning changes handled as adiabatic changes. Thus the non-linear problems may be linearized and the essential behaviour of the network deduced from the linear model.

We have assumed that the time evolves in discrete steps rather than continuously. This makes it possible to handle the problem using a transfer matrix. The continuum problem is very much more difficult to handle. However, one can give very powerful physical arguments to show that the essence of the phase transformations of continuum systems is contained in the discrete problem [1]. The lattice-gas model of the liquid-gas transition is a particular example which illustrates this point [5]. As yet no rigorous relation has been given relating the properties and behaviour of a continuous system to those of a discrete system but the existence of such a relationship is implied in many other physical systems.

The propagation of order in a spin system or the time evolution of a neural network can be treated using a transfer matrix. For a system of N neurons, each of which can be in one of two states there are 2^N possible configurations.

An element of the $2^N \times 2^N$ transfer matrix, P represents the probability of the system making a transition in one time period from one configuration α to another γ. We have shown that such an element can be represented by

$$\langle\gamma|P|\alpha\rangle = A \exp \{\beta \sum_{i,j} (s'_i V_{ij} s_j) + \sum_i s'_i H_i\}, \tag{1}$$

where

$$H_i = \sum_j (V_{ij}/2) - U_i \tag{2}$$

Here, A is a normalizing constant; β is the "thermal" factor, $(kT^*)^{-1}$; $s_j=+(-1)$ indicates the up (down) state of the jth spin in the α-configuration and s'_i the spin state in the γ- configuration; V_{ij} is the contribution to the post-synaptic potential of the i th neuron due to a synapse from the j th neuron; and, U_i is the threshold potential of the i th neuron. An essentially identical expression describes the probability of finding a column of spins in a configuration γ knowing that the adjacent column is in a configuration α, except that V_{ij} then represents the spin-spin interaction between two spins i and j; and H_i represents the contribution of an externally applied local magnetic field at the i th site.

In the spin problem it can be shown that long range order can only occur if the maximum eigen value of the matrix P is at least doubly degenerate. This occurs at the Curie point for a ferromagnetic Ising system and persists for all temperatures below this. Long range order implies that the configuration of spins at one end of the crystal is influenced by the configuration at the other. This can only occur at temperatures below the Curie point. For the neural network a similar argument applies. In order for the "state of the brain," (configuration of fired or not-fired neurons) to be dependent on the state of the brain at a time many time steps earlier it is again necessary for the eigen values of P to be at least doubly degenerate. This then is the condition for the existence of short term memory, namely the ability of the network to retain an active firing pattern over a period of many cycles. We refer to these as persistent states. The order parameter of the system is described by the eigen vectors of P which are associated with these maximum eigen values.

Long term memory is the ability of the network to recreate a firing pattern or sequence of patterns which previously had propagated through the network. We have been able to show [2] that if one assumes that the postsynaptic potential V_{ij} of a given synapse is slowly modified by the presence of repetitive "imprinting" signals, α to γ passing through the synapse so that

(3) $$V_{ij} = a(s_i')_\gamma \cdot (s_j)_\alpha$$

Where a is a positive constant and the subscripts α and γ refer to the patterns for which the set of s_j and s_i' are taken , then upon subsequent stimulus the persistent state which will evolve will be identical to that used during imprinting. The factorizable quadratic form chosen for (3) is based on the physiological fact that the vesicles of the j th axon which release the transmitter substance contribute one factor in determining the strength of the post synaptic potential, while the receptor site of the i th neuron contributes the other. This form gives the network the ability to respond to a stimulus in much the same way as does a hologram respond to the enciting reference beam. In preparing a hologram coherent light from a reference beam is mixed with light scattered from an object. If O(x) is the amplitude of the light from the object at point x on the photographic plate and R(x) the amplitude of the reference beam then, $(O(x) + R(x))^2$ is the total intensity at x. The degree of exposure at x is thus proportional to this factor. Expanding it we get $R(x)^2 + O(x)^2 + 2\,O(x).R(x)$. The first two terms give an essentially constant factor with all the information contained in the factorized quadratic term $2\,O(x).\,R(x)$. To view the hologram the developed plate is illuminated by the reference beam. The amplitude of the wave transmitted at x is then proportional to R(x) times the transmitance of the plate which is proportional to $(2.\ O(x)Rx + \text{const})$. This then gives $O(x)\ R(x)^2$ plus an uninteresting constant. $R(x)^2$ is essentially independent of x so the first term recreates the object beam O(x). This, in effect, is how the neural network responds. Given that the synaptic strength is described by the expression (3) then it acts very much as the transmissivity factor of the holographic plate. It is multiplied by the s_j's in (1). If the set of s_j's is identical to the imprinting pattern $(s_j)_\alpha$ it will act as the reference beam and through (3) recreate the "object" pattern $(s_i')_\gamma$. This through (1) will give a very large probability for a transition to the new pattern γ.

One other feature worth noting is the striking similarity between the behaviour of tightly interconnected groups or assemblies of neurons and magnetic domains. We have simulated on a digital computer the behaviour of such an assembly of four neurons and mapped out the phase diagram in the H, T* plane, where V_{ij} is different for each of the neurons but the sum H_i is constrained to be the same for all. As one might expect the diagram is symmetric with respect to the line H = 0. For T* less than T_c^* the assembly can exist in one of two states corresponding to the upper "firing state" (a large positive average value for $\sum_i s_i$) or the lower non firing state ($\sum_i s_i$ negative). An external excitatory or inhibitory signal acting like an applied magnetic field can flip the domain from one state to the other. This appears to be the means by

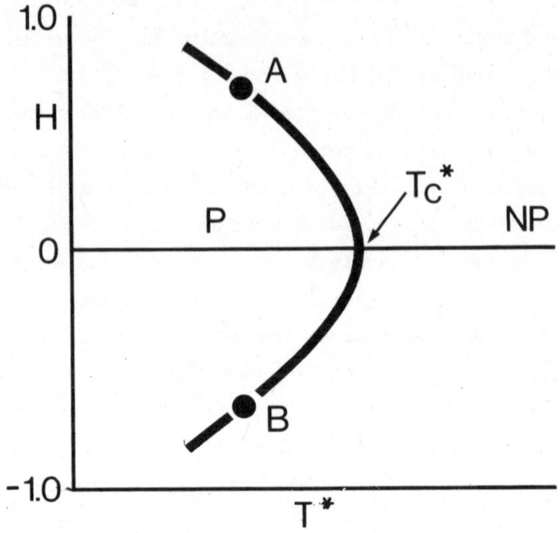

Figure 1. Phase diagram in H-T* plane showing the boundary between the persistent (P) and non-persistent phases (NP). Point A corresponds to the firing state (spin up) and B the non - firing state (spin down) of the neuron assembly.

which an external stimulus can be retained in short term memory long enough for physiological changes to occur to allow the information to be transferred to long term memory.

This is a brief overview of the model. Several important problems remain to be solved in the model itself and in devising means of how it could be tested experimentally. A theoretical problem of great importance is a determination of the representation of information in memory and of the storage capacity of the network. A start on this has been made by an exact solution of the "high temperature" limit of the model [3] but the more interesting "low temperature" behaviour remains to be solved. New experimental techniques are presently being developed which enable one to see or label those neurons involved in a particular processing activity.

I am indebted to Gordon Shaw with whom much of this work has been done and the the Research Corporation for partial funding of the work.

References

[1] Little, W. A., The Existence of Persistent States in the Brain, Math. Biosci. 19, 101-120 (1974).

[2] Little, W. A. and G. L. Shaw, A Statistical Theory of Short and Long Term Memory. Behav. Biol. 14, 115-133, (1975).

[3] Little, W. A. and G. L. Shaw, Analytic Study of the Memory Capacity of a Neural Network. Math Biosci. 39, 281-290 (1978).

[4] Shaw, G. L. and R. Vasudevan, Persistent States of Neural Networks and the Random Nature of Synaptic Transmission. Math Biosci. 21, 207-218 (1974).

[5] Lee, T. D. and C. N. Yang, Statistical Theory of Equations of State and Phase Transitions II, Lattice Gas and Ising Model, Phys. Rev. 87, 410 (1952).

EMBRYOGENESIS THROUGH CELLULAR INTERACTIONS

A. D. J. Robertson
Biological Research Corporation
Lexington, GA 30648

I. Introduction

I was asked to talk about percolation theory and its uses in biology, but this would be presumptuous since the inventor and leading exponent of the theory as far as applications go will (1). Professor Liggett, who follows me (2), and others at this meeting, are clearly more competent to describe this field and more interesting about its ramifications. I shall mention one use of the theory later, but otherwise take a more general approach to the idea of models in biology and their proper uses. I think this will be valuable because, as the fact of this meeting implies, many working as mathematicians or in the hard (lawful) sciences such as Physics would like, and, often, want to know how, to do biology. It is very difficult to tell them. This is because most of biology is not yet lawful; it is our task to make it so. We know that mathematics is required and we know that on the way biology will become a branch of physics, but it is hard to say how.

There has been great success with the smallest biological entities, genes and their molecular counterparts in the chromosome. This was because the gene could be isolated conceptually and its behavior clothed in a mathematical formalism. A different model, that for the structure of DNA, showed a potential isomorphism with the mathematical model (Mendel's Laws) for gene behavior (3). Thus, each reinforced the other (4). This is an important lesson: we know perfectly well that Mendel's Laws only apply to special idealized cases and that, therefore, they are in some sense not a good mathematical model. However, in a pragmatic sense they were perfect: once formulated they allowed a rational search for the carriers of the genes. It was shown that these were most likely to be the chromosomes because they behaved in conformity with Mendel's Laws. An ad hoc model accounting for all the aspects of inheritance of even simple characters would have been so complex as to be both meaningless and almost certainly wrong. In other words, we must not try to explain (or mimic) everything, but only to discover the ways in which matter interacts with matter to produce the growing complexity so characteristic of biological phenomena. There are two levels at which modelling is required: these are the more fundamental one of accounting for the origins and growth of the biological world; and the more local one of accounting for interactions between biological entities which have already formed, as for example in population biology. Of course, it is always possible, and indeed would seem quite 'biological', that the rules for more restricted, local behavior and interactions, might recapitulate or in some sense contain, those for the more fundamental process. I hope this is true because it is much easier to work at higher levels since the laws of physics are not sufficiently well understood that we can account for the formation, at least in terms of its structural stability, of a single atom. We are still reduced to the use of a series of metaphors describing, but incapable of accounting for, for example, the process of nucleosynthesis in a star. In the same way we can recount stories about the development of an embryo, but we cannot account for it, much less for its evolution from a molecular soup or the making of the soup itself. In this situation we must keep a firm grasp on the principles to be elucidated and must ignore, for the time being, complications. Once those principles are understood higher order effects will be easy to account for, just as we can now easily account for recombination frequencies between our once idealized genes.

With these points in mind I therefore want to mention some common pitfalls faced by modellers--and indeed by thinkers in general. These are all old, have all been understood for at least a few thousand years, and all still cause trouble. (This must mean that they arise from quite primitive behavioral mechanisms and that they reflect processes which have selective advantage, just as must apparently dangerous behavior like gambling. It takes an effort of will, not merely of the intellect, to surmount them.)

2. Fallacies

Cotes, in the preface to the second edition of Newton's 'Principia,' (5) described three classes of Natural Philosophers. There were those (e.g., Aristotle, the peripatetics) who . . . 'affirm that the several effects of bodies arise from the particular nature of those bodies. But whence it is that bodies derive these natures they do not tell us; and therefore tell us nothing." This result I call the "naming fallacy." Secondly, there were those who ". . . assume hypotheses as first principles of their speculations, although they afterwards proceed with the greatest accuracy from those principles" (e.g., Euclid). This behavior leads to what I call the 'Hyperborean fallacy." Finally, there were those Experimental Philosophers who assumed nothing and framed no hypotheses (note the change in meaning of this word today), but who, from some select phenoma . . ." deduce by analysis the forces of nature and the more simple laws of forces; and from thence by synthesis they show the constitution of the rest." Clearly this last is the ideal to which we all aspire.

The naming fallacy is pervasive: two examples spring to mind. The first is the identification of cell types in the embryo with specific macromolecular markers on the cell surface. Possession of these markers is held to allow not only intercellular recognition (or snubbing) but also to explain the achievement of a stable pattern of cellular differentiation. The question of how the possession of markers is arranged in the first place is usually not addressed. A discussion of this point of view is given by Moscona (6). Such 'explanation by naming' leads, of course, to a regression, very reminiscent of scholastic proofs of the existence of God. My other example leads to a progression. It comes from the assumption that visual perception can be explained by assuming cells 'higher and higher' in the visual system with increasingly specific properties. This point of view is discussed by Barlow (7); it is fair to say that it is held to widely varying extents by different authors--nonetheless the underlying fallacy is plain and is clearly of the naming sort.

Hyperborean fallacies are so-called because the Delians, presumably to explain the presence of offerings surreptitiously placed by their priests, believed that "certain unspecified sacred objects wrapped in wheat-straw are taken by the Hyperboreans to the confines of their land and handed to their neighbors, who duly hand them on from hand to hand and land to land until they reach Delos. Once, long ago they sent two of their own maidens, Hyperoche and Landike, all the way with the offerings, together with an escort of five men for their safety. But since these did not return they were indignant, and ever afterwards have employed the method described instead of sending envoys of their own. The maidens died and were buried in Delos, where Herodotus saw their tombs." (Quotation from Guthrie (8).) All evidence for the existence of the Hyperboreans was circumstantial, yet numerous scholars, to this day, have sought their identity.

I suppose that the most unpretentious paraphrase of the Hyperborean fallacy is 'jumping to conclusions'. Although it is, perhaps,invidious to single out an example the one I give contains so many other dubious arguments that I cannot resist the temptation. (Weinberger mentions another in his paper in this collection (9).)

In the introductory section of one of his books Chomsky says the following (10). "A theory of linguistic structure that aims for explanatory adequacy incorporates an account of linguistic universals, and it attributes tacit knowledge of these universals to the child. It proposes, then, that the child approaches the data with the presumption that they are drawn from a language of a certain antecedently well-defined type, his problem being to determine which of the (humanly) possible languages is that of the community in which he is placed. Language learning would be impossible unless this were the case." Note especially the last sentence. Perhaps an autistic child is one who thinks in Russian despite his parents insistence on speaking English.

My attitude to these fallacies, and to the confusion which arises from their use, might seem extreme were it not that so much called 'Science' depends on them. This is why I am sure that advances in mathematics always come before advances in the other sciences. A new generalization in mathematics simply opens a field of conjecture to all: it exposes a new part of knowledge (e.g., Non-Euclidean geometry and its reconversion of the then dynamic to a currently geometric Universe) or it allows the analysis, by the non-mathematician, of originally puzzling phenomena (e.g., the non-linear behavior of the action potential). A side benefit, of course, is that the proper use of mathematics requires precise definition. This is known, in microcosm, by anyone who has tried to solve a problem using a computer. To a biologist the experience is salutary, to say the least.

Even though all of us at this meeting no doubt understand the pitfalls I have described it is still possible, as a casual glance at The American Naturalist for the last four years shows, for famous biologists to argue over whether or not the definition of 'fitness' in terms of differential survival by 'natural selection' implies or involves a tautology; and, if so, whether or not experiments designed within the framework of the tautology can produce useful results. Anyone who doubts the merit of my case should read the above journal with care.

3. The Cellular Slime Molds

Cellular slime molds go through a life cycle which contains both free-living amoeboid individuals and multi-cellular 'pseudoplasmodia' (11). After hatching from a spore an amoeba of the best-known species, Dictyostelium discoideum, feeds on bacteria and other micro-organisms which it finds by chemotaxis in its damp environment. (This is an ideal species for an initial foray into experimental biology, for it can be cultured at room temperature on agar and thrives on most of its contaminants.) At 21°C, with abundant food, the cells divide every 3 1/2 hours. After the food is finished they undergo a period of differentiation then aggregate.

The hemispherical aggregate forms a tip, the whole looking like a breast in elevation, and secretes a mucopolysaccharide slime over its outer surface. The slime hardens into a membranous sheath, constraining the growth of the aggregate, caused by amoebae still entering from the periphery, to a more phallic erecting slimy column. This eventually becomes unstable, falls over and crawls off, when it is known, in the vernacular, as the slug. The slug migrates leaving a trail consisting of the collapsed sheath, which it continues to secrete. At the end of migration the slug compacts, rotates through 90° so that its tip becomes again superior, recapitulating its form during late aggregation and secretes internally a vertical cylindrical sheath of cellulose and other minor constituents whose radius is about one-third of that of the whole cellular mass. Cells from the tip pass into the tube, vacuolate and expand, dying, to make a stalk, lifting the remainder of the pseudoplasmodium above the substrate. About one-third of the cells form the stalk and a baseplate; the remainder differentiate into spores which are dispersed by air currents to start the life cycle anew in more favorable environments.

This process is extremely interesting to biologists because it includes examples of many of the developmental mechanisms thought to be important in Metazoan, and, particularly, vertebrate embryos. At the cellular level development can be broken down into the reiteration of six behaviors: cell growth, division, movement, secretion, contact formation and breaking, and differentiation (12). In some embryos, for at least part of development, these behaviors are locally autonomous. In others there is evidence for a more global control and harmonious coördination. Extreme examples of the first sort are called mosaic because loss of a part leads to a permanent loss of adult structure, while extremes of the second kind are known as regulative since loss of a part is compensated, leading to a perfect, though perhaps small, adult. Regulation, leading as it does to the production of form (or patterned differentiation) at least partly independent of absolute size, is felt to be one of the most profound mysteries of development, and it, too, is exhibited by the slime mold pseudoplasmodium. For example, if a slug is cut into thirds each part develops into a fruiting body with roughly correct proportions. Regulation, including the extreme case of regeneration of a missing limb so characteristic of Arthropods and Amphibians (i.e., groups with a well defined metamorphosis between larval and adult stages), has led to an extraordinary proliferation of models for the control of development (12). Fortunately these themselves regulate since they depend on only one or two principles (13)!

I became interested in the slime molds because they were known to aggregate using a chemotactic signal which was believed to function throughout development, possibly controlling morphogenesis in the pseudoplasmodial stages (11). At that time it also seemed probable that cyclic adenosine monophosphate (cAMP) was the aggregative attractant (14). These two features of the slime molds were extremely exciting, the first because the regulative properties of embryos imply the need for intercellular communication and here was a potentially tractable example of intercellular communication in a developmental context (15, 16) and the second because cAMP is almost ubiquitous as an intermediar in humorally controlled cell behavior. Thus (and of course this was pointed out by many (e.g., (14)) here was an apparently ideal system for illuminating one of the most profound problems of embryology with, already, a hint of a continuity of mechanism as would be expected in a conservatively evolving biosphere. The intriguing generalizations which immediately sprang to mind (15, 17) remain to be substantiated, but they appear more and more plausible (18, 19).

Faced with such potential bounty the biologically inclined physicist must model, while the mathematically incompetent biologist wishes he could. I was extremely fortunate to collaborate with Morrel H. Cohen (20, 21) in analyzing the slime mold aggregative system, in a way which subsequently facilitated experimental work. I shan't go through the modelling here; it was limited by lack of information--which indeed the process of modelling revealed--and some of the conclusions were inevitably wrong. Nonetheless the basic framework and the arguments for the form and function of the aggregativ signal were correct (22). Here I shall concentrate on a description of the signal and its functions and mention the role played in our analysis by percolation theory.

Since the aggregative signal is relayed and is coupled to cell movement time-lapse films can record its passage. In fact, signals are released above every five minutes, and are marked by inward waves of cell movement, which may have a circular or spiral form (23), producing patterns similar to those seen in galaxy formation and, similarly, revealing the presence of an attractive instability. The details of signalling and responding are now sufficiently well understood that they can be incorporated into a realistic simulation of aggregation as has been done by MacKay (24).

As I mentioned, the signalling system operates throughout development until the mature fruiting body has been formed (25, 26). While the responses to signalling after aggregation are complicated by additional mechanical constraints, and by a coupling to cellular differentiation, as well as to movement and the secretion of extra-cellular supporting materials, there is no evidence that the signal itself changes. This is extremely important because it not only suggests that complexity arises as the result of reiterated responses to a single, simple message, but is consistent with much of the behavior found when the dominant regions called organizers in Metazoan embryos induce a sequence of cellular responses.

The period of differentiation for aggregation (interphase) begins not simply as the result of starvation but of the specific lack of several amino acids in the extracellular medium. Evidence both from the behavior of wild-type cells and from sequences of differentiations delineated by inference from the properties of aggregateless (or poorly aggregating) mutants (27, 28) suggests the following. Some (perhaps all) cells leak cAMP which is detected by receptors already present, but in small numbers. Simultaneously, and independently, phosphodiestrase (PDE) which converts cyclic to linear AMP, is secreted into the extracellular medium. Since cAMP actually stabilizes the PDE molecule the extracellular PDE activity builds up as the PDE diffuses into the substrate much more slowly than the cAMP. (It is not known how much of the increase in PDE activity is due to new PDE molecules.) The detection of extracellular cAMP leads to an increase in adenylate cyclase activity within the cell as well as in cAMP binding capacity extracellularly. Field-wide signal propagation cannot begin until about six hours after the onset of interphase (29). Thus, a whole sequence of events is dependent on the initial release of cAMP; of course, the rise in intracellular cAMP concentration is a phylogenetically old response to starvation and prelude to differentiation, and, more specifically, to differential gene expression (30). It is noteworthy that we have not been able to advance any of the steps concerned with signalling artifically. The first movable step is the synthesis or exposure of the contact sites responsible for cell-to-cell adhesion front-to-back during aggregation (31). This is dependent on the propagation of the cAMP signal, and is required quickly once the signal propagates, but may be delayed for up to 24 hours when small populations of cells are prevented from aggregation.

The machinery for the pulsatile production of cAMP is therefore completed as a single complex: in other words, there can be no half-measures and signal propagation occurs only when the interaction between synthesis and destruction can reach a catastrophic instability.

Normal aggregation begins nine to ten hours after amoebae are deposited on non-nutrient agar. All the components of the signalling and response systems are therefore ready by the time the first cells begin to signal (32). These have been called autonomous cells. there is no evidence that they are a special sub-set of the amoebae. Most probably they are simply those cells with the lowest thresholds for the synthesis of cAMP in response to cAMP stimulation and are thus set off by the ambient cAMP; results of several authors (19, 33) suggest that a concentration of 10^{-9} M is sufficient to excite the most sensitive cells, and that the threshold itself, or the time taken to become most sensitive, is a random variable. Once the first cells start to signal in a sufficiently dense field the question becomes academic and signals are amplified and propagate. Above a critical density of cells wave propagation can be sustained; below it, it cannot, and aggregation begins purely by chemotaxis. When the medium is excitable and above critical density the signal can propagate for large distances, so aggregation territories are large and fruiting body density low. Thus, there is a bifurcation into regimes in which aggregation has different characteristics because the signalling mechanisms are different (34).

The critical density has turned out to be a surprisingly difficult, if not suspect, concept for many biologists. Herein lies a moral: even though the critical density arises naturally from the properties of the amoebae and can be demonstrated experimentally in several ways it is too abstract for many and is therefore assumed to be a modelling artifact. Unfortunately, suspicion of biological modelling is, on the average, well-founded and quite difficult to counter. I have even heard Hodgkin and Huxley's extraordinary feat of experimental and theoretical analysis dismissed as 'mere' curve-fitting.

In any case, the critical density arises because the range of the cAMP signal is fixed. The signal itself is all-or-nothing in nature, containing $\mathcal{N}$ molecules, and there is a threshold concentration, C*, for its stimulation. Signal range, R, is thus characterized by the parameter $C^*/\mathcal{N}$. For amoeba densities, N (cells/unit area of culture), below N*, the critical density, signal concentration decays below C* before a neighboring cell is reached. That is, on the average cells are more than R apart. For $N > N^*$ there exist percolation channels, extending throughout the field, within which amoebae are close enough together to allow signal relaying. Thus, as N passes through N* aggregation morphology changes drastically from one of many small aggregation territories, containing small fruiting bodies, to one of relatively few, large aggregates. If one carefully varies cell density in steps and plots the dependence of fruiting body number on cell density a sharp discontinuity emerges (35). The function at first rises, but, at the critical density, drops abruptly. If one is careless in plating out a sequence of cell densities a 'ringing' in the function--an abrupt increase in variance--is seen around the critical density. (Actually, this can be useful in locating a critical density crudely in a quick experiment which can then be repeated more carefully.) The critical density is easy to measure accurately, at least for D. discoideum, since it is easy to count large numbers of fruiting bodies.

At the critical density, since sensitive amoebae are distributed at random throughout the field, percolation theory must be used to calculate signal range: thus $\pi R^2 N^* = 4.5$ (36). It is only at the critical density that signal range can be measured so exactly and it is percolation theory which makes this possible. Signal range having been calculated, many more experiments are possible. For example, we wanted to know the proportion, $X_2(t)$, of cells, in a differentiating population of amoebae, capable of relaying a signal. This was found by Gingle in an ingenious series of exponents (35). He measured the times at which signal propagation could first be induced for different densities of amoebae, using a microelectrode releasing cAMP as an external stimulus. Thus he obtained a time at which the relaying-competent cells became sufficiently dense to relay the signal. At each onset time, t*, the fraction of relaying-competent cells in a field of total cell density N is $X_2(t^*) = N^*/N$. Repeating the experiment with fields of different cell densities, each with a characteristic t*, gives the function $X_2(t)$ and is a direct measure of a process of cellular differentiation. However, there is a complication: critical density is a function of total cell density through the dependence of signal range on PDE activity, itself a function of cell density. The PDE activity must be measured and its distorting effect calculated. To do this, Gingle, diluted populations of wild-type amoebae with cells of a mutant (Dl) which could not relay the signal, but which produced PDE. Thus, at a fixed total cell density he could find the ratio of wild-type to mutant densities at which critical density could be reached at its limiting time, that is when $X_2(t) = 1$, or about 10 hours. Since the wild-type amoebae alone contributed to the relayed signal it is their density which is being measured, and being varied independently of PDE activity.

It is noteworthy that the enzyme activities measured by this technique were much higher than those then in the literature. This, we found, was because the assays used were most unfavorable to recovering all the PDE activity. Direct measurements of PDE under conditions consistent with aggregation confirmed Gingle's results (22). The critical density technique (and thus percolation theory) has some quite surprising further applications. For example, suppose that a cell type from a vertebrate embryo aggregates (or does something else) and the behavior is controlled by an unknown propagated signal. The signal can be identified tentatively by discovering reagents--e.g., enzymes--which displace the critical density through their effects on signal range. Gingle applied this technique to cells from the early chick embryo, finding that they appeared to propagate a cAMP signal (37); this was confirmed by direct assay (38).

During D. discoideum aggregation the signal is coupled to movement, which involves actin and probably myosin, but nothing is known of the coupling, although, of course, it requires the presence of cAMP receptors, which could well be, on the intracellular side of the membrane, directly concerned in the contractile mechanism. The signal has other significant features (22). It is brief, at the level of the single cell, it is pulsatile, containing about 3×10^7 cAMP molecules, and it is self-limiting since there are both thresholds and lintels for signal production. At or above 3×10^{-7} M signal production is cut off. Thus, the cAMP receptor-synthetic system has an inherent periodicity when coupling across the cell membrane occurs. This suggests that the periodic signal is not driven by a separate oscillator, but can oscillate independently under the right conditions. Interestingly enough, chemotaxis shows a concentration dependence with a similar shape (39) but the space between threshold and lintel spans 5 or more orders of magnitude in cAMP concentration as compared to perhaps 2 1/2 for signal relaying. Of course, the suppression of chemotaxis represents receptor saturation rather than a cut-off in cAMP synthesis, so it is not surprising that the kinetics are different. A cut-off of cAMP synthesis or release is essential--otherwise the system would 'explode'.

As a sufficient density of amoebae is reached in the center of the aggregate the global properties of the signal change. A nipple-shaped tip is elevated above the center, away from the substrate, and it releases a continuous cAMP signal (40). Since the signal is continuous, and above the relaying threshold, all the remaining, more peripheral, amoebae are entrained and the tip continues to act as though it were an organizer, defining the developmental axis of the pseudoplasmodium for the remainder of morphogenesis. This property of the tip comes directly from the response function for cAMP reduced cAMP release. As soon as the ambient cAMP concentration can rise faster than it can be reduced--by PDE and diffusion--then cells in the region are stimulated to fire when they are sensitive and when the ambient concentration is within the sensitivity window. Since the signal is brief, and the signalling periods and thresholds are random variables, then firing of individual cells becomes unsynchronized and the tip region approximates a continuous source. Thus the tip defines a reflecting boundary, cAMP being able to diffuse away from it only towards the substrate. Close to the edge of the tip the cAMP concentration will be maximal; it will fall with distance from the tip surface until the balance between cAMP synthesis and removal is reversed, when a region in which periodic, entrained signals are possible forms. The beauty of the system is that, with no extra properties, the signalling system developed by amoebae before they aggregate evolves, when coupled to movement, to produce a spatial distribution of cells with an 'organizer', a developmental axis, and two regimes of steady cAMP concentration.

The later stages of development include formation and migration of the slug, the erection of the fruiting body and the differentiation of stalk and spore cells. I shall not discuss these in detail since good descriptions are available; I simply reiterate my conviction that the mechanical constraint of the slime sheath surrounding the slug allows migration by means of the coordinated movements of its constituent amoebae, and that similar, simple changes in mechanical properties will suffice to explain fruiting body erection. We already know that the signalling system operates throughout development, and there is good evidence that the two different regimes of cAMP concentration in the slug are responsible for the regulated pattern of stalk cell and spore formation, the stalk cells differentiating in the high cAMP concentration region (41, 42).

4. Observations on Embryos

Despite the bewildering variety of embryonic and adult phenotypes the tremendous intellectual crystallization which resulted from the study of comparative anatomy led, even before Darwin, to the realization that only a few simple plans of development exist. One is struck, when looking at the variety of methods for gastrulation, for example in Coelenterates, by the fact that the number is small and the range the same as in other Phyla (43). Obviously details vary tremendously, but there are only a few ways in which one geometry is easily transformed to another with the preservation of topology. That is, the ways of converting a hollow sphere to a tube with several layers and one or more orifices are, in fact, severely limited. Thus, while all are used, not too much reliance should be placed on the method of gastrulation as a phylogenetic marker, except possibly in comparisons between closely related orders. The problem then becomes one of explaining how the few common mechanisms might work and getting experimental evidence to support one (or more) of the explanations.

At a more general level the difference in early development between regulative and many mosaic embryos is that, as one might expect, in the mosaic case cytoplasmic localization is preserved by spiral cleavage (44, 45). Radial cleavage leads to a pattern of blastomeres relatively unstable to perturbation, a trait clearly correlated with the capacity for regulation; in spiral cleavage each daughter blastomere fits into a crevice between two others already present--another example of the geometrical stability of helices and their superior capacity for retaining information.

In regulative embryos cell movement becomes vitally important during geometrical transformations, for example gastrulation. I cannot give an exhaustive review, but again simply want to stress that directed cell movement plus mechanical constraints can account for much morphogenesis as was clearly shown by His a century ago (46). How are movements directed? Here, the major generalization is due to Spemann (47) who showed the pre-eminence of the organizer in defining, as a result of directed cell migration during gastrulation and subsequent inductive interactions, the embryonic axis. It is noteworthy that, without the mechanical constraint of a vitelline membrane, exogastrulation occurs, and that with experimentally induced changes in the mechanical constraints offered by the yolk multiple ingressions can occur. Thus, mechanical factors are again important for defining gross features of morphogenesis, while not necessarily responsible for initiating or controlling cellular movement and differentiations caused by subsequent inductive interactions.

Unfortunately, while there is much descriptive material, almost nothing is known about the causes of cell movement, or, indeed about the chemical (or other) nature of signals released by organizers. It was, therefore, a provocative discovery that the amphibian organizer could attract slime mold amoebae, and do so better than any other part of the embryo (48). While this experiment must be repeated, and while much more evidence will be required before we can say that organizers operate through the release of small molecules, the outlook is at least promising since the most significant feature of the signal released by organizers is that it is not species specific, but can cross the line between orders, classes and possibly even Phyla (15, 16). Thus, in analogy with the slime mold signal, the message is the same but the interpretation varies. A word of caution is in order: while this conclusion is consistent with the experimental data it is based on inference; so far, no one has been able to identify an inducing substance or to show, unequivocally, that organizers act through chemical signalling. Proof will require both the identification of the putative substances and the manipulation of their activity to produce predictable changes in development. In any case, such results led to some of our recent work and to a realization that there was considerable supporting evidence in the literature. I have reviewed some of this evidence (18), and for the moment I shall restrict myself to observations on the early chick embryo, and on cells derived from it.

In the chick embryo cell movementd during gastrulation and the formation of the organ rudiments have been shown to be associated with wave propagation suggesting that periodic signalling occurs and may be involved in the control of development. During the first two days of development cells isolated from the embryo share many properties with slime mold amoebae, in particular the capacity to amplify a cAMP signal. Most striking are the concentration dependence and brevity of the response. There is a threshold and lintel, the pulse contains up to 10^8 cAMP molecules, and is released within at most 5 sec, probably less than two. There is an extracellular PDE, both in the intact embryo and in the supernatant of disaggregated cells, and, of great significance, the PDE activity is enhanced by Lithium ions which are known to have teratogenic effects on many regulative embryos. We have found that some of the most easily observed effects, for example the inhibition of somite formation, can be reversed with PDE inhibitors, further evidence that a cAMP signalling system is important in the intact embryo (49). These results led to the discovery that the primary axis of the intact embryo could be diverted by a micro-electrode releasing cAMP, that this effect had a threshold of about 10^{-8} M and that individual cells were attracted to the micro-electrode with a lower threshold for response (19). In addition, a narrow range of cAMP concentrations enhances the agglutination of cells disaggregated from the one day embryo, suggesting again that a propagated cAMP signal is coupled to a cellular behavior, in this case contact formation (37). While these results are exciting it will take a long time to demonstrate whether or not cAMP, and possibly other small molecules, act to control cell behavior extracellularly in the chick embryo during normal development. Perhaps the best hope comes from my observation that primordial germ cells migrate prematurely in embryos exposed to PDE inhibitors and that they are attracted to micro-electrodes releasing cAMP just above the ventral surface of explanted embryos.

5. Conclusions

What can be learned from the examples given in my rather rambling talk? Firstly, I think that some conclusions about development in general are possible. I shall mention these and then finish with a short statement of what seem to me to be important gaps in our knowledge and understanding. The first suggestion comes from a question: "How did development evolve?"

This must go back to the origin of the Metazoans since development is found in all the Metazoan Phyla. This suggests that multicellular organisms must almost always arise from single cells by a developmental process rather than by, say,accumulation of some cells to form buds, as in Hydrozoans, which is in fact a surprisingly rare mechanism. Certainly the benefits of sexual reproduction would be hard to realize other than through the formation of a zygote. In any case the most likely explanation is that the mosaic and regulative patterns of development seen today are reminders, living fossils, of two of the ways in which the original Metazoans formed: cellularization of an individual and accumulation, by aggregation, of many similar cells. Clearly the second process must be regulative since it would be difficult to guarantee membership of the final organism to any individual cell, and it must involve signalling to allow aggregation. One may speculate that, as is seen in the slime molds, the signal detection apparatus existed as part of a food seeking device and became linked to intercellular communication as a basis for avoiding death by starvation. In the soil, there are indeed amoebae which simply cluster for 24 hours while the food supply regenerates itself: the predator-prey cycle is broken. It is easy to imagine how a more complex process leading to spore dispersal was added to this primitive pattern of aggregation, particularly since some soil amoebae can form cysts independently under adverse conditions (50) and this behavior has been found in one of our mutants of D. discoideum called P2.

Signalling itself becomes much more efficient when the signal is propagated as a pulse over relatively long distances, as in D. discoideum. A larger, more complex, organism can form, and control can be more precise. Those slime molds without a propagated signal have a very simple morphogenesis (e.g., D. minutum) (51) and there is evidence that those with branching fruiting bodies have two signal molecules, one used for aggregation, and the other, cAMP, for multicellular morphogenesis (52). While the identity of the aggregative attractant for two of these species, in the genus Polysphondylium, is not known some feel it is a small peptide, while Dr. J. Grutsch has told me that there is evidence that it is acetyl-choline or a very closely related molecule. The amoebae certainly respond dramatically to acetyl-choline and show cholinesterase activity, while they do not respond well to cAMP until after aggregation. However, cAMP leaks out early, as in D. discoideum, and is probably responsible for mediating the effects of starvation, setting differentiation for aggregation in train and, later on, for controlling stalk cell differentiation (53).

It is likely, therefore, that regulative development is controlled by extra-cellular signals and that the molecules used are those which both controlled aggregation early in evolution and do so today in simple organisms. Furthermore, the natural candidates for the molecular species are neural transmitters, small peptides and some hormones which are widespread in phylogeny. The more ubiquitous, but peculiar, a small molecule, the more likely it is to be used, which gives cyclic nucleotides an edge, but makes neural transmitters quite probable, as morphogens. A further advantage of cAMP is the large amount of energy released on hydrolysis, since a molecule with important signalling functions should be easy to destroy (54). In a rather similar statement of his beliefs Haldane (55) emphasized that steps in an extracellular signalling chain might become internalized, making the evolutionary history of a developmental control mechanism harder to unravel. He gives examples, and it is, indeed, easy to see how an originally extracellular mechanism could become more and more specialized during evolution with a gradual change from overt extracellular signalling to signalling between cells coupled by tight junctions, or simply coupled electrically.

There are now many publications containing data supporting the role of neural transmitters in early Vertebrate and higher Metazoan development; unfortunately the data provide associations only--for example between the phases of gastrulation and acetylcholine and serotonin activities in sea urchins (56). The difficulty is that authors rarely suggest a mechanism by which the molecule concerned acts--in other words no testable hypothesis emerges. For this reason I propose that the most profitable way of attacking this problem is a combination of visual observation, especially using time-lapse filming, and of screening tests which will reveal whether or not extracellular signals exist; if they do, what aspects of cellular behavior they are coupled to; and, finally, whether or not they are propagated. Such methods allow positive identifications of the molecules concerned. We have already applied them to cells from the early chick embryo. Two corollaries of such a program will be that an objective basis for the action of some teratogens will be found--for example, we have shown the Lithium ions enhance PDE activities and thus interfere with cAMP metabolism and signalling--and evidence for the nature of organizer activity will come as we learn to manipulate development of the intact embryo, for example with a micro-electrode source of a morphogen.

It is never obvious, except to hind-sight, which are the mathematical tools essential for advance. However, it is clear that an understanding of non-linear interactions in finite systems with complicated boundary conditions is needed in embryology. Thom's formalism might seem appropriate, yet it has had startlingly few successes, especially in the realm of biological morphogenesis where it should be most applicable. Perhaps the difficulty is again one of erecting testable hypotheses rather than post hoc descriptions of geometrical transformations. On the other hand, the mathematics of linear systems such as passive filters is well understood yet often not applied to biological systems where it is clearly appropriate. A good example comes from the slime molds. Calculations show, and common sense demands, that the signal from a single cell must be brief for effective communication. Its time constant must be shorter than those of functions it controls, such as movement, and shorter than characteristic diffusion times (20). Nonetheless, results from many laboratories appear to show signals which are minutes long--too slow by several orders of magnitude. The explanation is simple: cAMP output is assayed from populations of many (up to 10^8) cells in suspension and poorly synchronized. Samples are taken at various intervals of up to a minute. Well established theorems set clear limits to the brevity of a response which could be measured under these circumstances. In other words high frequencies are being smoothed out and lost in a series of low pass filters. Here is an example of undue faith in observation without sufficient examination of the methods used (22, 33).

Another clear danger that I have mentioned is naming things whose existence is not known independently. Positional information is an example. Giving it a name immediately conjurs up a vision of a coordinate system, already in place, which may be quite misleading. At the very least it becomes more difficult to notice alternatives. Indeed, the author of the term has stretched its meaning as counter-examples have emerged until, to the sceptic, all that is left is the name. This is not to say that we should not model: clearly we must have ways to order phenomena and to examine their interactions. After all our brains themselves make models of the environment which can be manipulated and experimented with internally (57). But this modelling process, like that of the successful scientist, does not depend on mere re-description but includes ways of ensuring that what we think, if not necessarily correct, is at least feasible.

References

1. J. R. Hammersley. This volume.

2. T. Liggett. This volume.

3. J. D. Watson and F. H. C. Crick. Nature 171 (1953), 737 and 964.

4. E. A. Carlson. The Gene: A Critical History. Saunders, Philadelphia, 1966, pp. 301.

5. R. Cotes. Page XX of the Preface to the 2nd edition of Newton's Principa, University of California Press, Berkeley; reprinted 1964.

6. A. A. Moscona, ed. The Cell Surface in Development. John Wiley, New York, 1974.

7. H. B. Barlow. In Current Problems in Animal Behavior, W. H. Thorpe and O. L. Zangwill, eds. Cambridge University Press, Cambridge, 1961, pp. 331-60.

8. W. K. C. Guthrie. The Greeks and Their Gods. Beacon Press, Boston, 1950.

9. H. F. Weinberger. This volume.

10. N. Chomsky. Quotation from p. 27 of Aspects of the Theory of Syntax. M.I.T. Press, Boston, 1965.

11. J. T. Bonner. The Cellular Slime Molds. Princeton, 1967.

12. A. Robertson and M. H. Cohen. Ann. Rev. Biophys. Bioeng. 1 (1972), 409.

13. A. T. Winfree. This volume.

14. T. M. Konijn, D. S. Barkley, Y. Y. Chang and J. T. Bonner. Am. Nat. 102 (1967), 225.

15. L. Wolpert. J. Theor. Biol. 25 (1969), 1.

16. A. Robertson. In How Animals Communicate, Thomas F. Sebeok, ed., Indiana University Press, 1977, pp. 33-44.

17. B. C. Goodwin and M. H. Cohen. J. Theor. Biol. 25 (1969), 49.

18. A. Robertson. J. Embryol. Exptl. Morphol. 50 (1979), 155.

19. A. R. Gingle and A. Robertson. J. Embryol. Exptl. Morphol. 53 (1979), 353.

20. M. H. Cohen and A. Robertson. J. Theor. Biol. 31 (1971a), 101.

21. M. H. Cohen and A. Robertson. ibid. (1971b), 119.

22. J. F. Grutsch and A. Robertson. Dev. Biol. 66 (1978), 285.

23. B. M. Shaffer. The Acrasina, Part I. Adv. Morphogen 2 (1962), 109.

24. S. A. Mackay. J. Cell. Sci. 33 (1978), 1.

25. J. T. Bonner. J. Exp. Zool. 110 (1949), 259.

26. J. Rubin and A. Robertson. J. Embryol. Exptl. Morphl. 33 (1975), 227.

27. M. Darmon, P. Brachet, and L. H. Pereria da Silva. Proc. Natl. Acad. Sci. USA 72 (1975), 3163.

28. R. P. Yeh, F. K. Chan, and M. B. Coukall. Dec. Biol. 66 (1978), 361.

29. A. Robertson, D. Drage, and M. H. Cohen. Science

30. I. Pastan and S. Adhya. Bact. Rev. 40 (1976), 527.

31. G. Gerisch, H. Fromm, A. Heusgen, and U. Wick. Nature 255 (1975), 547.

32. R. K. Raman, Y. Hashimoto, M. H. Cohen, and A. Robertson. J. Cell. Sci. 21 (1976), 243.

33. P. N. Devreotes and T. L. Steck. J. Cell. Biol. 80 (1979), 300.

34. Y. Hashimoto, A. Robertson, and M. H. Cohen. J. Cell. Sci. 19 (1975), 215.

35. A. Gingle. J. Cell. Sci. 20 (1976), 1.

36. V. K. Shante and S. Kirkpatrick. Adv. Phys. 20 (1971), 325.

37. A. R. Gingle. Dev. Biol. 58 (1977), 394.

38. A. Robertson, J. F. Grutsch, A. R. Gingle. Science 199 (1978), 990.

39. T. M. Konijn. Antibiotics and Chemotherapy 19 (1974), 96.

40. A. Robertson. The Biology of Brains, W. B. Broughton, ed., Blackwell, Oxford, 1974, pp. 1-10.

41. I. N. Feit, G. A. Fournier, R. D. Needleman, and M. Z. Underwood. Science 200 (1978), 439.

42. J. T. Bonner. Proc. Nat. Acad. Sci. USA 65 (1970), 110.

43. G. Reverberi. Experimental Embryology of Marine and Fresh Water Invertebrates. North Holland, Amsterdam, 1971.

44. J. N. Cather. Adv. Morphogen. 9 (1971), 67.

45. D. T. Anderson. Embryology and Phylogeny in Annelids and Arthropods. Pergamon, Oxford, 1973.

46. W. His. Unsere Korperform. Vogel, Leipzig, 1974.

47. B. M. Shaffer. The Acrasina, Part I. Adv. Morphogen. 2 (1962), 109.

48. V. Nanjundiah. Exptl. Cell. Res. 86 (1974), 408.

49. Unpublished data of A. R. Gingle, J. F. Grutsch and the author.

50. E. N. Willmer. Cytology and Evolution. Academic, London, 1970.

51. G. Gerisch. Curr. Top. Devl. Biol. 3 (1968), 157.

52. M. E. Jones and A. Robertson. J. Cell Sci. 22 (1976), 41.

53. J. F. Grutsch, private communication.

54. M. E. Jones. Aggregation in Polysphondylium. Ph.D. thesis, The University of Chicago, 1974.

55. J. B. S. Haldane. Annee Biol. 58 (1954), 89.

56. T. Gustafson and M. Toneby. Exptl. Cell. Res. 62 (1970), 102.

57. K. J. W. Craik. The Nature of Explanation. Cambridge University Press, Cambridge, 1943.

BIOLOGICAL INTERPRETATION OF A RANDOM CONFIGURATION MODEL FOR CARCINOGENESIS

Petre Tautu

German Cancer Research Center
Heidelberg

Since 1973 a new stochastic model for carcinogenesis has been developed (Tautu, 1974; Schürger and Tautu, 1976 a,b, 1977,1978) by introducing some biological assumptions rather different from those commonly considered in mathematical cancer models. It is the aim of this paper to give a detailed introduction to the biological assumptions leading to this model as well as to comment on some biological consequences inferred from the mathematical development.

1. Cell population structure. The first basic assumption is that there is a stability of the arrangements of cells in tissues which will be mathematically represented by a d-dimensional ($d \geq 1$) infinite square lattice on which our process of normal and cancerous growth and spread takes place. Cells are located at the nodes (sites) of this lattice, and thus each node can be specified as 'occupied' or 'vacant' (observe that the set of nodes is countably infinite). The dimension of our square lattice plays an important role. Our infinite square lattice system can be considered as the limiting case of a lattice system with a finite but very large number of components.

Lattice systems are already known in statistical mechanics and polymer chemistry, e.g. lattice gases, lattice crystals, etc. Roughly speaking, the main objective of statistical mechanics does not differ from our objective, as regards the definition of macroscopic variables or properties in terms of underlying microscopic features. The reader interested in possible relations between statistical mechanics and theoretical biology is referred to M.A.Garstens (1969) and W.M. Elsasser (1970). In most ecological applications a regular array of plants (Cochran, 1936; Freeman, 1953) is a rectangular two-dimensional lattice; furthermore, a cell membrane is often represented as a lattice graph (Hill and Kedem, 1966; Zand and Harary, 1972). An equivalent of the two-dimensional square lattice is the 'tessellation' known in the theory of cell automata, which is a two-dimensional space subdivided into square cells of equal size, like the squares of a checkerboard (Moore, 1962).

The use of a square lattice instead of a hexagonal one does not significantly influence the results. T.Williams (1971) who preferred the hexagonal lattice in his model for skin cancer, has noticed that the numerical differences are related to the degree of connectedness in moving from the triangular to the square or the hexagonal lattice.

The d-dimensional square lattice - which we denote by Z^d - is given by $Z^d=\{x|x=(x^1,\dots,x^d),\ x^i \text{ integer}, 1\le i\le d\}$, $d\ge 1$. The following assumption introduces a certain regularity in the spatial distribution of cells:

A1. The possible locations of cells are given by the nodes belonging to Z^d. No site of Z^d can be occupied by more than one cell at a time.

Assumption A1 expresses the rule of exclusion of multiple occupancy whose consequences as regards the emergence of order in a growing tissue will be examined in the sequel. The spatial connection between cells will be defined with the aid of the concept of 'neighbourhood'. Let us adopt the most simple definition according to which two sites, x and y, of Z^d are neighbours if they are unit distance apart (in this case, we write $x\sim y$).

Thus we have

A2. A site $x=(x^1,\dots,x^d)\in Z^d$ is a neighbour of a site $y=(y^1,\dots,y^d)\in Z^d$ iff $\|x-y\|=1$, where $\|\cdot\|$ denotes the Euclidean norm. The set of all neighbours of $x\in Z^d$ is denoted by $N_x=\{y|y\sim x\}$.

In a previous paper (Schürger and Tautu, 1976a) we have defined N_x by the aid of H, the neighbourhood index (stencil), which is used to define a correspondence between a cell $x\in Z^d$ and its neighbours. Let $H=\{\nu_1,\dots,\nu_m\}$, where $\nu_j=(h_j^1,\dots,h_j^d)$, $1\le j\le m$; h_j^i integer, $1\le i\le d$. Then $N_x=\{x+\nu_1,\dots,x+\nu_m\}$ where $x+\nu_j$ equals $(x^1+h_j^1,\dots,x^d+h_j^d)$. This is the obvious formalism employed in the theory of cell automata (see, e.g., Yamada and Amoroso, 1969; Aladyev, 1974). The definition of H has an important consequence: since the neighbours of each cell on the lattice are defined by the same H, the 'relative positions' of the neighbouring cells with respect to any cell on Z^d can be thought of as being the same on the whole lattice. We obtain a uniform interconnection pattern among the cells of this idealized tissue. The above spatial connection is generally meant to be equivalent with a functional connection between cells. For instance, neighbours of a cell at site $x\in Z^d$ are those from which the cell directly receives information (signals). Thus, the neighbourhood may be interpreted as the uniform spatial organization of a cell-to-cell communication. The contact between neighbouring cells (or some specialized molecules on their surfaces) is the present-day accepted mechanism for growth regulation and the transmission of positional information (e.g. McMahon, 1973).

If the neighbourhood relation is defined by using H, it is supposed to be symmetric, i.e., $x \in N_y$ implies $y \in N_x$, $x, y \in Z^d$. If the neighbourhood relation is an equivalence relation, it determines a partition $\{\Lambda_o, \Lambda_1, \Lambda_2, \ldots\}$ of Z^d, called a lamination of Z^d (Yamada and Amoroso, 1969). This partition decomposes the lattice structure into independent laminated substructures Λ_i acting in parallel. The consequences of this assumption might be important in neurobiology.

2. Growth mechanisms. The second basic assumption is that each cell on the lattice has a finite set of internal states $S=\{1,\ldots,s\}$. In this case, we identify the elements of S with the discrete phases of the cell cycle (i.e. G_1, S, G_2 and M phases). A cell begins its life in phase 1 and completes it at the end of state s (mitotic phase) when producing two daughters. The crucial control events for the regulation of growth seem to reside in the G_1 phase of a cell. In mid- or late G_1, the cell decides whether to initiate DNA synthesis and undergo division or to stop proliferation. This should be the sensitive phase for the action of many chemical carcinogens (e.g. Warwick, 1971; Bertram et al., 1975) whereas the DNA-synthesis phase seems to be sensitive for x-ray carcinogenic action (e.g. Terasima and Tolmach, 1961; Tolmach et al., 1971), and finally G_2, the sensitive phase for carcinogenic viruses (e.g. Basilico and Marin, 1966). A mathematical model for carcinogenesis must take into account a differential cell sensitivity to carcinogens which primarily initiates a non-homogeneity into the given cell population. The same holds also good for the action of cycle-specific antineoplastic drugs (e.g. Valeriote and van Putten, 1975), particularly the mitotoxic and oncotoxic agents in the classification of T.C.Hall (1975).

Let us now suppose that at the end of its lifespan (state $s \in S$) a cell located at site $x \in Z^d$ gives birth to two daughter cells. In accordance with assumption A1, one of these two new cells must leave its parental site. The behaviour of this movable cell is assumed to be the following:

A3. A daughter cell born at $x \in Z^d$ performs a positional shifting and is located at a site of the neighbourhood $\{y | y \sim x\}$ of x. Each possible site is chosen with probability $(2d)^{-1}$. Two possibilities can occur: (i) the site $y \in N_x$ is vacant and the daughter cell occupies it, or (ii) the site $y \in N_x$ is occupied and the daughter cell discards the inhabitant.

This assumption is complicated by the joint action of different biological processes. Firstly, it takes into consideration a 'stationary' cell motility (Sträuli, 1979), the daughter cell having the competence to move only to its original neighbouring places. It is known that a decrease of cell attachment is a regular event during cell division (e.g. Sträuli and Weiss, 1977). Such a movement can be paralleled with the 'local rearrangement' identified in developing systems by A.Robertson and M.H.Cohen (1972). The dislocation of the old inhabitant can be easily depicted in the following way: the spatial growth takes place on k (>1) parallel, stratified two-dimensional lattices; the first one may be thought of as a 'basal' lattice from which 'old' cells are pushed up on the next lattice. The cell flow goes forward to the last (k-th) lattice. The 'pushing up' process creates a transit cell subpopulation; this may be evidenced by analyzing, e.g., the proliferative cell populations of skin or small intestine (e.g. Iversen et al., 1968; Potten, 1978). For example, in the mouse small intestine, k should be equal to 10 (i.e., about 10 cell layers deep: Potten, 1978) and the dislocation process may occur at random, a hypothesis put forward by N.J.Nadler (1965) for esophagus (see also Eaves, 1973). However, for skin the dislocation process may be cycle-dependent: in their cybernetic model for the population kinetics of epidermal cells, O.H.Iversen and R.Bjerknes (1963) considered that the oldest cell in the G_1 phase adjacent to a mitotic cell is pushed up.

The killing action assumed in A3 is actually analogous to the complex process of cell loss which is a vital factor in cell kinetics analysis (Steel, 1968). The production of new cells is balanced by the loss of an approximately equivalent number of cells; besides the dislocation mechanisms, there also are intrinsic (e.g. aging, cell abnormalities) and extrinsic (e.g. host-immune defenses) mechanisms (Cooper et al., 1975). So far, as it concerns 'killing' itself, the term 'apoptosis' has been created (Kerr et al., 1972) in order to designate the events involving the break up of a cell followed by its ingestion and digestion taken by its neighbouring cells. Many normal as well as malignant cells produce and release proteases -

e.g. elastase, collagenase, plasminogen activator, etc. - which are meant to be causal factors of growth control and invasion. For more details the reader is referred to the book edited by E.Reich, D.B. Rifkin and E.Shaw (1975). In the cell automata theory the death of old elements is also supposed; for instance, to his recursive definition of construction of new elements, S.Ulam (1962) adds the rule of erasing all elements which are k generations old (mainly k=3). In constructing his cellular automaton designed as a 'game of life', J.H. Conway introduced the assumption that a living cell at x will die just in case $N_x>4$ or $N_x<3$ (Gardner, 1970,1971,1972; Baer and Martinez, 1974; Dresden and Wong, 1975).

Assumption A3 suggests the concept of 'invasivity'. Long ago, the belief that invasivity is an "outstanding" feature of malignancy has been unanimously accepted. Recent investigations, however, do not support this view (see Mareel, 1979), showing for instance, that there are various degrees of invasiveness (e.g., within the same tumour, invasive and non-invasive areas are present) and also a spatial condition (e.g. invasion in a three-dimensional culture but not in a monolayer). The biologists and biomathematicians have two important tasks: (i) give a clear, precise and unambiguous definition of invasivity in the context of spatial competition of multitype populations, (ii) find, define and prove the specific characteristics - if any - of malignant invasion. It was observed that slowly proliferating tumours may be highly invasive and that an increased mitotic rate is not a sign of invasivity (Rubio, 1978). These facts should apparently be in contradiction with our assumption A3 because of a suspected confusion between local (intercellular) invasion (=contact-induced spread) and colonization. Sometimes, however, invasion in the pathologist's meaning could imply a selective orientation and a velocity which cannot be explained by the spatial organization of growth. Our hypothesis considers only a neighbourhood interaction (competition) process having global proliferative effects. Moreover, the assumption that a daughter cell born at $x \in Z^d$ chooses all neighbour sites $y \in N_x$ with equal probability obstructs orientation or polarity effects. The process takes place in an isotropic, homogeneous microenvironment.

3. <u>Metamorphose mechanism (the natural history of a malignant clone)</u>. The third assumption in our model is that carcinogenesis might be a multi-stage process. This hypothesis is unanimously accepted although no precise definition of 'stage' is given. After a careful

analysis of the crude approach of a two-stage mechanism, e.g. "initiation-promotion", one can gather that in fact each of both stages is itself a multistage process. In a previous paper (Schürger and Tautu, 1978) we have illustrated the multi-stage carcinogenesis as a reversible chain of cell events:

$$\text{normality} \rightleftarrows \text{induction} \rightleftarrows \ldots \rightleftarrows \text{promotion} \rightleftarrows \ldots \rightleftarrows \text{malignancy.} \qquad (1)$$

Scheme (1) exemplifies two basic assumptions, namely (i) there is a linear sequence of discrete cell metamorphoses specifying the 'stages', (ii) each 'stage' may be reversible. We have

A4. The stages of metamorphoses of a target cell and its offspring from normal to malignant are represented by elements from a finite set $K=\{1,\ldots,k\}$. The elements of K are called 'types' or 'colours'. A cell of type $i \in K$ makes visible at the end of its mitotic phase $s \in S$ its metamorphosis to type $i+1$ or to type $i-1$. The division of this cell is symmetric so that both daughter cells will have the same type $i' \in \{i-1,i,i+1\} \cap K$.

Although assumption A4 is unambiguous, some biological comments are necessary.

(a) A4 gives no suggestion about the genetic/epigenetic events leading to a colour change. The set K is thought of as suitable for depicting the genetic instability and the resulting phenotypic heterogeneity. Consequently, the term 'metamorphosis' is chosen in order to avoid any genetic connotation (e.g. somatic mutation, mutation-like event, transformation, etc.).

(b) The linear scheme (1) could be accepted as the simplified approach to multi-stage carcinogenesis. It is an elementary reconstitution of a complicated process, which neglects other possible alternatives. Careful studies on hepatocarcinogenesis (Bannasch, 1978) show that a single type of precancer cells does not exist: most tumour cells develop from the clear glycogen storage cells, but other precursors (e.g. acidophilic and large X-cells) can have malignant offspring, too. For those biologists who represent differentiation as a continuous process (e.g. Seglen, 1976), our discretization in (1) might be inadequate. However, when observing a given differentiation, we must realize that it is rather arbitrarily defined by the observer. The differentiative program can be recognized only at the terminal stages.

(c) There is evidence to support the hypothesis of colour reversion. A genetic mechanism of reversion could be the explanation of the observation that some highly malignant, undifferentiated cells can give rise to differentiated cells that are no longer malignant (Comings, 1973). This reversion occurs in a damaged genetic network and, plausibly, can be achieved either by (i) bypassing the damaged step in the original path, (ii) restorating the original path, or (iii) by building a totally new path (Ts'o et al., 1977). In some instances, the reverted cell, while functioning 'normally', may be not the same in all respects as the original cell. The thesis that initiation is an irreversible stage appears to be indefensible in its absolute formulation. For other arguments regarding reversibility the reader is referred to the papers by Z.Rabinovitz and L.Sachs (1970a,b; 1972) and E.J.Stanbridge (1976).

Taking now into account the assumptions about the growth rules on a spatial structure and the assumptions about the metamorphosis mechanisms some additional implications are of significance:

(B1) Cancer cells occur and grow in the cell population which originated them. The growth and renewal rules of the normal cells in the population govern, at least partially and for a while, the incipient malignant subpopulation. How long the tumour heterogeneity can be partially explained by a differentiation scheme specific to the normal original tissue, the term 'tumour autonomy' is a misnomer.

(B2) Because of assumption A3 it is allowed to think that a cancer cell may disappear by the effect of the local process. This effect is also the result of the consecutive assumption of an 'equivalent' neighbourhood, that is, the hypothesis that the neighbouring cells interact with the same force constant. Thus, the habitual assumption that a 'black' site remains black is inconsistent with our system of hypotheses. R.Baserga (1976) came to an identical conclusion: "It is quite possible that although neoplastic transformations may occur continuously in the body every day, most of the initially transformed cancer cells are continuously eliminated".

(B3) The hint B2 leads to the following tentative definition: The total dose of an ultimate carcinogen is that dose (having a certain time pattern) which weakens the local cell interaction and increases the probability that some multi-stage metamorphosed cells can continuously grow together. The latent period is neither more nor less than the lifespan of a neighbourhood interaction process. It cannot be deduced from the exposure time to an external carcinogen.

(B4) Because every cell has a risk to be erased, its chance to establish a clone depends on the local interaction process. The opinion that tumours frequently develop as a clone from a single (or a few) malignant cell(s) should be re-examined (see e.g. Fialkow, 1976). A similar mechanism has been suggested for human atherosclerotic plaques (Benditt, 1977). Our discussion of assumption A3 gives informal arguments against the generality of the monoclonal hypothesis. Actually, in his 1976 survey, P.J.Fialkow suggested also that "multicellular origin should be found for tumours caused by cell-to-cell spread and transformation by oncogenic viruses and might also be detected for neoplasms strongly influenced by hormones".

(B5) Our assumption about the growth mechanisms do not consider the possibility that a 'dormant' cancer cell might exist: the set S of cell states does not include a resting ('quiescent') phase. All cells on Z^d are cycling and thus the considered cell assembly is proliferative. It was hypothesized (e.g. Fisher and Fisher, 1959) that a malignant or a precancerous cell may enter a 'dormant' state, possibly in the conversion stage. Normal cells, however, can also enter a quiescent G_o state and, on the contrary, neoplastic change may involve a decreased ability of some cells to enter that phase. Thus, if malignant 'dormant' cells exist, we must assume that G_o should consist - "like Dante's Inferno" (Baserga, 1976) - of different levels of depth.

(B6) The single mechanism of tumour growth compatible with our assumptions above is the modification (mostly by decreasing) of the length of cell cycle. The first report that malignant cells do not necessarily proliferate faster than do normal cells appeared in 1962 when R.Baserga and W.E.Kisieleski showed that there are certain normal cells in the adult animal whose cell cycle is shorter than that of even the fastest growing tumours.

(B7) The above assumptions on growth and metamorphosis mechanisms represent, in fact, the microscopic description of the temporal behaviour of each cell on the considered lattice. Our task is to deduce mathematically the behaviour of the 'idealized tissue' as a whole when a carcinogenic perturbation occurs, knowing the microscopic properties of the lattice components. A very tempting analogy is the behaviour of a ferromagnetic material where the bearing of each particle (located on a d-dimensional lattice) can be described by the position change of the electron's spin. The space S contains only two states, 'up' and 'down'. When this ferromagnet approaches a critical (temperature) point, the length of the nearest-neighbour influence grows rapidly. Smaller local fluctuations, however, create a new structure superimposed on that of the large lattice. Thus, an immense number of spins that are mostly 'up' may have within it an 'island' of spins that are mostly 'down', which in turn surrounds a small group of 'up' spins with an 'islet' of 'down' spins. The above mentioned Ising model is the rough analogue to the formation of areas, nodules, islands or foci (ostensibly reversible or irreversible) which have been observed in carcinogenesis (Bannasch, 1976; Farber, 1976). Without any reductionist implication, the analogy with the Ising model strongly suggests that space dimension and nearest-neighbour interactions are important explanatory tools in both cases. The main problem is the identification of those essential (relevant) biological properties which determine the critical behaviour of a given cell assembly. For some applications of the Ising model in molecular and cell biology the reader is referred to N.S.Goel (1968, 1972), C.J.Thompson (1968, 1972), W.T.Yap and H.A.Saroff (1971), G.S.Grest and A.K.Rajagopal (1974) and W.A.Little (1974).

4. Some possible modifications of A1-A4 and open biological questions. In this paragraph we briefly suggest possible variants and refinements of our assumptions A1-A4. It must be pointed out, however, that apparently minor modifications of some assumptions lead to the construction of new models. Also, one may conjecture that different biological assumptions can lead to equivalent mathematical models.

(C1) Replace in A1 the infinite lattice Z^d with a finite array $A \subset Z^d$. In a previous paper (Schürger and Tautu. 1976a) we introduced and studied this variant we called 'Model I'. The array A is chosen such that $A \cap N_x \neq \emptyset$, $x \in A$. Also A3 is modified by assuming that if a cell located at site $x \in A$ divides, a site $y \in N_x \cap A$ is chosen (all possible choices being equiprobable) for the movable daughter cell. The number of cells located on A is supposed to remain constant and thus, the spatial competition on A is intensified. It is perhaps interesting to mention the suggestion of A.B.Cairnie (1976) that homeostasis in the small intestine might be the result of a process resembling to a predator-prey process.

(C2) Replace the uniformity hypothesis concerning the neighbourhood index H by assuming space- and time-varying neighbourhoods. Such variable neighbourhoods seem to be the natural consequence of variable cell-to-cell communication due either to a nonhomogeneous diffusion of growth factors or to a heterogeneity of a substratum to which the cell is applied (Weiss, 1963; Vasiliev and Guelstein, 1966). Assumption C2 is a component in the construction of a pattern formation process.

(C3) Replace in A3 the assumption that all possible choices of sites are equally probable by the hypothesis that a selective (dominant or reversible) choice occurs. The result must be an oriented spread. Biologically speaking, assumption C3 may express either attraction-repulsion phenomena or the creation of an intrinsic cell polarity. Following E.W.Sinnot (1960), "polarity is simply the specific orientation of activity in space" (see also Lindenmayer, 1971). The splitting of a homogeneous population into two subpopulations is associated, in some cases, with the creation of opposite polarities (Garcia-Bellido and Ripoll, 1978). Fibroblasts growing in vitro constitute a good experimental model for studying orientation, because they are motile and the normal cells have a parallel orientation to their nearest neighbours. It is claimed (Elsdale, 1972) that these cells come to adopt that configuration which allows them the maximum exercise of their motility. If a perturbation factor is (randomly) introduced, e.g. occurrence of malignant cells, the nearest fibroblasts change their orientation. Computer simulations of a simple lattice model show pictures similar to an Ising process (Vasil'ev et al., 1973), but the global stability of orientation is maintained even in the presence of substantial disturbances (20-25% 'malignant' cells). Cells in solid tissues may have an internal polarity which should be responsible for asymmetric division (see C5 and C6).

(C4) Change the regular disposition of cells on lattice points by assuming inter-site locations: 'black' daughter cells are produced in excess, overcoming the cell loss mechanism, and they are able to adapt their shape to the local structure (see Sträuli, 1979: shape adaptation during locomotion).

(C5) Change the state space S by introducing a quiescent state 0, that is, assume that $S=\{0,1,2,...,s\}$. Biologically, this means the inclusion of the G_0 phase into the set of cell states. Thus, the cellular system will also contain non-cycling cells; they will be recruited into the proliferating pool during carcinogenesis (Baserga, 1976). The new assumption C5 entails essential modifications in the system of hypotheses commented in the above paragraphs. Actually, a new model is needed. We have to refine the growth mechanism in order to distinguish the interaction between the occupant of site x and that of the chosen site $y \in N_x$. The first change is to assume C3, that is, a site $y \in N_x$ is chosen with a probability depending on the state of its inhabitant. The case of two daughter cells at x entering G_0 simultaneously is a puzzling one, if we suppose that G_o cells have no 'killing' potency. We ought to require that the daughter cells never enter simultaneously G_o and that, in addition, the cycling daughter is movable. Whether this hypothesis is biologically consistent remains an open question.

In the theory of cell automata the quiescent (blank, passive, unexcitable, fiducial) state is a specially designated component in the definition of the cell automaton or the tessellation structure (Moore, 1962). It is stipulated that at t=0 all but a finite number of cells are in state 0. The 'level of depth' of a quiescent cell depends on its neighbours states (the idea is suggested by the construction given by A.W.Burks (1970)). When all neighbours of x are in G_o, the cell at x is also in G_o phase.

(C6) Change the set K of colours by introducing the pluripotent stem cell as 'white'. Change also scheme (1) by assuming a differentiation network where the common source for normal and malignant metamorphoses is the 'white' stem cell (or type 0). The hypothesis that the normal stem cell is the target cell in carcinogenesis gains ground in the last time (e.g. Pierce, 1976; Pierce et al., 1978; Mintz et al., 1978). As J.Cairns (1975) suggests, "the only mutations that accumulate during the life of an animal are, of course, those that arise in cells that themselves survive or contribute descendants that survive for the animal's lifetime".

Assumption C6 entails drastic modifications in A1-A4. The most important amendment is the lattice design. We assume that a type 0 cell can interact with cells of all other types but not with another cell of the same type. This corresponds to the expectation that the stem cells are restricted to limited territories (the stem cell 'niche' hypothesis: Schofield, 1978; Potten et al., 1979) so that they cannot easil compete with each other (Cairns, 1975). A possible representation of a niche for a stem cell located at $x \in Z^2$ should be a 'box' centered at x, that is, a planar area like the lamination Λ or rather the 'zone of influence' introduced by D.J.Gates and M.Westcott (1978). The cells contained in the box constitute the 'proliferative unit' (for skin see the scheme suggested by C.S.Potten (1976)) whose formal definition requires a re-statement of A2. The metamorphose mechanism, the "oncogeny" (Hauschka, 1961), particularly the generation of a 'malignant' stem cell by a 'normal' stem cell, is still obscure.

(C7) Replace Z^d by an Abelian group $G=(X/R)$, where X has 2d generators and R contains all relations other than the commutativity and the inverse iterations. Assumption A2 becomes as follows: A direct neighbour of $x \in G$ is any point $y \in G$ such that $y=xg$ for some $g \in X'$, where X' is X plus the group identity (Mylopoulos and Pavlidis, 1971). Thus $N_x=\{xg | g \in X'\}$.

Finally, we must give attention to two other points. Firstly, to the fact that some steps and variables in carcinogenesis are still unobservable. As a matter of fact, for this process there are state variables which are physically undistinguishable by our present experimental methods. It is the opportunity to remind here that in mechanical statistics or quantum mechanics any measurable quantity associated with a particle system can be thought of as an 'observable' (for the theory see, e.g., Gudder, 1979). This is not always possible in biology, particularly cell biology; besides, some measurable quantities may not be of interest for some models:"Whether one has quantified a phenomenon in an interesting way is determined both by the intuitive familiarity of the resulting concept and its fruitfulness in facilitating the construction of a successful theory" (Sneed, 1971). This is perhaps an explanation for the paucity of 'laws' in biology (Rosen, 1977). Clearly, "if we are to understand the essential nature of neoplastic development, we must begin to identify isolate, and study the different cell types that seem to be the key precursors from which the malignant neoplasm is ultimately derived" (Farber, 1973).

The second point evokes our difficulties in describing the temporal and spatial course of complex biological processes by using a single language. An example will make this idea more understandable. Suppose that we are to describe the growth of fibroblasts in vitro. If the process develops in the absence of collagen, then it can be described as growth in a two-dimensional space; the presence of collagen adds a new dimension and growth takes place in a three-dimensional space (multilayering). Moreover, in the absence of collagen, the structures are more variable until attaining a certain stability. The cells firstly form a patchwork pattern separated by distiguishable "frontiers", but in the final step they constitute a single uniform array. As T.Elsdale (1972) pointed out, pattern generation based on random cell movements and specific mutual constraints is a stochastic process. However, in order to describe it as a whole, we must improve its construction by coupling or compounding the original process with other processes (e.g. in this example, the diffusion of collagen, changeable neighbourhood relationships, and so on). This difficulty can be encountered in modeling morphogenesis (see e.g. Tautu, 1975), if this process is viewed as one where "unoriented and erratic cell groups ... will through a sequence of identified interactions give rise to an emergent pattern of a much higher degree of regularity and orientation" (Weiss, 1963).

Here, we concentrated only on assumptions dealing with carcinogenesis. However, with some simplifications in A3 and A4 (especially two cell states, 'white' and 'black'), the mathematical model can also describe other processes which require neighbourhood contacts, e.g. the spread of simple epidemics, the diffusion of innovations, the cell infection by virus particles or simplified morphogenesis. For some examples and mathematical treatments, the reader is referred to D. Mollison (1977), K.Schürger (1979), T.Hägerstrand (1967), A.L.Koch (1964), W.Schwöbel et al. (1966) and M.Eden (1961).

5. Short notes on previous work. In 1951, J.C.Fisher and J.H. Holloman interpreted the formation of cancer foci in terms of cell interactions. The main hypothesis is the existence of a critical size of malignant cells able to generate a tumour: in authors' words, "a single cancer cell isolated in normal tissue will not generate a tumour, whereas a clump of cancer cells will". The explanation given is that a putative malignant cell expresses its malignant characteristics (phenotype) when all its neighbours are malignant. Also, the occupancy by cell loss mechanism is suggested: if the neighbour of a malignant cell dies "for one reason or another", then the place is occupied by a malignant daughter cell. Thus, "repeated killing of neighbours ... can produce a supercritical colony from a single cancer cell".

An important result was obtained in 1958, when J.C.Fisher considered the growth of a group of cells ('clone') located in a circular region of radius r (area A). The growth of r at steady state is defined by $r=k(t-t_0)$, where t_0 is the time when the clone started growing, and k is a constant. Clearly, $A=\pi r^2=\pi k^2(t-t_0)^2$. The number N of cells in the clone will be proportional to its area, hence

$$N = c(t-t_0)^2, \qquad (2)$$

where c is a constant. "A law of this sort for the growth of non-malignant epithelial clones seems more reasonable than the exponential", commented Fisher. He assumed that a malignant cell has an advantage over its normal neighbours, and claimed that the growth of a cancer clone might be exponential.

It is interesting to note that in the theory of cell automata a result similar to (2) was obtained by E.F.Moore in 1962. His theorem says the following: If a finite two-dimensional self-reproducing automaton is capable of reproducing f(t) offspring by time t, then there exists a positive real number c such that $f(t) \leq ct^2$. Here, the exponent 2 represents the space dimensionality. E.F.Moore observes that this "Malthusian sort of argument" depends on the fact that there is a finite spread velocity of the 'primary' (nonquiescent) automaton, since each unit of time permits the finite area of the primary automaton to grow only one cell in each direction. The growth with a Poissonian rate on a two-dimensional lattice has been studied in 1965 by R.W. Morgan and D.J.A.Welsh. In their model, the "infected" objects have only two neighbours, so that they spread only through the positive quadrant. The authors conjectured that if M(t) denotes the expected number of objects infected at time t, then, for some finite constant c, $M(t)/t \to c$, as $t \to \infty$. This conjecture was proved by J.M.Hammersley (1966) using arguments from percolation theory, which, in general, is dealing with a certain class of (nonlinear) non-Markovian processes. Although these results are not directly concerned with tumour growth, they suggest a general idea about the restrictions of a growth process in space.

Starting from the same hypothesis of critical size, J.K.Wright and R.Peto (1969) tried to find an expression for the probability that a tumour will be formed provided a certain number m of precancerous ('activated') cells lie close enough together to interact in some way. The authors considered a cluster structure rather than a lattice: it is a group of cells whose spread is defined by

$$s_m = \sum_{i=1}^{m} |x_i - \bar{x}|^2,$$

where x_i is the position of the i-th cell in the cluster and

$$\bar{x} = m^{-1} \sum_{i=1}^{m} x_i .$$

Since the 'compactness' of such a cluster is unknown, the authors assume that there exists a series of 'critical spreads' $\sigma_2, \sigma_3, \sigma_4, \ldots$ such that a tumour will be formed in a tissue if, for some $m>1$, one can find m precancerous (activated) cells so close together that $s_m<\sigma_m$. This inequality - for some $m\geq m_o$, where m_o is the smallest number of cells that can interact to form a tumour - is the criterion which must be satisfied by any group of precancerous cells in a cluster with radius $\sqrt{s_m/m}$.

In 1968, N.T.J.Bailey published his stochastic model for the birth, death and migration of spatially distributed populations, where tumour growth in a three-dimensional lattice is mentioned as an example. His assumptions are different as concerns location and interaction: on a lattice site is located a clump of cells (a 'colony') and the interaction between the nearest neighbour clumps is assumed to be a migration process with constant rate. At each site the cells multiply according to a simple linear birth-and-death stochastic process. Initially, the population consists of an arbitrary number of cells (possibly infinitely many). N.T.J.Bailey suggested viral carcinogenesis as a possible application.

The stochastic model constructed by T.Williams and R.Bjerknes (1971a,b; 1972) actually describes a one-stage mechanism for carcinogenesis. In this model, one starts with a simple 'black' (cancer) cell which, together with all its descendants, is postulated to divide κ $(1<\kappa<\infty)$ times as fast as the 'white' (normal) cells which initially occupy all other sites of a hexagonal lattice. The quantity κ is called "heritable advantage" (Williams and Bjerknes, 1971b) or "carcinogenic advantage" (Williams and Bjerknes, 1972). The cell division times are assumed to be exponentially distributed, which renders the process Markovian. Thus, if the probability that a 'white' cell divides during a short time interval $(t,t+\Delta t)$ is Δt, the 'black' cell will divide during $(t,t+\Delta t)$ with probability $\kappa\Delta t$. In 1976, D.H.Green and D.Y.Downham assumed instead that the parameter of the (exponential) division time distribution of a cancer cell depends upon the number of its normal neighbours.

If the structural assumptions are identical with those above presented, in turn, the 'carcinogenic advantage' is, as a matter of fact, the simplest assumption which involves only a cell phase and one transformation. When $\kappa=\infty$, then only cancer cells divide (and the malignant type is thus irreversible); when $\kappa=1$, there is no carcinogenic advantage, and the cells are interchangeably 'white' or 'black'. The latter case - called an "invasion process" - was studied in 1973 by P.Clifford and A.Sudbury (see also Kelly, 1977), and the former, in the same year, by D.Richardson - as a percolation process with an exponential time coordinate distribution. It is not out of place to remark now that, in fact, we are in possession of different mathematical tools enabling us to study mathematically various growth processes in space - e.g. Markovian as well as non-Markovian ones, specific applications of automata and formal languages theories, pattern theory, etc. Our interest in some of them (see e.g. Tautu, 1975, 1976) stems from the idea that formal correspondences are logically indispensable for revealing some key stones of a scientific theory.

6. Blackening a three-dimensional lattice. It was noticed (Williams, 1971) that, for the Williams-Bjerknes model, the embedded Markov chain obtained by considering the total number of black cells at the time of the n-th jump actually is a one-dimensional simple random walk on $N=\{0,1,2,...\}$ with positive drift $(\kappa-1)/(\kappa+1)$. It follows from the gambler's ruin formula that the probability of absorption at the origin is $1/\kappa$ (in fact, this holds for $1\leq\kappa\leq\infty$). In this context, the reader is also referred to P.Clifford and A.Sudbury (1971); J.M.Hammersley (1963); H.Kesten (1963) as well as J.M.Hammersley and R.S.Walters (1963). It is perhaps the right place to emphasize that the first important results in the mathematical development of the Williams-Bjerknes model or related models were obtained with the aid of percolation theory and the theory of subadditive processes (e.g. Mollison, 1972,1977; Richardson, 1973; Kesten, 1973).

If $\kappa=1$, it follows from the above mentioned result concerning the drift that the 'blackening' process for the Williams-Bjerknes model is a symmetric random walk with 0 as an absorbing barrier. A well known theorem due to G.Pólya (1921) implies that on a one- and two-dimensional square lattice the symmetric (unrestricted) random walk will, with probability one, return, sooner or later (infinitely often) to its original position. On a three-dimensional square lattice, however, a symmetric (unrestricted) random walk has a positive probability to wander off to infinity, i.e. to 'escape'. We interpreted (Tautu, 1978) this theorem by suggesting that every multiplicative cell system might be 'invasive' in a three-dimensional space, if there is no 'absorber point' or absorbing barrier (case $\kappa=\infty$). For the invasion model ($\kappa=1$), it was claimed (Clifford and Sudbury, 1973) that the probability for any finite set of positions on a one- or two-dimensional rectangular lattice to be occupied by objects of the same colour tends to one as time increases. The situation for a three-dimensional square lattice will not be the same because the probability that two independent symmetric random walks on a three-dimensional square lattice do not meet is less than one. However, when both independent (symmetric) random walks are unrestricted, there is a constant (positive) intersection probability (Erdös and Taylor, 1960). For the intermediate case $1<\kappa<\infty$, F.P.Kelly (1977) conjectured that it resembles the case $\kappa=\infty$ in many ways.

The above elementary investigation about the spatial behaviour of the black random walk for the Williams-Bjerknes model suggests a precise meaning of the common expression 'escape from growth control'. In a three-dimensional idealized tissue the escape probability is always positive even in the immediate presence of some 'absorbers' (Watanabe, 1978). Any dividing cell, even a precancerous one, can be an absorber for a black random walk. The local interaction which is simply described by the hypotheses A1-A4 represents the chance mechanism that makes escape possible. One may, thus, imagine that 'escape from growth control' should be governed by certain probabilistic spatial rules similar to those mentioned above. A careful interpretation of cellular growth in vitro (possibly two-dimensional) and in vivo (three-dimensional) is suggested.

The random walk picture given above should not conceal another aspect of local interaction in space. If any dividing cell may be an absorber for the black random walk, one can, conversely, deduce that any black random walk cannot establish and spatially develop a clone independently of its neighbours. This hypothesis is in contrast to the opinion that a malignant tumour originates in a single (or a few) cancer cell(s). If black cells might develop independently of other coloured cells, there is some mathematical evidence of the clonal origin of a tumour. The answer to this problem was obtained in 1967 by P.S.Puri in the form of limit theorems concerning Markov branching processes in both discrete and continuous time. The author takes into consideration the existence of N independent time-homogeneous branching processes. If one of these N processes is supercritical, then the ultimate population will be composed only of the progeny of the supercritical process, given that the total population does not become extinct as $t\to\infty$. (It is necessary to mention that in many models for carcinogenesis - e.g. Neyman and Scott (1967) or Williams (1974) - it is assumed that the malignant population grows supercritically.) A consequence of Puri's theorem is that even in the case where none of the N independent branching processes is supercritical, the ultimate population will almost surely be the product of one and only one of the

N parents, given that the total population will survive as $t\to\infty$.

In general, such results cannot be obtained if we assume that the N branching processes will interfere with each other. We should like to point out the substantial differences in interpreting a growth process by a 'tree representation' which means independence of branches (see e.g. Nowell, 1976) or by a 'lattice representation' which means configuration-dependence. It seems plausible to think of the clone problem in cancer as a possible monoclonality in composition (depending on the metamorphosis network) but not in origin (depending on local interaction).

In the d-dimensional symmetric random walk picture, growth is defined by the number of different sites occupied by the walk during, say, the first $n\geq 1$ steps. According to a theorem given by A.Dvoretzky and P.Erdös (1951), this number, R_n^d, obeys the strong law of large numbers, for $d\geq 3$. (The reader should observe that according to another theorem in the same paper, the asymptotic behaviour of $E(R_n^d)$, as $n\to\infty$, is given by

$$E(R_n^d) = \begin{cases} \dfrac{\pi n}{\log n} + O\left(\dfrac{n\log\log n}{\log^2 n}\right) & , d=2 \\ cn + O(\sqrt{n}) & , d=3 \end{cases}$$

(c denoting a positive constant) which is essential for 'invasive' growth.) In pattern theory, a similar law of large numbers has been obtained by U.Grenander (1976).

A great deal has been done to obtain asymptotic results for spatial growth processes. D.Richardson (1973) showed that for some classes of processes with local interactions the normalized set of occupied sites asymptotically has a certain geometrical shape: a ball with respect to a certain norm. Richardson's result should be compared with the recent theorems derived by K.Schürger (1979) - particularly the "strong law of large configurations" - and with S.J.Willson (1977, 1978), motivated by crystal growth. It has been shown by U.Grenander (1976, p.313) that in some cases extremum principles force the configuration to assume a regular pattern: "It should be noted, however, both that the extrema need not be uniquely attained, and that the statements are all asymptotic in character. It is only for large configurations that a regular appearance is caused by the extremum principles." However, in order to better understand carcinogenesis by the aid of models based on assumptions A1-A4 or their variants, we need some non-asymptotic results and a topological pattern analysis.

References

ALADYEV,V.(1974) Survey of research in the theory of homogeneous structures and their applications. Math.Biosci. 22, 121-154.

ARMSTRONG,P.B.(1977) Cellular positional stability and intercellular invasion. BioScience 27, 803-809.

BAER,R.M. and MARTINEZ,H.M.(1974) Automata and biology. Ann.Rev. Biophys.Bioeng. 3, 255-291.

BANNASCH,P.(1976) Cytology and cytogenesis of neoplastic (hyperplastic) hepatic nodules. Cancer Res. 36, 2555-2562.

BANNASCH,P.(1978) Sequential cellular alterations during hepatocarcinogenesis. In: Rat Hepatic Neoplasia (P.M.Newberne, W.H. Butler eds.), 58-93. Cambridge: MIT Press.

BARRETT,J.C. and TS'O,P.O.P.(1978a) Relationship between somatic mutation and neoplastic transformation. Proc.NAS USA 75, 3297-3301.

BARRETT,J.C. and TS'O,P.O.P.(1978b) Evidence for the progressive nature of neoplastic transformation in vitro. Proc.NAS USA 75, 3761-3765.

BASERGA,R.(1976) Multiplication and Division in Mammalian Cells. New York: M.Dekker.

BASERGA,R. and KISIELESKI,W.E.(1962) Comparative study of the kinetics of cellular proliferation of normal and tumorous tissues with the use of tritiated thymidine. I.Dilution of the label and migration of labeled cells. J.Natl.Cancer Inst. 28, 331-339.

BASILICO,C. and MARIN,G.(1966) Susceptibility of cells in different stages of the mitotic cycle to transformation by polyoma virus. Virology 28, 429-437.

BAILEY,N.T.J.(1968) Stochastic birth, death and migration processes for spatially distributed populations. Biometrika 55, 189-198.

BENDITT,E.P.(1977) Implications of the monoclonal character of human atherosclerotic plaques. Amer.J.Pathol. 86, 693-702.

BERTRAM,J.S.,PETERSON,A.R. and HEIDELBERGER,C.(1975) Chemical oncogenesis in cultured mouse embryo cells in relation to the cell cycle. In Vitro 11, 97-106.

BOUCK,V. and DIMAYORCA,G.(1976) Somatic mutation as the basis for malignant transformation of BHK cells by chemical carcinogens. Nature 264, 722-727.

BRAMSON,M. and GRIFFEATH,D.(1979) On the Williams-Bjerknes tumour growth model.I. Preprint.

BURKS,A.W.(1970) Von Neumann's self-reproducing automata. In: Essays on Cellular Automata (A.W.Burks ed.), 3-64. Urbana: Univ. of Illinois Press.

CAIRNIE,A.B.(1976) Homeostasis in the small intestine. In: Stem Cells of Renewing Cell Populations (A.B.Cairnie, P.K.Lala, D.G.Osmond eds.), 67-77. New York: Academic Press.

CAIRNS,J.(1975) Mutation selection and the natural history of cancer. Nature 255, 197-200.

CLIFFORD,P. and SUDBURY,A.(1971) Some mathematical aspects of two-dimensional cancer growth. Symp.Tobacco Research Council, London, June 1971.

CLIFFORD,P. and SUDBURY,A.(1973) A model for spatial conflict. Biometrika 60, 581-588.

COCHRAN,W.G.(1936) The statistical analysis of the distribution of field counts of diseased plants. J.Roy.Statist.Soc. 3, Suppl. 49-67.

COMINGS,D.E.(1973) A general theory of carcinogenesis. Proc.NAS USA 70, 3324-3328.

COOPER,E.H.,BEDFORD,A.J. and KENNY,T.E.(1975) Cell death in normal and malignant tissues. Adv.Cancer Res. 21, 59-120.

DOWNHAM,D.Y. and GREEN,D.H.(1976) Inference for a two-dimensional stochastic growth model. Biometrika 63, 551-554.

DOWNHAM,D.Y. and MORGAN,R.K.B.(1973a) Growth of abnormal cells. Nature 242, 528-530.

DOWNHAM,D.Y. and MORGAN,R.K.B.(1973b) A stochastic model for a two-dimensional growth on a square lattice. Bull.Intern. Statist.Inst. 45, Vol.1, 324-331.

DRESDEN,M. and WONG,D.(1975) Life games and statistical models. Proc. NAS USA 72, 956-960.

DVORETZKY,A. and ERDÖS,P.(1951) Some problems on random walk in space. Proc.2nd Berkeley Symp.Math.Statist.Probab.,353-367.

EAVES,G.(1973) The invasive growth of malignant tumours as a purely mechanical process. J.Pathol. 109,233-237.

EDEN,M.(1961) A two-dimensional growth process. Proc.4th Berkeley Symp. Math.Statist.Probab., Vol.IV, 223-239.

ELSASSER,W.M.(1970) The role of individuality in biological theory. In: Towards a Theoretical Biology (C.H.Waddington ed.), Vol.3, 137-166. Edinburgh: Edinburgh Univ.Press.

ELSDALE,T.(1972) Pattern formation in fibroblast cultures, an inherently precise morphogenetic process. In: Towards a Theoretical Biology (C.H.Waddington ed.), Vol.4, 95-108. Edinburgh: Edinburgh Univ.Press.

ERDÖS,P. and TAYLOR,S.J.(1960) Some intersection properties of random walk paths. Acta Math.Acad.Sci.Hungar. 11, 231-248.

FARBER,E.(1973) Carcinogenesis - Cellular evolution as a unifying thread: Presidential address. Cancer Res. 33,2537-2550.

FARBER,E.(1976) Hyperplastic areas, hyperplastic nodules, and hyperbasophilic areas as putative precursor lesions. Cancer Res. 36, 2532-2533.

FIALKOW,P.J.(1976) Clonal origin of human tumors. Biochim.Biophys. Acta 458, 283-321.

FISHER,B. and FISHER,E.R.(1959) Experimental evidence in support of the dormant tumor cell. Science 130, 918-919.

FISHER,J.C.(1958) Multiple-mutation theory of carcinogenesis. Nature 181, 651-652.

FISHER,J.C. and HOLLOMON,J.H.(1951) A hypothesis for the origin of cancer foci. Cancer 4, 916-918.

FREEMAN,G.H.(1953) Spread of disease in a rectangular plantation with vacancies. Biometrika 40, 287-305.

GARCIA-BELLIDO,A. and RIPOLL,P.(1978) Cell lineage and differentiation in Drosophila. In: Genetic Mosaics and Cell Differentiation (W.J.Gehring ed.), 119-156. Berlin-Heidelberg-New York: Springer.

GARDNER,M.(1970) Mathematical games. The fantastic combinations of John Conway's new solitaire game "life". Scientific Amer. 223, Oct. 120-123.

GARDNER,M.(1971) Mathematical games. On cellular automata, self-reproduction, the Garden of Eden and the game "life". Scientific Amer. 224, Febr. 112-117.

GARSTENS,M.A.(1969) Statistical mechanics and theoretical biology. In: Towards a Theoretical Biology (C.H.Waddington ed.), Vol.2, 285-292. Edinburgh: Edinburgh Univ.Press.

GATES,D.J. and WESTCOTT,M.(1978) Zone of influence models for competition in plantations. Adv.Appl.Prob. 10, 499-537.

GREEN,D.H. and DOWNHAM, D.Y.(1976) An extension of the two-dimensional stochastic model of Williams and Bjerknes. Comm.9th European Meeting of Statisticians, Grenoble, 1976.

GOEL,N.S.(1968) Relaxation kinetics of denaturation of DNA. Biopolymers 6, 55-72.

GOEL,N.S.(1972) Cooperation processes in biological systems. Progress Theor.Biol. 2, 213-302. New York: Academic Press.

GRENANDER,U.(1976) Pattern Synthesis. Lectures in Pattern Theory, Vol.1. New York-Heidelberg-Berlin: Springer.

GREST,G.S. and RAJAGOPAL,A.K.(1974) Impurity and end effects on finite polymer chains. Ising model approach. J.Theor.Biol. 47, 43-56.

GUDDER,S.P.(1979) Stochastic Methods in Quantum Mechanics. New York: North Holland.

HALL,T.C.(1975) The cellular mechanisms of cancer chemotherapy: "The sevenfold path". In: Cancer Chemotherapy. Fundamental Concepts and Recent Advances, 295-309. Chicago: Year Book Med.Publ.

HAMMERSLEY,J.M.(1963) Long-chain polymers and self-avoiding random walks. Sankhyā Ser.A 25, 15-34.

HAMMERSLEY,J.M.(1966) First-passage percolation. J.Roy.Statist.Soc. Ser.B 28, 491-496.

HAMMERSLEY,J.M. and WALTERS,R.S.(1963) Percolation and fractional branching processes. SIAM J.Appl.Math. 11, 831-839.

HAUSCHKA,T.S.(1961) The chromosomes in ontogeny and oncogeny. Cancer Res. 21, 957-974.

HÄGERSTRAND,T.(1967) Innovation Diffusion as a Spatial Process. Chicago: Univ. of Chicago Press.

HILL,T.L. and KEDEM,O.(1966) Studies in irreversible thermodynamics. III. Models for steady state and active transport across membranes. J.Theor.Biol. 10, 399-441.

IVERSEN,O.H.,BJERKNES,R. and DEVIK,F.(1968) Kinetics of cell renewal, cell migration and cell loss in the hairless mouse dorsal epidermis. Cell Tissue Kinet. 1, 351-367.

KAUFFMAN,S.(1971) Differentiation of malignant to benign cells. J. Theor.Biol. 31, 429-451.

KELLY,F.P.(1977) The asymptotic behaviour of an invasion process. J.Appl.Prob. 14, 584-590.

KERR,J.F.R.,WYLLIE,A.H. and CURRIE,A.R.(1972) Apoptosis: A basic biologic phenomenon with wide-ranging implications in tissue kinetics. Brit.J.Cancer 26, 239-257.

KESTEN,H.(1963) On the number of self-avoiding walks. J.Math.Phys. 4, 960-969.

KESTEN,H.(1964) On the number of self-avoiding walks. J.Math.Phys. 5, 1128-1137.

KESTEN,H.(1973) Contribution to the discussion of the paper "Subadditive ergodic theory" by J.F.C.Kingman. Ann.Prob. 1, 903.

KITAGAWA,T.(1976) Sequential phenotype changes in hyperplastic areas during hepatocarcinogenesis in the rat. Cancer Res. 36, 2534-2539.

KOCH,A.L.(1964) The growth of viral plaques during the enlargement phase. J.Theor.Biol. 6, 413-431.

LINDENMAYER,A.(1971) Polarity, symmetry and development. Unpubl. manuscript.

LITTLE,W.A.(1974) The existence of persistent states in the brain. Math.Biosci. 19, 101-120.

MARQUARDT,H.(1974) Cell cycle dependence of chemically induced malignant transformation in vitro. Cancer Res. 34, 1612-1615.

MAREEL,M.M.K.(1979) Is invasiveness in vitro characteristic of malignant cells? Cell Biology Intern.Rep. 3, 627-640.

McMAHON,D.(1973) A cell-contact model for cellular position determination in development. Proc.NAS USA 70, 2396-2400.

MINTZ,B.,CRONMILLER,C. and CUSTER,R.P.(1978) Somatic cell origin of teratocarcinomas. Proc.NAS USA 75, 2834-2838.

MOLLISON,D.(1972) Conjecture on the spread of infection in two dimensions disproved. Nature 240, 467-468.

MOLLISON,D.(1974) Percolation processes and tumour growth (abstr.) Adv.Appl.Prob.6, 233-235.

MOLLISON,D.(1977) Spatial contact models for ecological and epidemic spread (with discussion). J.Roy.Statist.Soc.Ser.B 39, 283-326.

MOORE,E.F.(1962) Machine models of self-reproduction. Proc.Symp. Appl.Math. 14, 17-33.

MORGAN,R.W. and WELSH,D.J.A.(1965) A two-dimensional Poisson growth process. J.Roy.Statist.Soc.Ser.B 27, 497-504.

MYLOPOULOS,J.P. and PAVLIDIS,T.(1971) On the topological properties of quantized spaces.I.The notion of dimension. J.Assoc.Comp.Mach. 18, 239-246.

NEYMAN,J. and SCOTT,E.L.(1967) Statistical aspects of the problem of carcinogenesis. Proc.5th Berkeley Symp.Math.Statist.Probab., Vol.4, 745-776.

NOWELL,P.C.(1976) The clonal evolution of tumor cell populations. Science 194, 23-28.

PETERSON,A.R.,BERTRAM,J.S. and HEIDELBERGER,C.(1974) Cell cycle dependency of DNA damage and its repair in transformable mouse fibroblasts treated with N-methyl-N'-nitro-N-nitrosoguanidine. Cancer Res. 34, 1600-1607.

PIERCE,G.B.(1976) Origin of neoplastic stem cells. In: Progress in Differentiation Research (N.Müller-Bérat, C.Rosenfeld, D.Tarin, D.Viza eds.), 269-273. Amsterdam: North-Holland.

PIERCE,G.B.,SHIKES,R. and FINK,L.M.(1978) Cancer: A Problem of Developmental Biology. Englewood Cliffs: Prentice-Hall.

PÓLYA,G.(1921) Über eine Aufgabe der Wahrscheinlichkeitsrechnung betreffend der Irrfahrt im Strassennetz. Math.Ann. 84, 149-160.

POTTEN,C.S.(1976) Identification of clonogenic cells in the epidermis and the structural arrangement of the epidermal proliferative unit (EPU). In: Stem Cells of Renewing Cell Populations (A.B.Cairnie, P.K.Lala, D.G.Osmond eds.), 91-102. New York: Academic Press.

POTTEN,C.S.(1978) Epithelial proliferative subpopulations. In: Stem Cells and Tissue Homeostasis (B.J.Lord, C.S.Potten, R.J.Cole eds.), 317-334. Cambridge: Cambridge University Press.

POTTEN,C.S.,SCHOFIELD,R. and LAJTHA,L.G.(1979) A comparison of cell replacement in bone marrow, testis and three regions of surface epithelium. Biochim.Biophys.Acta 560, 281-299.

PURI,P.S.(1967) Some limit theorems on branching processes related to development of biological populations. Math.Biosci. 1, 77-94.

RABINOWITZ,Z. and SACHS,L.(1970a) Control of the reversion of properties in transformed cells. Nature 225, 136-139.

RABINOWITZ,Z. and SACHS,L.(1970b) The formation of variants with a reversion of properties of transformed cells. V.Reversion to a limited life-span. Intern.J.Cancer 6, 388-398.

RABINOWITZ,Z. and SACHS,L.(1972) The formation of variants with a reversion of properties of transformed cells.VI.Stability of the reverted state. Intern.J.Cancer 9, 334-343.

REICH,E.,RIFKIN,D.B. and SHAW,E.(eds.)(1975) Proteases and Biological Control. Cold Spring Harbor Lab.

RICHARDSON,D.(1973) Random growth in a tessellation. Proc.Camb.Phil. Soc. 74, 515-528.

ROBERTSON,A. and COHEN,M.H.(1972) Control of developing fields. Ann.Rev.Biophys.Bioeng. 1, 409-464.

ROSEN,R.(1977) Observation and biological systems. Bull.Math.Biol. 39, 663-678.

RUBIO,C.A.(1978) Rate of cell division in atypias and invasive carcinoma of the uterine cervix in the mouse. J.Natl.Cancer Inst. 61, 259-263.

SCHOFIELD,R.(1978) The relationship between the spleen colony-forming cell and the haemopoietic stem cell. A hypothesis. Blood Cells 4, 7-25.

SCHÜRGER,K.(1979) On the asymptotic geometrical behaviour of a class of contact interaction processes with a monotone infection rate. Z.Wahrsch. 48, 35-48.

SCHÜRGER,K. and TAUTU,P.(1976a) Markov configuration processes on a lattice. Rev.Roumaine Math.Pures Appl. 21, 233-244.

SCHÜRGER,K. and TAUTU,P.(1976b) A Markovian configuration model for carcinogenesis. Lecture Notes in Biomath. 11, 92-108.

SCHÜRGER,K. and TAUTU,P.(1977) A spatial stochastic model for carcinogenesis: A Markov configuration process. Invited paper, 10th European Meeting of Statisticians, Leuven.

SCHÜRGER,K. and TAUTU,P.(1978) Die Simulation eines mathematischen Modells der Krebsentstehung. IBM Nachrichten, Heft 242, 265-273.

SCHWÖBEL,W.,GEIDEL,H. and LORENZ,R.J.(1966) Ein Modell der Plaquebildung. Z.Naturforsch. 21, 953-959.

SEGLEN,P.O.(1976) Differones: A simplifying concept in differentiation. In: Progress in Differentiation Research (N.Müller-Bérat, C.Rosenfeld, D.Tarin, D.Viza eds.), 205-210. Amsterdam: North-Holland.

SINNOT,E.W.(1960) Plant Morphogenesis. New York: McGraw-Hill.

SNEED,J.D.(1971) The Logical Structure of Mathematical Physics. Dordrecht: D.Reidel.

STANBRIDGE,E.J.(1976) Suppression of malignancy in human cells. Nature 260, 17-20.

STEEL,G.G.(1968) Cell loss from experimental tumours. Cell Tiss.Kinet. 1, 193-207.

STRÄULI,P.(1979) The role of cell movement in tumor invasion: A general appraisal. In: Cell Movement and Neoplasia (M. de Brabander, M.Mareel, L. de Ridder eds.), 187-191. Oxford: Pergamon Press.

STRÄULI,P. and WEISS,L.(1977) Cell locomotion and tumor penetration. Report on a Workshop of the EORTC Cell Surface Project Group. Europ.J.Cancer 13, 1-12.

TAUTU,P.(1974) Random systems of locally interacting cells (abstr.) Adv.Appl.Prob. 6, 237.

TAUTU,P.(1975a) A stochastic approach to the French Flag problem (abstr.) Adv.Appl.Prob. 7, 262-263.

TAUTU,P.(1975b) A stochastic automaton model for interacting systems. In: Perspectives in Probability and Statistics (J.Gani ed.), 403-415. Applied Probability Trust.

TAUTU,P.(1976) Formal languages as models for biological growth. Lecture Notes in Biomath. 11, 127-134.

TAUTU,P.(1978) Blackening a d-dimensional lattice. Rev.Roumaine Math. Pures Appl. 23, 141-152.

TERASIMA,T. and TOLMACH,L.J.(1961) Changes in X-ray sensitivity of HeLa cells during the division cycle. Nature 190, 1210-1211.

THOMPSON,C.J.(1968) Models for hemoglobin and allosteric enzymes. Biopolymers 6, 1101-1118.

THOMPSON,C.J.(1972) Mathematical Statistical Mechanics. New York: MacMillan.

TOLMACH,L.J.,WEISS,B.G. and HOPWOOD,L.E.(1971) Ionizing radiation and the cell cycle. Fed.Proc. 30, 1742-1751.

TS'O,P.O.P.,BARRETT,J.C. and MOYZIS,R.(1977) The relationship between neoplastic transformation and the cellular genetic apparatus. In: The Molecular Biology of the Mammalian Genetic Apparatus (P.O.P.Ts'o ed.), Vol.2, 241-267. Amsterdam: North-Holland.

ULAM,S.M.(1962) On some mathematical problems connected with patterns of growth of figures. Proc. Symp.Appl.Math. 14, 215-224.

VALERIOTE,F. and VAN PUTTEN,L.(1975) Proliferation-dependent cytotoxicity of anticancer agents: A review. Cancer Res. 35, 2619-2630.

VASILIEV,J.M. and GUELSTEIN,V.J.(1966) Local cell interactions in neoplasms and in the foci of carcinogenesis. Progr.Exp.Tumor Res.8,26-65

VASIL'EV,N.B.,LEONTOVICH,A.M.,MARGOLIS,L.B.,PETROVSKAYA,M.B. and PYATETSKII-SHAPIRO,J.J.(1973) A model for maintenance of cell orientation by local interaction. Ontogenez 4, 412-415.

VERHAGEN,A.M.W.(1977) A three parameter isotropic distribution of atoms and the hard-core square lattice gas. J.Chem.Phys. 67, 5060-5065.

WARWICK,G.P.(1971) Effect of the cell cycle on carcinogenesis. Fed.Proc. 1760-1765.

WATANABE,H.(1978) Escape probability of a random walker on a lattice doped with absorbers. J.Chem.Phys. 69, 4872-4875.

WEISS,P.(1963) Cell interactions. In: Canadian Cancer Conference, Vol.5, 241-276. New York: Academic Press.

WILLIAMS,T.(1971) Working paper. Symp.Tobacco Research Council, London, June 1971.

WILLIAMS,T.(1974) Evidence for super-critical tumour growth (abstr.) Adv. Appl.Prob. 6, 237-238.

WILLIAMS,T. and BJERKNES,R.(1971a) Hyperplasia: the spread of abnormal cells through a plane lattice (abstr.) Adv.Appl.Prob. 3, 210-211.

WILLIAMS,T. and BJERKNES,R.(1971b) A stochastic model for the spread of an abnormal clone through the basal layer of the epithelium. Symp. Tobacco Research Council, London, June 1971.

WILLIAMS,T. and BJERKNES,R.(1972) Stochastic model for abnormal clone spread through epithelial basal layer. Nature 236, 19-21.

WILLSON,S.J.(1977) The growth of configurations. Math.Systems Theory 10, 387-400.

WILLSON,S.J.(1978) On convergence of configurations. Discrete Math. 23, 279-300.

WOLPERT,L.(1970) Positional information and pattern formation. In: Towards a Theoretical Biology (C.H.Waddington ed.), Vol.3, 199-230. Edinburgh: Edinburgh Univ. Press.

WRIGHT,J.K. and PETO,R.(1969) An elementary theory leading to non-linea dose-risk relationships for radiation carcinogenesis. Brit.J.Cancer 23, 547-553.

YAMADA,H. and AMOROSO,S.(1969) Tessellation automata. Inform.Control 14, 299-317.

YAP,W.T. and SAROFF,H.A.(1971) Application of the Ising model to hemoglobin. J.Theor.Biol. 30, 35-39.

ZAND,R. and HARARY,F.(1972) Biochemical and biophysical graphs.I. Lattice graphs as membrane models. Mimeographed paper.

TOPIC III

Spatial models in epidemiology and genetics

Discussion by the Editors

In his 1967 paper on the simulation of stochastic epidemics in two dimensions, N.T.J.Bailey pointed out that most real communities entail the spatial distribution of individuals as an essential ingredient, with susceptible individuals in the neighbourhood of an infective one. 'Neighbourhoods' may, however, be very complex and hard to define, while various forms of complicated social and geographical stratifications may also exist. Clearly, in the framework of our conference, spatial models in epidemiology represent a predestinated topic. In other meetings on spatial processes - e.g. Newcastle upon Tyne, 1974; Edinburgh, 1974; Swansea, 1975, or Bath, 1977 - models for epidemics have also been reported. The "geography of genes" is a second natural topic: the term 'cline' (introduced by J.Huxley in 1933) denotes a spatial gradient in the frequencies of different genotypes or phenotypes within a continuous population in equilibrium in a one-dimensional habitat.

The group of papers in this section contains two fundamental survey lectures given by N.T.J.Bailey and H.T.Weinberger. For additional information on threshold theorems the reader is referred to the paper by J.Hammersley (Topic VI) where he discusses this theorem in terms of percolation theory, introducing a theorem given by C. McDiarmid (see also the paragraph on lattice models in Bailey's survey). The generalization of this theorem to mixed Bernoulli percolation was quite recently published by J.Hammersley in Math. Proc.Camb.Phil.Soc. (1980). H.-P. Altenburg and W.J.Bühler consider an urn model as a stochastic model for epidemic spread, and K. Schumacher studies the asymptotic speed of propagation, the travelling-front solutions and the influence of delays on the minimal speed of propagation for the deterministic general epidemic model of Kermack and McKendrick. Two other papers deal with the "real purpose" of a mathematical theory of epidemics, that is, suggestion or facilitation of practical prevention and control: K.Bögel discusses practical problems and control of epidemics in wildlife, while H.Thieme gives sufficient mathematical conditions for stopping the spatial spread of rabies.

The genetical problems treated in this section are: genetic identity in an (infinite homogeneous) tree (S.Sawyer), clines (O.

Diekmann; P.C.Fife and L.A.Peletier) and the qualitative analysis of the solutions of Fisher's diffusion equation (F.Rothe). As P.Fife wrote in his book "Mathematical Aspects of Reacting and Diffusing Systems" (Lecture Notes in Biomath., Vol. 28, 1979), qualitative results for biological models are very important, since accurate quantitative results can only occasionally be expected. The main reason, remarked Fife, is that, here, biological models are much more idealized than in physics and chemistry: biological populations are never homogeneous, the environment is never uniform in time or space, and the population is never isolated from other influences. The non-mathematical conclusion is drawn by Hermann Hesse: "No permanence is ours; we are a wave / That flows to fit whatever form it finds."

ON THE SPATIAL SPREAD OF A PATTERN

H.-P. Altenburg
Universität Heidelberg, Fakultät f. Klin. Medizin Mannheim

and

W. J. Bühler
Universität Mainz, Fachbereich Mathematik

Summary

A simple process is considered for the spread of a pattern in a spatially distributed population. Expressions are given for the stochastic means, variances and covariances. Central limit theorems are obtained for the number of individuals that have the pattern, and the time needed for the pattern to reach the n-th subpopulation.

1. Introduction

The process that we are considering in this paper was originally constructed as an attempt to model the spread of an epidemic disease in a spatially distributed population. The mathematical assumptions that we needed led us to a model that is tractable, however overly simplified to be realistic as a model for the original purpose.

Let us assume that a population is divided into a denumerable number of subpopulations S_i $(i=0,1,\dots)$ and that $X_i(t)$ is the number of individuals in S_i at the time t, that have a certain characteristic (pattern) E. We assume

$$(1) \qquad X_i(0) = \begin{cases} 0 & i \neq 0 \\ 1 & i=0 \end{cases}$$

The probabilistic rate of transfer of E to an individual in S_i is proportional to the number $X_{i-1}(t)$, that is

$$(2) \qquad P\{X_i(t+\Delta t) - X_i(t) = 1 \mid X_0(t), X_1(t), \dots\} = \lambda X_{i-1}(t)\Delta t + o(\Delta t) \qquad i=1,2,\dots$$

where we set $\lambda = 1$ to simplify notation.

There are at any time in all subpopulations an infinite number of individuals, that do not yet possess the characteristic E and an individual cannot lose the characteristic. With the help of Bailey's "random variable technique" (Bailey, 1964) we can write down the partial differential equation for the common probability generating function $P(s_0,s_1,\ldots|t) = E\left[\prod_{i=0}^{\infty} s_i^{X_i(t)}\right]$ of the $X_i(t)$:

$$\frac{\partial P}{\partial t} = \sum_{i=0}^{\infty} (s_{i+1}-1)s_i\frac{\partial P}{\partial s_i} \tag{3}$$

$$P(s_0,s_1,\ldots|0) = s_0$$

2. Expectations and Variances

(3) can be transformed into a differential equation for the cumulant generating function

$$K(\theta_o,\theta_1,\ldots|t) = \log P(e^{\theta_o},e^{\theta_1},\ldots|t) \tag{4}$$

from which by differentiating with respect to the θ_i at $\theta_i=0$ we obtain the system (5) of differential equations for the $m_i(t) = E\left[X_i(t)\right]$.

$$\frac{d}{dt} m_i(t) = m_{i-1}(t), \qquad m_i(0)=0 \quad i=1,2,\ldots \tag{5}$$

$$\frac{d}{dt} m_0(t) = 0, \qquad m_0(0)=1$$

with the solution

$$m_i(t) = t^i/i! \qquad i=0,1,2,\ldots \tag{6}$$

If we apply the operator $\frac{\partial^2}{\partial\theta_i\partial\theta_j}$ to the differential equation of $K(\ldots,\theta_i,\ldots|t)$ and set all θ_i, θ_j equal to zero we also get a system of differential equations for the variances and covariances. Denoting the variance of $X_i(t)$ by $v_{ii}(t)$, and the covariance of $X_i(t)$ and $X_j(t)$ by $v_{ij}(t)$ we get the system of differential equations

$$\frac{dv_{ij}(t)}{dt} = 2(v_{i-1,j}(t) + v_{i,j-1}(t)) \qquad i,j=1,2,...$$

(7)

$$\frac{dv_{ii}(t)}{dt} = m_{i-1}(t) + 2(v_{i-1,i}(t) + v_{i,i-1}(t)) \qquad i=1,2..$$

with the initial conditions

(8) $$v_{ij}(0) = 0 \qquad i,j=0,1,2,...$$

Let $V_n(y,t) = \sum_{j=0}^{\infty} v_{nj}(t)y^j$ be the generating function of the $v_{nj}(t)$ and $V_n^*(y,s)$ its Laplace-transform. We get from the equations (7)

(9) $$\frac{\partial}{\partial t} V_n(y,t) = t^{n-1}y^n/(n-1)! + 2V_{n-1}(y,t) + 2yV_n(y,t) \qquad n=1,2..$$

with $V_n(y,0)=0$, $V_0(y,t)=0$. We obtain from (9) for the Laplace-transform $V_n^*(y,s)$:

(10) $$V_n^*(y,s) = \left[(y/s)^n + 2V_{n-1}^*(y,s)\right]/(s-2y)$$

whose solution

(11) $$V_n^*(y,s) = \sum_{i=0}^{n-1} y^{n-i}2^i s^{i-n}/(s-2y)^{i-1}$$

can again be inverted into

(12) $$V_n(y,t) = \sum_{i=0}^{n-1} y^{n-i}2^i \int_0^t (t-u)^{n-i-1}u^i e^{2yu}du/((n-i-1)!i!)$$

from which we easily get the expressions for the $v_{ij}(t)$ by the expansion of the series, for example for n=1,2,3:

(13) $$v_{1j}(t) = 2^{j-1}t^j/j! \qquad j=1,2,3,...$$

(14) $$v_{2j}(t) = (2t)^j/(4j!) + j2^jt^{j+1}/(j+1)! \qquad j=2,3,...$$

$$v_{3j}(t) = t^{j+2}j2^j/(j!(j+2)) + t^{j+1}(j-1)2^{j-1}/(j!(j+1))$$

(15)

$$+ t^j2^{j-3}/j! \qquad j=3,4,...$$

3. A related urn model

In this section we consider an urn model equivalent to the process $\{X_0(\tau_n), X_1(\tau_n), \ldots\}$, where τ_n is the moment of the n-th transfer. We label the balls with 0,1,2,... and at the beginning of the experiment there is one ball of the kind 0 in the urn. The store of balls is unlimited. We repeat the following experiment n times:

(a) a ball is randomly drawn out of the urn and its kind (for example k) is registered

(b) the ball is put back in the urn and a ball of the next kind (k+1) is added.

After the n-th drawing we have n+1 balls in the urn, of which 1 is of the 0, a_1 are of the kind 1,... and a_n are of the kind n. We always have $a_1 + \ldots + a_n = n$. Obviously $a_{k+j} = 0$ for j=1,2,... if $a_k = 0$ for any $k > 1$. The case $a_n > 0$ only occurs if $a_1 = \ldots = a_n = 1$. Let $Y_k(n)$ be the number of balls of the kind k after the n-th drawing and set

(16) $$A(a_1,\ldots,a_n) := n!P_n(a_1,\ldots a_n) = n!P\{Y_0(n)=1,\ldots,Y_n(n)=a_n\}$$

We obviously have

(17) $$P_n(a_1,\ldots,a_n) = \sum_{i=1}^{n} P_{n-1}(a_1,\ldots,a_{i-1},\ldots a_n)\ a_{i-1}/n\ .$$

From this we can easily derive a partial differential equation for the probability generating function $F_n(s_0,\ldots,s_n) = E\left[s_0^{Y_0(n)} \ldots s_n^{Y_n(n)}\right]$ namely

(18) $$F_n(s_0,\ldots s_n) = n^{-1}\sum_{i=1}^{n} \frac{\partial F_{n-1}}{\partial s_{i-1}} s_i s_{i-1}\ ,$$

which for general n is not easy to solve.

From (17) we are able to derive several rules which are helpful in computing the $A(a_1,\ldots,a_n)$:

(19a) $$A(a_1,\ldots,a_n) = \sum_{i=1}^{n} a_{i-1}A(a_1,\ldots,a_{i-1},\ldots,a_n) \qquad \text{with } a_0=$$

(19b) $$A(1) = A(1,1) = A(1,1,\ldots,1) = A(1,\ldots,1,k) = 1$$

(19c) $$A(1,\ldots,1,b_1,\ldots b_k) = A(b_1,\ldots,b_k)$$

Now we determine the expectations of the $Y_k(n)$. Let $C_k(n) = E\left[Y_k(n)\right]$.

Then from $C_k(n) = E\left[E\left[Y_k(n) \mid Y_j(\nu),\ \nu=0,1,2,\dots n-1;\ j=0,1,2,\dots\right]\right]$ we obtain

$$(20) \qquad C_k(n) = C_k(n-1) + C_{k-1}(n-1)/n$$

with the initial conditions $C_0(1) = C_1(1) = 1$, $C_k(1) = 0$ for $k > 1$. From (20) we get for the generating function of the $C_k(n)$, $D_n(t) = \sum_{k=0}^{n} C_k(n) t^k$ the expression

$$(21) \qquad D_n(t) = \prod_{k=1}^{n} (t+k)/k .$$

Except for the sign of k and for the factor 1/n! the polynomial in (21) is the generating function of the Stirling numbers of the first kind. Thus we write

$$(22) \qquad D_n(t) = \left(\sum_{k=0}^{n} s(n+1,k+1)t^k\right)/n!$$

where s(i,j) is the absolute value of the Stirling number of the first kind. So we have for the expectations

$$(23) \qquad C_k(n) = s(n+1,k+1)/n! \qquad\qquad n \geq 1,\ k \geq 0$$

and especially

$$(24) \qquad C_0(n) = 1,\ C_n(n) = 1/n!,\ C_{n-1}(n) = (n+1)/(2(n-1)!) .$$

In a similar way higher moments can be evaluated. In general this is rather messy. Denoting the second moments $E\left[Y_k(n)Y_j(n+l)\right]$ by $C_{kj}(n,l)$ in analogy to (20) we get by conditioning on the $Y_j(\nu)$, $\nu \leq n+l-1$, the relations

$$(25a) \qquad C_{k,j}(n,l) = C_{k,j}(n,l-1) + C_{k,j-1}(n,l-1)/(n+l) \qquad \text{for } l \geq 1$$

$$(25b) \qquad C_{k,j}(n,0) = C_{k,j}(n-1,0) + \tfrac{1}{n}C_{k-1,j}(n-1,0) + \tfrac{1}{n}C_{k,j-1}(n-1,0) \qquad \text{for } k \neq j$$

$$(25c) \qquad C_{k,k}(n,0) = C_{k,k}(n-1,0) + \tfrac{2}{n}C_{k,k-1}(n-1,0) + \tfrac{1}{n}C_{k-1}(n-1)$$

Now if B_n is the number of the ball, which is drawn at the n-th drawing $(n>2)$, we have

THEOREM 1 a) $P\{B_n = k\} = p_{kn} = s(n,k+1)/n!$

b) The random variable $Z_n = (B_n - E[B_n])/\sqrt{Var[B_n]}$ has an asymptotic standard normal distribution, i.e.

$$\lim_{n\to\infty} P\{Z_n \le x\} = \phi(x) = \frac{1}{\sqrt{2\pi}} \int_{-\infty}^{x} e^{-y^2/2} dy$$

c) $Z_n, Z_{n+1}, \ldots$ are asymptotic independent.

Proof:

a) Because of $p_{kn} = C_k(n-1)/n$ we get from (20) the recursive relation

$$p_{kn} = ((n-1)p_{k,n-1} + p_{k-1,n-1})/n \tag{26}$$

out of which we derive a product formula for the probability generating function $F_n(t) = E[t^{B_n}]$:

$$F_n(t) = \prod_{k=0}^{n-1} ((t+k)/(k+1)) = (\sum_{k=1}^{n} s(n,k)t^{k-1})/n! \; . \tag{27}$$

So we have

$$p_{kn} = s(n,k+1)/n! \tag{28}$$

b) According to (27) the distribution of B_n is that of a sum of n random variables: $B_n = L_0 + \ldots + L_{n-1}$, where $P\{L_i = 1\} = 1/(i+1) = 1 - P\{L_i = 0\}$ From this we get the assertion.

c) Consider the joint distribution of B_n and B_{n+1}. Then (25) can be used in a similar way as (20) in the case of one variable to represent the joint generating function $F_{n,1}$ as

$$F_{n,1}(t,s) = \sum_{\alpha=0}^{n-1} W_{\alpha,n} \prod_{k=0}^{n+1-1} f_{\alpha,n,k}(t,s) \tag{29}$$

where

$$W_{\alpha,n} = n^{-2} \prod_{k=\alpha+1}^{n-1} (k+2)/k = (n+1)n^{-1}(\alpha+1)^{-1}(\alpha+2)^{-1} \tag{30}$$

$$f_{\alpha,n,k}(t,s) = \begin{cases} (k+st)/(k+1) & \text{for } k < \alpha \\ (k+s+t)/(k+2) & \text{for } \alpha \le k < n \\ (k+s)/(k+1) & \text{for } n \le \alpha \end{cases} \tag{31}$$

The product in (29) for fixed α corresponds to pairs of random variables $(L_{\alpha,n,k}, L^*_{\alpha,n,k})$ taking values (0,0), (0,1), (1,0) and (1,1) only whose sums $(B_{\alpha,n}, B_{\alpha,n+1})$ satisfy the central limit theorem with $E[B_{\alpha,n}] \sim \log n$, $E[B_{\alpha,n+1}] \sim \log n$, $Var[B_{\alpha,n}] \sim \log n$, $Var[B_{\alpha,n+1}] \sim \log n$ and $Cov[B_{\alpha,n}, B_{\alpha,n+1}] = \sum_k Cov[L_{\alpha,n,k}, L^*_{\alpha,n,k}] = O(1) = o(\log n)$.
Since $\sum_{\alpha>K} W_{\alpha,n}$ tends to zero uniformly in n for $K \to \infty$ the conclusion remains valid for the mixed distribution in (29).

COROLLARY 2 $\lim_{n\to\infty} P\left\{ (B_n - \log n)/\sqrt{\log n} \leq x \right\} = \Phi(x)$

THEOREM 3 For fixed k, $Y_k(n)$ has an asymptotic normal distribution with expectation $(\log n)^k/k!$ and variance $(\log n)^k/k!$.

Proof (by induction in k):
a) k=1: We put

$$W_j = \begin{cases} 1 & \text{if a ball of the sort 1 is put back after the j-th drawing} \\ 0 & \text{otherwise} \end{cases}$$

But $P\{W_j = 1\} = 1/j = 1 - P\{W_j = 0\}$. So we have analogously to theorem 1 that $Y_1(n)$ has an asymptotic normal distribution with expectation log n and variance log n.
b) Conditionally on $Y_k(m)$, m=0,1,2,... $Y_{k+1}(n)$ is the sum of Bernoulli variables with success probabilities $Y_k(m)/(m+1)$, m=k+1,k+2,...n . Thus if $\sum Y_k(m)/(m+1)$ diverges, then $Y_{k+1}(n)$ is asymptotically normal with expectation and variance asymptotically equal to $\sum_{m=k+1}^{n} Y_k(m)/(m+1)$.
Replacing $Y_k(m)$ by $(\log m)^k/k!$ (approximately its expected value) yields

$$(32) \qquad \sum_{m=k+1}^{n} (\log m)^k/(k!(m+1)) \sim \int_{k+1}^{n} (\log x)^k/(k!x)\,dx \sim (\log n)^{k+1}/(k+1)!$$

c) The law of the iterated logarithm can be used to show that $Y_k(m)$ differs sufficiently little from its expected value as to leave the argument given in b) valid.

4. Limit distribution of $X_k(t)$

Now we return to our original process and we first consider the subpopulation in which a transfer is taking place at the time t. Denoting by H(t) the number of the subpopulation in which a transfer occurs at the time t, we have:

THEOREM 4 $\lim_{t\to\infty} P\left\{ (H(t) - t)/\sqrt{t} \le x \right\} = \phi(x)$

Proof:

We put $N(t) = \sum_{i=0}^{\infty} X_i(t)$ and $V(t) = N(t)e^{-t}$. Given $N(t)=n$, the distribution of $H(t)$ is equal to that of $A_n=B_n+1$. $V(t)$ converges with probability 1 (for $t \to \infty$) to a positive random variable V. Given $V=v$ we have $\log(N(t)e^{-t}) \to \log v$ $(t \to \infty)$ and so $\log N(t) = t + \log v + o(1)$. So it is $\mathrm{Var}[A_{N(t)}]/t \sim \log N(t)/t \to 1$ and $(E[A_{N(t)}] - t)/\sqrt{t} \sim (\log N(t) - t)/\sqrt{t} \to 0$. From this we conclude that

$$P\left\{\frac{H(t) - t}{\sqrt{t}} \le x\right\} = E\left[P\left\{\frac{H(t) - E[A_{N(t)}]}{\sqrt{\mathrm{Var}[A_{N(t)}]}} \le x\frac{t}{\sqrt{\mathrm{Var}[A_{N(t)}]}} + \right.\right.$$

$$\left.\left. + \frac{t - E[A_{N(t)}]}{\sqrt{\mathrm{Var}[A_{N(t)}]}} \;\middle|\; N(t)\right\}\right]$$

converges to $\phi(x)$ for $t \to \infty$.

THEOREM 5 For fixed $k \ge 1$ we have

$$\lim_{t\to\infty} P\left\{(X_k(t) - t^k/k!)/\sqrt{t^k/k!} \le x\right\} = \phi(x)$$

Proof:

The assertion follows analogously to theorem 4 from

$$\mathrm{Var}\left[Y_k(N(t))\right]/(t^k/k!) \sim \log^k(N(t) + 1) /(t^k/k!) \to 1$$

and

$$(E\left[Y_k(N(t))\right] - t^k/k!)/\sqrt{t^k/k!} \sim \log^k(N(t) + 1) - t^k/k!)/\sqrt{(t^k/k!)} \to 0$$

5. The distribution of time

In this last section we want to say something about the distribution of time T_n until the pattern reaches the n-th subpopulation. Obviously w have

(33) $$P\left\{T_n > t\right\} = P_n(1,1,..1,0;\ t)$$

where $P_n(\ldots ; t)$ is joint probability generating function of $X_0(t), X_1(t), \ldots, X_n(t)$. Analogously to (3) we get a partial differential equation for $P_1(s_o, s_1; t)$ with the solution $P_1(s_o, s_1; t) = s_o\exp(-(1-s_1)t)$, i.e. T_1 is distributed according to an exponential distribution. This implies the following procedure:

$\{B_k'\}$ is designated as the sequence of drawings in the urn model. Every B_k' we associate with a number b_k: $b_k = a_{j-1}$ if after the k-th drawing a ball of the sort j is put back in the urn.

Now we write T_n as the sum of K_n independent exponentially distributed random variables V_j, where the parameter of the exponential distribution is b_j $(j=1,\ldots K_n)$:

$$T_n = V_1 + \ldots + V_{K_n} \quad . \tag{34}$$

Here K_n is a random variable greater than or equal to n and $E[V_j] = 1/b_j$, $\mathrm{Var}[V_j] = 1/b_j^2$.

<u>LEMMA 6</u> The sequence of the numbers b_k contains 1 infinitely often (with probability one).

<u>Proof:</u>

In the urn we always have only one ball of the kind 0. Consequently the probability of drawing a ball of the kind 0 in the k-th drawing is equal to 1/k. Because of the divergence of the series $\sum_{k=1}^{\infty} 1/k$ the Borel-Cantelli lemma yields the conclusion.

<u>COROLLARY 7</u> The series $\sum_{k=1}^{\infty} 1/b_k$ and $\sum_{K=1}^{\infty} 1/b_k^2$ are divergent (w.p.1).

With these results we obtain the following main theorem of this section.

<u>THEOREM 8</u>

$$\lim_{n\to\infty} P\left\{ (T_n - C_{K_n}) / \sqrt{S_{K_n}^2} \leq x \;\middle|\; B_1', \ldots, B_{K_n}' \right\} = \Phi(x) \qquad \text{(w.p.1)}$$

where

$$C_{K_n} = \sum_{k=1}^{K_n} 1/b_k \ , \quad S_{K_n}^2 = \sum_{k=1}^{K_n} 1/b_k^2 \quad .$$

References

1. Athreya, K.B. / Ney, P. (1972): Branching Processes. Berlin - Heidelberg - New York, Springer

2. Bailey, N.T.J. (1964): The elements of stochastic processes with applications to natural science. New York, Wiley

3. Bailey, N.T.J. (1968): Stochastic birth, death and migration processes for spatially distributed populations. Biometrika 55, 189-198

4. Bühler, W.J. (1971): Generations and degree of relationship in supercritical Markov branching processes. Z. Wahrscheinlichkeitstheorie Verw. Gebiete 18, 141-152

5. Erdelyi, A. / Magnus, W. / Oberhettinger, F. / Tricomi, F.G. (1954): Tables of Integral Transforms. Bateman Manuscript Project. New York

6. Jordan, Ch. (1950): Calculus of Finite Differences. New York, Chelsea Publishing Company

SPATIAL MODELS IN THE EPIDEMIOLOGY OF INFECTIOUS DISEASES

Norman T. J. Bailey
Health Statistical Methodology Unit
World Health Organization, Geneva

CONTENTS

1. Introduction

We shall concentrate in this review on the population dynamics of infectious diseases, with special emphasis on the explicit handling of spatial components. Even the simplest nonspatial infectious disease models are moderately complex. The Law of Mass Action is used to describe the rate at which new infections occur in a community, i.e. making it proportional to the product of the numbers of both susceptible and infectious persons existing at any particular time. This makes the process nonlinear.

In small groups like families or classrooms, chance plays a major part in the spread of infection requiring the use of stochastic formulations. We expect deterministic approximations to be valid for large population sizes. But care is needed, as shown by Bartlett's demonstration of the importance of stochastic modelling to describe fade-out phenomena for measles in towns with populations as large as 250 000. When the complication of spatial structure is added, very complex systems can result. Nevertheless, a steadily increasing literature over the past twenty years has enabled both a growth of insight into the general nature of spatially structured infectious disease processes, and the development of practical methods of investigating actual public health problems.

Many high-priority applications arise in tropical countries where the populations are often spread out over a large number of small villages, automatically involving the spatial aspect. Moreover, many of the major infectious diseases are parasitic in nature. This means that we have to consider, not just one population with the person-to-person transmission of disease, but two or more populations harbouring different stages of the parasite. In malaria, for example, there is the human host and the mosquito vector, with the parasite obligatorily switching back and forth between the two. But, although eradicated from many areas of the globe, there has been a recent resurgence of this disease. Hundreds of millions of people are at risk, there are an estimated 150 million new cases every year, and in Africa alone about a million children under the age of five die from it every year.

The literature on the spatial analysis of infectious disease processes is both more extensive than one might imagine, and very variable in content. It ranges from highly abstruse theory to empirical descriptions of natural phenomena. I propose to review available material with a primarily mathematical content. In general this means describing approaches, assumptions, results and implications, without for the most part embarking on actual mathematical derivations. The latter are liable to involve considerable intricacy, and those who are interested should consult the publications cited for further details.

2. Classification of models

The broad field of coverage is large and heterogeneous. A classification for the different types of modelling undertaken will make reviewing the field easier, will introduce certain perspectives, and identify major trends and gaps in our knowledge. The typology used makes no claim to uniqueness, but is introduced solely as a matter of convenience.

First of all, there is a growing literature on general theoretical modelling. We expect such work to provide an intellectual background for the investigation of actual processes in the real world. It should clarify thought, as well as give insight into the underlying mechanisms of geographical spread. We may also become more aware of the dangers of research that yields neither theoretical insight nor practical control capabilities, but only generates sophisticated mathematical complexity for its own sake.

Two subdivisions may be noted. One handles the spatial aspect in terms of functions that compactly describe the spatial structure as a whole. The other avoids such simplifying functional assumptions, conceiving the total population as composed of a large number of inter-connected subgroups. The spatial aspect must then be introduced via an ad hoc correspondence between the subgroups and some chosen topographical pattern.

Secondly, there is the area of specific applications. Here we want models to describe with sufficient accuracy the behaviour of a given disease over a given geographical area. Good qualitative agreement is required between theory and observation, even though rigorous statistical fitting may not be possible. For example, in recent years in the USSR multiple-subgroup models have been used to predict how influenza, starting in one big town, may spread to any other big town. Again, much work has been done on describing the spread of rabies in Europe, using the computerized analysis of surveillance data, as well as on simulation studies to test methods of control.

The above two areas both involve major mathematical components. There is also a third important area, though we shall not review it in any detail. This is medical geography (see, for example, McGlashan, 1972), which includes spatial variations in health and disease, particularly as causatively related to environmental conditions. Although much of the quantitative work reported so far involves mainly descriptive data, we may expect a more powerful theoretical basis to emerge in due course. The development of a strong, mutually supportive, interaction between spatial epidemiological modelling and medical geography could have great practical consequences.

3. General theory: functional treatment of space

3.1 Elementary threshold theorems

The earliest, simplest and most elegant treatment of a spatial epidemic process is the derivation by David Kendall (in discussion on Bartlett, 1957) of a spatial threshold theorem closely resembling the nonspatial result of Kermack & McKendrick (1927). This is a good starting point, because of the relative simplicity of the treatment, and the use of equations similar to those in elementary nonspatial theory.

Basically, we have a homogeneously mixing population of x susceptibles, y infectives and z removed individuals who are isolated, dead, or recovered and immune. It is assumed that, with an infection-rate β and a removal-rate γ, there are $\beta xy\Delta t$ new infections and $\gamma y\Delta t$ removals in time Δt. The differential equations describing this process are

$$\left.\begin{aligned} dx/dt &= -\beta xy, \\ dy/dt &= \beta xy - \gamma y, \\ dz/dt &= \gamma y, \end{aligned}\right\} \qquad (3.1.1)$$

where $x + y + z = n$, a constant.

We define a relative removal-rate $\rho = \gamma/\beta$. From the second equation in (3.1.1) it is evident that no epidemic will even start to build up unless $\rho < x_0$, the initial number of susceptibles. This gives a genuine threshold. Kermack & McKendrick (1927) showed that if $x_0 = \rho + \nu$, where ν is small, then the total epidemic size is 2ν. Thus the number of susceptibles is reduced to $\rho - \nu$, i.e. as far below the threshold as it was originally above.

More generally, let the intensity of the epidemic be defined as the proportion of susceptibles finally contracting the disease, i.e.

$$i = z_\infty/n, \qquad (3.1.2)$$

where the outbreak is triggered off by a mere trace of infection, i.e. $x_0 \doteq n$. An exact treatment of (3.1.1) by Kendall (1956) shows that i is given by the unique positive root of

$$i = 1 - e^{-in/\rho}. \qquad (3.1.3)$$

The spatial generalization developed by Kendall redefines x, y and z as the relative proportions of susceptibles, infectives and removals, relevant to a small area dS surrounding a point P, where the density of individuals per unit area is a constant σ. The quantities x, y and z are now functions of both time and position, with $x + y + z = 1$. Equations (3.1.1) can then be generalized to

$$\left.\begin{aligned} \partial x/\partial t &= -\beta\sigma x\breve{y}, \\ \partial y/\partial t &= \beta\sigma x\breve{y} - \gamma y, \\ \partial z/\partial t &= \gamma y, \end{aligned}\right\} \qquad (3.1.4)$$

where $\breve{y}$ is a spatially weighted average of y as defined by

$$\tilde{y}(P,t) = \iint \lambda(PQ)y(Q,t)dS. \qquad (3.1.5)$$

In (3.1.4) the parameters β and γ are constants, and the explicit factor σ in the term $\beta\sigma x\tilde{y}$ keeps the infection-rate independent of density. We assume in (3.1.5) that dS is an areal element surrounding Q, and that $\lambda(PQ)$ is a suitably chosen non-negative weighting coefficient.

Suppose we start with an initial focus of infection, uniformly spread for example over a circle centred on the origin and having radius a, putting

$$\left.\begin{aligned} \varepsilon(P) &= \alpha, \ OP < a, \\ &= 0, \ OP > a. \end{aligned}\right\} \qquad (3.1.6)$$

Certain mild restrictions have to be introduced in order to make progress, such that $\lambda(PQ)$ vanishes if PQ is greater than some constant b. It can then be shown that as $t \to \infty$, x and z must respectively decrease and increase to limits X and Z, with $y \to Y \equiv 0$. In addition, Kendall made the reasonable conjecture that $Z(P)$ decreases steadily as $OP \to \infty$, at least for P sufficiently far from the origin. It can then be shown that $Z(P)$ must tend to a limit, ζ say, satisfying

$$\zeta = 1 - e^{-\sigma\zeta/\rho}. \qquad (3.1.7)$$

This always has a zero root. But there is a unique positive root as well if $\sigma > \rho$, in which case $Z(P)$ must converge to $\zeta > 0$. We then have a <u>pandemic</u>, in which the proportion of individuals contracting the disease is ultimately at least as great as ζ, however distant we are from the original focus. The quantity ζ is called the <u>severity</u> of the pandemic.

We thus have Kendall's Pandemic Threshold Theorem, which states that:

(i) there will be a pandemic if and only if the population density σ <u>exceeds</u> the threshold density $\rho = \gamma/\beta$;

(ii) when there is a pandemic its severity ζ is given by the unique positive root of equation (3.1.7).

There is thus a close resemblance between (3.1.7) for the severity of a pandemic and (3.1.3) for the intensity of a non-spatial epidemic.

An analogue of Kendall's threshold theorem has recently been obtained by Radcliffe (1976) for a host-and-vector type of infection where there are two populations of individuals exhibiting a kind of criss-cross infection: susceptible hosts are infected only by infectious vectors, and conversely. This corresponds to the infectious organism being obliged to spend two successive phases of its life-cycle in host and vector alternately.

An immediate extension of (3.1.1), followed by a spatialisation in terms of densities corresponding to (3.1.4), gives

$$\left.\begin{aligned} \partial x/\partial t &= -\beta\sigma x\tilde{y}', & \partial x'/\partial t &= -\beta'\sigma' x'\tilde{y}, \\ \partial y/\partial t &= \beta\sigma x\tilde{y}' - \gamma y, & \partial y'/\partial t &= \beta'\sigma' x'\tilde{y} - \gamma' y', \\ \partial z/\partial t &= \gamma y, & \partial z'/\partial t &= \gamma' y'. \end{aligned}\right\} \qquad (3.1.8)$$

Following arguments analogous to those in Kendall's original exposition, Radcliffe arrives at a Host-Vector Pandemic Threshold Theorem that states:

(i) there will be a pandemic if and only if the product of the two initial population densities $\sigma\sigma'$ exceeds the threshold given by the product of the two relative removal-rates $\rho\rho'$;

(ii) when there is a pandemic the severities in the two populations, ζ and ζ', are given by the unique non-zero solutions of

$$\zeta = 1 - \exp(-\beta\sigma\zeta'/\gamma'), \quad \zeta' = 1 - \exp(-\beta'\sigma'\zeta/\gamma). \qquad (3.1.9)$$

Radcliffe pointed out that this model would be most appropriate to a viral disease producing immunity in both hosts and vectors. In malaria we might consider the assumptions approximately valid by supposing that temporary immunity was conferred on the human hosts during the spread of the disease, and interpreting all the mosquito vector removals as deaths (since mosquitoes do not lose acquired infectiousness before they die).

The big advantage of such simplified investigations is that they crystallise out the essential mechanisms underlying complex processes in the real world. This provides theoretical insight and an impetus for further research, as well as reaching mathematically formulated conclusions with an immediate epidemiological interpretation. Attempts to introduce additional structure into such models increase the complexity of the mathematical work alarmingly. For example, in Thieme (1977) a model is specified involving the four principal epidemiological classes: susceptibles; infected individuals in a latent state who are not yet infectious; actively infectious infectives; and removed individuals. The infectious influence of an active infective can act at a distance, but its intensity is a function of the time elapsed since the receipt of the infection. There is no birth or immigration of susceptibles, though infectives can move around their initial location. Thieme concentrates on analysing the final state of the epidemic, paying particular attention to the question of whether in a large area the epidemic remains restricted to a small part or, alternatively spreads over the whole area.

The model is characterised by a nonlinear integral equation with a unique solution. This solution has a temporally asymptotic limit which is the minimal solution of another nonlinear integral equation, leading to a generalization of Kendall's Pandemic Threshold Theorem. However, the complexity of the mathematical analysis, and the subtlety of the necessary analytic assumptions, are such that it seems difficult to state the final threshold result succinctly in epidemiological terms. Further work may remedy this deficiency. More recent studies by Thieme (1979) on the asymptotic behaviour of population spread can also be applied to epidemic processes.

3.2 Diffusion models

Another kind of model, in some ways more specialized than Kendall's model above though capable of greater development in detail, is the diffusion representation of Bartlett (1956). We modify the nonspatial model of (3.1.1) by introducing a birth-rate μ providing a steady supply of new susceptibles, and keep the population constant by assuming a corresponding death-rate in the removal group. In general, this gives rise to endemic phenomena with an equilibrium state. As pointed out to me by Professor H. W. Hethcote (private communication), this model, originally due to H. E. Soper (1929), is not quite well-posed and needs certain minor adjustments. We therefore include in the model, not a steady rate of accession of new susceptibles, but equal birth- and death-rates, μ, per individual, for all groups, still assuming that all new births are susceptible. (Compare Dietz, 1976.)

Bartlett's approach is to generalise these equations, first for two population groups characterised by the variables $(\underline{x}_1, \underline{y}_1)$ and $(\underline{x}_2, \underline{y}_2)$, and then to an intuitively obvious extension to a spatially continuous distribution of population with $\underline{x}$ and $\underline{y}$ representing densities of susceptibles and infectives. The adjusted equations can accordingly be written as

$$\left.\begin{aligned} \partial x/\partial t &= -\beta x(y + \alpha\nabla^2 y) + \mu(n - x) + \theta\nabla^2 x, \\ \partial y/\partial t &= \beta x(y + \alpha\nabla^2 y) - (\gamma + \mu)y + \phi\nabla^2 y, \end{aligned}\right\} \qquad (3.2.1)$$

where $\underline{x} \equiv \underline{x}(\zeta,\eta,\underline{t})$, $\underline{y} \equiv \underline{y}(\zeta,\eta,\underline{t})$, and $\nabla^2 \equiv \partial^2/\partial\xi^2 + \partial^2/\partial\eta^2$, ξ and η being appropriately chosen spatial coordinates. The factor α refers to the spatial influence of infectives, and θ and ϕ to the migration-rates of susceptibles and infectives, respectively. We assume that all these influences are both local and isotropic: this accounts for the operator ∇^2. Looking at small first-order fluctuations in the usual way, we find that oscillatory solutions are possible, but all involve damping in time.

A particular application of this theory is to study the consequences of introducing a small focus of infection into a susceptible area. For the initial stages of such a process, which may be either epidemic (with $\mu = 0$) or endemic, $\underline{x}$ can be taken as approximately constant. We then need only the second equation of (3.2.1), which can be written as

$$\partial y/\partial t = Ay + B\nabla^2 y; \quad A = \beta x - \gamma - \mu, \; B = \phi + \alpha\beta x. \qquad (3.2.2)$$

This is a standard diffusion equation, and can be solved by using a Fourier transform of the spatial coordinates. The solution is

$$y = \{C/(2Bt)\} \exp\{At - (\xi^2 + \eta^2)/(4Bt)\}, \qquad (3.2.3)$$

where $\underline{C}$ can be determined from the initial conditions $\underline{y} = \underline{y}_0$, $\underline{t} = 0$. Note that if $\underline{x} < (\gamma + \mu)/\beta$, we have $\underline{A} < 0$ and the effect of the infection at any point is only transient. For a substantial epidemic result, we must have $\underline{x} > (\gamma + \mu)/\beta$, i.e. above the threshold value, with $\underline{A} > 0$.

Let $\underline{y}_R$ be the amount of infection outside a circle of radius $\underline{R}$. Then

$$y_R = \int_{\xi^2+\eta^2 \geq R^2} y\, d\xi\, d\eta = 2\pi C \exp\{At - (R^2/4Bt)\}. \qquad (3.2.4)$$

It follows from (3.2.4) that

$$R = 2(AB)^{\frac{1}{2}}\, t\left[1 - \{\log(y_R/2\pi C)\}/(At)\right]^{\frac{1}{2}}. \tag{3.2.5}$$

Hence as $t \to \infty$ we have $R \to 2(AB)^{\frac{1}{2}}t$. For any given value of y_R the circle of radius R grows at an approximately constant rate R/t when t is large. This rate is the velocity of propagation, V, and has the limiting value

$$V \equiv 2(AB)^{\frac{1}{2}} = 2\{(\beta x - \gamma - \mu)(\phi + \alpha\beta x)\}^{\frac{1}{2}}. \tag{3.2.6}$$

Such ideas are rather over-simplified, but the concept of foci of infection is extremely important. In the discussion to Soper's paper (1929), it was suggested by Dr Halliday that a measles epidemic starting in Glasgow took about 24 weeks to cover the city. This was the first indication that this kind of analysis was both possible and worthwhile. If the city could then be roughly regarded as a circle with a two mile radius, the observed velocity of propagation would be about 1/12 mile per week. Further approximate calculations are possible. To estimate the diffusion coefficient $\sigma = (2B)^{\frac{1}{2}} = V/(2A)^{\frac{1}{2}}$, we can put $\mu = 0$, and $\gamma = \frac{1}{2}$ (corresponding to an incubation period of two weeks). Bartlett (1956) suggested putting $\beta x/\gamma = 2$, i.e. a density of susceptibles equal to twice the relative removal-rate. In this case, $A \doteq \gamma = \frac{1}{2}$, giving some idea of the standard deviation involved in the diffusion process.

Radcliffe (1973) has developed an extension to cover the initial geographical spread of an infection involving both hosts and vectors, as in malaria with the parasite obligatorily alternating between man and mosquito. The mathematical treatment now deals with two populations of infectives, and a more complicated expression is obtained for the limiting velocity of propagation.

An application has also been made by Noble (1974) in a tentative analysis of data on the spread of plague (Black Death) in Europe in the 14th century. In this discussion, the insect vector is ignored, and person-to-person infection is assumed. This would be realistic in so far as a pneumonic form of the disease predominated. Some interesting numerical results are obtained.

Though diffusion approximations are important and useful, certain shortcomings have become apparent in recent years (see Mollison, 1972b, 1977; Daniels, 1975, 1977). The method adopted above, leading to the result in (3.2.6), involves an invalid use of the central limit theorem. A more searching analysis can, however, be made using the spatial contact approach described below in Section 3.4. Generally speaking, the advancing wave ultimately settles down to a certain minimum velocity c_0: the diffusion approximation leads to a spuriously low value of c_0, the correct value depending on the exact form of the contact distribution envisaged.

3.3 Stochastic models

It is natural to enquire about developing stochastic versions of the deterministic models described above. Infection processes are inherently probabilistic in nature, and stochastic considerations can have important consequences in quite large populations. Even in a nonspatial situation considerable complications can arise. We start by looking at a typical model involving infection and removal, plus the accession of fresh susceptibles, the population containing r susceptibles and s infectives. Let the probability of (r,s) occurring at time t be $p_{rs}(t)$. During time Δt the transitions are given by

$$\left.\begin{array}{rlcl} \text{(i)} & \Pr\{(r,s) \to (r-1,\ s+1)\} & = & \beta rs\Delta t, \\ \text{(ii)} & \Pr\{(r,s) \to (r,\ s-1)\} & = & \gamma s\Delta t, \\ \text{(iii)} & \Pr\{(r,s) \to (r+1,\ s)\} & = & \mu\Delta t, \end{array}\right\} \qquad (3.3.1)$$

where β and γ are the usual infection- and removal-rates, respectively, and μ is an immigration-rate for fresh susceptibles. Standard methods allow us to write down immediately the partial differential equation for the associated probability-generating function, $\Pi(z,w,t) = \sum p_{rs}(t)z^r w^s$,

$$\frac{\partial\Pi}{\partial t} = \beta(w^2 - zw)\frac{\partial^2\Pi}{\partial z\partial w} + \gamma(1 - w)\frac{\partial\Pi}{\partial w} + \mu(z - 1)\Pi. \qquad (3.3.2)$$

This is intractable as it stands, but a number of approximations and modifications have yielded valuable insights into observable epidemiological phenomena (see Bailey, 1975). Generalization to a spatial distribution of infection is fairly easily done in formal terms, but explicit manipulation cannot as yet be taken very far.

Following Bartlett (1956), we replace the probability-generating function $\Pi(z, w, t)$, containing the two auxiliary variables z and w, by a probability-generating functional $\Pi\{z(\mathbf{a}), w(\mathbf{a}), t\}$, containing the two spatial functions $z(\mathbf{a})$ and $w(\mathbf{a})$, where the vector $\mathbf{a} \equiv (\xi,\eta)$. In principle, the "point" stochastic process can be studied by investigating the functional Π. In practice, there are severe difficulties in making progress without major simplifying assumptions.

We first introduce migration-rates θ and ϕ for susceptibles and infectives, respectively. If the process is temporally homogeneous, then γ and μ are constants, but β, θ and ϕ are functions of position, i.e. the distance between any two relevant points indicated by $\mathbf{a}$ and $\mathbf{b}$. If the process is also isotropic, only the modulus $|\mathbf{a} - \mathbf{b}|$ will be involved.

We further simplify the discussion by looking only at the initial stages of an epidemic, with an approximately constant number of susceptibles $n(\mathbf{a}) = n$, say. Thus Π is a functional of $w(\mathbf{a})$ only. The process is now multiplicative, and the 'backward' equation is simpler in form than the 'forward' one implied above. Suppose that we decide to study the consequences of introducing a single case of infection into a spatially distributed population of susceptibles. Let $\Pi(\mathbf{v})$ be the appropriate functional. It can be shown that it satisfies the equation.

$$\begin{aligned}\partial\Pi(\mathbf{v})/\partial t = \gamma\{1 - \Pi(\mathbf{v})\} &+ n\int\beta(\mathbf{u} - \mathbf{v})\,\Pi(\mathbf{v})\,\{\Pi(\mathbf{u}) - 1\}\,d\mathbf{u} \\ &+ \int\phi(\mathbf{v} - \mathbf{u})\,\{\Pi(\mathbf{u}) - \Pi(\mathbf{v})\}\,d\mathbf{u}. \qquad (3.3.3)\end{aligned}$$

This is still moderately formidable, but the chance of extinction $p_t(\mathbf{v})$ at time t can be obtained by putting $w(\mathbf{a}) \equiv 0$. If we make the further **homogeneity assumption that**

$$p_t(\underset{\sim}{v}) = p_t(\underset{\sim}{u}) = p_t, \tag{3.3.4}$$

then (3.3.3) reduces to

$$\partial p_t/\partial t = \gamma(1-p_t) + n\, p_t(p_t-1)B; \quad B = \int\beta(\underset{\sim}{u})d\underset{\sim}{u}. \tag{3.3.5}$$

This is identical with the equation for the chance of extinction in a simple birth-and-death process with birth- and death-rates $\underline{nB}$ and γ, with

$$p_t = \gamma\{e^{(nB-\gamma)t} - 1\}/\{nBe^{(nB-\gamma)t} - \gamma\}. \tag{3.3.6}$$

Naturally, these results apply only to the beginning of the process in question, and it is not clear what the implications are for long-term stability.

Expressions for moment-densities can also be found. Suppose we want the mean density of infectives given by

$$m(\underset{\sim}{a},t)\, d\underset{\sim}{a} \equiv E\{ds(\underset{\sim}{a},t)\}, \tag{3.3.7}$$

where $\underline{ds(\underset{\sim}{a},t)}$ is the stochastic number of infectives at time $\underline{t}$ in $d\underset{\sim}{a}$. We can readily obtain an appropriate partial differential equation for $\underline{m(\underset{\sim}{a})}$, which can first be integrated with respect to ξ and η using a Fourier transform, and then integrated with respect to time. With local and isotropic infection and diffusion we recover the deterministic results of Section 3.2.

A generalization of this stochastic approach has also been made by Radcliffe (1973) to the corresponding host-and-vector model. General solution has not been possible, but, making assumptions similar to the above, Radcliffe shows that once the process of infection starts from a focus, it does not die out but leads to a spreading wave of infection.

It should, however, be pointed out that the criticisms mentioned at the end of Section 3.2 in connexion with the inadequacy of diffusion approximations also apply here. Bartlett (1960, Ch. 8) considered the properties of the second-order density

$$m_2(\underset{\sim}{a}, \underset{\sim}{b}, t)\, d\underset{\sim}{a}\, d\underset{\sim}{b} \equiv E\{ds(\underset{\sim}{a}, t)\, ds(\underset{\sim}{b}, t)\}, \tag{3.3.8}$$

and suggested that the velocities of propagation in all individual realisations of the stochastic model would ultimately coincide. Daniels (1977) has, however, shown in a more searching analysis that there are three different ranges for the velocity of propagation with quite distinct properties. Bartlett's argument really applies only to the lower range, while the actual minimum velocity $\underline{c_o}$ in fact lies in the middle range.

Finally, we should mention the simplifying approach of McNeil (1972), which appeals to notions of geographical spread but does not incorporate them explicitly. Working in the context of a simple epidemic, involving infection without recovery, he considers the general idea of an infection spreading deterministically from a focus with the rate of spread proportional to the length of the circular boundary. This leads to the use of a modified nonspatial stochastic model in which the chance of infection in a small interval of time is proportional to the <u>square root</u> of the product of the numbers of infectives and susceptibles. A good deal of subsequent analysis is possible, involving both deterministic diffusion approximations as well as stochastic arguments (see Bailey, 1975).

3.4 Spatial contact models

Some years after his Pandemic Threshold Theorem, Kendall (1965) returned to the problem of introducing a spatial factor into epidemic models. To clarify the issues involved, he confined attention to a one-dimensional population uniformly distributed along a line, and looked for waves of infection travelling through the linear community.

Specifically, he assumed an overall population density of σ, with $\sigma x(s,t)$, $\sigma y(s,t)$ and $\sigma z(s,t)$, as the local densities of susceptibles, infectives and removals, respectively, at distance s from the origin and at time t. The appropriate analogue of (3.1.4) is

$$\left.\begin{aligned} \partial x/\partial t &= -\beta\sigma x\tilde{y}, \\ \partial y/\partial t &= \beta\sigma x\tilde{y} - \gamma y, \\ \partial z/\partial t &= \gamma y, \end{aligned}\right\} \qquad (3.4.1)$$

which is identical in form with (3.1.4), but now referred to the single spatial coordinate s. If we absorb the constant $\beta\sigma$ into the time-scale and write $\rho = \gamma/\beta$, as usual, the first two, independent equations of (3.4.1) appear as

$$\partial x/\partial t = -x\tilde{y}; \quad \partial y/\partial t = x\tilde{y} - (\rho/\sigma)y. \qquad (3.4.2)$$

To simplify the treatment Kendall proposed two modifications. The first was to use a diffusion approximation to the spatial average $\tilde{y}$ given by

$$\tilde{y} \doteq y + k(\partial^2 y/\partial s^2). \qquad (3.4.3)$$

The second was to look for a travelling wave solution putting

$$x(s,t) \equiv x(s - ct), \quad y(s,t) \equiv y(s - ct), \qquad (3.4.4)$$

where $c > 0$ is the velocity with which the wave travels, the form of the wave being preserved.

A useful substitution is then to put $S = x - ct$, as a single argument for x and y. Reasonable boundary conditions can then be easily imposed. Using the variable S to locate the phenomenon relative to the frame of axes travelling at velocity c with the wave, we can substitute (3.4.3) in (3.4.2) and change the independent variable from t to S. This gives

$$\left.\begin{aligned} c(dx/dS) &= x\{y + k(d^2y/dS^2)\}, \\ c(dy/dS) &= -x\{y + k(d^2y/dS^2)\} + (\rho/\sigma)y. \end{aligned}\right\} \qquad (3.4.5)$$

By careful phase-plane analysis Kendall showed that no waves were possible unless $\sigma > \rho$, in which case no waves could have a smaller velocity than the limiting value

$$c_0 = 2k^{\frac{1}{2}}\beta\sigma\{1 - (\rho/\sigma)\}^{\frac{1}{2}}, \qquad (3.4.6)$$

where the time scale appropriate to (3.4.1) has now been restored, though all velocities greater than or equal to c_0 would be possible. This constitutes a threshold theorem for waves.

Kendall pointed out that similar methods had been used by Kolmogorov, Petrovsky & Piscounoff (1937) in their investigation of the wave of advance of an advantageous gene. The latter had shown that an initial discontinuity of arbitrary form, but of spatially restricted extent, would generate a wave ultimately travelling with the minimum velocity. It seemed likely that an epidemic starting with a local concentration of infectives would generate two waves travelling in opposite directions towards $\underline{s} = \pm\infty$ having the asymptotic velocity $\underline{c}_o$.

Kendall's work led directly to the extensive studies of Mollison. The latter made a thorough investigation of the deterministic simple epidemic (i.e. involving infection only, with no removal) for a continuous population distributed along a line (Mollison, 1968; 1970; 1972a, b), with many generalizations and extensions, including stochastic situations (Mollison, 1972b, 1977, 1978). A considerable amount of highly intricate mathematical argumentation is involved, and the interested reader should consult the works cited. However, in spite of the complexity, the models are very clearly defined, the resulting theorems are completely intelligible, and the practical interpretations in epidemiological terms are not difficult to envisage.

Mollison's linear deterministic simple epidemic model is defined as follows. The population is supposed to be uniformly spread along a line with density σ. At any particular location $\underline{s}$ there are only two types of individuals, the susceptible with proportion $\underline{x}(\underline{s},\underline{t})$ and the infective with proportion $\underline{y}(\underline{s},\underline{t})$, where $\underline{x}(\underline{s},\underline{t}) + \underline{y}(\underline{s},\underline{t}) = 1$. There is no removal or recovery from the infective state. And it is assumed that, at time $\underline{t}$, the infectious influence of infectives at $\underline{s} - \underline{u}$ on susceptibles at $\underline{s}$ is given by $\beta\underline{x}(\underline{s},\underline{t})\underline{y}(\underline{s}-\underline{u},\underline{t})d\underline{V}(\underline{u})$, where V is the contact distribution.

The basic equation for the propagation of infection can be taken from the second line of (3.4.1) with $\gamma = 0$, i.e.

$$\partial y/\partial t = \beta\sigma\tilde{y}(1-y); \quad \tilde{y} = \int y(s-u,t)dV(u). \tag{3.4.7}$$

This convolution, taken over the whole population space, is clearly rather general in form. Suppose we consider first the special form of the contact distribution given by the double negative exponential

$$dV(u) = \tfrac{1}{2}\alpha e^{-\alpha|u|}du, \quad -\infty < u < \infty. \tag{3.4.8}$$

Mollison (1972a) proves there exists a minimal velocity $\underline{c}_o$, given by

$$c_o = (3^{3/2}/2)(\beta\sigma/\alpha), \tag{3.4.9}$$

such that waveforms exist only for velocities equal to or greater than $\underline{c}_o$, and that for each such velocity there exists a unique waveform. (Note that Mollison's treatment has α and β interchanged: we have used the above definition to retain uniformity in our own notation.) In other words, there is a velocity spectrum given by the interval $[\underline{c}_o,\infty)$.

Mollison (1972b) has shown that, for contact distributions with negative-exponentia bounded tails, upper bounds greater than or equal to some $\underline{c}_*$ can be found, with the minimal velocity corresponding (at least) to all initial conditions involving a bounded set of infectives. When $\underline{V}$ is strictly negative exponential, $\underline{c}_*$ and $\underline{c}_o$ coincide. Thus, for negative-exponentially bounded $\underline{V}$, the diffusion approximation and the exact equation are in qualitative agreement: both suggest a single velocity of special importance, with all higher velocities being possible. However, it also emerged that, for all other contact distributions with greater spread, there existed no velocity spectrum. From this it follows that fairly restrictive conditions are required for diffusion approximations to be valid, and they must accordingly be treated with considerable caution.

In the case of the negative-exponential contact distribution given by (3.4.8), the use of Kendall's local diffusion approximation gives a slightly lower value for $\underline{c}_o$. The variance in (3.4.8) is $2\alpha^{-2}$, so in (3.4.3) Kendall's approximation requires $\underline{k} = \alpha^{-2}$. Substituting in (3.4.6), and letting $\rho \to 0$ for application to the simple epidemic without removal, gives $\underline{c}_o = 2(\beta\sigma/\alpha)$. Here the factor on the right is 2, compared with $3^{3/2}/2 \doteqdot 2.6$ in (3.4.9).

While the above work concentrates on the one-dimensional problem, extensions to two dimensions can easily be envisaged though are more complicated (Kendall, 1965; Mollison, 1972b). Again, what about epidemics <u>with</u> removal? As pointed out by Mollison (1972b) one would expect an epidemic with removal to be bounded in some sense by a simple epidemic, since it would be implausible that removal should speed up an outbreak.

Of greater interest are stochastic treatments. Thus Mollison (1972b) carefully defines a simple stochastic epidemic in which the population consists of discrete individuals, with a constant number σ living at each integer point of the real axis. Using a slightly artificial 'germ' model, it is assumed that cross-infection between any infective-susceptible pair is a Poisson process with parameter $\alpha(\underline{n})$, where $\underline{n}$ is the separation of the pair. Each individual emits germs in a Poisson process with parameter $\sigma\sum_{n=-\infty}^{\infty}\alpha(\underline{n})$, where we define the probability density on the integers given by $\underline{v}(\underline{n}) = \alpha(\underline{n})/\alpha$. Thus $\underline{v}(\underline{n})$ corresponds to the contact distribution $\underline{dV}(\underline{s})$ used for the continuously distributed population of the deterministic model. The net result is that the condition for a finite rate of propagation is that the variance of $\underline{V}$ must be finite.

Mollison (1972b, 1977) also carried out a number of interesting simulations. When the variance of $\underline{V}$ is infinite, the epidemic progresses by 'a series of "great leaps forward" which get increasingly out of hand.' But there is an intermediate case in which $\underline{V}$ has finite variance but the tails are not exponentially bounded. Simulations here show an alternation of great leaps forward and periods of steady progress.

A number of further extensions and generalizations have also been made in

Mollison's (1977) review.   All aspects of spatial contact models are covered, with special reference to ecological and epidemic spread, giving an integrated and consolidated account of the existing 'state of the art'.   A further paper concentrates on very broadly defined Markovian contact processes (Mollison, 1978), which include non-linear processes such as simple and general epidemics as special cases.

Investigations have also been made by Daniels (1975, 1977), adopting various approximation procedures including saddlepoint methods (Daniels, 1960).   Several results obtained verify the conclusions previously reached by Mollison who used more general, difficult and rigorous topological arguments.   Not only are the approximate solutions found relatively easily, but the behaviour of the epidemics are properly described and the critical velocities are correctly calculated.   Perhaps a more vigorous use of such applied mathematical methods would be worth encouraging.   Heuristic investigations of many difficult problems might produce results of practical value fairly quickly, with rigorous confirmation or disproof following later.

Finally, the investigations of Diekmann (1978) on thresholds and travelling waves must be mentioned.   A very general model is introduced, involving a variable period of infectiousness, which leads to nonlinear integral equations of mixed Volterra-Fredholm type.   Results, relating to thresholds and the velocities of travelling waves, are to some extent analogous to the findings of Mollison described above.

3.5   Lattice models

While the generality possible in many models is in principle very powerful, near-intractability has led some writers to investigate models for which the spatial structure is more specific in the hopes of achieving explicitly constructible solutions. This expectation has, however, been only partially realised.   One of the biggest simplifications is to retain some patterning of spatial relationships, without going all the way in the direction of multi-site models, by assuming that the individuals in a discrete population are positioned at the vertices of a two-dimensional lattice.

One of the first models of this kind was due to Morgan & Welsh (1965).   An initial population of susceptibles is distributed over the vertices $(\underline{x},\underline{y})$ of a square lattice, $\underline{x}$ and $\underline{y}$ being confined to the non-negative integers.   The epidemic process is started by supposing the individual at the origin $(0,0)$ to be infected from elsewhere at $\underline{t} = 0$.   The infection is allowed to spread stochastically to nearest neighbours according to a Poisson process, but only in the directions of increasing $\underline{x}$ and $\underline{y}$.   Several theoretical results were proved, and certain conjectures were derived from the study of computer simulations.   Hammersley (1966) subsequently generalised the situation and proved some of the conjectures in terms of first-passage percolation theory.   A more extensive theoretical discussion is given by Hammersley & Welsh (1965). The latter give the example of a large orchard planted with trees at the vertices of a square lattice. We may suppose that an infected tree can infect only the four nearest neighbours. If the probability of an infected tree spreading the disease to a neighbouring susceptible tree is p, we may ask what is the probability $P_N(p)$ that the disease will spread to more than N other trees? We want to choose the spacing between the trees so that p is below the appropriate critical value for the lattice, so that the chance of a large number of trees being affected is negligible. An additional degree of reality can be introduced by incorporating a time-distribution for the spread of infection. First-passage percolation theory deals with the time at which infection first passes beyond some given region.

Unfortunately, the literature on percolation processes has never been adapted to deal convincingly with concrete epidemiological phenomena. And some of the assumptions of percolation theory, in which the stochastic mechanisms are a property of the medium, are at variance with the usual epidemiological picture of disease transmission. In addition, Hammersley has indicated in his presentation to the Conference that many percolation-theoretic results require populations of order 10^6 per dimension, thus tending to rule out applications to human diseases spread over two dimensions. There is also the recent work of Gertsbakh (1977), in which random graph theory is investigated as a possible basis for the study of, amongst other things, certain epidemic processes. This whole subject would be worth investigating further to see if practical insights could be obtained into specific disease problems, with results being couched in terms that could be understood by epidemiologists.

Lattice models lend themselves to computerised descriptions, and are easily used for simulation studies. Thus much of the theoretical work of Bartlett (1957, 1960, 1961), was complemented by computer studies which allowed an examination of the spatial aspect in addition to the inevitable time factor. Bartlett used a 6 × 6 grid of cells, each cell being individually treated as a nonspatial sub-population. The spatial element was introduced by migration between cells having a common boundary. Bartlett showed theoretically that, to a first approximation, the critical community size would be uninfluenced by spatial factors unless there was a diffusion of individuals. The simulations indicated further that, even with a moderate degree of diffusion, results did not differ substantially from nonspatial models. This tended to show that the spatial aspect was less important than previously supposed.

A more purely spatial simulation model was investigated by Bailey (1967), who considered a population of susceptibles located at the vertices of a square lattice bounded by the lines $x = \pm k$, $y = \pm k$ (for integral k), the total population being of size $n = (2k + 1)^2$. Calculations were mainly based on $k = 5$, $n = 121$ and $k = 11$, $n = 441$. It was assumed that the epidemic was started off by a single infective individual at the origin, with the infection spreading in a probabilistic way to the eight nearest neighbours. A fixed generation-time was chosen, characterised by a constant latent period and very short subsequent infectious period. The chance of 'adequate contact' was defined by the probability $\underline{p}$. Both 'simple' epidemics with no recovery, and 'general' epidemics with recovery and immunity, were considered.

For $\underline{p} = 1$, the simple epidemic clearly spreads deterministically with a square with sides $\underline{x} = \pm \underline{g} = \underline{y}$ being covered by the $\underline{g}$'th generation. Results can be summarised by the average epidemic curves for individuals in subpopulations on the perimeters of squares centred on the origin. The graphical presentations showed epidemics spreading in steady waves away from the origin. When $\underline{p}$ is small, e.g. 0.2, the wave is more spread out the further one moves from the initial focus. But with $\underline{p}$ large, e.g. 0.8, the wave moves out to the boundary more or less unchanged in form. For general epidemics the behaviour is rather different. When $\underline{p} = 0.2$ outbreaks are more or less localised around the origin, but with $\underline{p} = 0.8$ the resulting spread is practically indistinguishable from the corresponding simple epidemic. There is clearly a threshold effect, the critical value of $\underline{p}$ being in the neighbourhood of 0.3.

Similar models have also been independently investigated for the study of infected plaques by Schwöbel, Geidel & Lorenz (1966), who worked mostly with a hexagonal lattice (though similar results were found for square and triangular lattices as well) and also observed the phenomenon of an approximately linear rate of spread.

Again, the process of two-dimensional carcinogenic growth, studied by Williams & Bjerknes (1971, 1972), is also relevant. It is supposed that the generative mass in the epithelium is situated in the basal layer, above which are differentiated cells. Whenever a basal cell divides the daughter cells displace a randomly chosen neighbouring cell upwards. Cells may be normal or abnormal, the latter leading to tumours. Cancer cells multiply κ times as fast as normal cells, and a clone of cancer cells gradually invades the whole basal layer and the tissue above it. If the time scale $\underline{t}$ is changed to $\tau = \kappa\underline{t}$, and we let $\kappa \to \infty$, a continuous-time, two dimensional, simple epidemic process is obtained. Thus the theory of the Williams-Bjerknes model could have implications for epidemiological spread.

In particular, it was observed empirically from the computer printouts that, independently of κ, the periphery of the abnormal clone presented a greater crinkliness than expected, corresponding to a dimensionality of 1.1. The conjecture that this result might hold generally was disproved by Mollison (1972c), but the limited validity actually observed might have some practical implications for the examination of histological specimens.

Additional investigations, both theoretical and empirical, have been made of the Williams-Bjerknes model by Downham & Morgan (1973 a,b) and Downham & Green (1976). Further theoretical studies have also been made on a wide class of random growth processes in tessellations and honeycombs by Richardson (1973). Many of the previously discussed processes appear as special cases.

4. General theory: multiple-site models

4.1 Deterministic models

In the previous section we considered models in which the spatial element has a structure that can be represented compactly in functional form. This simplification provides insight into general qualitative behaviour, but lacks realism since actual geographical areas can be highly heterogeneous. An alternative is to employ models that simply involve several subpopulations located at different sites. The subpopulations, each homogeneously mixing within itself, are mutually interacting. It should be noted that we are not committed to a spatial interpretation: the different subpopulations might be different social groups having different degrees of contact within and between them.

A few results for the general deterministic epidemic model were given many years ago by Wilson & Worcester (1945) in the case of two subgroups. Later, Rushton & Mautner (1955) discussed the case of an arbitrary number of groups, m, for the simple epidemic involving no recovery. Thus, if the i'th subpopulation, of size n_i, contains $x_i(t)$ susceptibles at time t, and if β_{ij} is the cross-infection rate between the i'th and j'th subpopulations, we have

$$dx_i/dt = -x_i\sum_{j=1}^{m}\beta_{ij}(n_j-x_j); \quad i=1, \ldots, m, \tag{4.1.1}$$

where β_{ii} is the internal infection-rate for the i'th subpopulation. Rushton & Mautner examined various special cases, and obtained a formal general solution to the whole set of equations when $n_i = n$, $\beta_{ii} = \beta$, and $\beta_{ij} = \beta q$ (all $i \neq j$). General qualitative conclusions included the observation that, corresponding to certain epidemiological observations, there was a tendency for the overall epidemic curve to fall more slowly than it rose, as compared with the single population situation for which the curve is symmetrical. See also Watson (1972) described in the following section.

A considerable amount of deterministic multiple-site modelling has been done over a number of years by the Soviet School of mathematical epidemiology (see Baroyan *et al*. 1967 and later; Rvachev, 1968 and later). This has however been developed *ad hoc* in the applied context of influenza prediction and control in the USSR, as is more appropriately reviewed in Section 5.3.

4.2 Stochastic models

Stochastic models in nonlinear situations are notoriously difficult to handle in full detail. This is especially true for epidemic models, even in their simplest form incorporating only infection without subsequent recovery. Thus Haskey (1957) dealt with a simple stochastic epidemic model involving just two separate subpopulations. He obtained explicit results only when the infection-rates within groups and between groups were all equal. A simplification of these extremely complex derivations was later obtained by Becker (1968).

A general stochastic multiple-site model was introduced by Neyman & Scott (1964). This incorporates specific realistic features that had been previously ignored, and uses a position-dependent branching process model with a discrete-time parameter corresponding to the incubation period. The number of susceptibles infected by a given infective is made to depend on the latter's position. An individual who becomes infected at any point is allowed to move away and exert his infectious influence from some other location. Although careful attention is paid to mathematical rigour, a serious restriction is introduced by the assumption that the stock of susceptibles is not reduced by each new infection. Thus the model could be approximately valid at the start of an epidemic, but would become progressively less so as time passed. Conclusions about the total epidemic size cannot therefore be drawn satisfactorily. Indeed, this is pointed out by Neyman & Scott themselves in noting that immunising ten per cent. of the susceptible population leads to an expected epidemic size of only nine cases, a conclusion obviously at variance with epidemiological observation!

The recursion methods introduced by Severo (1967) can also be applied to the type of modelling first developed by Haskey. Generalisations by Severo (1969) deal with stochastic cross-infection between several groups. Further results were later obtained by Kryscio (1972). These recursion methods are however all highly algorithmic in character, and do not readily reveal the qualitative behaviour of the processes in question. More recently, Capasso (1976) has made an elaborate mathematical investigation of the stochastic two-subpopulation situation, establishing a number of rigorously defined theorems.

The most easily appreciated work is that of Watson (1972). Basically, we envisage an extension of the general stochastic epidemic model, though ignoring the accession of new susceptibles. Let the total population of size N be divided into m distinct subpopulations C_i of size N_i $(1 \leq i \leq m)$, having X_i susceptibles and Y_i infectives. We define the infection-rate for susceptibles in C_i due to infectives in C_j as β_{ij} $(1 \leq i, j \leq m)$, and the removal-rate in C_i as $\gamma_i (1 \leq i \leq m)$. An appropriate partial differential equation for the associated multivariate probability-generating function for susceptibles and infectives is easily written, but so far appears to be intractable.

Some general progress is possible by first restricting the discussion to 'equivalent' classes defined by

$$N_i \equiv N_o; \quad \gamma_i \equiv \gamma; \quad \beta_{ii} \equiv \beta; \quad \beta_{ij} = q\beta, \; i \neq j, \tag{4.2.1}$$

where $0 \leq q \leq 1$. This model then corresponds to the general stochastic epidemic except for the assumption of homogeneous mixing. Suppose we write $X = \sum_i X_i$ and $Y = \sum_i Y_i$ for the total numbers of susceptibles and infectives, respectively, at time t. The basic transitions during Δt are

$$\left.\begin{aligned} &\text{(i)} \quad \Pr\{(X,Y) \to (X-1,\, Y+1)\} = \textstyle\sum_{i,j} \beta_{ij} X_i Y_j \,\Delta t, \\ &\text{(ii)} \quad \Pr\{(X,Y) \to (X,\, Y-1)\} = \textstyle\sum_i \gamma_i Y_i \,\Delta t, \end{aligned}\right\} \tag{4.2.2}$$

corresponding to (3.3.1) without arrival of any new susceptibles. Since $\underline{X}_i \leq \underline{N}_o$, we easily find, using (4.2.1), that

$$\textstyle\sum_{i,j} \beta_{ij} X_i Y_j \,\Delta t \leq Q\beta N_o Y \,\Delta t; \qquad Q = 1 + (m-1)q. \tag{4.2.3}$$

We now envisage a linear birth-and-death process $\underline{Y}_u(\underline{t})$, having birth-rate $\underline{Q}\beta\, \underline{N}_o$ and death-rate γ, which is an upper bounding process for $\underline{Y}\ (\underline{t})$ and approximating the latter in the early stages. It follows therefore that a major outbreak is possible only if $\underline{Q}\beta\, \underline{N}_o > \gamma$. In this case there will be an outbreak affecting most subpopulations if $\underline{q}$ is sufficiently large, while only a few subpopulations will be affected if $\underline{q}$ is small. Otherwise, with $\underline{Q}\beta\, \underline{N}_o < \gamma$, only a minor outbreak is to be expected in the initially infected group. These qualitative distinctions between a 'generalized outbreak', a 'restricted outbreak', and a 'localized outbreak', could have some practical relevance to insight and control at the public health level.

Watson showed how the equations for the deterministic analogue, regarded as an approximation to the stochastic model, could be solved numerically to give information on the severity of outbreaks in different subpopulations. But such a model could not reflect the full behaviour of the stochastic process, especially in regard to extinction.

Another line of development in the direction of obtaining explicitly computed solutions has recently been opened up by Lynne Billard (1976) in which she uses a matrix-blocking technique to simplify the recursive approach of Severo. This method had previously been applied with some success in a nonspatial situation. Further development is required, and the question of whether insight can be obtained into the structure and qualitative behaviour of the dynamic processes involved also needs to be investigated.

5. Epidemiological applications

5.1 Preliminary remarks

It is evident from the foregoing discussion of the primarily mathematical aspects of the spatial spread of infectious diseases that the whole subject is extremely complex. The existence of neat and tidy practical applications is hardly to be expected. Nevertheless, the basic ideas can be of considerable importance in providing general insight, as well as indicating what techniques might be of practical value. Three areas have been chosen to illustrate the potentialities. The first is Bartlett's work, which has always, in both spatial and nonspatial investigations, been discussed in relation to available data. The main emphasis has been on measles, outbreaks of which have been well documented over the years. The second area is the Soviet work by Baroyan and his colleagues on the control of influenza in the USSR. This work has been pursued in the context of practical public health decision-making and introduces concepts not to be found elsewhere. Thirdly, there has been a great deal of effort put into the surveillance and control of the animal disease rabies, which is of considerable danger to man. Theoretical work in this area is not extensive, but a lot of empirical studies have been made, including computerised simulations.

If the very difficult subject of the spatial spread of infectious disease is not to become overloaded with purely mathematical technicalities, far more work is needed to increase the range of practical applications. Not only do we need better methodologies for predicting the spread of disease in highly heterogeneous populations and geographical regions, but more knowledge is required about *when* spatial aspects must be included and when they can be safely neglected.

5.2 Measles

A series of important contributions to our understanding of the population dynamics of measles were made by Bartlett over the years 1953-60 (see Bailey, 1975). These were particularly related to both endemic and recurrent epidemic disease. Previous deterministic theory could explain the occurrence of periodic oscillations, but unfortunately only in a *damped* form - thus contradicting familiar epidemiological observations. Bartlett showed how a stochastic formulation could lead to an *undamped* series of recurrent outbreaks, and this led to an analysis of the concept of 'critical community size', above which oscillations would tend to be maintained indefinitely, ar below which the infection would be liable to fade out. Theory was carefully combinec with suitable simulation studies, and compared with extensive data on the incidence o measles in communities of varying sizes in both the UK and the USA. A figure of about 250 000 represented the actual critical level of gross community size for measles

Most of Bartlett's work was in fact nonspatial. But it provided an important context for a number of studies on the implications of introducing a spatial aspect. We noted in Section 3.5 that so far as the overall qualitative phenomena were concerned the spatial factor had little influence on critical size.

In most developed countries measles is not a serious health hazard, and is not a high priority for the practical application of mathematical modelling. However, in many developing countries, especially in Africa, the disease is an important cause of childhood mortality, and receives special attention from multiple immunization programmes, such as those promoted by WHO. Subsequent evaluation of these field programmes may have to take account of the population dynamics of the diseases involved. New theoretical investigations may then be required for models representing the communities concerned. It is not obvious in advance that spatial factors will be found unimportant, either for measles or the other diseases involved. I believe, therefore, that the subject of spatial modelling in the context of developing country immunization programmes is worthy of further investigation.

For recent discussions of <u>nonspatial</u> undamped oscillations, see Dietz (1976).

5.3   Influenza

The Soviet work on influenza started with Baroyan & Rvachev in about 1967 and has developed over a ten-year period (see Bailey, 1975, Ch. 19). This work began in the context of public health control. It contains certain specific features and modifications which, though important, make the basic ideas more difficult to understand.

We begin by looking at a simple generalization of the deterministic model in (3.1.1) to a multi-site situation which includes migration as the mechanism of contact between subpopulations. Suppose that there are $\underline{G}$ subpopulations. In the $\underline{i}$'th group at time $\underline{t}$ there are $\underline{x}_i$ susceptibles, $\underline{y}_i$ circulating infectives, $\underline{u}_i$ ill and isolated individuals, and $\underline{z}_i$ recovered and immune persons. Thus

$$x_i + y_i + u_i + z_i = n_i, \quad i = 1, \ldots, G. \qquad (5.3.1)$$

We make an explicit distinction between those who are infected, but circulating, and those who are ill and isolated, since the former are subject to migration but the latter are not. Let us concentrate on morbidity, rather than mortality. We assume a universal infection-rate within groups of β, and a removal-rate γ, as before. The number of new infections in the $\underline{i}$'th group in $\Delta\underline{t}$ is thus $\beta\underline{x}_i\,\underline{y}_i\Delta t$, and the number of removals is $\gamma\underline{y}_i\Delta t$. The latter are observed new cases, and join the ill and isolated group, for which we assume a recovery-rate ω. In $\Delta\underline{t}$ therefore $\omega\underline{u}_i\Delta\underline{t}$ individuals move to the recovered and immune class. Next, we assume a migration-rate μ_{ij} for the migration from group $\underline{i}$ to group $\underline{j}$ for all classes of individuals except those who are ill and isolated.

The basic dynamic equations are evidently

$$\left.\begin{aligned} dx_i/dt &= -\beta x_i\, y_i + \textstyle\sum_{j=1}^{G}(\mu_{ji}\, x_j - \mu_{ij}\, x_i), \\ dy_i/dt &= \beta x_i\, y_i + \textstyle\sum_{j=1}^{G}(\mu_{ji}\, y_j - \mu_{ij}\, y_i) - \gamma y_i, \\ du_i/dt &= \gamma y_i - \omega u_i, \\ dz_i/dt &= \omega u_i + \textstyle\sum_{j=1}^{G}(\mu_{ji}\, z_j - \mu_{ij}\, z_i). \end{aligned}\right\} \qquad (5.3.2)$$

In general n_i is a variable quantity, but we can hold the overall population number, N, constant by setting $\sum n_i = N$.

The Soviet work introduces both simplifications and complications. It is assumed that the contact-rate *per* individual is constant, as when the rate at which an individual makes contacts is more or less independent of the total size of the community in which he lives. We must then replace β in (5.3.2) by β/n_i. It is also assumed that the class of u_i ill and isolated persons can be ignored. A new element is introduced into the treatment of the migration factor by supposing that the actual rate of transfer from group i to group j is σ_{ij}. This applies to all individuals that are not ill and isolated, so in general $\sigma_{ij} = \mu_{ij}(n_i - u_i)$, or $\sigma_{ij} \doteq \mu_{ij}\, n_i$, where the ill class can be ignored in the overall dynamics.

The basic model of Baroyan *et al.* (1967 and later) makes a more sophisticated allowance for an age-dependent recovery-rate, in contrast with the exponential decay implicit in the foregoing formulation. Suppose that $F(\tau)$ is the distribution function for the duration of infection, so that $F(\tau)$ is the proportion of newly infected persons who are infectious up to a time τ measured from the point of infection. An auxiliary variable $\phi_i(t, \ell)$ is introduced, indicating that the number of individuals in group i at time t who became infected in the time-interval $(\ell, \ell + \Delta\ell)$ is $\phi_i(t, \ell)\, \Delta\ell$ (for $\ell < t$), whether these individuals have recovered by time t or not. It follows that

$$x_i(t) = x_i(0) - \int_0^t \phi_i(t, \ell)\, d\ell, \qquad (5.3.3)$$

$$y_i(t) = \int_0^\infty \phi_i(t, t-\tau)\{1-F(\tau)\}\, d\tau, \qquad (5.3.4)$$

where (5.3.3) says that the number of susceptibles remaining at time t equals the initial number less all the infections that have taken place, and (5.3.4) says that the number of infectives at time t is given by the total number of previous infections which have not yet terminated. A maximum duration of infection T may be assumed with $F(\tau) \equiv 1$ for $\tau \geqslant T$.

The number of infections in t is

$$\phi_i(t, t) = \beta x_i\, y_i/n_i = \{\beta x_i(t)/n_i\} \int_0^\infty \phi_i(t, t-\tau)\{1-F(\tau)\}\, d\tau, \qquad (5.3.5)$$

and the rate of change of susceptibles is

$$\frac{dx_i(t)}{dt} = -\,\phi_i(t, t) + \sum_{j=1}^{G}\left(\frac{\sigma_{ji}\, x_j(t)}{n_j} - \frac{\sigma_{ij}\, x_i(t)}{n_i}\right). \qquad (5.3.6)$$

In addition, $\phi_i(\underline{t}, \underline{\ell})\, \Delta\underline{\ell}$ is affected only by migration, so

$$\frac{\partial \phi_i(t, \ell)}{\partial t} = \sum_{j=1}^{G} \left(\frac{\sigma_{ji}\phi_j(t, \ell)}{n_j} - \frac{\sigma_{ij}\phi_i(t, \ell)}{n_i} \right). \qquad (5.3.7)$$

Equations (5.3.5) to (5.3.7), for $\underline{i}$ = 1, ..., $\underline{G}$, suffice to determine the behaviour of the whole system, given appropriate initial conditions.

The function $\underline{F}(\underline{t})$ was estimated approximately from epidemiological data. To start with, it was assumed that $\sigma_{ij} = \underline{k}\; \underline{n}_i\; \underline{n}_j$, where $\underline{k} \doteq 2^{-32}$, as determined from official inter-city transportation statistics. On the basis of a computerised solution of (5.3.5) to (5.3.7), the expected courses of secondary epidemics in 128 cities were predicted, using only data for an initial epidemic in Leningrad. Results were obtained for 23 cities where data were available, showing close agreement between theoretical predictions and observed data in respect of size, time of onset, and maximum point of outbreak.

These results are impressive. But some statisticians say that the statistical material published is insufficient to convince them; while some virologists say that the whole influenza situation is biologically so complex that it must defy any attempt to provide a mathematical basis for prediction. However, the recent work of Spicer (1979) on influenza in the UK is very encouraging. The controversy is important and further investigation is urgently needed. The fundamental notion of using the transportation structure to help quantify the migrational aspect is a big advance. Moreover, there is an ever-present danger of a world-wide pandemic arising from some new mutant strain of influenza virus. This warrants a major study of the possibility of building a world-wide model to help plan ahead for the intervention and control that would be needed.

5.4   Rabies

Human cases of rabies normally occur through contact with infected animals, and animal reservoirs of the disease constitute a permanent threat in many parts of the world. In some countries the domestic dog is the chief source of danger, but in Europe the fox is the main reservoir of infection. Increasing publicity has been given to the slow but steady spread of wild-life rabies across central Europe during the past 30 years. France was not reached until 1975 and Italy not until 1977. The United Kingdom is still free at the time of writing. During the five-year period 1972-76 there were some 82 000 laboratory-confirmed animal cases in countries in the European Region of WHO. Post-exposure treatment was given to more than a million people, despite which there were over 600 human deaths. The advancing wave of infection gives rise to considerable public anxiety, and has substantial economic implications as well. Improved understanding of the mechanisms of the spread of rabies is required for both surveillance and control. While theoretical work provides a useful background for developing the population dynamics of rabies, specific applications are uncommon.

Many data have been collected in recent years and the broad quantitative picture has been well established, e.g. Toma & Andral (1970) and WHO (1978 a, b). More detailed investigations, using sophisticated methods of computerised spatio-temporal analysis have been made by Moegle et al. (1974) and Bögel et al. (1976). These latter approaches are essentially empirical in nature, though they provide a basis for regular surveillance and lead to suggestions for the public health control. The presentation of the two-dimensional spread of rabies in time has been studied by Sayers, Mansourian, Phan Tan & Bögel (1977), using computerised signal and pattern analysis with anisotropic filtering and the estimation of descriptive statistical parameters. The routine use of such methods, coupled with visual computer displays, could materially assist surveillance. None of the work models the underlying mechanisms of rabies spread. Nevertheless, it is worth considering because of its direct public health relevance, and the possibility of stimulating further investigations into rabies dynamics that might have valuable practical consequences.

Recently, however, Berger (1976) has developed a lattice type model involving random distributions of foxes over the grid, as well as incorporating a stochastic spread of infection. He concluded that in practice the spread of rabies in areas with high fox population densities could not be controlled by vaccination only, and additional measures such as hunting and gassing would be required.

Subsequently, Lambinet et al. (1978) developed a deterministic spatial model, which for continuous spatial coordinates (x, y) leads to a pair of integro-differential equations for the densities of healthy and infected foxes. No analytic solution seems possible, but extensive computerised simulations were carried out after suitable discretization of the basic equations. The basic formulation is complicated, but from a computational point of view there were no undue difficulties. Most parameters were considered to be known approximately by ecologists, while the more elusive effective contact-rate was estimated by adjusting it empirically to give results in broad qualitative agreement with known facts. The authors recognise the existence of certain over-simplications and the need to estimate the contact-rate more definitively. Nevertheless, they foresee the possibility of further modifications and developments, leading to validation of the model and ultimate application to public health control through the presentation and study of alternative scenarios of prophylaxis.

6. Medical geography

Finally, some comments must be added on the relevance of spatial models of infectious disease spread to the whole corpus of knowledge comprising medical geography. This is a widely based subject covering the spatial distribution of all kinds of diseases, both communicable and noncommunicable, especially as related to geographical and environmental factors, infectious disease dynamics being clearly of importance here. Conversely, a closer acquaintance with the work of medical geographers might be of considerable value to epidemiological modellers. The public health control of disease would be improved by integrating activities in the overlapping areas of public health decision-making, epidemiology, medical geography and applied biomathematical modelling.

A useful introduction to the scope of medical geography is the book edited by McGlashan (1972). Part V deals with Disease Diffusion and contains four separate important contributions by Brownlea, Haggett, Hunter and Tinline. For an excellent review of some aspects of model building and spatial pattern analysis in human geography, see Cliff & Ord (1975). However, their emphasis is on stationary spatial processes, and they rely on fitting linear time series rather than adopting the dynamic modelling technique we have concentrated on. The approach of Ripley (1977) would be helpful in avoiding these limitations. The literature also contains a large number of descriptive studies of the spread of infectious diseases, by epidemiologists, geographers and biomathematicians, that could be legitimately classified under different headings. For example, Splaine, Lintott & Angulo (1974) deal with the computerised contour mapping of variola minor in Brazil; Hugh-Jones (1976) discusses the spread of foot-and-mouth disease in actual dairy herds in the UK through the use of a simulation model based on a space-time grid; McGlashan (1977) investigates the spread of viral hepatitis in Tasmania through a planar graph approach; etc.

It would take us too far afield to attempt to review this wider area. But the works quoted above, together with their attached references and bibliographies, should enable interested readers to become better acquainted with the field of medical geography. Closer collaboration between disease modellers, epidemiologists and medical geographers is essential, in my view, for the development of realistic quantitative instruments that can be used by public health decision-makers for predicting the scenarios associated with different available control strategies.

REFERENCES

The following bibliography contains all sources mentioned in the text, and includes for convenience a few additional items as well. The list has been made as complete as possible for specifically mathematical references, though there may be many omissions. I should be glad to receive information about further relevant material. So far as epidemiological applications are concerned, the field is much wider and it is impossible to attempt a comprehensive list here. Similarly, only selected references to the vast subject of medical geography can be incorporated.

ARONSON, D. G. (1977). The asymptotic speed of propagation of a simple epidemic. In Nonlinear Diffusion, Research Notes in Mathematics, 14, 1-23. (W. E. Fitzgibbon, III & H. F. Walker, Eds.) London: Pitman.

ATKINSON, C. and REUTER, G. E. H. (1976). Deterministic epidemic waves. Math. Proc. Camb. Phil. Soc., 80, 315-330.

BAILEY, N. T. J. (1967). The simulation of stochastic epidemics in two dimensions. Proc. Fifth Berkeley Symp. Math. Statist. & Prob., 4, 237-257. Berkeley and Los Angeles: University of California.

BAILEY, N. T. J. (1975). The Mathematical Theory of Infectious Diseases. London: Griffin.

BAROYAN, O. V., BASILEVSKY, U. V., ERMAKOV, V. V., FRANK, K. D., RVACHEV, L. A. and SHASHKOV, V. A. (1970). Computer modelling of influenza epidemics for large-scale systems of cities and territories. (Working paper for WHO Symposium on Quantitative Epidemiology, Moscow, 23-27 Nov. 1970. In English and in Russian. Summarised in Baroyan et al., 1971.)

BAROYAN, O. V., GENCHIKOV, L. A., RVACHEV, L. A. and SHASHKOV, V. A. (1969). An attempt at large-scale influenza epidemic modelling by means of a computer. Bull. Int. Epid. Assoc., 18, 22-31.

BAROYAN, O. V. and RVACHEV, L. A. (1967). Deterministic epidemic models for a territory with a transport network. Kibernetika, 3, 67-74. (In Russian.)

BAROYAN, O. V. and RVACHEV, L. A. (1968). Some epidemiological experiments carried out on an electronic computer. Vestnik Akad. Med. Nauk, 23 (5), 32-34. (In Russian.)

BAROYAN, O. V., RVACHEV, L. A., BASILEVSKY, U. V., ERMAKOV, V. V., FRANK, K. D., RVACHEV, M. A. and SHASHKOV, V. A. (1971). Computer modelling of influenza epidemics for the whole country (USSR). Adv. Appl. Prob., 3, 224-226.

BAROYAN, O. V., RVACHEV, L. A., FRANK, K. D., SHASHKOV, V. A. and BASILEVSKY, U.V. (1973). Mathematical and computer modelling of influenza epidemics in the U.S.S.R. Vestnik Akad. Med. Nauk, 28 (5), 26-30. (In Russian.)

BAROYAN, O. V., RVACHEV, L. A. and IVANNIKOV, Yu. G. (1977). The modelling and Prediction of Influenza Epidemics in the USSR. Moscow. (In Russian.)

BAROYAN, O. V., ZHDANOV, V. M., SOLOVIEV, V. D., ZAKSTELSKAYA, L. Ya., RVACHEV, L. A., URBAKH, Yu. V., ERMAKOV, V. V. and ANTONOVA, I. V. Prospects of machine modelling of influenza epidemics for the territory of the USSR. Zh. Mikrobiol. Epidemiol. Immunobiol., 49 (5), 3-11. (In Russian.)

BARTLETT, M. S. (1954). Processus stochastiques ponctuels. Ann. Inst. Poincaré, 14 (Fasc. 1), 35-60.

BARTLETT, M. S. (1956). Deterministic and stochastic models for recurrent epidemics. Proc. Third Berkeley Symp. Math. Statist. & Prob., 4, 81-109. Berkeley and Los Angeles: University of California.

BARTLETT, M. S. (1957). Measles periodicity and community size. J. Roy. Statist. Soc., A, 120, 48-70.

BARTLETT, M. S. (1960). Stochastic Population Models in Ecology and Epidemiology. London: Methuen.

BARTLETT, M. S. (1961). Monte Carlo studies in ecology and epidemiology. Proc. Fourth Berkeley Symp. Math. Statist. & Prob., 4, 39-55. Berkeley and Los Angeles: University of California.

BARTLETT, M. S. (1978). Stochastic Processes (3rd edn). Cambridge University Press.

BECKER, N. G. (1968). The spread of an epidemic to fixed groups within the population. Biometrics, 24, 1007-1014.

BERGER, J. (1976). Model of rabies control. Lecture Notes in Biomathematics, 11, 74-88.

BILLARD, Lynne (1976). A stochastic general epidemic in m sub-population. J. Appl. Prob., 13, 567-572.

BÖGEL, K., MOEGLE, H., KNORPP, F., ARATA, A., DIETZ, K. and DIETHELM, P. (1976) Characteristics of the spread of a wildlife rabies epidemic in Europe. Bull. Wld Hlth Org., 54, 433-447.

BROWNLEA, A. A. (1972). Modelling the geographic epidemiology of infectious hepatitis. In Medical Geography (ed. N. D. McGlashan), 279-300. London: Methuen.

CAPASSO, V. (1976). A stochastic model for epidemics in two interacting regions of a large population. Bollettino U.M.I. (5) 13-B, 216-235.

CLIFF, A. D. and ORD, J. K. (1975). Model building and the analysis of spatial patterns in human geography. J. Roy. Statist. Soc., B, 37, 297-328.
DANIELS, H. E. (1960). Approximate solutions of Green's type for univariate stochastic processes. J. Roy. Statist. Soc., B, 22, 376-401.
DANIELS, H. E. (1975). The deterministic spread of a simple epidemic. In Perspectives in Probability and Statistics (ed. J. Gani), 373-386. London: Academic Press.
DANIELS, H. E. (1977). The advancing wave in a spatial birth process. J. Appl. Prob., 14, 689-701.
DAVIDSON, J. (1976). Velocities of propagation for general spatial epidemics. (M.Sc. Thesis: University of Edinburgh).
DIEKMANN, O. (1978). Thresholds and travelling waves for the geographical spread of infection. J. Math. Biol., 6, 109-130.
DIETZ, K. (1976). The incidence of infectious disease under the influence of seasonal fluctuations. Lecture Notes in Biomathematics, 11, 1-15.
DOWNHAM, D. Y. and GREEN, D. H. (1976). Inference for a two-dimensional stochastic growth model. Biometrika, 63, 551-554.
DOWNHAM, D. Y. and MORGAN, R. K. B. (1973a). Growth of abnormal cells. Nature, 242, 528-530.
DOWNHAM, D. Y. and MORGAN, R. K. B. (1973b). A stochastic model for a two-dimensional growth on a square lattice. Bull. ISI, 45 (1), 324-331.
GERTSBAKH, I. B., (1977). Epidemic processes on a random graph: some preliminary results. J. Appl. Prob., 14, 427-438.
HAGGETT, P. (1972). Contagious processes in a planar graph: an epidemiological application. In Medical Geography (ed. N. D. McGlashan), 307-324. London: Methuen.
HAMMERSLEY, J. M. (1966). First-passage percolation. J. Roy. Statist. Soc., B, 28, 491-496.
HAMMERSLEY, J. M. and WELSH, D. J. A. (1965). First-passage percolation, subadditive processes, stochastic networks and generalized renewal theory. In Bernoulli-Bayes-Laplace (ed. J. Neyman and L. M. LeCam), 61-110. Berlin: Springer.
HASKEY, H. W. (1957). Stochastic cross-infection between two otherwise isolated groups. Biometrika, 44, 193-204.
HUGH-JONES, M. E. (1976). A simulation spatial model of the spread of foot-and-mouth disease through the primary movement of milk. J. Hyg. Camb., 77, 1-9.
HUNTER, J. M. (1972). River Blindness in Nangodi, Northern Ghana: A hypothesis of cyclical advance and retreat. In Medical Geography (ed. N. D. McGlashan), 261-278. London: Methuen.
KELKER, D. (1973). A random walk epidemic simulation. J. Amer. Statist. Assoc., 68, 821-823.
KENDALL, D. G. (1965). Mathematical models of the spread of infection. In Mathematics and Computer Science in Biology and Medicine, 213-225. London: H.M.S.O.
KERMACK, W. O. and McKENDRICK, A. G. (1927). Contributions to the mathematical theory of epidemics (Part I). Proc. Roy. Soc., A, 115, 700-721.
KOLMOGOROFF, A. N., PETROVSKY, I. and PISCOUNOFF, N. (1937). Etude de l'équation de la diffusion avec croissance de la quantité de matière et son application à un problème biologique. Bull. Univ. État. Moscou, A 1 (6), 1-25.
KRYSCIO, R. J. (1972). The transition probabilities of the extended simple stochastic epidemic model and the Haskey model. J. Appl. Prob., 9, 471-485.
LAMBINET, D., BOISVIEUX, J.-F., MALLET, A., ARTOIS, M. and ANDRAL, L. (1978). Modèle mathématique de la propagation d'une épizootie de rage vulpine. Rev. Epidém. Santé Publ., 26, 9-28.
McGLASHAN, N. D. (Ed.)(1972). Medical Geography. London: Methuen.
McGLASHAN, N. D. (1977). Viral hepatitis in Tasmania. Soc. Sci. & Med., 11, 731-744.
McNEIL, D. R. (1972). On the simple stochastic epidemic. Biometrika, 59, 494-497.
MOEGLE, H., KNORPP, F., BÖGEL, K., ARATA, A., DIETZ, K. and DIETHELM, P. (1974). Zur Epidemiologie der Wildtiertollwut. Zbl. Vet. Med., B, 21, 647-659.

MOLLISON, D. (1968). Waves in an epidemic without removal. (Smith's Prize Essay: Cambridge University.)
MOLLISON, D. (1970). Spatial propagation of simple epidemics. (Ph.D. Thesis: Statistical Laboratory, Cambridge University.)
MOLLISON, D. (1972a). Possible velocities for a simple epidemic. Adv. Appl. Prob., 4, 233-257.
MOLLISON, D. (1972b). The rate of spatial propagation of simple epidemics. Proc. Sixth Berkeley Symp. Math. Statist. & Prob., 3, 579-614. Berkeley and Los Angeles: Univ. Calif.
MOLLISON, D. (1972c). Conjecture on the spread of infection in two dimensions disproved. Nature, 240, 467-468.
MOLLISON, D. (1977). Spatial contact models for ecological and epidemic spread. J. Roy. Statist. Soc., B, 39, 283-326.
MOLLISON, D. (1978). Markovian contact processes. Adv. Appl. Prob., 10, 85-108.
MORGAN, R. W. and WELSH, D. J. A. (1965). A two-dimensional Poisson growth process. J. Roy. Statist. Soc., B, 27, 497-504.
NEYMAN, J. and SCOTT, E. (1964). A stochastic model of epidemics. In Stochastic Models in Medicine and Biology (ed. J. Gurland), 45-85. Madison: University of Wisconsin Press.
NOBLE, J. V. (1974). Geographical and temporal development of plagues. Nature, 250, 726-729.
RADCLIFFE, J. (1973). The initial geographical spread of host-vector and carrier-borne epidemics. J. Appl. Prob., 10, 703-717.
RADCLIFFE, J. (1976). The severity of a viral host-vector epidemic. J. Appl. Prob., 13, 791-794.
RICHARDSON, D. (1973). Random growth in a tessellation. Proc. Camb. Phil. Soc., 74, 515-528.
RIPLEY, B. D. (1977). Modelling spatial patterns. J. Roy. Statist. Soc., B, 39, 172-192.
RUSHTON, S. and MAUTNER, A. J. (1955). The deterministic model of a simple epidemic for more than one community. Biometrika, 42, 126-132.
RVACHEV, L. A. (1968). Computer modelling experiment on large scale epidemics. Dokl. Akad. Nauk, SSSR, 180 (2), 294-296. (In Russian.)
RVACHEV, L. A. (1971). A computer experiment for predicting an influenza epidemic. Dokl. Akad. Nauk, SSSR, 198 (1), 68-70. (In Russian.)
RVACHEV, L. A. (1972). Modelling medico-biological processes in a community in terms of the dynamics of continuous media. Dokl. Akad. Nauk, SSSR, 203 (3), 540-542. (In Russian.)
SAYERS, B. McA., MANSOURIAN, B. G., PHAN TAN, T. and BÖGEL, K. (1977). A pattern-analysis study of a wildlife rabies epizootic. Med. Inform., 2, 11-34
SCHWÖBEL, W., GEIDEL, H. and LORENZ, R. J. (1966). Ein Modell der Plaquebildung. Zeit. Naturf., 21, 953-959.
SEVERO, N. C. (1967). Two theorems on solutions of differential-difference equations and applications to epidemic theory. J. Appl. Prob., 4, 271-280
SEVERO, N. C. (1969). A recursion theorem on solving differential-difference equations and applications to some stochastic processes. J. Appl. Prob., 6, 673-681.
SOPER, H. E. (1929). Interpretation of periodicity in disease-prevalence. J. Roy. Statist. Soc., 92, 34-73.
SPICER, C. C. (1979). The mathematical modelling of influenza epidemics. Brit. Med. Bull., 35, 23-28.
SPLAINE, M., LINTOTT, A. P. and ANGULO, J. J. (1974). On the use of contour maps in the analysis of spread of communicable disease. J. Hyg. Camb., 73, 15-26.
THIEME, H. R. (1977). A model for the spatial spread of an epidemic. J. Math. Biol., 4, 337-351.
THIEME, H. R. (1979). Asymptotic estimates of the solutions of nonlinear integral equations and asymptotic speeds for the spread of populations. Z. angew. rein. Math. (At press.)

TINLINE, R. R. (1972). Lee wave hypothesis for the initial pattern of spread during the 1967-8 foot and mouth epizootic. In Medical Geography (ed. N. D. McGlashan), 301-306. London: Methuen.

TOMA, B. and ANDRAL, L. (1970). La rage vulpine en France. Cahiers méd. vét., 39, 98-155.

WATSON, R. K. (1972). On an epidemic in a stratified population. J. Appl. Prob., 9, 659-666.

WIERMAN, J. C. (1979). The front velocity of the simple epidemic. J. Appl. Prob., 16, 409-415.

WILLIAMS, T. and BJERKNES, R. (1971). Hyperplasia: the spread of abnormal cells through a plane lattice. Adv. Appl. Prob., 3, 210-211.

WILLIAMS, T. and BJERKNES, R. (1972). Stochastic model for abnormal clone spread through epithelial basal layer. Nature, 236, 19-21.

WILSON, E. B. and WORCESTER, J. (1945). The spread of an epidemic. Proc. Nat. Acad. Sci., Wash., 31, 327-332.

WORLD HEALTH ORGANIZATION (1978a). Rabies in Europe. Wkly Epidem. Rec., 53 61-63, 69-71, 77-78.

WORLD HEALTH ORGANIZATION (1978b). World Health (October 1978).

POPULATION CONTROLLED SPREAD AND INTENSITY OF DISEASES IN WILDLIFE

K. Bögel
World Health Organization
Geneva, Switzerland

This paper will be devoted to some special problems which should be considered by those who would like to study epidemiological conditions in wildlife with mathematical procedures and models.

Firstly I will talk about movement and migration patterns of wild animals as we observe them in behavioural studies on single animals.

Secondly I will discuss aspects of population dynamics, in particular the threshold phenomenon of minimum population densities for maintaining chains of infection and wavelike reduction and recovery of populations. If time permits I would also refer to the complex problems of multi-host reservoirs of infectious agents.

In the past years much of our knowledge has been derived from studies on wildlife rabies. In the following this disease will, therefore, be used several times as an example.

Of course, the spread of a disease is influenced by many factors as explained by Dr Bailey. In general we have, however, to consider three conditions in nature concerning the movement of animals actively carrying the infective agent, namely the contacts between animals as neighbours confined to their territories which may or may not overlap (slowly moving herds may represent a particular condition within this category of relatively short distances between cases of infection). Another type of movement is often found, seasonally dependent, in the same animal species, namely moderate distance movement to find sexual partners whereas long distance movement, generally called migration, can be observed when families break up or climatic and food conditions cause the animals to look for new territories.

The spread of epidemics observed over periods of one or more years is therefore often based on the three different movement patterns of populations and within populations. This makes modelling of epidemiological conditions very difficult, particularly in respect of diseases carried by wild animals.

The second aspect we have to consider in epidemics in wildlife is the population density and dynamics. Underlying factors are the availability of food - often with seasonal changes - competition of species for food, fertility and productivity rates, predator-prey animal relationship, hunting or population management by man, measures of disease control, and last but not least the causative agents of diseases and epidemics.

It is understood that the maintenance of a chain of infection, if not vertically transmitted almost like a genetic factor, requires a certain density of the host population, below which the contact rates are insufficient. Where one animal is able to transmit the disease only to less than 1.0 animal the disease can not spread far into populations or will die out. Populations therefore have not to be totally eliminated for complete disappearance of a disease. It is the subject of ecologists and epidemiologists to find the critical mass level of population density and to see whether and where this level can be reached by population reduction without seriously destroying the ecological balance in nature. Here all those who develop tools for modelling and quantification have wide areas of application and considerable responsibility for health of man, animals and the environment.

In our models considering reduced population densities we have always to include the reducing factors and their impact which is density dependent as well as the counteracting population pressure responsible for the recovery of a reduced population.

Here we come to a very difficult assessment of the actual conditions in nature. We know of the regular interactions between populations and infectious agents.

The wavelike increase and decrease of the disease frequency is recorded for a number of diseases, such as rabies and respiratory or intestinal virus infections. The principle is always the same, namely the reduction and increase of numbers of susceptible animals. This may be due to increasing and decreasing herd immunity or to mortality and population recovery.

In the case of rabies, mortality factors are decisive for the wavelike epidemics in the wild. How can we measure the actual conditions? We have found three indicators which, when brought together could be used for some estimates and the construction of models.

(1) the observation of the epidemic. If peaks of disease frequency are observed every four generations, one could guess that it takes 3-4 generations for complete recovery of reduced populations.

(2) the population turnover, assuming a steady population over several generations.[1]

A model of the steady population can be deduced from some population data, namely,

- the ratio of male and female adults
- the ratio of adult females over productive females
- the mean number of newborn, live animals per productive female. This is in fact the litter size.

From these three figures one can estimate the population increase through the addition of the new generation, which must be compensated for by mortality factors of the same order of magnitude. Overlapping pregnancy periods in a population during the course of the year, as we find in the mouse population, complicate this model a bit but are not major obstacles.

The turnover model for populations with one breeding season became relatively simple. From the investigation of fox populations in some study areas we obtained even some data on the various mortality factors in that system. Thus out of 3 animals of the maximum population after birth of the new fox generation 2 are removed within the 12 months until the next whelping season and of these 1.2 are taken by hunting, according to hunting statistics, and 0.8 are lost due to diseases and accidents.

(3) Besides epidemiological waves and standard conditions of population turnover, we need for our population dynamics a third complex of information, namely data on the population density. Since absolute data of density are generally not available, indicators for relative density serve the purpose. In rabies, hunting statistics have been used. Trapping indices, track counts in snow, etc. are used for foxes and other animal species. We first examine whether the indicators of population density match with the observed epidemic waves. If the result is positive we obtain from the density indicator some idea of the population reduction and increase. The increase must comply with the possible changes of the animal population turnover.

All three observations: population density indicator, model of turnover and epidemiological intervals can be applied to a nomogram. This graphical technique[2] is based on the following information on fox populations and wildlife rabies.

- We know the population turnover under standard conditions.
- We observe 3-4 year cycles of the epidemic.
- We obtain some indication on population recovery from hunting statistics.

The nomogram considers 3 dimensions:

(a) years required for complete population recovery
- epidemic observation
- density indicator

(b) mean recovery rate in per cent
- population turnover for possible maximum increase from generation to generation

(c) starting level of reduced population (before recovery)

By critical comparison of all available data the slope reflecting most probably the natural conditions can be drawn on the nomogram.

From this slope the population increase in the first recovery year can be read and a table be prepared for this first year's increase. This can also be called the population pressure of the first generation counteracting the reduction factors. The years in the nomogram may be replaced by generations on the Y-axis for other species and the slopes of the nomogram may be made on the basis of natural logarithmic growth in case of overlapping or many generations to be observed. The technique is, however, generally applicable to combine scarce information from various types of observation in a mathematical, logical and comprehensible way. The result can however not be more accurate than the logical combination of the observations, which add to each other in a non-quantifiable way and therefore result in an estimate, not more.

A third aspect of my presentation shall draw attention to the mechanics of spread if more than one single species is responsible for maintaining the chain of infection. The role of arthropod populations in major protozoal and viral diseases of man and animals provide many examples. Also in many complicated patterns of tapeworm infection, with one or two intermediate hosts in addition to the carrier of the mature worm, we have to consider the interaction of dynamics and densities of several populations.

However, let us think of a reservoir in which different animal species have the same function as active transmitters of the disease or as passive host. These different species contribute in different quantities to the reservoir and to the chain of infection. In the area of fox rabies in Europe we obtained some evidence for the existence of short chains of infections in martens. Such chains may on an average last for 3-4 months. Of course, the badger, sharing his den with the fox, is a frequent victim of rabies and this animal may contribute to the chain of infection. Rate and intensity of virus excretion through the saliva are lower in rabid badgers than in foxes and even lower in rabid martens than in rabid badgers. Nevertheless, these species add to the total vector potential of an area. This has implications for the control of a disease or the self-extinction of a disease through population reduction and must be considered in mathematical models. Our mathematical approaches have to reflect that the maintenance of a chain of infection and its interruption depend on the total number of vector or transmitting units per km^2.

References

1. Lloyd, H. G., Jensen, B., van Haaften, J. L., Niewold, F. J. J., Wandeler, A., Bögel, K., Arata, A. A.: Annual Turnover of Fox Populations in Europe, Zbl. Vet. Med. B, 23, 580-589 (1976)

2. Bögel, K., Arata, A. A., Moegle, H. and Knorpp, F.: Recovery of Reduced Fox Populations in Rabies Control, Zbl. Vet. Med. B, 21, 401-412 (1974)

Author's address: Dr K. Bögel, Veterinary Public Health, World Health Organization, 1211 Geneva 27/Switzerland

CLINES IN A DISCRETE TIME MODEL IN POPULATION GENETICS

O. Diekmann

Mathematisch Centrum
2e Boerhaavestraat 49
1091 AL Amsterdam
The Netherlands

1. INTRODUCTION

The perhaps simplest deterministic model in population genetics takes the form of the difference equation

$$(1.1)\qquad u_{n+1} = g(u_n),$$

where

$$(1.2)\qquad g(u) = \frac{\alpha u^2+\beta u(1-u)}{\alpha u^2+2\beta u(1-u)+\gamma(1-u)^2},$$

for some $\alpha,\beta,\gamma \in (0,1)$. This model originated from the works of Mendel, Fisher, Wright and Haldane and it describes the influence of selection on the genetic composition of a diploid population which consists of synchronized, nonoverlapping generations and whose members are distinguished according to the genotype with respect to one diallelic locus. In fact u_n represents the fraction of alleles of one type, say a, amongst the total number of alleles in the n-th generation. The parameters α, β and γ, the so-called survival fitnesses, are by definition the fractions of the neonates of the three genotypes aa, aA and AA which reach the reproductive stage. Finally, the model assumes that mating occurs at random.

One may verify, by some elementary but amusing calculations, that $g'(u) > 0$ for $u \in [0,1]$. Consequently the dynamics of (1.1) on the invariant set $[0,1]$ can be described in all detail and one easily arrives at a classification in terms of the parameters α, β and γ (see HADELER [8] and WEINBERGER [16]).

Since the dynamics of (1.1) are so simple, we are in an ideal situation for investigating the consequences of spatial dependence, if we bring this somehow into the model. So let us now, following WEINBERGER [16] and others, assume that the population is continuously distributed in a habitat Ω, which we think of as a closed subset of $\mathbb{R}^m$, and that the life cycle is given by the following scheme:

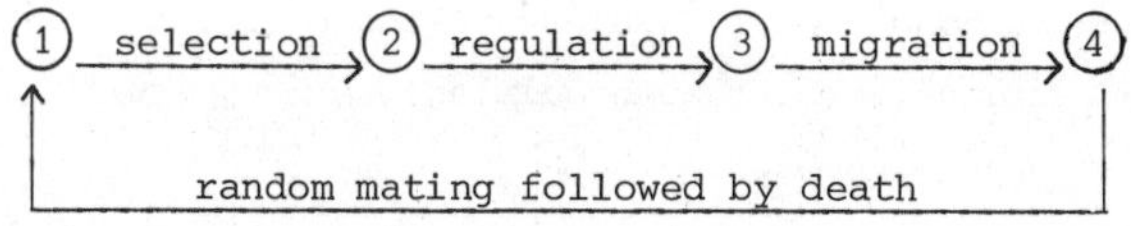

The chance that an arbitrary individual will survive the transition from stage 1 to stage 2 depends on both genotype and position. In the next two steps the genotype is irrelevant. First the total population size is cut down to the local carrying capacity $C(x)$ and subsequently individuals may migrate from their original position to a new one near by. In fact we suppose that we know a function $K_1 = K_1(\xi,\eta)$, the forward migration kernel, with the property that the number of those who migrate from a neighbourhood $\omega(y)$ of y to a neighbourhood $\omega(x)$ of x is given by

$$\int_{\omega(x)} \int_{\omega(y)} K_1(\xi,\eta)C(\eta)\,d\eta\, d\xi.$$

Since then the total population density at x after migration is given by

$$\int_{\Omega} K_1(x,\eta)C(\eta)\,d\eta$$

we conclude that the fraction of those in $\omega(x)$ who came from $\omega(y)$ is given by

$$\int_{\omega(x)} \int_{\omega(y)} K_2(\xi,\eta)\,d\eta\, d\xi,$$

where K_2, the so-called backward migration kernel, is given by

$$K_2(\xi,\eta) = \frac{K_1(\xi,\eta)C(\eta)}{\int_{\Omega} K_1(\xi,\zeta)C(\zeta)\,d\zeta} .$$

We observe that K_2 describes how frequencies, like u, transform as a consequence of migration, and that we can express K_2 explicitly in K_1 as a result of the regulation mechanism which brings about that the total population density before and after migration is independent of the generation. Finally, the life cycle is completed by the production of offspring after random mating, followed by the death of the old generation.

The analogue of equation (1.1) is given by

$$u_{n+1}(x) = \int_{\Omega} g(y,u_n(y))K_2(x,y)\,dy, \tag{1.3}$$

where $g(y,u)$ is the function defined in (1.2), but now with α, β and γ being functions of y. We shall restrict our attention to the special case $\Omega = \mathbb{R}$ and $K_2(x,y) = k(x-y)$. If α, β and γ are constant, then, roughly speaking, the asymptotic behaviour is locally the same as it was in the space independent situation (however, there are some interesting new phenomena like travelling waves and an asymptotic speed of propagation; see WEINBERGER [16]). Here we concentrate on the case where α, β and γ do depend on position (for instance, as a result of variations in temperature, humidity or soil composition) and we are interested in nonconstant stable equilibrium solutions of (1.3). Thus, in other words, we investigate whether both alleles can be preserved in the population as a consequence of an inhomogeneous environment.

The effect of an inhomogeneous environment has been analysed in the context of other, related, mathematical models. For diffusion equations the problem has been studied by HALDANE [9], STATKIN [15], NAGYLAKI [12], FLEMING [7], FIFE & PELETIER [6], PELETIER [14] and others. Problems in discrete time and discrete space have been analysed by, for instance, KARLIN & RICHTER-DYN [10] and NAGYLAKI [13]. A reference pertaining to the present model is DOWNHAM & SHAH [4].

2. EXISTENCE, UNIQUENESS AND GLOBAL ASYMPTOTIC STABILITY OF A CLINE

In this section we shall study the difference equation

$$(2.1)\qquad u_{n+1} = Tu_n, \qquad u_0 \text{ is given,}$$

where $u_n \in C = C(\mathbb{R};[0,1])$ and where the operator $T\colon C \to C$ is given by

$$(2.2)\qquad (T\phi)(x) = \int_{-\infty}^{\infty} g(y,\phi(y))k(x-y)\,dy.$$

Throughout we assume that k and g satisfy:

$$H_k\colon\ k \in L_1(\mathbb{R});\quad k \geq 0;\quad \int_{-\infty}^{\infty} k(x)\,dx = 1;$$

$$k(-x) = k(x);\quad k \text{ has a compact support.}$$

$$H_g^1\colon\ g\colon \mathbb{R}\times[0,1] \to [0,1] \text{ is a } C^1\text{-mapping;}$$

$$g(x,0) = 0 \text{ and } g(x,1) = 1,\ \forall x \in \mathbb{R};$$

$$\frac{\partial g}{\partial u} > 0 \text{ on } \mathbb{R}\times[0,1].$$

Further assumptions on g will be introduced later.

From these assumptions we infer that the operator T has fixed points ϕ_0 and ϕ_1 defined by $\phi_0(x) = 0$, $\phi_1(x) = 1$, $\forall x \in \mathbb{R}$. These we call trivial fixed points. They correspond to a situation where one of the alleles is absent from the population. In principle, we want to obtain answers to the following questions. Does there exist a nontrivial fixed point of T? If so, is it stable as an equilibrium solution of the difference equation (2.1)? If so, what is its domain of attraction (i.e., how are the functions u_0 characterized for which the sequence u_n defined by (2.1) converges towards the fixed point as $n \to \infty$)? We remark that in this context monotone nontrivial fixed points are called clines.

An important consequence of the assumptions, notably of the nonnegativity of k and the monotonicity of g with respect to u, is that the operator T is monotone (or, order-preserving): if $\phi \geq \psi$ then $T\phi \geq T\psi$ (here and in the following $\phi \geq \psi$ means $\phi(x) \geq \psi(x)$, $\forall x \in \mathbb{R}$). This observation suggests that we might try to construct suitable upper and lower solutions and subsequently use monotone iteration. In order to obtain candidates for comparison functions we first pay attention to linear convolution inequalities.

LEMMA 1. *Let* $\nu > 1$. *There exists a continuous, nonnegative function* q *with compact, nonempty support, such that*

$$\nu(q * k) \geq q.$$

For a constructive proof of this lemma see WEINBERGER [16] or [2; Lemma 3]. We point out that the support of q becomes larger as ν decreases to 1. This can be concluded from the following result, which is due to ESSÉN [5; Theorem 3.1].

LEMMA 2. *Let* ψ *be a bounded continuous function which satisfies* $\psi * k \geq \psi$. *Then* ψ *is constant.*

Our second assumption on g:

H_g^2: there exist for $i = 1,2$, $x_i \in \mathbb{R}$, $v_i \in (0,1)$, $\nu_i > 1$, such that
(i) $g(x,u) \geq \nu_1 u$ for $x \geq x_1$ and $0 \leq u \leq v_1$,
(ii) $g(x,u) \leq 1 - \nu_2(1-u)$ for $x \leq x_2$ and $1 - v_2 \leq u \leq 1$.

expresses that the allele a is protected far to the right whereas the allele A is protected far to the left. Since translation, multiplication and convolution commute, it is clear that the functions $\underline{\psi}$ and $\overline{\psi}$ defined by

$$\underline{\psi}(x;\delta,\xi) = \delta q(x-\xi),$$

$$\overline{\psi}(x;\delta,\xi) = 1 - \delta q(x+\xi),$$

satisfy $T\underline{\psi} \geq \underline{\psi}$ and $T\overline{\psi} \leq \overline{\psi}$ if we choose for q a solution of the inequality $\min\{\nu_1,\nu_2\}(q*k) \geq q$ (cf. Lemma 1) and if we choose $\delta \leq (\max q)^{-1}\min\{v_1,v_2\}$ and $\xi \geq$ **$\max\{x_1 - \min(\text{support } q), \max(\text{support } q) - x_2\}$. Moreover, for δ small and/or ξ large, $\underline{\psi} \leq \overline{\psi}$. Hence,**

$$\phi_0 \leq \underline{\psi} \leq T\underline{\psi} \leq \ldots \leq T^n\underline{\psi} \leq \ldots \leq T^n\overline{\psi} \leq \ldots \leq T\overline{\psi} \leq \overline{\psi} \leq \phi_1. \tag{2.3}$$

Consequently both $T^n\underline{\psi}$ and $T^n\overline{\psi}$ converge, as $n \to \infty$, to a limit (in fact, as one can deduce from the Arzela-Ascoli theorem, uniformly on compact subsets). Invoking Lebesgue monotone convergence theorem we conclude that the limit is a fixed point of T. Subsequently (2.3) shows that this fixed point is nontrivial. With the definition

$$X = \text{set of nontrivial fixed points of } T,$$

we can formulate this result as follows.

THEOREM 3. (Existence). X *is nonempty*.

Now suppose for a while that we know that $X = \{w\}$, a singleton, then both $T^n\underline{\psi}$ and $T^n\overline{\psi}$ converge to w as $n \to \infty$. But likewise the monotonicity of T implies that for any u_0 with $\underline{\psi} \le u_0 \le \overline{\psi}$ the sequence u_n defined by (2.1) converges to w as $n \to \infty$. In the next lemma we extend the applicability of this idea by showing that, except for ϕ_0 and ϕ_1, any $u_0 \in C$ yields after a finite number of iterations with T a function which can be sandwiched between $\underline{\psi}$ and $\overline{\psi}$ if δ is chosen sufficiently small.

LEMMA 4. *Let* $u_0 \in C$ *and* $u_0 \neq \phi_0, \phi_1$, *then there exist* $m = m(u_0)$ *and* $\varepsilon = \varepsilon(u_0)$ *such that* $\underline{\psi} \le u_m \le \overline{\psi}$ *provided* $\delta \le \varepsilon$.

PROOF. We shall give a proof only for the case that, for some $\eta > 0$, $[-\eta,\eta] \subset$ support k. Define $J_n = \{x \mid u_n(x) > 0\}$ and let $[\xi_1,\xi_2]$ be a nonempty interval contained in J_0. Then $[-n\eta + \xi_1, \xi_2 + n\eta] \subset J_n$ and consequently, for fixed ξ, support $\underline{\psi}(\cdot;\xi,\delta) \subset J_n$ for n sufficiently large, say $n \ge m$. Hence $u_m \ge \underline{\psi}$ if we take $\delta \le \varepsilon = (\max q)^{-1}\min\{u_m(x) \mid x \in \text{support } \underline{\psi}\}$. Essentially the same argument applied to $1 - u_n(x)$ shows the correctness of the right inequality. In the general case the proof is based on the same idea, but one has to consider properties of the support of iterated convolutions of k with itself, cf. [3; Lemma 2.1]. □

COROLLARY 5. (Uniqueness implies global asymptotic stability). *Suppose* $X = \{w\}$. *Let* $u_0 \in C$, $u_0 \neq \phi_0,\phi_1$. *Then* $\lim_{n\to\infty} u_n = w$.

Motivated by this result we shall now concentrate on finding conditions on g which guarantee that X is a singleton indeed. Our next assumption

$$H_g^3 : \frac{\partial g}{\partial x} > 0 \text{ on } \mathbb{R} \times (0,1),$$

expresses that the environment changes in a monotone way. Under this assumption the operator T has the additional property that it leaves the set of nondecreasing functions invariant: if ϕ is nondecreasing then so is $T\phi$. Starting from the same function q as before we define nondecreasing functions $\underset{\sim}{\psi}$ and $\tilde{\psi}$ by

$$\underset{\sim}{\psi}(x;\delta,\xi) = \delta \sup\{q(x-\eta) \mid \eta \ge \xi\},$$

$$\tilde{\psi}(x;\delta,\xi) = 1 - \delta \sup\{q(x+\eta) \mid \eta \ge \xi\}.$$

Again assumption H_g^2 implies that $T\underset{\sim}{\psi} \ge \underset{\sim}{\psi}$ and $T\tilde{\psi} \le \tilde{\psi}$ if δ is small and ξ is large.

LEMMA 6. *Either* X *consists of one increasing function or* X *contains at least two such functions* w_1 *and* w_2 *with* $w_1 \ge w_2$.

PROOF. Let $w \in X$ be arbitrary. By some rather technical arguments one can show that $\lim_{x\to+\infty} \inf w(x) \geq v_1$ and $\lim_{x\to-\infty} \sup w(x) \leq 1 - v_2$, where v_1 and v_2 are defined in H_g^2 (this part of the proof is essentially the same as the proof of Theorem 5.3 in [1], so we omit it). Hence $\underset{\sim}{\psi} \leq w \leq \tilde{\psi}$ if we choose δ and ξ properly. Consequently the sequences $T^n\tilde{\psi}$ and $T^n\underset{\sim}{\psi}$ converge monotonically to monotone elements w_1 and w_2 of X and $w_1 \geq w \geq w_2$. So if $w_1 = w_2$ then $w = w_1 = w_2$ for any $w \in X$, whereas $w_1 \neq w_2$ yields precisely the other alternative of the lemma (note that H_g^3 implies that any nondecreasing element of X is in fact increasing). □

On account of Lemma 6 it is sufficient to show uniqueness among increasing fixed points of T to conclude uniqueness in general. So let w_1 and w_2 be two increasing fixed points of T such that $w_1 \geq w_2$. We implicitly define a nonnegative function $h: \mathbb{R} \to \mathbb{R} \cup \{+\infty\}$ as follows:

$$\begin{aligned} w_1(x) &= w_2(x+h(x)) && \text{if } w_1(x) \in \{w_2(\xi)\,|\,\xi \in \mathbb{R}\}, \\ h(x) &= +\infty && \text{otherwise.} \end{aligned}$$

Since w_1 and w_2 are increasing, h is uniquely defined and continuous. Next, let the number $\bar{h}$ be defined by

$$\bar{h} = \sup\{h(x)\,|\,-\infty < x < \infty\}.$$

It is our intention to show that $\bar{h} = 0$ and the following result is a first step in that direction.

LEMMA 7. If $0 < \bar{h} < \infty$ then $h(x) < \bar{h}$ for all $x \in \mathbb{R}$.

PROOF.

$$\begin{aligned} w_2(x+h(x)) &= w_1(x) \\ &= \int_{-\infty}^{\infty} g(y,w_1(y))k(x-y)\,dy \\ &= \int_{-\infty}^{\infty} g(y-\bar{h},w_1(y-\bar{h}))k(x+\bar{h}-y)\,dy \\ &= \int_{-\infty}^{\infty} g(y-\bar{h},w_2(y-\bar{h}+h(y-\bar{h})))k(x+\bar{h}-y)\,dy \\ &< \int_{-\infty}^{\infty} g(y,w_2(y))k(x+\bar{h}-y)\,dy \\ &= w_2(x+\bar{h}). \end{aligned}$$

□

Our final assumption:

H_g^4: there exist $x_3 \in \mathbb{R}$ and $\mu, \rho \in (0,1)$ such that

(i) $g(x,u) > u$ for $u \in (0,1)$ and $x \geq x_3$,

$\frac{\partial g}{\partial u}(x,u) \leq \rho$ for $u \in [1-\mu, 1]$ and $x \geq x_3$,

(ii) $g(x,u) < u$ for $u \in (0,1)$ and $x \leq -x_3$,

$\frac{\partial g}{\partial u}(x,u) \leq \rho$ for $u \in [0,\mu]$ and $x \leq -x_3$,

is intended to get hold on the behaviour of $h(x)$ for $x \to \pm\infty$. The assumption implies that, in the absence of migration, the alleles A and a are doomed to disappear from the population far away to the right and to the left, respectively. Also it implies that for any $w \in X$, $w(-\infty) = 0$ and $w(+\infty) = 1$ (cf. the proof of Lemma 6), and consequently $h(x) < \infty$ for all $x \in \mathbb{R}$.

THEOREM 8. (Uniqueness). $X = \{w\}$, a singleton.

PROOF. Let w_1, w_2, h and $\bar{h}$ be as above and suppose $\bar{h} > 0$. Then, on account of Lemma 7, either $\bar{h} = \lim_{x\to+\infty} \sup h(x)$ or $\bar{h} = \lim_{x\to-\infty} \sup h(x)$. Let us assume that the first alternative holds. We define numbers $\tilde{x}$, $\bar{x}$, b and $\tilde{h}$ as follows:

$$\tilde{x} = \inf\{x \mid w_i(\xi) > 1 - \mu \text{ for } \xi \geq x \text{ and } i = 1,2\},$$

$$\bar{x} = \max\{x_3, \tilde{x}\},$$

$$b = \max(\text{support } k),$$

$$\tilde{h} = \max\{h(x) \mid \bar{x} - b \leq x \leq \bar{x}\},$$

and we choose a number $d \in (\tilde{h}, \bar{h})$. Hence $h(x) < d$ for $\bar{x} - b \leq x \leq \bar{x}$ and there exists $x_0 > \bar{x}$ such that $h(x_0) > d$. So if we define a function v by $v(x) = w_1(x) - w_2(x+d)$ then $v(x) \leq 0$ for $\bar{x} - b \leq x \leq \bar{x}$ and $v(x_0) > 0$. Since both w_1 and w_2 are fixed points of T we can write:

$$v(x) = \int_{-b}^{b} \{g(x-\xi, w_1(x-\xi)) - g(x+d-\xi, w_2(x+d-\xi))\}k(\xi)\,d\xi$$

$$\leq \int_{-b}^{b} \{g(x-\xi, w_1(x-\xi)) - g(x-\xi, w_2(x+d-\xi))\}k(\xi)\,d\xi.$$

Furthermore, defining sets $\Sigma(x)$ by

$$\Sigma(x) = \{\xi \mid w_1(x-\xi) \geq w_2(x+d-\xi)\}$$

and taking $x \geq \bar{x}$, we deduce

$$v(x) \leq \rho \int_{\Sigma(x)} v(x-\xi)k(\xi)d\xi$$
$$\leq \rho \sup\{v(\xi) \mid \xi \geq \bar{x} - b\}$$
$$= \rho \sup\{v(\xi) \mid \xi \geq \bar{x}\}.$$

Since $\rho < 1$ this implies $\sup\{v(\xi) \mid \xi \geq \bar{x}\} = 0$ which is, however, in contradiction with $v(x_0) > 0$. Similarly one can exclude the second alternative, $\bar{h} = \lim_{x\to-\infty} \sup h(x)$. Hence our assumption $\bar{h} > 0$ must be false and we conclude that $w_1 = w_2$. This in turn implies, by Lemma 6, the conclusion of the theorem. □

An example of a function g satisfying $H_g^1 - H_g^4$ is obtained by taking $\alpha(x) = 1 + c(x)$, $\beta(x) = 1$ and $\gamma(x) = 1 - c(x)$ in (1.2), where $c'(x) > 0$ and $-1 < c(-\infty) < 0 < c(+\infty) < 1$.

Our results show that spatial variation in survival fitness can lead to a cline with strong stability properties. The same conclusion appears from the work on diffusion equations and from that on discrete habitat models. So this work adds to the robustness of that conclusion.

Also we hope that this paper shows some of the flavour of working with convolution equations. Recently these have shown up in various mathematical models from population dynamics (see, for example, [1,11,16]) and it looks as though this will happen increasingly in the near future.

ACKNOWLEDGEMENT

The author has benefitted by stimulating discussions with J.P. Pauwelussen and T. Nagylaki.

REFERENCES

[1] DIEKMANN, O., Thresholds and travelling waves for the geographical spread of infection, J. Math. Biol. 6 (1978) 109-130.

[2] DIEKMANN, O., Run for your life. A note on the asymptotic speed of propagation of an epidemic, J. Diff. Equ. 33 (1979) 58-73.

[3] DIEKMANN, O. & H.G. KAPER, On the bounded solutions of a nonlinear convolution equation, Nonlinear Analysis, Theory, Methods & Applications 2 (1978) 721-737.

[4] DOWNHAM, D.Y. & S.M.M. SHAH, A sufficiency condition for the stability of an equilibrium, Adv. Appl. Prob. 8 (1976) 4-7.

[5] ESSÉN, M., Studies on a convolution inequality, Ark. Mat. 5 (1963) 113-152.

[6] FIFE, P.C. & L.A. PELETIER, Nonlinear diffusion in population genetics, Arch. Rat. Mech. Anal. 64 (1977) 93-109.

[7] FLEMING, W.H., A selection-migration model in population genetics, J. Math. Biol. 2 (1975) 219-233.

[8] HADELER, K.P., Mathematik für Biologen, (Berlin, Springer, 1974).

[9] HALDANE, J.B.S., The theory of a cline, J. Genetics 48 (1948) 277-284.

[10] KARLIN, S. & N. RICHTER-DYN, Some theoretical analysis of migration selection interaction in a cline: a generalized two range environment, p. 659-706 in: S. Karlin & E. Nevo (eds.), Population Genetics and Ecology (New York, Academic Press, 1976).

[11] MOLLISON, D., Spatial contact models for ecological and epidemic spread, J. Roy Statist. Soc. B 39 (1977) 283-326.

[12] NAGYLAKI, T., Conditions for the existence of clines, Genetics 80 (1975) 595-615.

[13] NAGYLAKI, T., Selection in One- and Two-Locus Systems, Lect. Notes in Biomath. 15 (Berlin, Springer, 1977).

[14] PELETIER, L.A., The mathematical theory of clines, p. 295-308 in: P.C. Baayen, D. van Dulst & J. Oosterhoff (eds.), Proceedings of the Bicentennial Congress of the Wiskundig Genootschap, Part II, MC Tract 101 (Amsterdam, Mathematisch Centrum, 1979).

[15] SLATKIN, M., Gene flow and selection in a cline, Genetics 75 (1973) 733-756.

[16] WEINBERGER, H.F., Asymptotic behaviour of a model in population genetics, p. 47-96 in: J.M. Chadam (ed.), Nonlinear Partial Differential Equations and Applications, Lect. Notes in Math. 648 (Berlin, Springer, 1978).

CLINES INDUCED BY VARIABLE MIGRATION

P. C. Fife
Department of Mathematics
University of Arizona
Tucson, Arizona 85721

L. A. Peletier
Department of Mathematics
University of Leiden
Leiden, Netherlands

Clines are spatial gradients in the genetic composition of a population in equilibrium. They have been studied a great deal by population geneticists ([1-5], for example) in particular within the context of the simple scalar equation

$$\frac{d^2u}{dx^2} + s(x)f(u) = 0 \tag{1}$$

(and other contexts as well, of course), where u is some gene frequency, f a selective pressure mechanism tending to drive the population to certain preferred states s a measure of the selective intensity, and the second derivative a term designed to account for spatial migration. This model can be criticised on various accounts. Nevertheless, it does arise in a systematic and logical manner [6] from heuristic assumptions which are acceptable, for some situations, to experienced people in the field.

We suppose $f(0) = f(1) = 0$, and for our purposes define clines to be solutions $u(x)$ on the real line satisfying $u(-\infty) = 0$, $u(\infty) = 1$, which are stable, in some weak but reasonable sense, as equilibrium solutions of the corresponding nonlinear diffusion equation. (Similar results hold for finite intervals; we do not report them here.) Past analyses of clines (such as [1-5,7-9]) have been entirely devoted to the case when the "state" $u = 0$ is "favored" for large negative x, and $u = 1$ is "favored" for large positive x. (If $s(x)\int_0^1 f(u)du < 0$, we say 0 is favored; the opposite for 1. When s is constant, traveling wave solutions of the corresponding diffusion equation exist, for which $u(x,t)$ approaches the favored state at each x.)

We now suppose the state $u = 0$ is favored for <u>all</u> x ($\int f du < 0, s > 0$) and ask whether it is possible for clines to exist. It is shown that if s is allowed to vary in certain ways, while remaining positive, a cline can indeed exist, provided the equation is bistable ($u = 0$ and $u = 1$ are both stable solutions of (1)). This is contrary to the case when there is no x-variation, for then clines do not exist.

Moreover, we generalize (1) to allow for a variable migration rate, and find that this variation may also be sufficient for the existence of a cline. The generalization is

$$\frac{d}{dx}\left(D(x)\frac{du}{dx}\right) + s(x)f(u) = 0, \tag{2}$$

which arises in natural situations.

Various sufficient conditions can be given for the existence of clines; generally, D or s should take on smaller values in a finite zone separating two communities which are more or less in states $u = 0$ and $u = 1$, respectively. The intuitive explanation is that the effect of such a variation in D is that the decreased diffusion in the intermediate zone makes for more effective isolation of the communities and so may provide for their coexistence.

Eq. (2) may be reduced to one of the form (1) by means of the rescaling

$$\hat{x} = \int_0^x d\xi/D(\xi).$$

In this note we therefore restrict attention to equations of the form (1), although some results are more appropriately given directly in terms of D and s in (2).

We always assume that $0 < s_0 < s(x) < s_1$, $\int_0^1 f(u)du < 0$, $f(0) = f(1) = 0$, $f'(0) < 0$, $f'(1) < 0$, and that for some $a \in (0,1)$, $f(u) < 0$ for $u \in (0,a)$ and $f(u) > 0$ for $u \in (a,1)$.

Conditions for nonexistence. It is well known that a cline cannot exist if s is constant. More generally, the following holds, where

$$\gamma = \frac{\int_a^1 f(u)du}{-\int_0^a f(u)du} \qquad (\text{so } 0 < \gamma < 1):$$

Theorem 1. If $\dfrac{\inf[s(x)]}{\sup[s(x)]} > \gamma$, then no cline exists.

This condition is sharp. Moreover, not only must (infs)/(sups) be large enough, but also the variation of s must be abrupt enough:

Theorem 2. Let $s_1(x) \in C^1(-\infty,\infty)$, and $s_\varepsilon(x) = s_1(\varepsilon x)$. There exists an $\varepsilon_0 > 0$ such that if $0 < \varepsilon < \varepsilon_0$ and $s(x) = s_\varepsilon(x)$, then there exists no cline.

Conditions for existence. Here we insert an extra positive parameter λ into (1):

$$\frac{d^2u}{dx^2} + \lambda s(x)f(u) = 0 \qquad (3)$$

Let $\theta(x,\xi)$ be any nonnegative bounded function which vanishes for $|x| > \xi$ and is chosen so that $\sup\theta = 1$. For some positive μ and ξ, let

$$s(x) = [1 + \mu\theta(x,\xi)]^{-1}.$$

Theorem 3. If $\lambda\xi^2 > \frac{a^2}{8}(\int_a^1 f(r)dr)^{-1}$, then there is a number μ^* such that if $\mu > \mu^*$, (3) has a cline solution.

Of course, μ being large assures us that the "dip" in s will be low enough and abrupt enough.

Let us now consider the more general equation

$$\frac{d^2u}{dx^2} + h(x,u) = 0$$

where for each fixed x, h has the properties of f assumed above. Various general conditions on h can be given which suffice for the existence of a cline. We shall not detail them here. A typical such condition involves, in addition to minor stipulations, the assumption that $h(x,u) \geq s(x)f(u)$ where

$$s(x) < \gamma, \quad x \in [0,B],$$
$$s(x) \leq 1, \quad x \in [B, B+T]$$
$$s(x) \geq 1, \quad x \in [B+T,\infty) .$$

Here the constants γ , B (the extent of the barrier region), and T (the extent of the transitional region between the barrier and the community in state $u = 1$) have to satisfy certain inequalities. For example, T cannot be too large.

The proofs of the existence results consist in constructing a supersolution $\bar{u}$ and a subsolution $\underline{u}$ with $\underline{u} < \bar{u}$ and both satisfying the required boundary conditions at $\pm\infty$. Stability is to be understood in a certain technical sense. In particular, the allowed perturbations should not take u outside the range $\underline{u} \leq u < \bar{u}$. With the sub- and super solutions at hand, an extension of a result of Matano [10] then assures us of the existence of a solution, stable in the above sense, between them; this, by our definition, is a cline.

Results of the type announced here were conjectured to be true by S. Levin (personal communication and [11]). In two-dimensional problems, the effect of the shape of the domain on existence of clines was explored in the interesting paper of Matano [10]. We thank S. Levin and C. Holland for bringing this to our attention. The results in the present note will appear later in more detailed form and with proofs.

Literature

1. J. B. S. Haldane, The theory of a cline, J. Genetics 48, 277-284 (1948).

2. R. A. Fisher, The wave of advance of advantageous genes, Ann. of Eugenics 7, 355-369 (1937).

3. M. Slatkin, Gene flow and selection in a cline, Genetics 75, 733-756 (1973).

4. T. Nagylaki, Conditions for the existence of clines, Genetics 80, 595-615 (1975).

5. T. Nagylaki, Clines with asymmetric migration, Genetics 88, 813-827 (1978).

6. P. Fife, Mathematical Aspects of Reacting and Diffusing Systems, Lecture Notes in Biomathematics No. 28, Springer-Verlag (1979).

7. C. Conley, An application of Wazewski's method to a nonlinear boundary value problem which arises in population genetics, J. Math. Biol. 2, 241-249 (1975).

8. W. H. Fleming, A selection-migration model in population genetics, J. Math. Biol. 2, 219-233 (1975).

9. P. C. Fife and L. A. Peletier, Nonlinear diffusion in population genetics, Arch. Rational Mech. Analysis 64, 93-109 (1977).

10. H. Matano, Asymptotic behavior and stability of solutions of semilinear diffusion equations, Publ. Res. Inst. Math. Sci. (Kyoto), to appear (1979).

11. S. A. Levin, Non-uniform stable solutions to reaction-diffusion equations: applications to ecological pattern formation, Proceedings of the International Symposium on Synergetics, Schloss Elmau, Bavaria, Springer-Verlag (1979).

ASYMPTOTIC BEHAVIOR OF THE SOLUTIONS OF THE FISHER EQUATION

F. Rothe
Lehrstuhl für Biomathematik
Universität Tübingen
Auf der Morgenstelle 28
D-7400 Tübingen

Abstract

For Fisher's diffusion model which describes the advance of an advantageous gene, the following two questions are discussed: Existence of travelling fronts and convergence to travelling fronts. There are striking differences between the heterozygote intermediate and the heterozygote inferior case.
In the first case, there exists a continuum of travelling fronts with different velocities which occur asymptotically for different initial data. In the second case there exists only one travelling front which is very stable.

Introduction

Various models for moving fronts in population genetics and epidemics have been studied in recent years. This note deals with the classical Fisher model for the spread of an advantageous gene [5]. We briefly outline the biological background.
Consider a homogeneous diploid population with two alleles A and a at one gene locus. Let the variable $p \in [0,1]$ measure the probability that gene A occurs. Like R.A. Fisher in [5], we assume that the population is in Hardy-Weinberg equilibrium. Thus the probabilities of the genotypes AA, Aa and aa are given by

$$p^2, \; 2p(1-p), \; (1-p)^2.$$

Let the numbers

$$f \qquad 1 \qquad g$$

measure the fitness of the different genotypes.
We consider a population with a continuous sequence of generations spread out in an infinite one-dimensional territory. Thus p is a function of the two variables $(x,t) \in R \times [0,\infty)$.

case	I	II	III	IV
	$1 \leq f$ $0 \leq g \leq 1$ $(f,g) \neq (1,1)$	$1 < f$ $1 < g$	$1 \leq g$ $0 \leq f \leq 1$ $(f,g) \neq (1,1)$	$0 \leq f < 1$ $0 \leq g < 1$
comparison of fitness	aa≤Aa≤AA aa<AA	Aa<AA Aa<aa	AA≤Aa≤aa AA<aa	AA<Aa aa<Aa
name adopted by Weinberger	heterozygote intermediate	heterozygote inferior	heterozyg. intermed.	heterozyg. superior
graph of the function F(p)				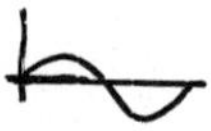
stable equilibria	1	0 , 1 (bistable)	0	α
unstable equilibria	0	α	1	0,1

Fig.1

In Fisher's model, the time evolution of the population is governed by the diffusion equation

$$p_t = p_{xx} + F(p) \qquad (1)$$

which must be supplemented by the initial condition

$$p(x,0) = \varphi(x) \qquad \text{for all } x \in R\ , \qquad (2)$$

there the function F(p) is the third order polynomial

$$F(p) = p(1-p)(1-g-(2-f-g)p) \quad . \qquad (3)$$

To begin with, consider the ordinary differential equation

$$p_t = F(p) \quad . \qquad (4)$$

As shown in Fig. 1 there occurs different qualitative behavior for different values of the fitness parameters f and g. The names for the different situations I, II, III, IV are taken from Aronson and Weinberger [1] .

In cases II and IV, the function F(p) has the zero $\alpha = \frac{1-g}{2-f-g}$ in the interval (0,1).
Let us turn back to the diffusion model (1). In the classical paper [7] Kolmogoroff et.al. consider the case I and take as initial data

$$\varphi(x) = \begin{cases} 1 & x < 0 \\ 0 & x > 0 \end{cases} \quad . \qquad (5)$$

In this case there occurs a front moving to the right which, for large time, has about constant speed and shape. Similar behavior can be expected for more general initial data.
Thus, it is natural to consider the following two problems along with the diffusion model:

(A) Existence of travelling fronts. By travelling fronts we mean solutions of (1) which have the form

$$p(x,t) = \phi(x-ct) \qquad (6)$$

and satisfy the condition

$$\lim_{x\to-\infty} \phi(x) = 1 \quad , \quad \lim_{x\to\infty} \phi(x) = 0 \quad , \quad 0 \leq \phi(x) \leq 1 \text{ for all } x \in R. \tag{7}$$

(B) Convergence to travelling fronts in some sense for a larger class of initial data.

Exchanging the genes A and a, case III can be reduced to case I, therefore case III can be omitted.
Problems (A), (B) can also be treated for more general source functions F(p).

A) Existence of travelling fronts

This problem really involves only ordinary differential equations. It can be solved by phase plane methods. We give some important results; for details see [4,6,7].

Case I (monostable, heterozygote intermediate)
As a natural generalization, we impose the following condition on the function F:

(Ia) $F \in C^1 [0,1]$

(Ib) $F(0) = F(1) = 0, \quad F'(0) > 0, \quad F'(1) < 0$

(Ic) $F(p) > 0 \quad$ for $p \in (0,1)$.

Theorem 1: (Hadeler, Rothe [6])

There exists a minimal velocity $c_o > 0$. For velocities $c \geq c_o$ there exist travelling fronts which are unique up to translation. For velocities $c < c_o$ there exist no travelling fronts.
The minimal velocity c_o obeys the inequality

$$F'(0) \leq \frac{c_o^2}{4} \leq \max_{0\leq p\leq 1} \frac{F(p)}{p} \tag{8}$$

If F(.) is the third order polynomial (3), the minimal velocity c_o is explicitely given by

$$c_o = \begin{cases} 2\sqrt{1-g} & \text{for } f + 3g \leq 4 \\ \dfrac{f-g}{\sqrt{2(f+g-2)}} & \text{for } f + 3g \geq 4 \end{cases} . \tag{9}$$

In the latter case, also the minimal front can be written down explicitely:

$$\phi(x) = \left(1 + \exp \sqrt{\frac{f+g-2}{2}}\; x\right)^{-1} \qquad \text{for} \quad f + 3g \geq 4 \quad . \tag{10}$$

Define the velocity c^* by

$$F'(0) = \frac{c^{*2}}{4} \quad .$$

The following definition is taken from Stokes [10] .

<u>Definition:</u> The front with minimal velocity is called a <u>pushed</u> front if $c_o > c^*$ and a <u>pulled</u> front is $c_o = c^*$.
These names can be explained as follows:

For pulled fronts, the velocity of propagation $c_o = 2\sqrt{F'(0)}$ is determined by the leading edge of the population distribution for $x \to \infty$, $p \to 0$. Hence one can imagine that the leading edge *pulls* the front. On the other hand, for pushed fronts, the velocity of propagation is determined not by the behavior of the leading edge of the distribution, but by the whole wave shape. Hence one can imagine that the front is pushed.
If F(.) is the third order polynomial (3), we have a pulled front for $f + 3g \leq 4$ and a pushed front for $f + 3g > 4$, as can be seen from formula (9).

<u>Case II</u> (bistable, heterozygote inferior)
It is natural to impose the following conditions on the function F:

(IIa) $F \in C^1\,[0,1]$

(IIb) $F(0) = F(1) = 0, \quad F'(0) < 0, \quad F'(1) < 0$

(IIc) There exists some $\alpha \in (0,1)$ such that
$F(p) < 0 \quad$ for $p \in (0,\alpha)$
$F(p) > 0 \quad$ for $p \in (\alpha,1)$

<u>Theorem 2:</u> Then there exists a unique front velocity c_o, and a monotone travelling front unique up to translation.

If F(.) is the third order polynomial (3), the front and the front velocity are explicitely given by

$$\phi(x) = \left(1 + \exp \sqrt{\frac{f+g-2}{2}}\; x\right)^{-1} \qquad \text{and} \tag{11}$$

$$c_o = \frac{f-g}{\sqrt{2(f+g-2)}} \quad . \tag{12}$$

In case IV there exist no travelling fronts.
So furtheron we will be only concerned with cases I and II.

B. Convergence to travelling fronts

Case II

First consider case II which is much simpler and completely solved by Fife & McLeod [4] .

Theorem 3:

Let the function F(.) satisfy (IIa), (IIb), (IIc) and take initial data $\varphi(x)$ which satisfy

$$0 \leq \varphi(x) \leq 1 \quad \text{for all } x \in R \tag{13}$$

and

$$\liminf_{x \to -\infty} \varphi(x) > \alpha , \qquad \limsup_{x \to \infty} \varphi(x) < \alpha . \tag{14}$$

For some constants $x_o \in R$, $K, \omega > 0$, the solution p of the diffusion equation (1) satisfies

$$|p(x,t) - \phi(x-c_o t-x_o)| \leq Ke^{-\omega t} \quad \text{for all } x \in R,\ t > 0.$$

The proof makes use of the Ljapunov-functional

$$\mathcal{L}(p) = \int_{-\infty}^{+\infty} e^{c_o x} \left[\tfrac{1}{2}p_x^2 - V(p)\right] dx$$

with $V(p) = \int_0^p F(p')dp'$.

Case I:

Let the function F satisfy (Ia), (Ib), (Ic). In this case, the asymptotic behavior is much more complicated, because there exists a whole continuum of travelling fronts for all velocities $c \geq c_o$.
Let $\phi(x;c)$ be the travelling front with velocity c. The important characteristic of the front is the asymptotic behavior for $x \to \infty$.
There occurs exponential decay:

$$\phi(x;c) = ae^{-\lambda(c)x}(1+o(1)) \qquad \text{for } x \to +\infty$$

where the decay rate is

$$\lambda(c) = \begin{cases} c/2 + \sqrt{c^2/4 - F'(0)} & \text{if } c = c_o \\ c/2 - \sqrt{c^2/4 - F'(0)} & \text{if } c > c_o \end{cases} .$$

Weaker results are given by Aronson and Weinberger [1] and Rothe [8]. We skip them and give the stronger results of Uchiyama [12]. Concerning initial data $\varphi(x)$ we need the following assumptions:

$$\liminf_{x \to -\infty} \varphi(x) > 0 \tag{15}$$

$$\lim_{x \to \infty} \frac{\varphi(x+y)}{\varphi(x)} e^{\lambda y} = 1 \qquad \text{for all } y \in R . \tag{16}$$

<u>Theorem 4:</u> (Uchiyama [12])

Assume the initial data $\varphi(x)$ satisfy (13), (15), (16) with

$$\lambda > c_o/2 - \sqrt{c_o^2/4 - F'(0)} .$$

Let the function m(t) be given by

$$m(t) = \sup \left\{ x \mid u(x,t) = \tfrac{1}{2} \right\} . \tag{17}$$

Then m(t) is well defined for t large enough and

$$\lim_{t \to \infty} u(\,x + m(t), t) = \phi(x; c_o) \quad \text{uniformly for } x \geq \text{const.}$$

On the other hand, if the initial data satisfy (13), (15), (16) with

$$\lambda = \lambda(c) \leq c_o/2 - \sqrt{c_o^2/4 - F'(0)}$$

for some velocity $c \geq c_o$, then

$$\lim_{t \to \infty} u(\,x + m(t), t) = \phi(x; c) \quad \text{uniformly for } x \geq \text{const.}$$

Similar results can be shown for pairs of diverging fronts. As an example, we mention the following

Corollary 5

Let the initial data satisfy (13) and have compact support.
Define

$$m^*(t) = \inf\left\{ x \mid u(x,t) = \frac{1}{2} \right\} .$$

Then

$$\lim_{t\to\infty} |u(x,t) - \phi(x-m(t);c_o)| = 0 \qquad \text{for } x \geq 0$$

and

$$\lim_{t\to\infty} |u(x,t) - \phi(-x+m^*(t);c_o)| = 0 \qquad \text{for } x \leq 0 .$$

Theorem 4 and the Corollary 5 give only results concerning the "shape" of the wavefront. It is a much more delicate problem to get good estimates for m(t), which measures the position of the wavefront.

Concerning the higher speed fronts, one gets good results by sharpening Condition (16) about the asymptotic behaviour of the initial data.

Theorem 6: (Uchiyama)

Let the initial data satisfy (13), (15) and let

$$\lim_{x\to\infty} \varphi(x)\, e^{\lambda x} > 0 \text{ exist} \tag{18}$$

where $\lambda = \lambda(c)$ corresponds to some speed $c \geq c_o$.

Then m(t)-ct is bounded and there exists $y \in R$ such that

$$\lim_{t\to\infty} u(x+xt-y,t) = \phi(x;c) \text{ uniformly for } x \geq \text{const.}$$

It is still harder to give estimates of m(t) if the front with minimal speed occurs asymptotically.
The following two Theorems 7,8 from Uchiyama [12] and Rothe [9] cover some cases.

Theorem 7: (Uchiyama)

Let the function F satisfy

$$F(p) \leq F'(0)p \qquad \text{for all } p \in (0,1) .$$

(Then by estimate (8) $c^* = c_o$, hence the minimal front is a pulled front).

Let the initial data satisfy (13) and

$$\varphi(x) = 0 \qquad \text{for all } x \geq \text{const.}$$

Then m(t) satisfies the estimate

$$c^*t - \frac{3}{2c^*}\log t + \text{const} \leq m(t) \leq c^*t - \frac{3}{2c^*}\log t + O(\log\log t).$$

Similar estimates are given by Bramson [2,3] .

Theorem 8: ((Rothe [9])

Let the function F be such that the minimal front is a pushed front, e.g. $c^* < c_0$.
Let the initial data satisfy (13), (15) and

$$|\varphi(x)e^{\lambda_- x}| \leq \text{const} \qquad \text{for all } x \in R$$

where

$$\lambda_- = c_0/2 - \sqrt{c_0^2/4 - F'(0)} \quad .$$

Then $m(t) - c_0 t$ is bounded and there exist constants $y \in R$, $K, \omega > 0$ such that

$$|u(x,t) - \phi(x - c_0 t - y; c_0)| \leq Ke^{-\omega t} \qquad \text{for all } x \in R.$$

A corresponding result holds for pairs of diverging fronts.

Corollary 9

Let the function F be such that the minimal front is a pushed front, e.g. $c^* < c_0$.
Let the initial data satisfy (13) and

$$|\varphi(x)|e^{\lambda_- |x|} \leq \text{const.}$$

Then there exist constants $y, z \in R$, $K, \omega > 0$ such that

$$|u(x,t) - \phi(x - c_0 t - y; c_0)| \leq Ke^{-\omega t} \quad \text{for } x \geq 0$$

$$|u(x,t) - \phi(-x + c_0 t - z; c_0)| \leq Ke^{-\omega t} \quad \text{for } x \leq 0 \quad .$$

It is interesting to take for F again the original third order polynomial (3).

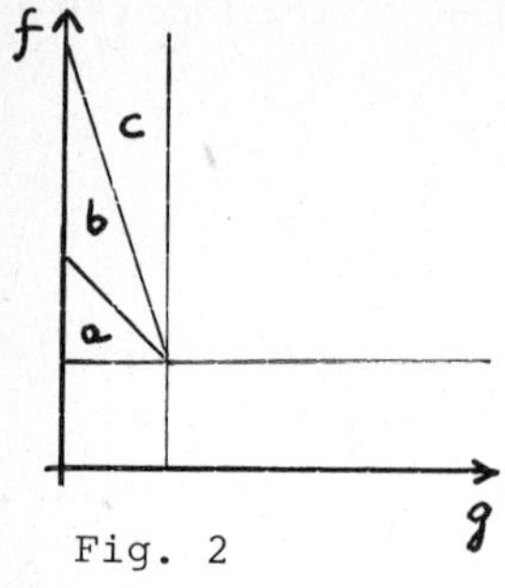

Fig. 2

Theorem 7 can be applied if $f + g \leq 2$, which corresponds to region a. Theorem 8 can be applied for the pushed front. By Theorem 1 this occurs if $f + 3g > 4$, hence in region c. In region b the exact asymptotical behavior of m(t) is not known.

References

[1] D.G.Aronson, H.F.Weinberger: Nonlinear diffusion in population genetics, combustian and nerve propagation. Partial diff. Equations and related Topics, ed. J.A.Goldstein. Lecture Notes in Math. 446, 5-49, New York, Springer 1975

[2] M.D.Bramson: Maximal displacement of branching Brownian motion. Communications on Pure and Applied Math. 31, 531-581 (1978)

[3] M.D.Bramson: Minimal displacement of branching random walk. Zeitschrift f. Wahrscheinlichkeitstheorie und verwandte Gebiete 4 5, 89-108 (1978)

[4] P.C.Fife, S.B.McLeod: The approach of solutions of nonlinear diffusion equations to travelling fronts solutions. Archi ve for Rat. Mech. and Anal. 65 (4), 335-361 (1977)

[5] R.A.Fisher: The advance of advantageous genes, An. Eugenics (1937) 355-369

[6] K.P.Hadeler,F.Rothe: Travelling fronts in nonlinear diffusion equations. Journ. of Math.Biol. 2, 251-263 (1975)

[7] A.N.Kolmogorov, I.G. Petrovskii, N.S.Piskunov: A study of the equation of diffusion with increase in the quantity of matter and its application to a biological problem, Bjul. Moskovskogo Gos. Univ. 1,7 1-26 (1937)

[8] F.Rothe:Convergence to travelling fronts in semilinear parabolic equations, Proceedings of the Royal Society of Edinburgh 80A, 213-234 (1978)

[9] F.Rothe: Convergence to pushed fronts ,Rocky Mountain Journal of Mathematics, to appear

[10] A.N.Stokes: On two types of moving fronts in quasilinear diffusion Math.Bioscience 31, 307-315 (1976)

[11] A.N.Stokes: Nonlinear diffusion waveshapes generated by possibly finite initial disturbances. Journ. of Math. Anal. and Appl. 61, 2, 370-381 (1977)

[12] K. Uchiyama: The behavior of solutions of some nonlinear diffusion equations for large time, Journ. of Math. of Kyoto Univ. 18, 3, 453-508 (1978)

RANDOM WALKS AND PROBABILITIES OF GENETIC IDENTITY IN A TREE

Stanley Sawyer
Department of Mathematics
Purdue University
West Lafayette, IN 47907 U.S.A.

1. Introduction

The purpose here is to discuss isotropic random walks on a set of graphs called infinite homogeneous trees, and to describe a population genetics model on these graphs such that migration between generations is represented by an isotropic random walk. The population model is of "Stepping Stone" type, which are used in Biology to model selectively neutral gene flow in a stationary interbreeding population. Here the main influences are local migration and the consolidative effects of random finite offspring distributions. In one or two dimensions, these random effects cause expanding waves which can mimic the "waves of advance" of epidemiological models or genetic models with selection. In three or more dimensions, or in infinite trees, there are still waves of advance but they do not exclude other types behind the "wave" (see §3 below).

The material on isotropic random walks is adapted from Sawyer (1978) See Sawyer (1976), (1977), (1978), (1979) for most of the material on Stepping Stone models, and Cavalli-Sforza and Bodmer (1971), Chapter 8 for typical applications to biological populations.

2. Trees

Let T_{a+1} be an infinite tree of connectivity $a+1$; i.e. an infinite connected graph with no non-trivial closed loops in which every node belongs to exactly $a+1$ edges. In particular each node has $a+1$ "nearest neighbors". Thus T_2 is the one-dimensional Euclidean graph. Henceforth we assume $a+1 > 2$.

Since T_{a+1} has no non-trivial closed cycles, each pair of nodes x,y is connected by a unique (except for orientation or retracing of steps) path of $d = d(x,y)$ edges. We define an <u>isotropic random walk</u> on T_{a+1} as a Markov chain $\{X_n\}$ such that the transition function

$$\mathrm{Prob}[X_{n+1} = y | X_n = x] = p(x,y) \tag{2.1}$$

is invariant with respect to isomorphisms of T_{a+1}, which is equivalent to

$$p(x,y) = A(d), \quad d = d(x,y). \tag{2.2}$$

Let

$$p_n(x,y) = \text{Prob}[X_n = y | X_0 = x] = \sum_z p_{n-1}(x,z)p(z,y).$$

The results below will depend on the asymptotics of $p_n(x,y)$, and of the potential

$$R(s,x,y) = \sum_{n=1}^{\infty} s^n p_n(x,y). \tag{2.3}$$

The corresponding results for Euclidean spaces are usually obtained by Fourier-analytic methods. For example, in one dimension, assume $p(x,y) = p(x-y) = p(y-x)$ and define

$$\phi(y) = \sum e^{i\theta x} p(x) = \sum y^x p(x) \quad \text{for} \quad y = e^{i\theta}. \tag{2.4}$$

Then $p_n(x,y) = p_n(x-y) = p_n(y-x)$, where

$$p_n(x) = \frac{1}{\pi}\int_0^{\pi} \phi(e^{i\theta})^n \cos x\theta d\theta = \frac{1}{2\pi i}\int_C \phi(y)^n y^x \frac{dy}{y} \tag{2.5}$$

$$R(s,x,0) = \frac{1}{\pi}\int_0^{\pi} \frac{\cos(x\theta)}{[1-s\phi(y)]} d\theta = \frac{1}{2\pi i}\int_C \frac{y^x}{[1-s\phi(y)]}\frac{dy}{y}$$

where $C = \{y: |y| = 1\}$, assuming $x \neq 0$ in the equation for $R(s,x,0)$. Here $\phi(e^{i\theta})$ is twice-continuously differentiable in θ if $E[d(X_1,X_0)^2] < \infty$, but does not analytically continue beyond $|y| = 1$ unless $\sum A(d)\rho^d < \infty$ for some $\rho > 1$.

To extend these arguments to trees, we use the fact that T_{a+1} is essentially a symmetric space of a p-adic matrix group (or p-adic "Lie group"), at least if a is a prime power. Since matrix groups are highly non-Abelian, one cannot hope to expand all functions $p(x)$ on T_{a+1} as in (2.4)-(2.5) in terms of one-dimensional functions. The isotropic functions are a class which one can, with the objects corresponding to $\{\cos(\theta x)\}$ called <u>spherical functions</u>. A combinatorial approach due to Cartier (1973) avoids the restriction that a be a prime power.

The resulting theory is as follows. For each $a > 1$, there exist monic polynomials $\{P_d(x): d = 0,1,2,\cdots\}$ with $P_d(t)$ of degree d, such that $P_d(2\sqrt{a}\cos\theta)$ corresponds to $\cos(d\theta)$ in the following sense:

$$f(t) = \sum_{d=0}^{\infty} A(d)P_d(t), \quad p(x,y) = A[d(x,y)]$$

always converges uniformly for $|t| \leq 2\sqrt{a}$, and

$$f(t)^n = \sum_{d=0}^{\infty} A_n(d)P_d(t), \quad p_n(x,y) = A_n[d(x,y)].$$

Set $g(y) = f[\sqrt{a}(y+y^{-1})]$, so $g(e^{i\theta}) = f(2\sqrt{a}\cos\theta)$. Then if $d = d(x,y) > 0$ and $C = \{y: |y| = 1\}$

$$p_n(x,y) = \frac{a\cdot a^{-d/2}}{2\pi i}\int_C g(y)^n y^d \frac{1-y^2}{a-y^2}\frac{dy}{y} \tag{2.6}$$

$$R(s,x,y) = \frac{a\cdot a^{-d/2}}{2\pi i}\int_C \frac{y^d}{[1-sg(y)]}\frac{1-y^2}{a-y^2}\frac{dy}{y}.$$

Unlike as in (2.5), $g(y)$ and (2.6) always continue analytically into the region $a^{-1/2} < |y| < \sqrt{a}$; the sharpest asymptotic results are obtained by continuing to $|y| = a^{-1/2}$.

Typical results depend on two constants

$$\beta = \sum_{d=1}^{\infty} [d - \frac{2a}{a^2-1}(1-a^{-d})]A_d^* \tag{2.7}$$

$$R = \sum_{d=0}^{\infty} a^{-d/2}[1+\frac{a-1}{a+1}d]A_d^*$$

where $A_d^* = \text{Prob}[d(X_1,X_0) = d] = a^{d-1}(a+1)A_d$ $(d > 0)$, $A_0^* = A_0$. Thus $\sum A_d^* = 1$. Except in trivial cases, $0 < \beta \leq \infty$ and $0 < R < 1$. Fix $x,y \in T_{a+1}$, and let $\{X_n\}$ be as before. Then

Theorem 2.1. Given $X_0 = x$, $d(X_n,y) \sim \beta n$ almost surely. Thus $\{X_n\}$ is transient in a rather strong sense. Now assume $\{d: A(d) \neq 0\}$ contains both even and odd integers, and $E[d(X_1,X_0)^2] < \infty$.

Theorem 2.2. $p_n(x,y) \sim C_1 n^{-3/2}R^n$ as $n \to \infty$.

Theorem 2.3. $R(1-u,x,y) = C_2 a^{-d}[e^{-du/\beta} + o(1)]$ as $d = d(x,y) \to \infty$ uniformly for $0 \leq u \leq 1$.

See Gerl (1977), (1978) for other results for random walks on trees in addition to the papers quoted in Sawyer (1978). Bougerol (1979) has results analogous to Theorems 2.2-2.3 for isotropic random walks on the symmetric spaces of continuous Lie groups, and also uses spherical functions and Fourier analysis.

3. Stepping Stone models

We assume an infinite population distributed on the nodes of an infinite lattice such as the d-dimensional lattice J^d or T_{a+1}. Each node or "colony" has a carrying capacity of $2N$ individuals and is usually at that carrying capacity. For definiteness, we assume that individuals are initially all of distinct "colors".

Stepping Stone models can be set up with discrete time and discrete generations, or with continuous time and continual birth and deaths. In the latter case births and deaths occur simultaneously and the model is similar to the "voter model" of Holley and Liggett or the Williams-Bjerknes tumor model (without "selection"). For definiteness, we restrict ourselves to discrete generations. Almost all the results to follow are known in continuous time as well.

At the beginning of each generation, each "parent" has a large number of juvenile offspring. The parents disappear, and the juveniles migrate independently according to (2.1). In each colony, exactly 2N juveniles, in general of parentage from varying colonies, are chosen at random to form the surviving adults of the next generation. Each juvenile is the same color as his parent unless there has been a mutation (an event of probability u), in which case his color is of a type never seen before. Thus, after many generations, individuals are of the same color if and only if they have an ancestor in common, without intervening mutations. For definiteness, we assume $E[d(X_1,X_0)^2] < \infty$ and $\{d: A(d) \neq 0\}$ has even and odd integers.

One measure of consolidation is

$$I(t,x,y) = \text{Prob. that two different individuals chosen randomly at } x \text{ and } y \text{ at time } t \text{ are the same color.} \tag{3.1}$$

Always $I(t,x,y) \to I(x,y)$ as $t \to \infty$. If the mutation $u > 0$, $I(x,y) < 1$, and $I(x,y)$ can be obtained essentially as a ratio of potentials (2.3). Various asymptotics for large $x-y$ and small u are known and used. If $u = 0$, $I(t,x,y) \to 1$ in J^1 and J^2 and $I(x,y) < 1$ for J^d, $d \geq 3$, and all T_{a+1}.

Another measure is the random variable

$$N(t) = \text{number of individuals in generation } t \text{ of the same color as an individual chosen at random from colony } x_0 \text{ [fixed in advance] at time } t. \tag{3.2}$$

This should model a typical surviving class in the t-th generation; it is of course biased towards the larger classes. In the following, assume $u = 0$ unless otherwise noted.

<u>Theorem 3.1</u>. (Sawyer (1977), (1978)). In J^d, $E[N(t)] \sim C\sqrt{t}$ $(d = 1)$, $\sim Ct/(\log t)$ $(d = 2)$. In J^d $(d \geq 3)$ or T_{a+1}, $E[N(t)] \sim Qt$ where $Q = 1 - I(x_0,x_0)$.

In J^d for $d \leq 2$, the individuals in any preassigned bounded set of colonies are of a uniform color with probability converging to one as $t \to \infty$. Thus the blocks measured by $N(t)$ tend to be solid. The blocks are roughly of radius $\sqrt{t}$. However, any given block eventually dissipates and becomes extinct. In J^d $(d \geq 3)$ or T_{a+1}, the expected block sizes are larger, but they are not solid.

Theorem 3.2. In J^d $(d \geq 2)$ or T_{a+1}, for all λ

$$\lim_{t\to\infty} \mathrm{Prob}\left[\frac{N(t)}{E[N(t)]} \leq \lambda\right] = \int_0^\lambda e^{-2y} 4y\,dy. \tag{3.3}$$

The result for J^d $(d \geq 2)$ is in Sawyer (1979). The argument extends to T_{a+1} once one notes that for any two nodes $x, y \in T_{a+1}$, there exists a graph isomorphism of T_{a+1} which is also an involution which interchanges x and y. A recent result of Bramson and Griffeath (personal communication) strongly suggests that (3.3) does hold in one dimension for nearest neighbor migration, but with a different limiting distribution.

As a measure of the distribution of numbers within blocks, let $N(t,A)$ = same as (3.2), except that we only count individuals in those colonies x with $d(x,x_0) \leq A$. Then

Theorem 3.3. (Sawyer (1978)). In T_{a+1}

$$\lim_{t\to\infty} \frac{E[N(t,2\beta t\theta)]}{E[N(t)]} = \min\{\theta,1\} \tag{3.4}$$

for β as in (2.7) and all θ, $0 \leq \theta < \infty$.

See Sawyer (1977), (1978) for the analogs of Theorem 3.3 in J^d, and results analogous to Theorems 3.1-3.3 for $t = \infty$, $u > 0$, with limits as $u \to 0$. In that case the limiting distributions in (3.3) and (3.4) turn out to be exponential.

REFERENCES

Bougerol, P. (1979). Comportement asymptotique des puissances de convolution d'une probabilite sur un espace symetrique. Ms.

Cartier, P. (1973). Harmonic analysis on trees. Proc. Sympos. Pure Math. Amer. Math. Soc. 26, 419-424.

Cavalli-Sforza, L., and Bodmer, W. (1971). The genetics of human populations. W. H. Freeman, Chapter 8.

Gerl, P. (1977). Irrfahrten auf F_2. Monat. Math. 84, 29-35.

Gerl, P. (1978). Ein gleichverteilungssatz auf F_2. Ms.

Sawyer, S. (1976). Results for the Stepping Stone model for migration in population genetics. Annals of Prob. 4, 699-728.

Sawyer, S. (1977). Rates of consolidation in a selectively neutral migration model. Annals of Prob. 5, 486-493.

Sawyer, S. (1978). Isotropic random walks in a tree. Zeit. Wahrschein. 42, 279-282.

Sawyer, S. (1979). A limit theorem for patch sizes in a selectively neutral migration model. Jour. Appl. Prob. 16, No. 3 (in press).

TRAVELLING-FRONT SOLUTIONS FOR INTEGRODIFFERENTIAL EQUATIONS II

Konrad Schumacher
Institut für Biologie II
Universität Tübingen

Abstract Biological spread is considered in the light of models governed by integrodifferential equations of the form $w_t = f(w, \mu * w)$:Asymptotic speed of propagation, existence of travelling fronts, dependence of the minimal speed on delays.

1. Introduction

The three class epidemic model of Kermack & Mc.Kendrick is given by the following system of integrodifferential equations

$$
\begin{aligned}
u_t &= -a\,u\,\overline{v} \\
v_t &= a\,u\,\overline{v} - b\,v \\
w_t &= b\,v
\end{aligned}
\tag{1,1}
$$

where u, v, and w are functions of the time t and one space variable $x \in \mathbb{R}$ denoting the densities of the susceptible, infective, and removed individuals. The infection rate $a > 0$ and the recovery-rate $b > 0$ are assumed to be constant. The symbol $\overline{v}$ stand for a weighted average of v which will be specified.

In [2] existence of travelling-front solutions and in [1] the asymptotic behaviour of the solutions of (1.1) have been analysed for the case $\overline{v} = k * v$, where

$$(k * v)(t,x) := \int_{-\infty}^{+\infty} v(t, x-\eta)\,k(\eta)\,d\eta \quad , \; k \in \mathcal{L}^1(\mathbb{R}, \mathbb{R}_+) \quad . \tag{1,2}$$

Taking into account finite periods of incubation and infectiveness and more than one space-dimension the results of [1,2] have been generalized in [3,4,10] on the basis of models in terms of mixed Volterra-Fredholm integral equations.

However there is still another aspect from which the system (1.1) with $\overline{v}$ given by (1,2) seems to be unrealistic. If, for example, the individuals do not migrate while the transmission of the disease is caused by a "cloud" of randomly moving pathogenic agents the density of which at the point $\xi \in \mathbb{R}$ for the time s being proportional to $v(s,\xi)$, one should $\overline{v}$

replace by

$$(1.3)\qquad \overline{v}(t,x) = \int_0^{\infty}\int_{-\infty}^{+\infty} v(t-\tfrac{|\eta|}{\rho}, x-\eta)\, k_o(\eta,\tfrac{|\eta|}{\rho})\, k_1(\rho)\, d\eta\, d\rho \quad .$$

Here $k_1(\rho)$ describes the distribution of the velocities of the pathogenic agents and $k_o(\eta,\tau)$ measures their effectiveness as a function of their age τ and the covered distance η. If k_o and k_1 are nonnegative and Lebesgue measurable such that

$$(1.4)\qquad \int_0^{\infty}\int_{-\infty}^{+\infty} k_o(\eta,\tau)\; k_1(\tfrac{|\eta|}{\tau})\frac{|\eta|}{\tau^2}\; d\eta\, d\tau = 1 \quad ,$$

then (1.3) can be rewritten in the form ($\tau := |\eta|/\rho$)

$$(1.5)\qquad \overline{v}(t,x) = (\mu * v)(t,x) := \int_0^{\infty}\int_{-\infty}^{\infty} v(t-\tau, x-\eta)\, d\mu(\eta,\tau) \quad ,$$

where μ is the normed Lebesgue-measure on $\mathbb{R}\times\mathbb{R}_+$ with the density

$$(1.6)\qquad d\mu(\eta,\tau) = k_o(\eta,\tau)\; k_1(\tfrac{|\eta|}{\tau})\frac{|\eta|}{\tau^2}\; d\eta\, d\tau.$$

In the following let μ be an arbitrary normed Lebesgue-measure on $\mathbb{R}\times\mathbb{R}_+$ ($\mu(\mathbb{R}\times\mathbb{R}_+) = 1$) and let $\overline{v}$ be given by (1.5). It may be of interest to look for solutions of (1.1) existing for all time such that

$$(1.7)\qquad \lim_{t\mapsto-\infty} w(t,x) = \lim_{t\mapsto-\infty} v(t,x) = 0 \ , \ \lim_{t\mapsto-\infty} u(t,x) = 1 \ , \text{for every } x\in\mathbb{R},$$

i.e. solutions coming from an asymptotic state of the population where all individuals are susceptible. Under these hypotheses the system (1.1) is reduced to the single integrodifferential equation (c.f. [1,2,7])

$$(1.8)\qquad w_t = f(w, \mu * w) \quad , \text{ where}$$

$$(1.9)\qquad f(r,s) := -b\, r + b\,(1 - \exp(-\tfrac{a}{b}\, s)) \ .$$

Another example being of interest for the biological theory results from the Fisher-equation $w_t = D\, w_{xx} + F(w)$ by replacing the diffusion-term w_{xx} by a migration-term of the form $-w + \mu * w$. Then the resulting integrodifferential equation has again the form (1.8) with

$$(1.10)\qquad f(r,s) := -\,d\, r + F(r) + d\, s \qquad , d > 0 \ .$$

The analysis of a deterministic homogeneous one-dimensional spread-process which is autonomous with respect to space and time can be based upon the more general approach

$$(1.11)\qquad w_t = f(w, \mu * g(w)) \qquad .$$

However if $g' \neq 0$ on the range of w coming into question we arrive again at (1.8) by the substitution $\tilde{w} := g(w)$. In the following we will discuss some aspects of a spread-process modelled by (1.8) that have not yet been covered by the theory in [7].

2.Hypotheses on f and μ

Throughout this article we will suppose the following hypotheses:

(h1) There exists $p > 0$ such that $f:[0,p]^2 \mapsto \mathbb{R}$ satisfies $f(0,0)=0$ and $f(p,p) \leq 0$.

(h2) f satisfies the global Lipschitz-condition

$$|f(\tilde{r},\tilde{s}) - f(r,s)| \leq L\ (|\tilde{r}-r| + |\tilde{s}-s|) \quad \text{for all } \tilde{r},r,\tilde{s},s \in [0,p]\ .$$

(h3) $f(r,s\uparrow)$ holds for all $r,s \in [0,p]$, i.e. f is non-decreasing with respect to its second argument for every fixed first argument.

(h4) The partial derivatives $a_0 := \frac{\partial}{\partial r} f(r,s)_{r=s=0}$ and $a_1 := \frac{\partial}{\partial s} f(r,s)_{r=s=0}$ exist such that for some $K > 0$ and $\delta \in (0,1]$

$$|f(r,s) - a_0 r - a_1 s| \leq K(r^{1+\delta} + s^{1+\delta}) \quad \text{for all } r,s \in [0,p]\ .$$

(h5) μ is a finite Lebesgue-measure on $\mathbb{R}\times\mathbb{R}_+$ such that $\mu(\mathbb{R}\times\mathbb{R}_+)=1$ and $\int_0^\infty \int_{-\infty}^{\infty} \exp(\lambda s - \beta\eta)\, d\mu(\eta,s) < \infty$ for all $\lambda \in \mathbb{R}$ and $\beta \geq 0$.

The hypothesis (h5) can be weakened but it is sufficient with regard to the application, for instance if μ has compact support. (h1)-(h4) hold for the source term f of (1.9). In the case of (1.10) one requires $d > 0$ and $F:[0,p] \mapsto \mathbb{R}$ should be Lipschitz-continuous, differentiable in 0, and satisfy $F(0)=0$ and $F(p) \leq 0$. The cybernetical meaning of the hypothesis (h3) is that for a spread-process the correlation between the production-rate w_t of w at (t,x) and the values of w at the deviating coordinates $(t-s,x-\eta)$ $(s \geq 0, \eta \in \mathbb{R}, (s,\eta) \neq (0,0))$ should be positive and not negative. Selfinhibition however, i.e. negative correlation between the production-rate w_t and the value of w at the same point (t,x), may occur. It seems that this property, called quasimonotony, is essential for the realization of travelling fronts and it is characteristic not only for the simple epidemic modell given by (1.8) and (1.9) but also for the more general epidemic models considered in [3,4,10]. It enables us to apply comparison techniques.

3.Mathematical method and lemmata

From [9] and the hypotheses (h1)-(h3) and (h5) we obtain the following existence and uniqueness result for the initial value problem of (1.8). The symbol supp μ designates the support of the measure μ in $\mathbb{R}\times\mathbb{R}_+$.

3.1 Lemma Let $T \geq 0$ be given such that supp $\mu \subset \mathbb{R}\times[0,T]$ ($T:=\infty$ if such a real number T does not exist). Then for every continuous function

$w_o : (-T,0]\times\mathbb{R} \mapsto [0,p]$ there exists a unique solution $w:(-T,\infty)\times\mathbb{R}\mapsto[0,p]$ of (1.8) such that $w=w_o$ on $(-T,0]\times\mathbb{R}$.

The main instrument is the following comparison-principle basing on the Lipschitz-condition and the quasimonotony of the source-term f.

3.2 Lemma Let $w_i : (-T,\infty)\times\mathbb{R}\mapsto[0,p]$,i=1,2,be continuous functions with the following properties:

a) $w^1 \leq w^2$ holds for $(t,x)\in(-T,0]\times\mathbb{R}$.

b) The left partial derivatives $\frac{\partial}{\partial t-}(w^1(t,x)-w^2(t,x))$ exist for $(t,x)\in(0,\infty)\times\mathbb{R}$.

c) $\frac{\partial}{\partial t-}(w^1(t,x)-w^2(t,x)) \leq f(w^1(t,x),(\mu * w^1)(t,x))-f(w^2(t,x),(\mu * w^2)(t,x))$ holds for $(t,x)\in(0,\infty)\times\mathbb{R}$.

Then $w^1 \leq w^2$ follows for $(t,x)\in(-T,\infty)\times\mathbb{R}$.

Proof From [6,Satz 2] we obtain the following conclusion:

Let $\tilde{w} : (-T,\infty)\times\mathbb{R}\mapsto\mathbb{R}$ be continuous and bounded above,where $\tilde{w}\leq 0$ if $(t,x)\in(-T,0]\times\mathbb{R}$.If $(t,x)\in(0,\infty)\times\mathbb{R}$,suppose that the left derivative $\partial\tilde{w}(t,x)/\partial t-$ exists and that for some $N>0$ the following implication holds:

$$(3.1) \qquad \tilde{w}(t,x)>0 \implies \frac{\partial}{\partial t-}\tilde{w}(t,x) \leq N \sup\{\tilde{w}(s,y) \mid t-T<s\leq t, y\in\mathbb{R}\} .$$

Then $\tilde{w}\leq 0$ for $(t,x)\in(-T,\infty)\times\mathbb{R}$.

To apply this result take $w:=w^1-w^2$.It remains to verify the implication (3.1).For this reason assume $\tilde{w}>0$ for some $(t,x)\in(0,\infty)\times\mathbb{R}$.From (h2),(h3),and the hypothesis c) then follows at the point (t,x):

$$\frac{\partial}{\partial t-}\tilde{w} \leq f(w^1,\mu * w^2+\mu * \tilde{w})-f(w^1,\mu * w^2)+f(w^1,\mu * w^2)-f(w^2,\mu * w^2) \leq$$
$$\leq L(\max\{0,\mu * \tilde{w}\}+\tilde{w}) \leq 2L\sup\{\tilde{w}(s,y) \mid t-T<s\leq t, y\in\mathbb{R}\} .$$

This proves the lemma.

The notations sub- and supersolution of (1.8) are defined as follows:

3.3 Definition A continuous function $\underline{w}$ $(\overline{w})$: $\mathbb{R}^2\mapsto[0,p]$ is called sub-(super-)solution of (1.8) if $\underline{w}$ $(\overline{w})$ has a left partial derivative with respect to t for all $(t,x)\in\mathbb{R}^2$ such that

$$(3.2) \qquad \frac{\partial}{\partial t-}\underline{w}(t,x) \leq f(\underline{w}(t,x),(\mu * \underline{w})(t,x)) \qquad (\text{"}\geq\text{" for } \overline{w} \text{ instead of } \underline{w})$$

for all $(t,x)\in\mathbb{R}^2$.

Sub- and supersolutions existing for all time are sufficient for our purpose.From the lemma 3.2 and the definition 3.3 we obtain:

3.4 Lemma Let $\underline{w}$ and $\overline{w}$ be sub- and supersolutions of (1.8) and let w be a solution of (1.8) such that $\underline{w}\leq w\leq\overline{w}$ for all $(t,x)\in(-T,0]\times\mathbb{R}$. Then $\underline{w}\leq w\leq\overline{w}$ for all $(t,x)\in(-T,\infty)\times\mathbb{R}$ follows.

The following lemma which parallels some results in [5,ch.4] for diffusion-reaction equations is the main instrument to prove existence of travelling fronts and to estimate the asymptotic speed of propagation.

3.5 Lemma For any $c \in \mathbb{R}$ suppose $\underline{w}$ ($\overline{w}$) is a sub-(super-)solution of the form $\underline{w}(t,x)=\underline{v}(x+ct)$ ($\overline{w}(t,x)=\overline{v}(x+ct)$). Let w be the solution of (1.8) due to the initial data $w=\underline{w}$ ($w=\overline{w}$) for $(t,x) \in (-T,0]\times\mathbb{R}$. Then $v(t,y):= =w(t,y-ct)$ fulfils $v(t\uparrow,y)$ ($v(t\downarrow,y)$) for all $(t,y) \in \mathbb{R}^2$. The limit $\hat{v}(y):= = \lim_{t\to\infty} v(t,y)$ satisfies $\underline{v} \le \hat{v}$ ($\hat{v} \le \overline{v}$) and $\hat{w}(t,x):=\hat{v}(x+ct)$ solves (1.8) for all $(t,x) \in \mathbb{R}^2$. In addition $\hat{v}(y\uparrow)$ follows if $\underline{v}(y\uparrow)$ ($\overline{v}(y\uparrow)$) on $\mathbb{R}$.

Proof It is sufficient to sketch the proof for the case of the sub-solution. For given $\delta > 0$ define

$$w^{\delta}(t,x) := w(t+\delta, x-c\delta) \quad , \ (t,x) \in (-T,\infty)\times\mathbb{R} \ .$$

Then from lemma 3.4 and the choice of the initial data we obtain

$$(3.3) \qquad w^{\delta}(t,x) \ge \underline{w}(t+\delta,x-c\delta)=\underline{v}(x+ct)=\underline{w}(t,x)=w(t,x), \ \text{if } (t,x)\in(-T,0]\times\mathbb{R}.$$

Since w^{δ} again solves (1.8), (3.3) and the lemma 3.2 yield

$$(3.4) \qquad w^{\delta}(t,x) \ge w(t,x) \quad \text{for } (t,x) \in (-T,\infty)\times\mathbb{R}.$$

As $\delta > 0$ was arbitrary, (3.4) implies

$$(3.5) \qquad v(t\uparrow,y) \quad \text{for all } (t,y) \in (-T,\infty)\times\mathbb{R}.$$

Since v is bounded, the limit-function $\hat{v}$ exists on $\mathbb{R}$ and is Lebesgue - measurable. Because of $\underline{v} \le v$ ($v \le \overline{v}$) with aid of lemma 3.4, we conclude $\underline{v} \le \le \hat{v}$ ($\hat{v} \le \overline{v}$). Now let us prove that $\hat{w}(t,x):=\hat{v}(x+ct)$ solves (1.8) for all real values of t and x.

We have $\hat{w}(t,x)=\lim_{s\to\infty} v(s,x+ct)=\lim_{s\to\infty} w(s,x+c(t-s))=\lim_{\lambda\to\infty} w(t+\lambda,x-c\lambda)$ and this is a reasonable definition for all $(t,x)\in\mathbb{R}^2$. Thus by Lebesgue's convergence theorem (note $0 \le \hat{w} \le p$) we conclude

$$(\mu * \hat{w})(t,x) = \int_0^{\infty}\int_{-\infty}^{\infty} \lim_{\lambda\to\infty} w(t+\lambda-\tau, x-c\lambda-\eta)\, d\mu(\eta,\tau) = \\ = \lim_{\lambda\to\infty} (\mu * w)(t+\lambda, x-c\lambda) \ .$$

Hence $\mu * \hat{w}$ is Lebesgue-measurable. Again from Lebesgue's convergence theorem and (h1) follows for all $(t,x)\in\mathbb{R}^2$:

$$(3.6) \qquad \hat{w}(t,x)-\hat{w}(0,x) = \lim_{\lambda\to\infty} (w(t+\lambda,x-c\lambda)-w(\lambda,x-c\lambda)) = \\ = \lim_{\lambda\to\infty}\int_0^t f(w(s+\lambda,x-c\lambda),(\mu * w)(s+\lambda,x-c\lambda))\,ds = \\ = \int_0^t f(\hat{w}(s,x),(\mu*\hat{w})(s,x))\,ds \ ,$$

showing that $w(.,x)$ and therefore $(\mu * \hat{w})(.,x)$ and $f(\hat{w}(.,x),(\mu*\hat{w})(.,x))$ are continuous in $\mathbb{R}$ for every fixed $x \in \mathbb{R}$. Thus (3.6) implies that $\hat{w}$ solves (1.8) for all $(t,x)\in\mathbb{R}^2$.

Now assume that $\underline{v}$ is non-decreasing. Then, for any $h > 0$, $\underline{w}^h(t,x):=$

$=\underline{w}(t,x+h)$ is again a subsolution satisfying $\underline{w}^h \geq \underline{w}$ on $\mathbb{R}^2$. As $w^h(t,x) := =w(t,x+h)$ has the initial function $\underline{w}^h$, the comparison-principle implies $w^h \geq w$ for all $(t,x) \in (-T,\infty) \times \mathbb{R}$. Hence $\hat{v}(y+h) = \lim_{s\to\infty} w(s,y+h-cs) \geq \hat{v}(y) = \lim_{s\to\infty} w(s,y-cs)$ follows for all $y \in \mathbb{R}$ finishing the proof.

It should be mentioned that the application of the lemma 3.5 for the case c=0 only yields that $\hat{w}$ is Lebesgue-measurable. In order to ensure the continuity of $\hat{w}$ in this case we require further properties of f and μ, for example that μ has a $\mathcal{L}^1$-density.

The lemma 3.5 reduces the existence-problem for travelling-wave solutions of (1.8) to the problem of finding nontrivial sub- and supersolutions of the travelling-wave type. With another technique basing upon monotone iteration it can be shown [7] that this reduction is also possible for the more general equation (1.11), provided that f satisfies a Lipschitz-condition for its first argument uniformly in the second and f and g are continuous and have the properties $f(r,s\uparrow)$ and $g(r\uparrow)$. This generalizaton may be of interest if influence-functions like $g(r)=\sqrt{r}$ occur, where the reduction of (1.11) to (1,8) under the hypotheses (h1)-(h3) is in general impossible.

It should be further noticed that the existence results for non-trivial solutions of the equation

$$(3.7) \qquad z = \nu * g(z) \quad (z\colon \mathbb{R} \mapsto [0,p] \text{ measurable}, \nu \text{ Lebesgue-measure on } \mathbb{R})$$

in [3] appear as special cases of the present theory. Namely, $z=\nu * w$ solves (3.7) if w is a travelling-wave solution of (1.8) with $c:=0, f(r,s):= =-r+g(s)$, and μ being the Lebesgue-measure on $\mathbb{R}\times\mathbb{R}_+$ with support in $\mathbb{R}\times\{0\}$ such that $\mu(I\times\{0\}) := \nu(I)$ for every real interval I.

Looking at the hypothesis (h4) the behaviour of the solutions of (1.8) in the neighborhood of the trivial solution w=0 is dominated by the linearized equation

$$(3.8) \qquad Z_t = a_o Z + a_1 \mu * Z \quad .$$

The travelling-wave solutions $Z(t,x)=V(x+ct)$ of (3.8) are linear combinations of the functions $V_{n,m}(y) = y^n \exp(\omega_m y)$, $y=x+ct$, where the complex number ω_m is a zero of the multiplicity n of the characteristic equation

$$(3.9) \qquad 0 = -\omega c + a_o + a_1 \int_0^\infty \int_{-\infty}^\infty \exp(-\omega cs - \omega\eta)\, d\mu(\eta,s) \quad ,$$

the right hand side of which is analytic on $\mathbb{C}$. From a standard analysis one obtains (c.f. [7]):

<u>3.6 Lemma</u> Let $A,B \in \mathbb{R}$ be given such that $B > 0$ and $A+B > 0$. Then if $\mu(\{0,0\}) < 1$ the function

$$\rho_{c,A,B}(\omega) := -\omega c + A + B \int_0^\infty \int_{-\infty}^\infty \exp(-(\omega cs + \omega\eta))\, d\mu(\eta,s)$$

has the following properties:

a) $\rho_{c,A,B}(\omega)$ is a strict convex function of $\omega \in \mathbb{R}$.

b) There exists $c^*(A,B) \in \mathbb{R}$ such that $\rho_{c,A,B}(\omega)$ has exactly two simple real zeros $0 < \beta_{c,A,B} < \tilde{\beta}_{c,A,B}$, if $c > c^*(A,B)$, and a double zero $\beta^*(A,B) > 0$, if $c = c^*(A,B)$.

c) There exists $\varepsilon > 0$ such that $\rho_{c,A,B}(\omega)$ has two simple conjugate complex zeros $\beta_{c,A,B} \pm i\pi\gamma_{c,A,B}$ if $c^*(A,B) - \varepsilon < c < c^*(A,B)$, where $\lim_{c \mapsto c^*} \beta_{c,A,B} = \beta^*(A,B)$ and $\lim_{c \mapsto c^*} \gamma_{c,A,B} = 0$.

One finds that $c^*(A,B) > 0$ holds if the measure μ is symmetric with respect to the space, i.e. $d\mu(\eta,s) = d\mu(-\eta,s)$ for all $\eta \in \mathbb{R}$ and $s \geq 0$ (c.f. [7]). The analysis of ρ_c is essential for the construction of sub- and supersolutions of (1.8) having the travelling-wave form. Some important examples of such functions are shown in the following lemma.

3.7 Lemma Assume (h1)-(h5) and let $\bar{a}_o \in \mathbb{R}$ and $\bar{a}_1 > 0$ be given such that

(3.10) $$f(r,s) \leq \bar{a}_o r + \bar{a}_1 s \quad , \text{ for all } r,s \in [0,p] .$$

(for instance, $\bar{a}_o := a_o + Kp^\delta$, $\bar{a}_1 := a_1 + Kp^\delta$ can be chosen) Then one gets :

a) If $\bar{a}_o + \bar{a}_1 > 0$ and $c \geq c^*(\bar{a}_o,\bar{a}_1)$, then $\bar{w}(t,x) := \bar{v}(t,x)$, where

(3.11) $$\bar{v}(y) := \min\{p, R\exp(\beta y)\} \quad , \quad \beta := \beta_{c,\bar{a}_o,\bar{a}_1} \quad ,$$

is a supersolution for every $R > 0$.

b) If $\bar{a}_o + \bar{a}_1 > 0$, $c = c^*(\bar{a}_o,\bar{a}_1)$, and supp μ is compact, then there exists $q_o \in \mathbb{R}$ such that, for every $q \leq q_o$, $\bar{w}(t,x) := \bar{v}(x+ct)$, where

(3.12) $$\bar{v}(y) := \begin{cases} p & -q \leq y \\ p\exp(\beta y)(-y-q+\exp(\beta q)), & y \leq -q \end{cases} \quad , \quad \beta = \beta^*(\bar{a}_o,\bar{a}_1)$$

is again a supersolution.

c) If $a_1 > 0$, $a_o + a_1 > 0$, and $c > c^*(a_o,a_1)$, then there exist numbers $\varkappa > 0$ and $M > 0$ such that $\underline{w}(t,x) := \underline{v}(x+ct)$, where

(3.13) $$\underline{v}(y) := \max\{0, p\exp(\beta y)(1 - M\exp(\varkappa y))\} \quad , \quad \beta = \beta_{c,a_o,a_1} \quad ,$$

is a subsolution.

d) If $a_1 > 0$, $a_o + a_1 > 0$, $c = c^*(a_o,a_1)$, and supp μ is compact, then for every $q \in \mathbb{R}$ there exist $\xi_o > 0$ and $\lambda > 0$ such that $\underline{w}(t,x) := \underline{v}(x+ct)$, where

(3.14) $$\underline{v}(y) := \begin{cases} 0 & y \geq 0 \\ \xi\exp(\beta y)(-y-q+q\exp(\lambda y)) & , y \leq 0 \end{cases} \quad , \quad \beta = \beta^*(a_o,a_1)$$

is a subsolution for all $\xi \in (0,\xi_o]$.

e) If $a_1 > 0$ and $a_o + a_1 > 0$, there exists $\varepsilon > 0$ such that for every velocity $c \in (c^*(a_o,a_1) - \varepsilon, c^*(a_o,a_1))$ real numbers $\xi_o > 0$, $\beta > 0$, and $\gamma > 0$ can be chosen such that the function $\underline{w}(t,x) := \underline{v}(x+ct)$, where

$$(3.15) \qquad \underline{v}(y) := \begin{cases} 0 & , y \le 0 \\ \delta \exp(\beta y)\sin(\pi\gamma y) & , 0 \le y \le 1/\gamma \\ 0 & , y \ge 1/\gamma \end{cases}$$

is a subsolution for all $\delta \in (0, \delta_o]$.

Proof The statements a) and c) have been shown in [7,8].The argumentation in the cases b) and d) is the same as in [11,Theorem 5] .The proof of the statement e) is similar to that of [1,Lemma 1-3].Namely (h4) implies the existence of sufficiently small numbers $\tilde{\alpha}, \tilde{\delta} > 0$ such that $\tilde{\alpha} < a_1$, $2\tilde{\alpha} < a_o + a_1$, and

$$(3.16) \qquad f(r,s) \geqq (a_o - \alpha_o)r + (a_1 - \alpha_1)s \quad ,$$

if $0 < \alpha_i \le \tilde{\alpha}$ (i=1,2) and $r,s \in [0,\tilde{\delta}]$ .For given $\gamma_o > 0$ the lemma 3.6 c) and the fact that $c^*(.,.)$ is a continuous nondecreasing function of its arguments yield the existence of $\varepsilon > 0$ such that for every c in the interval $c^*(a_o,a_1) - \varepsilon < c < c^*(a_o,a_1)$ and $\alpha_i \in (0,\tilde{\alpha}]$ (i=1,2) with the property $c < c^*(a_o - \alpha_o, a_1 - \alpha_1)$ the function $\rho_{c,a_o-\alpha_o,a_1-\alpha_1}$ has two simple conjugate complex zeros,the real part $\beta := \beta_{c,a_o-\alpha_o,a_1-\alpha_1}$ of which is positive and the imaginary part $\gamma := \gamma_{c,a_o-\alpha_o,a_1-\alpha_1}$ fulfils $0 < \gamma < \gamma_o$.If $\gamma_o > 0$ was sufficiently small,the analogous argumentation as in [1] employing the equation $\rho_{c,a_o-\alpha_o,a_1-\alpha_1}(\beta \pm i\pi\gamma) = 0$ shows that the function $\underline{v}$ of (3.15) solves the inequality

$$(3.17) \; \max\left\{c \frac{d}{dy+}\underline{v}(y), c \frac{d}{dy-}\underline{v}(y)\right\} \le (a_o - \alpha_o)\underline{v}(y) + (a_1 - \alpha_1)\int_0^{\infty}\int_{-\infty}^{\infty} \underline{v}(y-\eta)\,d\mu(\eta - cs, s).$$

From (3.16) and (3.17) follows the statement e) if $\delta_o \in (0, \tilde{\delta}]$ is small.

After these preparations it is rather easy to analyse the spread-process modelled by (1.8).

4.Asymptotic speed of propagation

Let w be a solution of (1.8) such that $w \not\equiv 0$.The concept of asymptotic speed developed by D.G.Aronson and H.F.Weinberger (c.f.[1,8,10,11]) authorizes us to call the number

$$\tilde{c} := \inf\left\{c \in \mathbb{R} \mid \lim_{t\to\infty} w(t, x-ct) = 0 \text{ for every } x \in \mathbb{R}\right\} \in [-\infty, \infty]$$

the asymptotic speed of propagation due to the solution w.In the following theorem estimates of $\tilde{c}$ are given generalizing known estimates in the case of Fisher's equation.With regard to the interpretation of (1.8) as a spread process it is sufficient to take into consideration only initial functions vanishing rather quickly for $x \mapsto -\infty$.Generalizations of the upper bounds of c also for non-autonomous equations are put down in [8].

<u>4.1 Theorem</u> Suppose $\mu(\{0,0\}) < 1$ and let $\bar{a}_0 \in \mathbb{R}$ and $\bar{a}_1 > 0$ be given according to (3.10). Then the following statements hold:

<u>a)</u> If $\bar{a}_0 + \bar{a}_1 \le 0$ and $f(s,s) < 0$ for $0 < s \le p$, every solution w of (1.8) satisfies

(4.1) $$\lim_{t \to \infty} w(t,x) = 0 \text{ ,uniformly for } x \in \mathbb{R} \quad (\tilde{c} = -\infty).$$

<u>b)</u> Suppose $\bar{a}_0 + \bar{a}_1 > 0$ and define $\bar{c}^* := c^*(\bar{a}_0, \bar{a}_1)$, $\bar{\beta}^* := \beta^*(\bar{a}_0, \bar{a}_1)$. Let w be a solution of (1.8) the initial function w_0 of which fulfils

(4.2) $$\sup_{x \in \mathbb{R}, -T < t \le 0} w_0(t,x) \exp(-\bar{\beta}^*(x + \bar{c}^* t)) < \infty ,$$

where $T \ge 0$ is chosen according to lemma 3.1. Then $\tilde{c} \le \bar{c}^*$ and

(4.3) $$\lim_{t \to \infty} \sup_{x \le r} w(t, x - ct) = 0 \text{ ,for every } r \ge 0 \text{ and } c > \bar{c}^* .$$

<u>c)</u> Suppose $a_1 > 0$, $a_0 + a_1 > 0$, $\operatorname{supp} \mu \subset \mathbb{R} \times [0,T]$ with $T < \infty$, and define $\underline{c}^* := c^*(a_0, a_1)$. Let w be a solution of (1.8) due to a continuous initial function w_0 satisfying

(4.4) $$\varliminf_{x \mapsto \infty} w_0(t,x) > 0 \text{ ,uniformly for } -T \le t \le 0.$$

Then $\tilde{c} \ge \underline{c}^*$ and

(4.5) $$\varliminf_{t \to \infty} \inf_{x \ge -r} w(t, x - ct) > 0, \text{ for every } r \ge 0 \text{ and } c < \underline{c}^* .$$

<u>Proof</u> <u>a)</u> One finds that $\bar{w}(t,x) := \begin{cases} p & \text{,if } -T \le t \le 0 \\ q(t) & \text{,if } t \ge 0 \end{cases}$,where q solves the differential equation $q'(t) = f(q(t), q(t))$ for $t \ge 0$, is a supersolution. Because of $\lim_{t \to \infty} q(t) = 0$, the statement follows from the lemma 3.4.

<u>b)</u> Let R denote the value of the supremum in (4.2). Then the lemma 3.7 a) yields that

(4.6) $$\bar{w}(t,x) := \min\{p, R \exp(\bar{\beta}^*(x + \bar{c}^* t))\} \quad , \quad (t,x) \in \mathbb{R}^2,$$

is a supersolution satisfying $w_0 \le \bar{w}$ for $(t,x) \in (-T,0] \times \mathbb{R}$. Thus the lemm 3.4 finishes the proof and (4.3) is a consequence from $0 \le w \le \bar{w}$ and the definition of $\bar{w}$ in (4.6).

<u>c)</u> With regard to the definition of c it is sufficient to prove (4. For this reason let $r \ge 0$ and $c < \underline{c}^*$ be given. Choose a continuous function $\hat{w}_0 : [-T,0] \times \mathbb{R} \mapsto [0,p]$ such that, for all $(t,x) \in [-T,0] \times \mathbb{R}$, $\hat{w}_0 \le w_0, \hat{w}_0(t,x\uparrow)$, and $\hat{w}_0(t,x) \ge \hat{\xi} > 0$ for all $x \ge \hat{x}$ and $t \in [-T,0]$. With aid of (4.5) such a function $\hat{w}_0$ exists for sufficiently small $\hat{\xi}$ and large $\hat{x}$. If $\hat{w}$ designates the corresponding solution of (1.8), the comparison-principle 3.2 yields

(4.7) $$\hat{w}(t,x) \le w(t,x) \quad \text{and} \quad \hat{w}(t,x\uparrow), \text{ for all } (t,x) \in [-T, \infty) \times \mathbb{R}.$$

By lemma 3.7 e) let $\hat{c} \in [c, \underline{c}^*)$, $\xi \in (0, \hat{\xi}]$, $\beta > 0$, and $\gamma > 0$ such that $\underline{w}(t,x) := \underline{v}(x - \hat{y} + \hat{c}t)$ with $\hat{y} := x + |\hat{c}|T$ and $\underline{v}$ given by (3.15) is a subsolution of (1.

satisfying $\max_{y \in \mathbb{R}} \underline{v}(y) \leq \hat{\xi}$ .The construction of $\underline{w}$ yields

$$(4.8) \qquad \hat{w}_o(t,x) \geq \underline{w}(t,x) \quad ,\text{if } (t,x) \in [-T,0] \times \mathbb{R}.$$

Namely $\underline{w}(t,x)=0$ holds,if $x \leq \hat{y}-\hat{c}t$,and $\hat{w}_o(t,x) \geq \hat{\xi} \geq \xi$,if $x \geq \hat{y}-\hat{c}t$,because of $\hat{y}-\hat{c}t \geq \hat{y}-|\hat{c}|T = \hat{x}$ .Hence by the comparison lemma

$$(4.9) \qquad \hat{w}(t,x) \geq \underline{w}(t,x) \quad ,\text{if } (t,x) \in [-T,\infty) \times \mathbb{R} ,$$

follows.Together with (4.7) the inequality (4.9) yields for all values $t \geq (r+\hat{y}+(2\gamma)^{-1})/(\hat{c}-c)$ the estimate

$$\inf_{x \geq -r} w(t,x-ct) \geq \inf_{x \geq -r} \hat{w}(t,x-ct) = \hat{w}(t,-r-ct) = \hat{w}(t,-r-\hat{c}t+(\hat{c}-c)t)$$
$$\geq \hat{w}(t,\hat{y}-\hat{c}t+(2\gamma)^{-1}) \geq \underline{w}(t,\hat{y}-\hat{c}t+(2\gamma)^{-1}) = \underline{v}(\tfrac{1}{2\gamma}) > 0,$$

which finishes the proof.

The reason for the hypothesis $T < \infty$ underlying the part c) of the theorem 4.1 is that in the case of infinite delay the construction of the subsolution would require to replace the hypothesis (4.4) by the hypothesis $\lim_{t\to\infty} \inf_{x \geq \tilde{x}} w_o(t,x-(\underline{c}^*-\tilde{\varepsilon})t) > 0$ for a certain $\tilde{x} \in \mathbb{R}$ and $\tilde{\varepsilon} > 0$. However this hypothesis and the hypothesis (4.2) cannot be satisfied for the same function w_o.Thus the sophisticated problem wether the statement c) of the theorem 4.1 has a generalization for the case of infinite delay which yields an inclusion of $\tilde{c}$ remains open.

For the modified Fisher equation with the source-term f of (1.10), the theorem 4.1 generalizes a well known estimate for the minimal speed of propagation due to the original autonomous Fisher-equation,namely:

$$(4.10) \quad c^*(-d+F'(0),d) \leq \tilde{c} \leq c^*(-d+l,d) \qquad , \; l := \sup_{0<r\leq p} \frac{1}{r} F(r) ,$$

provided that $F(0)=0, F'(0) > 0$,and $F(p)=0$ for a certain $p > 0$.Of course, if $F'(0)=l$,(4.10) is a formula for the computation of $\tilde{c}$ (A good bibliography for the analysis of Fisher's equation is found in [5]).

5.Travelling-front solutions

The notation "travelling-front solution" that will be used is the following:

5.1 Definition A solution w of (1.8) is called travelling-front solution for the velocity c,if $w(t,x)=v(x+ct)$ for all $(t,x) \in \mathbb{R}^2$ and $v(y) \in [0,p]$ satisfies $\lim_{y\to-\infty} v(y)=0$ and $\underline{\lim}_{y\to\infty} v(y) > 0$.

The opposite case $\lim_{y\to\infty} v(y)=0$ and $\underline{\lim}_{y\to-\infty} v(y) > 0$ can be reduced to the former by the substitution $\tilde{x}:=-x, \tilde{t}:=t$,and $\tilde{c}:=-c$. As far as existence, non-existence,and uniqueness modulo translation of travelling-front solutions is concerned, the following theorem holds:

5.2 Theorem Assume (h1)-(h5), $a_1 > 0$, $a_o + a_1 > 0$, $\mu(\{(0,0)\}) < 1$, and let (h4) be replaced by the stronger hypothesis

(5.1) $\quad -K(r^{1+\delta} + s^{1+\delta}) \leq f(r,s) - a_o r - a_1 s \leq 0$ for all $r,s \in [0,p]$.

In addition let $p > 0$ be the smallest positive zero of the equation $f(s,s)=0$. Then the following is true:

a) For every $c > c^*(a_o,a_1)$ there exists a travelling-front solution w such that v (c.f. the definition 5.1) is non-decreasing, $\lim_{y\to-\infty} e^{-\beta y} v(y) = q > 0$ exists with $\beta := \beta_{c,a_0,a_1}$, and $\lim_{y\to\infty} v(y) = p$.

b) If $c < c^*(a_o,a_1)$, there exists no travelling-front solution.

c) If $c = c^*(a_o,a_1)$ and μ has compact support, there exists a travelling-front solution w such that v is non-decreasing, $\lim_{y\to-\infty} |y|^{-1} e^{-\beta^* y} v(y) = q > 0$ exists with $\beta^* := \beta^*(a_o,a_1)$, and $\lim_{y\to\infty} v(y) = p$.

d) Suppose that, in addition, f has Hölder-continuous first order partial-derivatives in $[0,p]^2$ and define

(5.2) $\quad \tilde{a}_o := \max_{r,s\in[0,p]} \frac{\partial}{\partial r} f(r,s) \; (\geq a_o)$, $\quad \tilde{a}_1 := \max_{r,s\in[0,p]} \frac{\partial}{\partial s} f(r,s) \; (\geq a_1)$.

Then for every $c > \tilde{c}^* := c^*(\tilde{a}_o, \tilde{a}_1)$ $(\geq c^*(a_o,a_1))$ there exists modulo translation a unique "regular" travelling-front solution of (1.8), i.e. a solution w such that $\lim_{y\to-\infty} v(y) e^{-\lambda y}$ exists and is positive for a certain $\lambda > 0$.

Proof a) The assertion follows from the lemma 3.5 and the lemma 3.7 with aid of the subsolution $\underline{w}$ and the supersolution $\overline{w}$ due to

$$\underline{v}(y) := \max\{0, p\, e^{\beta y}(1 - M e^{\varkappa y})\}, \quad \overline{v}(y) := \min\{p, p\, e^{\beta y}\}.$$

(Note that $\overline{v}$ is non-decreasing) That $\lim_{y\to\infty} v(y) = p$ holds for every non-decreasing travelling-front solution has been shown in [7].

b) Assume $w(t,x) = v(x+ct)$ is a travelling-front solution, where $c < c^* := c^*(a_o,a_1)$. Let $\rho_o > 0$ and $y_o \in \mathbb{R}$ such that $v(y) \geq \rho_o$ if $y \geq y_o$. The lemma 3.7 e) yields the existence of $\hat{c} \in (c, c^*)$, $\gamma > 0$, $\beta > 0$, and $\rho \in (0, \rho_o]$ such that $\underline{w}(t,x) := \underline{v}(x + \hat{c}t - y_o)$ with $\underline{v}$ given by (3.15) is a subsolution. For all $(t,x) \in (-\infty,0] \times \mathbb{R}$ we have $w(t,x) = v(x+ct) \geq \rho_o$ if $x+ct \geq y_o$ and $\underline{w}(t,x) = 0$ if $x+ct \leq y_o$ showing $w \geq \underline{v}$ for all $(t,x) \in (-\infty,0] \times \mathbb{R}$ if ρ is sufficiently small. Hence by the lemma 3.4 we conclude $w \geq \underline{w}$ for all $(t,x) \in \mathbb{R}^2$, or equivalently $v(y) \geq \underline{v}(y - y_o + (\hat{c}-c)t)$ for all $(t,y) \in \mathbb{R}^2$. This conclusion contradicts the travelling-front property $\lim_{y\to-\infty} v(y) = 0$ as $v(y) \geq \underline{v}(\gamma/2) > 0$ for every $y \in \mathbb{R}$ follows if $t = (\gamma/2 + y_o - y)/(\hat{c} - c)$ is chosen.

c) The statement follows from the lemma 3.5 and the lemma 3.7, b), d).

Namely because of $\beta^* = \beta^*(\bar{a}_o, \bar{a}_1)$ the number $\rho > 0$ can be chosen such that the functions $\underline{v}$ of (3.14) and $\bar{v}$ of (3.12) satisfy $\underline{v} \leq \bar{v}$ on $\mathbb{R}$.

d) A proof is given in [7].

The compactness hypothesis for supp μ in the case $c=c^*$ can be weakened by imposing additional assumptions on the measure μ which will be neglected here. The uniqueness-problem requires further analysis. It seems that every travelling-front solution is regular in the case $c > c^*$ such that the regularity-hypothesis would be superflous.

The epidemic-model, where f is given by (1.9), has the nice property that $\tilde{a}_o = a_o = -b$ and $\tilde{a}_1 = a_1 = a > 0$ hold such that $\tilde{c}^* = c^*$ and the uniqueness-problem is open only for the case $c = c^*$. The same conclusion for the modified Fisher-equation however, where f is given by (1.10), would require the additional hypothesis $F'(0) = \max_{0 \leq s \leq p} F'(s)$. Whether this hypothesis is necessary in the non-trivial case $\mu(\{(0,0)\}) < 1$ seems to be still an open question.

6. Dependence of the minimal speed on delays

It may be of interest how much delays in the transfer-mechanism of the spread-process control the minimal speed of propagation. To analyse this effect let the spread-process be modelled by (1.8), where f and μ satisfy (h1)-(h5), $a_1 > 0$, and $a_o + a_1 > 0$. Let $c^* = c^*(a_o, a_1)$ denote its minimal speed of propagation and let c_m denote the corresponding speed if the measure μ is replaced by the averaged measure μ_m with support in $\mathbb{R} \times \{0\}$, defined by

$$\mu_m(I \times \{0\}) := \int_{s \geq 0} \int_{\eta \in I} d\mu(\eta, s) \quad \text{, for every interval } I \subset \mathbb{R}.$$

Then one expects $c^* \leq c_m$ if $c_m \geq 0$ and $c^* \geq c_m$ if $c_m \leq 0$, which means that delays in the transfer-mechanism cannot increase the absolut value of the minimal speed of propagation no matter whether the front is travelling forward or backward. This fact is established in the following theorem:

5.1 Theorem $\quad |c^*| \leq |c_m|$ holds.

Proof Let $\rho^{(m)}$ denote the characteristic function due to μ_m. Then the definition of ρ in the lemma 3.6 yields

$$\rho_{c_m}(\lambda) \leq \rho_{c_m}^{(m)}(\lambda) \quad \text{, for all } \lambda \geq 0, \quad (A := a_o, B := a_1)$$

if $c_m \geq 0$, and

$$\rho_{c_m}(\lambda) \geq \rho_{c_m}^{(m)}(\lambda) \quad \text{, for all } \lambda \geq 0,$$

if $c_m \leq 0$.Because of $\partial \rho_c(\lambda)/\partial c \leq 0$ for all $\lambda \geq 0$,the smallest number c (which is equal to c^*) such that ρ_c has for the first time a positive zero,cannot be greater(smaller) than c_m ,if $c_m \geq 0$ (if $c_m \leq 0$).This proves the theorem.

With regard to the application one requires a quantitative analysis how much delays in the tranfer-mechanism decrease $|c^*|$.For this reason let us suppose $d\mu(\eta,s)=d\mu(-\eta,s)$ for all $\eta \in \mathbb{R}$ and $s \geq 0$,which means that the transfer is symmetric with regard to the space.Furthermore we assume that μ is concentrated in a sufficiently small neighborhood of (0,0) such that the moments

$$\mu_{ij} := \int_0^\infty \int_{-\infty}^\infty s^i \eta^j \, d\mu(\eta,s) \qquad , \; i,j \in \{0,1,2,\dots\} ,$$

can be neglected if $|i| + |j| \geq 3$.By the symmetry we have $\mu_{ij}=0$ if j is odd.Thus Taylor-expansion shows that the transcendental equations

$$0 = -\beta^* c^* + a_o + a_1 \int_0^\infty \int_{-\infty}^\infty \exp(-\beta^*(\eta+c^*s))\, d\mu(\eta,s)$$

$$0 = -c^* - a_1 \int_0^\infty \int_{-\infty}^\infty (\eta+c^*s)\exp(-\beta^*(\eta+c^*s))\, d\mu(\eta,s) \; ,$$

from which c^* is determined,can be replaced by the second-order system

$$0 = -\beta'c'+a_o+a_1-a_1\mu_{1o}\beta'c'+\tfrac{1}{2}a_1\mu_{o2}\beta'^2+\tfrac{1}{2}a_1\mu_{2o}\beta'^2c'^2$$

$$0 = -c'-a_1\mu_{1o}c'+a_1\mu_{o2}\beta'+a_1\mu_{2o}\beta'c'^2$$

from which c' can be eliminated.One obtains:

$$(6.1) \qquad c' = \sqrt{\frac{2\, a_1(a_o+a_1)\mu_{o2}}{(1+a_1\mu_{1o})^2 - 2\mu_{2o}\, a_1(a_o+a_1)}} \; .$$

Assuming $\mu_{2o} \ll \mu_{1o} \ll 1$ the formula (6.1) yields with the abbreviations

$$(6.2) \qquad D := \tfrac{1}{2} a_1 \mu_{o2} \qquad \text{"mean diffusion-rate"}$$

and

$$(6.3) \qquad l := a_o + a_1 \qquad \text{"total rate of increase"}$$

the following first order approximation of c^* obtained from the expansi[on] of μ by moments:

$$(6.4) \qquad \hat{c}^* = \frac{\sqrt{4\, l\, D}}{1+2D(\mu_{1o}/\mu_{o2})} \; .$$

If $\mu_{1o}=0$,the formula (6.4) is the formula for the minimal speed of propagation due to the autonomous Fisher-equation if $F'(0)=\max_{0<r\leq p} F(r)/r = l$ (c.f.[5]).

Finally let us apply the formula (6.4) to the epidemic model,where f is given by (1.9) and μ by (1.6) and $a > b$ holds.We assume that k_o is independent of τ which means that the effectiveness of the pathogenic agents depends only on the covered distance.Writing $k_o(\eta)$ instead of

$k_o(\eta,\tau)$ we obtain by an easy calculation from (6.4) (if $\int_0^\infty k_1(\rho)d\rho = 1$)

$$(6.5) \qquad \hat{c}^* = \frac{\sqrt{2\,a\,(a-b)\,V}}{1 + a\,\bar{\lambda}\,/\,\bar{\rho}} ,$$

where $V := \int_{-\infty}^{\infty} \eta^2 k_o(\eta)d\eta$ and $\bar{\lambda} := \int_{-\infty}^{\infty} |\eta|\, k_o(\eta)d\eta$ denotes the mean distance covered by the pathogenic agents and $\bar{\rho} := (\int_0^\infty \rho^{-1} k_1(\rho)d\rho)^{-1}$ their (harmonic) mean velocity. (6.5) shows that $\hat{c}^*$ is dominated by $\bar{\rho}$ if, for fixed distribution k_o, $\bar{\rho}$ becomes small.

References

1. D.G.Aronson,The asymptotic speed of propagation of a simple epidemic, W.E.Fitzgibbon(III) & H.F.Walker,Nonlinear diffusion,Research Notes in Mathematics,Pitman 1977.
2. C.Atkinson & E.H.Reuter,Deterministic epidemic waves,Mathematical Proceedings of the Cambridge Philosophical Society (1976),80,315-330.
3. O.Diekmann,On a nonlinear integral equation arising in mathematical epidemiology,Mathematical Centre Report TW 170/77,Amsterdam 1977.
4. O.Diekmann & H.G.Kaper,On the bounded solutions of a nonlinear convolution equation,Mathematical Centre Report TW 172/77,Amsterdam 1977.
5. P.C.Fife,Mathematical aspects of reacting and diffusing systems,Lecture Notes in Biomathematics 28,Springer-Verlag (1979).
6. R.Redheffer & W.Walter,Das Maximumprinzip in unbeschränkten Gebieten für parabolische Ungleichungen mit Functionalen,Math.Ann.226,155-170 (1977)
7. K.Schumacher,Travellimg-front solutions for integrodifferential equations I ,preprint.
8. K.Schumacher,Obere Schranken für die Ausbreitungsgeschwindigkeit bei parabolischen Funktionaldifferentialgleichungen,preprint.
9. K.Schumacher,Existence and continuous dependence for functional-differential equations with unbounded delay,Arch.Rat.Mech.Anal. 67, 315-334 (1978).
10. H.Thieme,Asymptotic estimates of the solutions of nonlinear integral equations and asymptotic speed for the spread of populations,Journal für die reine und angewandte Mathematik,306(1979),94-121.
11. H.F.Weinberger,Asymptotic behavior of a model in population genetics, in Nonlinear Partial Differential Equations ans Applications,J.M. Chadam (ed.),Proceedings,Indiana 1976-1977,Lecture Notes in Mathematics 648,1978.

SOME MATHEMATICAL CONSIDERATIONS OF HOW TO STOP THE SPATIAL SPREAD OF A RABIES EPIDEMIC

Horst R. Thieme
Universität Heidelberg, SFB 123
Im Neuenheimer Feld 294
D-6900 Heidelberg, West Germany

In the last thirty years an epidemic of silvatic rabies has spread over central Europe. Although other species are involved in this epidemic, foxes have been found to be its main carriers such that the epidemic's development and spread is widely determined by the ecology and the control of fox populations (see [8], [9; I, III].). The observations made during two outbreaks (1964-65, 1969-70) of silvatic rabies in South Jutland (Denmark) permit to conclude that the reduction of the fox density in a protective belt in front of the epidemic wave can prevent the disease from penetrating into a particular area (see [6], [9; I, III].). In this article we derive a deterministic mathematical model which supports this conclusion and shows how, at least in principle, one can determine sufficient conditions for the depth of the protective belt and the level of reduction, such that the penetration of the epidemic is stopped.

For information concerning foxes and rabies we refer the reader to the references [2], [3], [6], [8], [9], and the literature cited there. The author thanks J. Berger (Hamburg), K. Bögel (WHO, Geneva), and K. Dietz (Tübingen) for instructive conversations and information material.

1. ECOLOGICAL AND EPIDEMIOLOGICAL ASSUMPTIONS

We assume that the fox population settles a plane habitat; for simplicity we consider the whole two-dimensional real plane $\mathbb{R}^2 = \{x = (x_1,x_2); x_1,x_2 \in \mathbb{R}\}$. We suppose that foxes exhibit a territorial behaviour which has a one-day rhythm (see [3], [8].).

We neglect that a small percentage of foxes is permanently itinerant and that rabid foxes may cease to stick to their territories. Further we neglect that male foxes may leave their territories in the rutting season looking for a mate and that juvenile foxes disperse in autumn looking for territories of their own.

(i) In presuming territorial behaviour we associate a centre of activity with every fox. We assume that the migration of a fox is isotropic with respect to its activity centre, i.e. that a fox whose activity centres at the point $x \in \mathbb{R}^2$ visits the point $y \in \mathbb{R}^2$ at a daily rate which only depends on the distance $|x-y|$ between x and y.

(ii) Further we associate a radius R of daily activity with every susceptible fox (susceptible: capable of catching the disease) such that the fox does not move further away from its activity centre than a maximum distance R. The activity radius can vary with the day and with the fox under consideration at a probability distribution $dP(R)$, but we neglect that this distribution may vary with the age of the fox and with the season.

We introduce the notion of an activity radius for the following reason: In a deterministic model describing spatial phenomena and using continuous space variables one considers spatial densities such that, in general, the number of individuals which visit a particular area on the day under consideration is no integer; in particular it may even not come up to one individual. In nature, however, this number is an integer. Thus it seems reasonable to assume in the model that a susceptible fox is not infected on the day under consideration, if the (generally non-integer) number of rabid foxes which visit its activity circle does not come up to one fox.

(iii) Following this idea we insert into the model the rate at which a susceptible fox actually comes into contact with rabid foxes (without being necessarily infected) on the day under consideration. We assume that this rate depends monotone increasing on the (generally non-integer) number of rabid foxes visiting the activity circle of the susceptible fox, and that it tends to zero, if this number tends to zero (see (6),..., (10).).

(iv) The daily migrations of rabid and susceptible foxes are described by non-negative integral kernels h and k. $h(|x-y|)$ indicates the daily rate at which a rabid fox whose activity centres at x visits the point y. $k(R,|x-y|)$ indicates the daily rate at which a susceptible fox whose activity centres at x and has the radius R visits the point y. Obviously

(1) $$k(R,r) = 0 , \quad \text{for} \quad r > R.$$

We assume that

(2) $$k_o(R) = \int_{\mathbb{R}^2} k(R,|y|)dy \leq 1 , \quad \text{for all} \quad R \geq 0.$$

$k_o(R)$ can be interpreted as the daily activity of a susceptible fox which has the activity radius R. If $k_o(R) = 1/2$, e.g., such a fox is active half the day.

(v) We assume that an immediate contact is necessary for an infection, because the rabies virus is contained in the saliva of the rabid fox and is normally transmitted by bite. (Scavenging and the contact with infected urine, however, might also be a route of infection.) Thus the rate at which a rabid fox infects a suceptible fox is proportional to their contact rate. The daily rate at which a rabid fox whose activity centres at y contacts a susceptible fox whose activity centres at x and has the radius R is given by

$$(3) \qquad K(R,x,y) = \int_{\mathbb{R}^2} h(|y-z|)\, k(R,|x-z|)dz \; .$$

(vi) Once being infected a fox does not become infective immediately after the infection, but only after an incubation period of minimum length $\sigma > 0$. For simplicity we identify rabid and infective foxes [9, II], although sometimes foxes may transmit the disease one or two days before they show the first symptoms [8].

(vii) After a period of maximum length $\tau > \sigma$ has elapsed since the infection, the rabid fox ceases to be infectious and at the same time dies.

(viii) In spring the birth of young foxes refills the reservoir of susceptible foxes. For this reason we introduce the notion of an epidemic year which lasts from one spring to the next and consider the epidemic seperately for every epidemic year.

(ix) We only consider the case that the control operations (gassing, vaccination, e.g.) are performed at the beginning of the epidemic year immediately after the young foxes were born. We neglect that these operations take some time and that the foxes might react on operations which eliminate a part of the population by expanding their territories or occupying territories whose former occupants have been killed. We assume that the control operations mainly affect the susceptible population such that, on the whole, only the initial distribution of the susceptible population is affected. We introduce a function $\gamma : \mathbb{R}^2 \to [0,1]$ called the control function such that $\gamma(x)$ indicates the rate at which susceptible foxes escape from the control operations, at the point x.

(x) For the epidemic year under consideration, we split up the population into the classes of susceptible, incubating, and infective foxes. Let S and I denote the densities of the susceptible and infective foxes such that, for $t \geq 0$ and any open set B of $\mathbb{R}^2$, the integrals

$$(4) \qquad \beta \int_B S(t,x)dx \qquad \text{and} \qquad \beta \int_B I(t,x)dx$$

indicate the numbers of susceptible and infective foxes at time t whose activities centre in B. Hereby $t = 0$ is the beginning of the epidemic year under consideration just after the young foxes were born and the control operations were performed. β is some scaling factor.

(xi) We assume that in the years before the outbreak of the epidemic the fox population was uniformly distributed over the habitat and had exhausted its carrying capacity. Thus let $N > 0$ be the constant density which the population attained in every pre-epidemic spring after the young foxes were born. According to (ix) and (x), this implies that

$$(5)\qquad S(0,x) \leq \gamma(x)\, N\ , \quad \text{for } x \in \mathbb{R}^2,$$

in the epidemic year under consideration.

2. DERIVATION OF THE MODEL

It is convenient to choose the difference $\tau-\sigma$ as unit of time. $\tau-\sigma$ indicates the maximum length of the infectious period (see (vi) and (vii).).

The development of the susceptible population

Since we have assumed that the foxes have a one-day rhythm, we let $\tau-\sigma = \rho$ days and consider the decrease $S(t,.) - S(t+1/\rho,.)$ of the density of the susceptible population from the tth day to the subsequent day. We assume this decrease to be proportional to the density of the susceptible population on the tth day and to the expected number of infective foxes which a susceptible fox contacts on the tth day (see (v).). Since $1/\rho$ is small (the maximum length of the infective period is about one week), we replace $S(t,.) - S(t+1/\rho,.)$ by $-\dot{S}(t,.)/\rho$ with the dot denoting the time derivative. Thus we obtain the following differential equation:

$$(6)\qquad -\dot{S}(t,x)/S(t,x) = \alpha\rho \int_0^\infty \int_{\mathbb{R}^2} I(t,y)\, K(R,x,y)\, Q(R,t,x)\, P(dR)dy,$$

for $t \geq 0$, $x \in \mathbb{R}^2$. P is the probability distribution of the radius of daily activity (see (ii).). By (3), the integral

$$\int_{\mathbb{R}^2} I(t,y)\, K(R,x,y)\, dy$$

indicates the expected number of infective foxes which contact a susceptible fox whose activity centres at x and has the radius R, on the tth day, on the condition that this susceptible fox actually comes into contact with infective foxes on the tth day. The probability that this latter event occurs is denoted by $Q(R,t,x)$. Thus the integral on the right side of (6) in fact indicates the number of infective foxes which contact a susceptible fox whose activity centres at x. The proportionality factor α indicates the rate at which the contact with an infective fox leads to an infection.

According to assumption (iii) we have inserted into equation (6) the rate $Q(R,t,x)$ at which a susceptible fox whose activity centres at x and has the radius R actually comes into contact with infective foxes on the tth day. According to (iii), this rate is a function of the number of infective foxes which visit the activity circle of the susceptible fox. Hence, by (iv),

$$(7) \qquad Q(R,t,x) = g\Big(\int_{|x-y|<R} \int_{\mathbb{R}^2} I(t,z)\, h(|y-z|)\, dydz \Big) .$$

According to (iii) we assume that g is a monotone increasing function from $[0,\infty)$ to $[0,1]$ which satisfies

$$(8) \qquad g(r) \to 0 \quad \text{for} \quad r \to 0 .$$

We remark that the insertion of Q into equation (6) is the critical point in which (6) differs from related equations in other models (see,e.g., [1], [4], [5], [7].). The function g can be given, e.g., by

$$(9) \qquad g(r) = H(r - 1/\beta) ,$$

with β being the scaling factor in (4) and H the Heaviside function $H(r) = 0$, if $r < 0$, and $H(r) = 1$, if $r \geq 0$. (9) means that a susceptible fox whose activity centres at x and has the radius R has no contact with infective foxes, if the (generally non-integer) number of infective foxes visiting its circle of activity does not come up to one fox. If (4) does not indicate the absolute number of foxes, but indicates the total weight of the foxes whose activities centre in the set B, then another example is given by

$$(10) \qquad g(r) = \int_0^\infty H(r - s/\beta)\, \mu(s) ds ,$$

with μ being the probability density of the body weight of one fox. We remark that the function g given by (9) is discontinuous, whereas the g given by (10) is continuous.

<u>The development of the infective population</u>

The number of infective foxes can be estimated according to (vi) and (vii). Since any fox which is infective at time t was infected in the time interval $[t-\tau, t-\sigma]$, the number of infective foxes at time t does not exceed the difference of susceptible foxes at time $t-\tau$ and at time $t-\sigma$. We have to take into account, however, that the number of susceptibles is discontinuously changed at the beginning of the epidemic year by the birth of young foxes and the control operations. Let $\tilde{S}$ denote the density of the susceptible foxes of the previous epidemic year, such that, for $s \leq 0$, $x \in \mathbb{R}^2$, $\tilde{S}(s,.)$ indicates the density of susceptible foxes $-s$ time units before the beginning of the present year. In this case $\tilde{S}(0,.)$ gives the density of the susceptible foxes just before the young foxes are born and the control operations are performed, whereas $S(0,.)$ gives the density of the susceptibles just after these events. It is an evident assumption that $\tilde{S}(.,x)$ monotone decreases on $(-\infty,0]$. We obtain that

$$I(t,x) \leq \tilde{S}(t-\tau,x) - \tilde{S}(t-\sigma,x) , \quad \text{if } 0 \leq t \leq \sigma ;$$

(11) $$I(t,x) \leq \tilde{S}(t-\tau,x) - \tilde{S}(0,x) + S(0,x) - S(t-\sigma,x) , \quad \text{if } \sigma \leq t \leq \tau ;$$

$$I(t,x) \leq S(t-\tau,x) - S(t-\sigma,x) , \quad \text{if } t \geq \tau .$$

The annual severity of the epidemic

The size we are really interested in is the number of foxes which are infected in the epidemic year under consideration. This size provides a measure of the annual severity of the epidemic. Thus we define

(12) $$u(t,x) = S(0,x) - S(t,x)$$

for $t \geq 0$, $x \in \mathbb{R}^2$. $u(t,.)$ indicates the density of the foxes which were infected from the beginning of the epidemic year to time t. For convenience we set

(13) $$u(t,x) = 0 , \quad \text{for } t < 0, x \in \mathbb{R}^2 .$$

Further we set

(14) $$u_o(x) = \tilde{S}(-\tau,x) - \tilde{S}(0,x) , \quad \text{for } x \in \mathbb{R}^2 .$$

By the monotone decrease of $\tilde{S}(.,x)$ and $S(.,x)$ we obtain from (11) that

(15) $$I(t,x) \leq u_o(x) + u(t-\sigma,x) ,$$

for $t \geq 0$, $x \in \mathbb{R}^2$. By solving equation (6) we obtain that

(16) $$u(t,x) = S(0,x) \left[1 - \exp(-\alpha\rho v(t,x)) \right] ,$$

with

(17) $$v(t,x) = \int_0^t \int_0^\infty \int_{\mathbb{R}^2} I(s,y)\, K(R,x,y)\, Q(R,s,x)\, ds dP(R) dy .$$

Since the right side of inequality (15) monotone increases in t, we obtain from (7) and (15) that, if $0 \leq s \leq t$, $x \in \mathbb{R}^2$, $R \geq 0$,

(18) $$Q(R,s,x) \leq V\big(u(t-\sigma,.)\big)(R,x) ,$$

with

(19) $$V(v)(R,x) = g\Big(\int_{|x-y|<R} \int_{\mathbb{R}^2} [\, u_o(z) + v(z) \,]\, h(|y-z|)\, dy dz \Big) ,$$

for $v \in M_+$. M_+ denotes the Borel measurable functions from $\mathbb{R}^2$ to $[0,\infty)$. One easily derives from (11) that

$$\int_0^t I(s,y)ds \leq u_o(y) + \int_{t-1}^t u(s-\sigma,y)ds , \tag{20}$$

for $t \geq 0$, $y \in \mathbb{R}^2$. We recall that $\tau-\sigma = 1$ by our time scaling.

From (16),..., (20), and from (5) we obtain the following inequality for the density $u(t,.)$ of foxes which were infected from time 0 to time t:

$$u(t,.) \leq Z\left(\int_{t-1}^t u(s-\sigma,.)ds , u(t-\sigma,.)\right) \quad \text{on } \mathbb{R}^2 , \tag{21}$$

for $t \geq 0$, and

$$u(t,.) \equiv 0 \quad \text{on } \mathbb{R}^2 , \text{ for } t < 0, \tag{22}$$

with

$$Z(v,w)(x) = \gamma(x) N \left[1 - \exp\left(-\alpha\rho \int_0^\infty U(v)(R,x) V(w)(R,x) dP(R)\right) \right] , \tag{23}$$

for $x \in \mathbb{R}^2$, $v,w \in M_+$. V is defined by (19). U is defined by

$$U(v)(R,x) = \int_{\mathbb{R}^2} [u_o(y) + v(y)] K(R,x,y) dy , \tag{24}$$

for $v \in M_+$, $x \in \mathbb{R}^2$, $R \geq 0$. We recall that K is indicated by (3). The inequality (21) and the equalities (19), (22), (23), (24) make up our final model.

3. RESULTS

Let us now turn back to our original problem. We assume that, before the first epidemic year under consideration, the epidemic has not yet reached the right half-plane $\{x_1 > 0\} = \{x = (x_1,x_2) \in \mathbb{R}^2, x_1 > 0\}$. We want to know whether it is possible to prevent the epidemic from penetrating too far into the right half-plane by performing control operations in a strip $\{0 < x_1 < r\} = \{x = (x_1,x_2) \in \mathbb{R}^2; 0 < x_1 < r\}$ of finite depth $r > 0$. We cannot expect, of course, to prevent the epidemic from slightly penetrating into the right half-plane, because to this end we would have to eliminate or to vaccinate all susceptible foxes in a sufficiently deep strip, which is practically impossible. Before we state and interpret our results, we collect our mathematical assumptions.

ASSUMPTIONS. a) N is a positive constant.

b) h and k are (Borel measurable) functions from $[0,\infty)$ to $[0,\infty)$ and from $[0,\infty)^2$ to $[0,\infty)$, respectively. k satisfies (1) and (2).

c) P is a probability measure on the Borel measurable subsets of $[0,\infty)$.

d) g is a monotone increasing function from $[0,\infty)$ to $[0,1]$ such that $g(r) \to 0$ for $r \to 0$.

e) $u_o = \tilde{S}(-\tau,.) - \tilde{S}(0,.)$ is a (Borel measurable) function from $\mathbb{R}^2$ to $[0,\infty)$.

THEOREM. Let $\lambda > 0$ be such that

$$\int_0^\infty e^{\lambda R}\, dP(R) < \infty \quad \text{and} \quad \int_{\mathbb{R}^2} e^{\lambda y_1} h(|y|)dy < \infty .$$

Then there exists a (Borel measurable) control function $\gamma : \mathbb{R}^2 \to (0,1]$ having the following properties:

a) $\gamma(x) < 1$ only in a strip $\{0 < x_1 < r\}$ with some $r > 0$.

b) $\inf \gamma(\mathbb{R}^2) > 0$.

c) If

(25) $$u_o(x) := \tilde{S}(-\tau,x) - \tilde{S}(0,x) \leq N e^{-\lambda x_1},$$

for $x \in \mathbb{R}^2$, then, for all solutions u of (21) and (22),

$$u(t,x) \leq N e^{-\lambda x_1},$$

for $t \geq 0$, $x \in \mathbb{R}^2$.

REMARK. The function γ can be determined by the following inequality

(26) $$1/2 \geq \alpha\rho N \gamma(x)\, \Phi(\lambda) \int_0^\infty \Psi(\lambda,R)\, W(\lambda,R,x)\, dP(R) ,$$

with

$$\Phi(\lambda) = \int_{\mathbb{R}^2} e^{\lambda y_1} h(|y|)dy \quad \text{and} \quad \Psi(\lambda,R) = \int_{\mathbb{R}^2} e^{\lambda y_1} k(R,|y|)dy$$

and

$$W(\lambda,R,x) = g\left(2N e^{-\lambda x_1} \Phi(\lambda) \int_{|y|<R} e^{\lambda y_1} dy \right),$$

for $R \geq 0$, $x \in \mathbb{R}^2$, $x_1 > 0$.

Sketch of the proof: From the assumptions and from Lebesgue's theorem of dominated convergence follows that there is some γ satisfying a), b), and (26). (26) implies that the function

$$v(x) = N e^{-\lambda x_1}, \quad \text{for } x \in \mathbb{R}^2,$$

satisfies

$$v \geq Z(v,v) \quad \text{on } \mathbb{R}^2,$$

if (25) is satisfied. Since the operator Z monotone increases in both arguments, we can show for any solution u of (21) and (22) that

$$u(t,x) \leq v(x) \quad \text{for } t \geq 0, x \in \mathbb{R}^2,$$

by considering the intervals $[(n-1)\sigma, n\sigma]$ inductively for $n \in \mathbb{N}$.

Interpretation of the theorem: In the first year under consideration the assumption (25) is obviously satisfied, since the epidemic has not yet reached the right half-plane. As the theorem shows, one can, in every epidemic year, use the same control operations which have the following properties:

a) The control operations need to be performed only in a strip of finite depth.

b) Nowhere the density of susceptible foxes needs to be reduced to zero.

c) If the control operations are performed in the spring of every epidemic year, the number of foxes at x which are infected in the whole year can be estimated by

$$N e^{-\lambda x_1},$$

with the same $\lambda > 0$ for every year. In particular assumption (25) is satisfied in the subsequent year.

Thus the epidemic never penetrates too far into the right half-plane.

Final remarks: At least in principle, from (26), a control function γ can be determined the application of which can halt the epidemic. We do not calculate any example here, because the model and our estimates are not yet optimal. The derivation of a more complex and more realistic model and of more precise estimates is beyond the scope of this article. Further there is still not enough information available concerning the parameters of the model.

Finally we should draw attention to the following point: In nature, if the fox population is not completely eliminated in a sufficiently deep belt (which is practically impossible), one can never exclude that, by a chain of unfortunate events, a few rabid foxes might penetrate into the area beyond the belt and there trigger a new epidemic. Our theorem indeed allows a few rabid foxes to penetrate into the area beyond the belt, but, by the crucial assumption (iii) and the insertion of Q into equation (6), our model involves that the fox population can tolerate a sufficiently low density of infective foxes without a severe outbreak of rabies, even if the density of susceptible foxes is high. Without including this implication our model could not make any protective belt, in which the fox population is not completely eliminated, halt an epidemic. The alternative to our (deterministic) approach consists in deriving a stochastic rabies model which provides probabilities at which protective belts prevent the epidemic from penetrating into a particular area. Despite the deficiences we mentioned above we think that our model illustrates the use of protective belts in rabies control and provides some motivation to derive better models and sharper estimates and to determine the intrinsic parameters.

REFERENCES

1 BAILEY, N.T.J.: The Mathematical Theory of Infectious Diseases and its Applications. London: Griffin 1975

2 BÖGEL, K.; MOEGLE, H.; KNORPP, F.; ARATA, A.; DIETZ, K.; DIETHELM, P.: Characteristics of the spread of a wildlife rabies epidemic in Europe. Bull. WHO 54, 433 - 447 (1976)

3 BURROWS, R.: Wild Fox. Newton Abbot: David and Charles 1968

4 DIEKMANN, O.: Thresholds and travelling waves for the geographical spread of infection. J. math. Biol. 6, 109 - 130 (1978)

5 LAMBINET, D.; BOISVIEUX, J.-F.; MALLET, A.; ARTOIS, M.; ANDRAL, L.: Modèle mathématique de la propagation d'une épizootie de rage vulpine. Rev. Epidém. et Santé Publ. 26, 9 - 28 (1978)

6 MÜLLER, J.: The effect of fox reduction on the occurrence of rabies. Observations from two outbreaks of rabies in Denmark. Bull. Off. Int. Epizoot. 75, 763 - 776 (1971)

7 THIEME, H.R.: A model for the spatial spread of an epidemic. J. math. Biol. 4, 337 - 351 (1977)

8 TOMA, B.; ANDRAL, L.: Epidemiology of fox rabies. Adv. Virus Res. 21, 1 - 36 (1977)

9 WANDELER, A.; WACHENDÖRFER, G.; FÖRSTER, U; KREKEL, H; SCHALE, W.; MÜLLER, J.; STECK, F.: Rabies in wild carnivores in central Europe. I. Epidemiological studies. Zbl. Vet. Med. B 21, 735 - 756 (1974). II. Virological and serological examinations. idem, 757 - 764 (1974). III. Ecology and biology of the fox in relation to control operations. idem, 765 - 773 (1974)

SOME DETERMINISTIC MODELS FOR THE SPREAD OF GENETIC AND OTHER ALTERATIONS

Hans F. Weinberger
University of Minnesota
Minneapolis, MN 55455 U.S.A.

1. Introduction. Because there are several recent surveys of aspects of the theory of growth and spread of mutant genes, populations, and epidemics, including at least six by participants in this symposium [5, 11, 13, 19, 38, 39, 51] , I will only present some ideas on these subjects and illustrate them with some mathematical results which have been of interest to me. In particular, I will only be concerned with homogeneous habitats, and will not discuss clines or the effects of boundaries.

There is some misunderstanding of the ideas behind deterministic models, of what can be expected of such models, and of their relations to stochastic models. Many biologists seem to feel that because the measured outcomes of their experiments contain some fluctuations, it is not possible to apply deterministic models, so that the only useful mathematical models of biological processes are stochastic ones.

I shall begin the discussion of deterministic biological models by examining the analogous but better understood models for the flow of a gas.

If one thinks of a gas as a set of particles (molecules or atoms) which exert certain forces on each other and obey Newton's laws, one obtains a very accurate deterministic model. (We shall neglect quantum mechanical and relativistic effects.) Given the initial position and velocity of each particle, the model predicts the future behavior of the gas. Such a model is however, quite useless because all the initial conditions simply cannot be measured.

What one can measure is the force exerted by various amounts (masses) of a particular gas on certain walls of various containers at various temperatures. From such experiments, one deduces relations between a small set of macroscopic variables. If these relations are combined with Newton's laws one can divide a large volume of gas into small subsets and obtain rules for predicting how these small volumes of gas move in space and time. A second step is to assume that such rules apply for arbitrarily small volumes, to introduce concepts of continuum theory such as density, pressure, and mean velocity, and to pass to the limit as the control volumes approach zero. This limit process yields the usual partial differential equations of gas dynamics. It is made plausible by the fact that the density of particles is so great that even extremely small volumes contain extremely many particles. The model is justified by the fact that predictions made in terms of quantities which are measured agree well with the measured results of experiments as long as the experiment does not involve extreme conditions such as speeds on the order of thermal speeds or orifices whose size is of the order of the mean free path. It is this kind of deterministic model which one seeks in various biological situations.

It is the program of statistical mechanics to connect the first kind of deterministic model with the second by showing that it is extremely unlikely that a set of particles obeying Newton's laws will behave in any way other than that predicted by the equations of gas dynamics. In this process only the initial conditions can be chosen at random, since Newton's laws determine the subsequent behavior. Thus the final state depends in a highly nonlinear fashion on the initial probability distribution. This leads to great mathematical difficulties, which have not yet been overcome.

In analogous fashion, one could consider a population and its environment as a (rather complex) set of elementary particles which satisfy Newton's laws and some mutual force relations. Such a deterministic description is again useless because the initial data cannot be measured. We make this fact apparent by noting that since the weather certainly affects behavior, such a description would include, among other things, a means of predicting the weather.

We shall present here a few simple deterministic models of biological systems. The purpose of such a model is to predict some interesting qualitative or quantitative aspects of a system in terms of a reasonably small set of parameters which remains more or less fixed in time and from sample to sample and which is capable of being measured with some, though not perfect, accuracy. The fluctuations in the parameters are reflected in fluctuations of their measured values.

One of the interesting problems is to estimate the magnitude of the possible effect of such fluctuations. A system in which small changes in the parameters produce small changes in the outcome is said to be robust. Robustness is an important property of a model if it is to be of any use. More particularly, one needs to show that the quantitative or qualitative predictions made from the model are essentially the same for all possible values of the experimentally determined parameters. Thus the robustness of a model depends upon the magnitude of the fluctuations in the data, upon the prediction to be made, and upon what errors in the prediction are tolerable.

Roughly speaking, a robust model is one in which the data fluctuations are relatively small. Such quantities as the number of elephants in a three-meter square or the number of rabbits born in a three-meter square in one second tend to vary between one and zero, so that the fluctuations are not relatively small. The trick of introducing a continuum theory by allowing volumes or time intervals to approach zero, which is so successful in gas dynamics, is likely to fail in biological models because of the much smaller densities involved. Thus many deterministic biological models must involve functions of discrete rather than continuous variables. This fact may require the mathematician to create new tools to deal with new kinds of problems, but then that is what a mathematician is supposed to be good for.

The usefulness of creating a deterministic model by discretizing space and time does depend upon the desired applications. An epidemic model which deals with 100-

kilometer squares and one-year intervals may not be very useful in devising a strategy to keep an outbreak of rabies from spreading from the Odenwald into Heidelberg. For this purpose one may have to take into account not only the stochastic nature of the phenomena in a small vicinity but also detailed information about the location of the river, hills, and houses, and about the habits of the local dogs and foxes.

A stochastic model is designed to take account of fluctuations in the parameters which determine the system. If one can solve the deterministic problem for each possible set of parameter values and if the probability distribution of these sets of parameter values is known, one obtains, at least in principle, the solution of the stochastic model. However, this explicit solution is rarely possible, and the distribution of the parameters is not usually completely known. Because most growth models of interest are nonlinear, a knowledge of finitely many moments of the probability distribution does not, in general, determine even the average value of the solution.

If one can show that the deterministic model is so robust that all fluctuations which can occur do not affect the result significantly, one knows that the deterministic model will give essentially the same answer as the stochastic model. Thus a robust deterministic model is almost as good as a stochastic one, provided the fluctuations in the stochastic model are suitably restricted and the concept of robustness is suitable for dealing with the problem.

One can, of course, consider any deterministic model as a stochastic one in which the variance of the parameters happens to be zero. Thus it is intrinsically easier to solve a deterministic problem than the corresponding stochastic one. One can therefore expect to be able to solve at least some deterministic problems whose stochastic analogues are too complicated to solve. As we shall see, the chief defect of the models we shall consider is that they are too simple to take into account sufficiently many significant factors. Thus a more complicated model may well give better predictions than a simpler one, even though the former is deterministic and the latter stochastic.

If one knows the solution of a stochastic problem, one can find the solution of the deterministic one by letting the variance of the data go to zero. On the other hand we have seen that the solution of a deterministic problem together with a suitable proof of robustness leads to strong information about a related stochastic problem. Determinists and probabilists have much to learn from each other, and hopefully this conference will promote the interchange of pertinent knowledge. The difficulty in analyzing good models for biological processes is such that we need all the techniques we can find.

2. Models for spread in population genetics.

A model for the spread of an advantageous gene in a population was proposed by R.A. Fisher in 1937 [15]. He considered a diploid population in which there are two forms (alleles) of one particular gene. The alleles are denoted by a and A, so that there are three genotypes aa, aA, and AA. If the populations of these genotypes are sufficiently dense, one can define continuous population densities $\rho_{aa}(x,t)$, $\rho_{aA}(x,t)$, and $\rho_{AA}(x,t)$ of these three genotypes. The gene fraction $p(x,t)$ is defined as the fraction

$$p = \frac{2\rho_{AA} + \rho_{aA}}{2\rho_{AA} + 2\rho_{aA} + 2\rho_{aa}}$$

of all the genes at the given locus which are of the allelic form A. Fisher proposed the equation

$$\frac{\partial p}{\partial t} = f(p) + D\frac{\partial^2 p}{\partial x^2} \tag{2.1}$$

with $f(p) = p(1-p)$ to describe the time development of the gene fraction due to the fact that aA individuals are more fit than aa individuals and AA individuals are even more fit, and the fact that individuals migrate. As Fisher realized, this model is somewhat crude. The diffusion term was simply added to an earlier model which neglected spatial spreading. It is clear that one should really have a coupled system of three such equations for the three densities ρ_{aa}, ρ_{aA}, and ρ_{AA} (see e.g. [2]). The single Fisher equation (2.1) can only be derived from these by some limiting arguments (see [2], [13, §§ 2.2 - 2.3], [19, § II.2.1]). We shall show below how to obtain an analogous discrete model.

In order to study the behavior of solutions of (2.1), Fisher looked for travelling wave solutions of the form $W_c(x-ct)$ with $W_c(-\infty)=1$ (all A) and $W_c(+\infty) = 0$ (all a). He found that there is a positive number c^* with the property that such a wave exists for every $c \geq c^*$ and none exists for $c < c^*$. Fisher conjectured that the advantageous gene A should spread into a region occupied only by aa individuals at the rate c^*.

An important case of this conjecture was soon verified by Kolmogoroff, Petrowsky, and Piscounoff [30]. They showed that if $p(x,t)$ is the solution of (2.1) with

$$p(x,0) = \begin{cases} 1 & \text{for } x < 0 \\ 0 & \text{for } x > 0 \end{cases} \tag{2.2}$$

if $f(0) = f(1) = 0$ and $f'' < 0$, and if $m(t)$ is defined as that value of x such that

$$p(m(t),t) = \frac{1}{2},$$

then as t approaches infinity, $m(t)/t$ approaches c^* and $p(x+m(t),t)$

approaches $W_{c^*}(x)$, where $W_{c^*}(x-c^*t)$ is the travelling wave of lowest speed. This result was extended to various monotone initial data $p(x,0)$ by Kanel [21,22] and by Kametaka [20].

Aronson and the author [2] showed that if the allele A is initially confined to a bounded set, then

$$(2.3) \qquad \lim_{t\to\infty} p(x+ct,t) = \begin{cases} 1 & \text{for } |c| < c^* \\ 0 & \text{for } |c| > c^* \end{cases} .$$

Uchiyama [48,49] has shown that, in fact, it is still true in this case that $p(x+m(t),t)$ approaches a travelling wave of speed c^* . He showed that $m(t) = c^*t - \frac{3}{2c^*}\log t + O(\log t)$. Bramson [6] independently obtained the stronger result $m(t) = c^*t - \frac{3}{2c^*}\log t + O(1)$ by using stochastic tools.

If $f(0) = f(1) = 0$, $\int_0^1 f(u)\,du > 0$, and there is an $\alpha \in (0,1)$ such that f is negative for $0 < p < \alpha$ and positive for $\alpha < p < 1$ the Fisher equation (2.1) represents the spread of allele A such that the heterozygotes aA are less fit than the two homozygotes aa and AA . For this case there is a unique speed c^* for which there is a wave solution $W_{c^*}(x-c^*t)$ with $W_{c^*}(-\infty) = 1$, $W_{c^*}(\infty) = 0$. In this case it was shown by Fife and Mc Leod [14] that when $p(x,0)$ vanishes outside a bounded set, $p(x,t)$ uniformly approaches the sum $W_{c^*}(x-c^*t-a) + W_{c^*}(c^*t-x-b) - 1$ of a wave travelling to the right and a wave travelling to the left. Uchiyama [48,49] and Rothe [43] have found similar results for some functions $f(p)$ which are positive for all p in the interval $(0,1)$.

For Fisher's equation $\frac{\partial p}{\partial t} = f(p) + D\nabla^2 u$ in two or more dimensions, Aronson and the author [3] have obtained the analogue of (2.3), which allows the wave front thickness to grow linearly in t . No analogue of Uchiyama's convergence in shape result or the Fife-McLeod result is as yet known.

In the spirit expounded in the introduction, we shall now show how a discrete analogue of the Fisher model (2.1) can be derived. Our model has the form of a stepping stone model [23, 29, 33, 39].

We take a two-dimensional habitat and divide it into congruent squares by a system of equidistant perpendicular grid lines. We introduce rectangular coordinates in such a way that the centers of the squares have the coordinates (i,j) , $i,j = 0, \pm 1, \pm 2, \ldots$. We also choose a unit of time and look at integral values $n = 0, 1, 2, \ldots$ of the time.

We suppose that at time zero each of the squares contains a large number of gametes of the individuals under study, each of which contains one gene at each locus. Let the ratio of gametes of type A to all the gametes in the square centered at (i,j) be denoted by $p(i,j)$. These gametes are paired in a random manner to form young diploid individuals, which undergo selective mortality. We do not examine the detailed mechanism of the selection process, but we assume that there are "relative fitness functions" $f_{AA}(p)$, $f_{aA}(p)$, and $f_{aa}(p)$ with the property that the numbers of survivors to maturity of the three genotypes in the square (i,j) are $p_o(i,j)^2 f_{AA}(p_o(i,j))$, $2p_o(i,j)[1-p_o(i,j)]\, f_{aA}(p_o(i,j))$, and $[1-p_o(i,j)]^2\, f_{aa}(p_o(i,j))$, respectively. The relative fitness functions are to be determined experimentally. Because the habitat is homogeneous, they are the same for all the squares. We have assumed that the original number of gametes is so large that the number of survivors does not depend on the initial number of gametes but only on the gene fraction. We have also assumed that each square is so large that the number of individuals which cross its boundary during the maturation process is negligible.

We now assume that after the maturation process there is a migration process which is described by saying that a fraction $K(k-i,\ell-j)$ of the individuals of each genotype in the square centered at (i,j) migrate to the square centered at (k,ℓ) . (The functional form of the migration kernel K comes from the homogeneity of the habitat.) Then the gene fraction at (k,ℓ) after the migration is given by

$$(2.4)\qquad p_1(k,\ell) = \frac{\sum_{i,j} K(k-i,\ell-j)\, p_o\,[p_o f_{AA}(p_o) + (1-p_o) f_{aA}(p_o)]}{\sum_{i,j} K(k-i,\ell-j)[p_o^2\, f_{AA}(p_o) + 2p_o(1-p_o) f_{aA}(p_o) + (1-p_o)^2 f_{aa}(p_o)]}$$

where $p_o = p_o(i,j)$ in each term of the sum. Our final assumption is that each of the newly arrived individuals at (k,ℓ) produces the same number of gametes and dies. Then $p_1(k,\ell)$ represents the gene fraction in the square centered at (k,ℓ) at the beginning of the next cycle.

If one thinks of $p_o(i,j)$ as a function of the integer variables i and j , the right-hand side of (2.4) represents a transformation Q of one such function into another:

$$(2.5)\quad Q[p](k,\ell) \equiv \frac{\sum_{i,j} K(k-i,\ell-j)\, p\,[p f_{AA}(p) + (1-p) f_{aA}(p)]}{\sum_{i,j} K(k-i,\ell-j)[p^2 f_{AA}(p) + 2p(1-p) f_{aA}(p) + (1-p)^2 f_{aa}(p)]},$$

$p = p(i,j)$ in each term of the sum.

Then (2.4) can be written in the form $p_1 = Q[p_o]$. If we denote the gene fraction of the gametes in the square centered at (i,j) a time n by $p_n(i,j)$ and if we suppose that the life cycles do not change with time, we obtain the recursion

$$(2.6)\qquad p_{n+1} = Q[p_n] \qquad n = 0,1,2,\ldots$$

The operator Q can be measured experimentally, provided $K(k,\ell)$ is zero except when $k^2+\ell^2$ is not too large; that is, provided the size of the squares is sufficiently large relative to the maximum distance of migration during the unit of time. If not all our hypotheses are valid, Q may not be representable in the form (2.5). For example, the unit of time may need to consist of several life cycles in order that climatic variations or various other fluctuations will be averaged out. We shall suppose that the fluctuations in the experimentally determined values of Q are relatively small.

The operator Q has some simple properties. Since we do not allow the possibility of mutation into the allelic types a or A, Q must take the functions $p \equiv 0$ and $p \equiv 1$ into themselves:

$$Q[0] = 0 \qquad Q[1] = 1 \; . \tag{2.7}$$

Since p is a gene fraction, we must have $0 \le Q[p] \le 1$ whenever $0 \le p \le 1$. We shall make the somewhat stronger hypothesis that:

$$\begin{aligned} &r(i,j) \ge p(i,j) \quad \text{for all} \quad i,j \quad \text{implies} \\ &Q[r](i,j) \ge Q[p](i,j) \quad \text{for all} \quad (i,j) \end{aligned} \tag{2.8}$$

An operator with this property is said to be order preserving. (Another name is monotone, but this is also used for a quite different property.) If Q is of the form (2.5), the condition (2.8) follows from the rather plausible assumption that increasing the gene fraction $p(i,j)$ in one square (i,j) increases the number of alleles of type A and decreases the number of alleles a among the individuals which mature in the square.

The homogeneity of the domain is reflected in the fact that Q is translation invariant. That is, if we define the translation operator

$$T_{k\ell}[u](i,j) = u(i-k, j-\ell) \; ,$$

then

$$T_{k\ell}[Q[u]] = Q[T_{k\ell}[u]]$$

for any (k,ℓ).

We observe that

$$Q[p]-p = \frac{\sum\limits_{i,j} K(k-i,\ell-j)p[1-p][p\{f_{AA}(p)-f_{aA}(p)\}+(1-p)\{f_{aA}(p)-f_{aa}(p)\}]}{\sum\limits_{i,j} K(k-i,\ell-j)[p^2 f_{AA}(p)+2p(1-p)f_{aA}(p)+(1-p)^2 f_{aa}(p)]} \; .$$

It is clear from this form that if $f_{AA}(p) > f_{aA}(p) > f_{aa}(p)$ for all p, then $Q[p] \ge p$, with equality everywhere if and only if $p \equiv 0$ or $p \equiv 1$. For obvious reasons, this situation is known as the heterozygote intermediate case. In fact, if the relative fitnesses f_{AA}, f_{aA}, and f_{aa} are independent of p, the inequality $Q[p] \ge p$ for all p implies that $f_{AA} \ge f_{aA} \ge f_{aa}$. However, the relative fitnesses may well depend upon the genetic composition of the competing juveniles, and the inequality $p\{f_{AA}(p)-f_{aA}(p)\}+(1-p)\{f_{aA}(p)+f_{aa}(p)\} \ge 0$, which implies heterozygote intermediate behavior, can be satisfied even when $f_{aA} \equiv 0$ so that the heterozygote combination is lethal.

If the relative fitnesses are independent of p, then the inequality $f_{aA} > f_{AA} > f_{aa}$ implies that $Q[p] \ge p$ if $p \le \Pi_1$ everywhere and $Q[p] \le p$ if

$p \geq \Pi_1$, everywhere where

$$\Pi_1 = \frac{f_{aA} - f_{aa}}{2f_{aA} - f_{aa} - f_{AA}} ,$$

which is between zero and one. This is the heterozygote superior case. The heterozygote inferior case $f_{aA} < f_{aa} < f_{AA}$ leads to the inequalities $Q[p] \leq p$ for $p \leq \Pi_1$ and $Q[p] \geq p$ for $p \geq \Pi_1$, where Π_1 is defined by the same formula.

If one wishes to use the Fisher model (2.1) but discretize the time while leaving the space continuous, one simply defines $p_n(x) = p(x,n)$. Then p_n satisfies the recursion (2.6) where Q is defined by setting $Q[u](x) = p(x,1)$ where $p(x,t)$ is the solution of the initial value problem

$$\frac{\partial p}{\partial t} = f(p) + D\frac{\partial^2 p}{\partial x^2}$$

$$p(x,0) = u(x) \quad .$$

This recursion is even valid if one takes account of seasonal or cyclic variations by allowing f and D to vary periodically in time with period 1. The order preserving property of Q is now an immediate consequence of the comparison theorem for parabolic equations, which, in turn follows by applying the maximum principle to the difference of two solutions. (See, e.g., [16,40].) We note, however, that the solution operator Q associated with the continuous time model (2.1) allows infinite speeds of migration.

If one is dealing with a stationary stochastic model, one can obtain a deterministic model of the form (2.6) by setting

$$Q[p_0](k,\ell) = E\{\text{gene fraction in } (k,\ell) \text{ at time } 1 \mid \text{gene fraction in each square } (i,j) \text{ at time } 0 = p_0(i,j),\ i,j = 0, \pm 1, \pm 2, \ldots\}$$

The order-preserving property now follows from what Liggett called the population monotone property of the process in his lecture [32]. As Kurtz shows in his lecture [31], if one starts with a continuous-time stationary stochastic process, one obtains a stochastic process with small fluctuations when the unit of time is sufficiently large. However, one of the most prevalent of what Robertson [42] has called hyperborean fallacies is the belief that every discrete time process with small fluctuations must come from a continuous-time stationary process. One of the symptoms of this fallacy is the fact that such processes inevitably assign positive probabilities to arbitrarily large speeds of migration and to arbitrarily large fluctuations, which are biologically impossible. Moreover, the elementary continuous-time processes that one can treat tend to be so simple that many important factors must be left out of consideration. The approach of considering Q as an empirically determined quantity allows the effects of these factors to be included, even without identifying the factors themselves.

The experimental determinations of the operator Q will usually fluctuate, although these fluctuations are hopefully small. This raises the question of the robustness of the various properties of the solution. A property is said to be robust if it is changed little when the operator Q is replaced by a nearby operator $\tilde{Q}$. (The concepts "changed little" and "nearby" must, of course by defined precisely.)

Since Q does fluctuate, one can also think of (2.6) as a stochastic equation

$$p_{n+1} = Q_n[p_n]$$

where Q_n is an operator-valued random variable with certain statistical properties. This is a nonlinear version of a problem which was discussed by Kesten [28] in his lecture at this conference. If this problem can be solved, it gives a stronger result than robustness. It assigns probabilities to various deviations from the solution of the "mean equation" (2.6). Such an answer is clearly more powerful. Its only drawback is that it frequently cannot be found.

3. Properties of order preserving models.

We shall present here some mathematical ideas which give a unified view of much of what is known about the models in the preceding and following sections.

We shall deal with a recursion of the form

$$p_{n+1} = Q[p_n] \; . \tag{3.1}$$

We are given a subset $\mathscr{H}$ (the habitat) of a Euclidean space of one or more dimensions. We identify a point of $\mathscr{H}$ with the vector $\vec{x}$ from the origin and we assume that if $\vec{x}$ and $\vec{y}$ are in $\mathscr{H}$, the same is true of $-\vec{x}$ and of $\vec{x}+\vec{y}$. For example, $\mathscr{H}$ may be the real line, the Euclidean two-space, or the set of points in the Euclidean two-space with integer coordinates (i,j). We can identify the latter with the unit squares centered at these points.

If $\vec{y}$ is a point of $\mathscr{H}$, we define the translation operator $T_{\vec{y}}$ by saying that for any function $p(\vec{x})$ defined on $\mathscr{H}$

$$T_{\vec{y}}[p](\vec{x}) = p(\vec{x}-\vec{y}) \; .$$

Let Q be an operator which takes every nonnegative continuous bounded function $p(\vec{x})$ defined on $\mathscr{H}$ into another such function. We shall assume that Q has the following properties, which are patterned on those of the operators Q in the preceding section.

(3.2)

(i) Q is order preserving; that is, if $p(\vec{x}) \geq r(\vec{x})$ for all $\vec{x}$ in $\mathscr{H}$, then $Q[p](\vec{x}) \geq Q[r](\vec{x})$ for all $\vec{x}$ in $\mathscr{H}$.

(ii) Q is translation invariant; that is, $Q[T_{\vec{y}}[p]] = T_{\vec{y}}[Q[p]]$ for every $\vec{y}$ in $\mathscr{H}$.

(iii) $Q[0] = 0$.

(iv) Q is continuous; that is, if the sequence p_k converges to p uniformly on each bounded subset of $\mathscr{H}$ as $k \to \infty$, then $Q[p_k]$ converges to $Q[p]$ at each point of $\mathscr{H}$.

(v) Q is compact; that is, if the sequence p_k is uniformly bounded, then it has a subsequence p_{k_i} such that $Q[p_{k_i}]$ converges uniformly on each bounded subset of $\mathscr{H}$.

We remark that if $\mathscr{H}$ is a discrete set such as the set of points with integer coordinates, then all functions are continuous, all convergent sequences converge uniformly on bounded sets, and (v) is a consequence of (iv).

In any biological situation one can expect the migration distance to be bounded by some constant A :

(3.3) $p(\vec{x}) = r(\vec{x})$ for $|\vec{x}| \leq A$ implies that $Q[p](0) = Q[r](0)$.

This condition is not satisfied by the inverse of the Fisher equation and can be weakened considerably.

The principal tool in deriving all our results is the following proposition.

PROPOSITION. Let R be an operator with the order preserving property (3.2.i). If the sequences v_n and w_n satisfy the inequalities

$$v_{n+1} \geq R[v_n]$$

and

$$w_{n+1} \leq R[w_n]$$

for $n = 0,1,2,\ldots$ and if $v_o \geq w_o$, then $v_n \geq w_n$ for all positive n .

Proof. This result is proved by a simple induction argument. Suppose that $v_n \geq w_n$. Then because of the order preserving property of R

$$v_{n+1} \geq R[v_n] \geq R[w_n] \geq w_{n+1} \quad .$$

Thus the inequality $v_o \geq w_o$ implies that $v_1 \geq w_1$, which implies that $v_2 \geq w_2$, and so forth.

One important application of this proposition is

COROLLARY 3.1. Let Q^+ and Q^- be operators which take the class of bounded continuous nonnegative functions into itself and which have the property that

$$Q^-[p] \leq Q[p] \leq Q^+[p]$$

for all such function. Then if p_n , p_n^+ , and p_n^- are the solutions of the recursions

$$p_{n+1} = Q[p_n] \ ,$$

$$p_{n+1}^+ = Q^+[p_n^+] \ ,$$

and

$$p_{n-1}^- = Q^-[p_n^-]$$

with $p_o^+ = p_o^- = p_o$, the inequalities

$$p_n^- \leq p_n \leq p_n^+$$

are valid for all n .

Thus we have a way of bounding p_n above and below, from which robustness can be deduced. Note that it is not necessary that Q^+ and Q^- have the properties (3.2). They may, for example, be obtained from the largest and smallest of the fluctuating measurements of Q .

Kendall [26] pointed out that when the initial values p_o are small, a deterministic theory may miss the possibility of the extermination of the gene A for which a stochastic theory predicts a positive probability. For example, if a single individual is born with a new mutant allele A which makes its bearer in every way more fit, any reasonable deterministic theory will predict that this allele A will eventually replace the original allele a in the population. If, however, the individual is killed by lightning before he has a chance to reproduce, the allele A disappears. This possibility is taken care of in Corollary 3.1 by the statement that the worst-case prediction Q^- takes some non-zero functions into the zero function.

It follows from the translation invariance of Q that Q takes constants into constants. That is, the effect of migration cancels when there is no spatial variation.

We shall make the additional assumption that there is a constant $\Pi_1 \leq \infty$ such that

$$Q[\alpha] > \alpha \quad \text{for} \quad 0 < \alpha < \Pi_1 \ , \qquad (3.4)$$

$$Q[\Pi_1] = \Pi_1 \quad \text{if} \quad \Pi_1 < \infty \ .$$

In the population genetics model this condition is satisfied in the heterozygote intermediate $(\Pi_1 = 1)$ and heterozygote superior cases.

If we wish to discuss wave propagation on the set of points of two-space with integer coordinates, we must observe that this set is not rotationally symmetric. Hence the propagation speeds must depend upon direction.

THEOREM 3.1. For each unit vector $\vec{\xi}$ there is a propagation speed $c^*(\vec{\xi}) \in (-\infty,+\infty]$ with the property that for every finite $c \geq c^*(\vec{\xi})$ there is a nonincreasing function $W_c(s)$ with $W_c(-\infty) = \Pi_1$, $W_c(+\infty) \equiv 0$ and such that the function $p_n(\vec{x}) = W_c(\vec{x}\cdot\vec{\xi}-cn)$ satisfies the recursion (3.1). (A solution of this form is called a travelling wave with speed c in the direction ξ.) If the condition (3.3) is valid, then $|c^*(\vec{\xi})| \leq A$.

To discuss the propagation of an initial mutation we define the convex set

$$S = \{\vec{x} \mid \vec{x}\cdot\vec{\xi} \leq c^*(\vec{\xi}) \text{ for all unit vectors } \vec{\xi}\} .$$

The sequence $p_n(\vec{x})$ is assumed to satisfy the recursion (3.1).

THEOREM 3.2. Suppose $p_o(\vec{x}) = 0$ outside some bounded set and that $p_o \leq \alpha < \Pi_1$. Then if S' is any open set which contains S ,

$$\limsup_{n\to\infty} \max_{n^{-1}\vec{x}\notin S'} p_n(\vec{x}) = 0 .$$

THEOREM 3.3. For any $\gamma > 0$ there is a constant r_γ such that if $p_o > \gamma$ on a ball of radius r_γ , and if S" is any closed subset of the interior of S ,

$$\liminf_{n\to\infty} \min_{n^{-1}\vec{x}\in S''} p_n(\vec{x}) \geq \Pi_1 .$$

Proofs of some special cases of these results can be found in [52] and [53]. For some results for unbounded S see [54].

The proofs of the general results will be published elsewhere. The point of these theorems is that the qualitative behavior of solutions of the Fisher equation does not depend on the exact form of the equation, but is valid for the whole class of models of the form (3.1) as long as the conditions (3.2) and (3.4) are satisfied.

It is easily seen from Corollary 3.1 that if $Q^-[p] \leq Q[p] \leq Q^+[p]$ for all nonnegative p and if Q^- and Q^+ satisfy (3.2), then for each $\vec{\xi}$ the corresponding wave speeds satisfy

$$c_-^*(\vec{\xi}) \leq c^*(\vec{\xi}) \leq c_+^*(\vec{\xi}) .$$

In order to obtain explicit bounds for c^* we first consider a linear operator L which takes the class of nonnegative bounded continuous functions on $\mathcal{H}$ into itself and which satisfies the hypotheses (3.2).

To satisfy (3.4) we assume that $L[1] > 1$. The travelling waves for a linear operator are of the form

$$p_n(\vec{x}) = e^{-\lambda(\vec{x}\cdot\vec{\xi} - cn)}$$

where λ is a positive constant. The condition that this function satisfy $p_{n+1} = L[p_n]$ is that

$$\begin{aligned} e^{\lambda c} &= e^{\lambda\vec{x}\cdot\vec{\xi}} L[e^{-\lambda\vec{y}\cdot\vec{\xi}}](\vec{x}) \\ &= L[e^{-\lambda(\vec{y}-\vec{x})\cdot\vec{\xi}}](x) \\ &= L[e^{-\lambda\vec{y}\cdot\vec{\xi}}](0) , \end{aligned}$$

provided the right-hand side is finite. Thus the wave speed for such an L is

$$c^*(\vec{\xi}) = \inf_{\lambda > 0} \{ \frac{1}{\lambda}\log L[e^{-\vec{y}\cdot\vec{\xi}}](0)\} . \tag{3.5}$$

A somewhat more sophisticated argument shows that this same formula also gives the wave speed for the nonlinear operator

$$L[\min\{p(\vec{y}), \delta\}]$$

where δ is any positive constant. This fact is useful in obtaining a lower bound Q^- for a given operator Q.

The formula (3.5) shows that the wave speed of the family of operators $t\,L[\min\{p(\vec{y}), \delta\}]$ is continuous in t for $t > 0$. Consequently we obtain the following theorem.

THEOREM 3.4. If L is a linear operator with the property that for every positive ϵ there is a $\delta > 0$ such that the inequality $Q[p] \leq L[p]$ holds for all nonnegative p and

$$Q[p] \geq (1-\epsilon) L[p]$$

when $0 \leq p \leq \delta$, then for each unit vector $\vec{\xi}$ the wave speed $c^*(\vec{\xi})$ of Q is given by the formula (3.5).

In this case, the statement of Theorem 3.2 holds for any bounded nonnegative p_0 without any restriction on its size.

If, moreover, the operator L has the property that if $p \not\equiv 0$ then for any radius r there is an integer k such that $L^k[p]$ is positive on some ball of radius r, then the statement of Theorem 3.3 holds with $r_\gamma = 0$ for all $\gamma > 0$.

The proof of Theorem 3.4 is much simpler than those of Theorems 3.2 and 3.3. For example, the stronger version of Theorem 3.2 is an immediate consequence of applying the Proposition to the sequences $v_n = e^{-\lambda[\vec{x}\cdot\vec{\xi} - nc]}$ and $w_n = p_n$.

The last statement of the Theorem means that as long as p_0 is not identically zero, p_n increases to at least Π_1 on a growing sequence of sets. This is called the hairtrigger effect [2,3]. The condition on L for this to happen is usually satisfied. Almost all explicit formulas for c^* are based on Theorem 3.4. The only exception of which I am aware is a result of Hadeler and Rothe [17] on the Fisher equation with a cubic polynomial.

The formula (3.5) also arises as a propagation speed for subadditive processes in the work of Hammersley [18].

4. Models for the growth and spread of a population.

The Fisher equation

$$\frac{\partial p}{\partial t} = D\nabla^2 p + f(p) \tag{4.1}$$

can also be used (and, in fact, with somewhat more justification) as a model for the growth and spread of the density p of an isolated population. The function $f(p)$ represents the net rate of growth when the population density is p. Clearly $f(0) = 0$.

For the spatially independent case the model with $f(p) = rp$, or at least the corresponding exponential growth law, goes back at least to Euler. (See [34], 3rd. ed., vol. II, p.27.) Malthus [34] pointed out that a linear model does not take account of the fact that the scarcity of resources limits the population. Verhulst [50] suggested that there should be a carrying capacity Π_1 such that $f(p) < 0$ for $p > \Pi_1$ and $f(p) > 0$ for $p < \Pi_1$. He treated several possible functions including the logistic function $f(p) = rp\ (1 - p/\Pi_1)$.

The results about Fisher's equation such as the existence of travelling waves and the asymptotic behavior of solutions can be transferred directly to the growth model (4.1).

A discrete model for population growth in a population with non-overlapping generations is easily formulated. We again break the habitat into the set of squares centered at the points (i,j) with integer coordinates. We let $p_o(i,j)$ be the initial population in the square centered at (i,j), and we define an operator Q which takes nonnegative bounded integer-valued functions of the integer variables (i,j) into functions of the same kind in such a way that the population $p_1(k,\ell)$ at time 1 in the square centered at (k,ℓ) is given by

$$p_1 = Q[p_o] .$$

If the function p_n gives the population of the various squares at time n, we have, assuming that the conditions are the same in each time period, the recursion

$$p_{n+1} = Q[p_n] . \tag{4.2}$$

If no spontaneous generation occurs, the operator Q has the property

$$Q[0] = 0 .$$

If the habitat is homogeneous, Q will be translation invariant. It is reasonable to expect that Q is also order preserving, although this need always not be the case. Thus, Q often has the properties (3.2), and (3.4) so that the population p_n has the properties stated in Theorem 3.2 and 3.3 of Section 3.

The graph of $Q[\alpha]$ versus α for constants α is called the reproduction curve [41]. It is found that when some populations become too large, they degrade the habitat to such an extent that $Q(\alpha)$ actually decreases with increasing α.

In such situations, Q is certainly not order preserving. The following idea of Thieme [47] can be used to show that Theorems 3.2, and 3.3 may remain valid in such a case. It follows from the Proposition of the preceding section that if

$$Q[p] \leq Q^+[p] \quad \text{for all} \quad p$$

with Q^+ order preserving, then Theorem 3.2 remains valid for solutions of (4.1) when $c^*(\vec{\xi})$ is replaced by the wave speed c_+^* which corresponds to Q^+. Similarly, if $Q[p] \geq Q^-[p]$ for all p, where $Q^-[\alpha] > \alpha$ on some interval $(0,\Pi_1^-)$ then Theorem 3.3 is valid when c^* is replaced by the speed c_-^* which corresponds to Q^- and Π_1 is replaced by Π_1^-. In particular, if there is an operator L which satisfies the hypotheses (3.2) and (3.4), if $L[1] > 1$, and if for each positive ϵ there is a $\delta > 0$ such that

$$(1-\epsilon)\, L[\min(p,\delta)] \leq Q[p] \leq L[p] ,$$

then even though Q may not satisfy the hypotheses (3.2), the statements of Theorems 3.2, 3.3, and 3.4 are valid, with Π_1 in Theorem 3.3 replaced by the δ which corresponds to any ϵ such that $(1-\epsilon)L[1] > 1$.

If the generations overlap and interact, one must take the age structure of the population into account. Continuum models for populations with age structure but no spatial dependence were introduced by Sharpe and Lotka [44] and by McKendrick [35, pp. 121-124]. We present here a recent model of Thieme [46] in which spatial spread is also considered.

Let $u(\vec{x},t)$ represent the rate of production (per unit area per unit time) of the gametes of an asexual population at the point $\vec{x}$ at time t. Thieme supposes that the gametes pair to produce immature individuals who compete with each other but do not migrate until, a certain time later, adults of density $g(u)$ emerge and begin to reproduce and migrate.

Let $P(\tau)$ represent the rate of gamete production by the individuals of age τ. Suppose that one also knows that, as a result of death and migration, a fraction $K(|\vec{y}|,\tau)$ of the adults which started their adult life at $\vec{x}-\vec{y}$ at time $t-\tau+\sigma$ arrives at $\vec{x}$ at time t. Then one obtains the integral equation

$$u(\vec{x},t) = \int_0^\infty \int A(|\vec{y}|,\tau)\, g(u(\vec{x}-\vec{y},t-\tau))\, d\vec{y}\, d\tau$$

where $A(\vec{y},\tau) = P(\tau)K(|\vec{y}|,\tau)$. If the function $u(x,t)$ for $t \leq 0$ (that is, the birth rate in the past) is known explicitly, this integral equation determines $u(\vec{x},t)$ for $t > 0$. More generally, Thieme writes the model in the form

$$u(\vec{x},t) = \int_0^t \int A(|\vec{y}|,\tau)\, g(u(\vec{x}-\vec{y},t-\tau))\, d\vec{y}\, dt + m(\vec{x},t) \quad \text{for} \quad t > 0 , \tag{4.3}$$

where the bounded nonnegative function $m(\vec{x},t)$ is prescribed or can be found in terms of what is known at $t = 0$.

Under the hypotheses that

$$g'(u) \geq 0 ,$$

$$0 \leq g(u) \leq g'(0)\, u ,$$

$$\lim_{u \to \infty} \frac{g(u)}{u} = 0 ,$$

$$g'(0) \int_0^\infty \int A(\vec{y},\tau)\, d\vec{y}\, d\tau > 1 ,$$

Thieme [46] showed that when there is an x_o such that $m(\vec{x},t) = 0$ for $|\vec{x}| \geq x_o$, the solution $u(\vec{x},t)$ has propagation properties very much like those described in Theorem 3.4 with the propagation speed c^* given by

$$(4.4)\quad c^* = \inf \{c \mid g'(0) \int_0^\infty \int A(|\vec{y}|,\tau)\, e^{\lambda(y_1 - c\tau)}\, d\vec{y}\, d\tau = 1 \text{ for some } \lambda > 0\} ,$$

regardless of the direction $\vec{\xi}$.

The same result in the special case $A(|\vec{y}|,\tau) = H(\tau)V(|\vec{y}|)$ was found independently and simultaneously by O. Diekmann [9]. Diekmann [8] also proved the existence of travelling waves for $c \geq c^*$ for this case. Some properties of these waves were studied by Diekmann and Kaper [10].

In order to display the relation of the equation (4.3) with the machinery of Section 3 we introduce the operator $Q[v,m]$ from bounded nonnegative functions $v(\vec{x},t)$ and $m(\vec{x},t)$ defined for $t \geq 0$ to bounded nonnegative functions $w(\vec{x},t)$ defined for $t \geq 0$ by saying that $w = Q[v,m]$ if

$$w(\vec{x},t) - \int_0^{1-s} \int A(|\vec{y}|,\tau)\, g(w(\vec{x}-\vec{y},t+\tau))\, d\vec{y}\, d\tau$$

$$= \int_{1-s}^\infty \int A(\vec{y},\tau)\, g(v(\vec{x}-\vec{y},t+\tau-1))\, d\vec{y}\, d\tau + m(\vec{x},t) \quad \text{for} \quad 0 \leq t < 1 ,$$

$$w(\vec{x},t) = v(\vec{x},t-1) \quad \text{for} \quad t > 1 .$$

Then if we define

$$u_n(\vec{x},t) = \begin{cases} u(\vec{x},n-t) & \text{for} \quad 0 \leq t \leq n \\ 0 & \text{for} \quad t > n \end{cases}$$

and

$$m_n(\vec{x},t) = \begin{cases} m(\vec{x},n+1-t) & \text{for} \quad 0 \leq t \leq n+1 \\ 0 & \text{for} \quad t > n+1 \end{cases}$$

we see that the integral equation (4.2) is equivalent to the recursion

$$u_{n+1} = Q[u_n, m_n]$$

$$u_o = 0 .$$

The operator Q is easily seen to be order preserving in both of its variables and invariant with respect to $\vec{x}$ - translations but not t - translations. The proofs of Thieme and Diekmann are closely related to the proof of Theorem 3.4. A proof of the analogues of Theorems 3.2 and 3.3 would permit one to remove the condition $g'(u) \leq g'(0)u$.

Thieme [47] introduced the idea mentioned above to eliminate the hypothesis that $g(u)$ is increasing, so that non-negative reproduction curves can also be treated in this model.

Once the solution $u(\vec{x},t)$ of (4.3) is known, the age distribution $p(\vec{x},a,t)$ of the population for $a \leq t$ can be found from the formula

$$p(\vec{x},a,t) = \int K(|\vec{y}|,a)\, g(u(\vec{x}-\vec{y}, t-a))\, d\vec{y} .$$

Since $u(\vec{x},t) = \int_0^\infty P(a)\, p(\vec{x},a,t)\, da$, one can also replace (4.3) by an integral equation for p.

It is not difficult to write a discrete version of a generalization of this model. If $p_n(i,j,k)$ represents the number of individuals of age k in the square centered at (i,j) at time n, then the function p_{n+1} is determined by the function p_n:

$$p_{n+1} = Q[p_n] .$$

Note that $Q[p_n](i,j,0)$ represents the number of births at (i,j), while $Q[p_n](i,j,k)$ for $k > 0$ is determined by the migration and death patterns of the population. The operator Q is translation invariant in the variables (i,j) but not in k.

It would be useful to find conditions under which the results of Section 3 can be extended to such a model.

5. Models for epidemics.

An epidemic may be thought of as the growth of a pathogenic population whose habitat consists of the members of a host population. The locations, or at least the residences, of the hosts are assumed to be fixed. The population of pathogenic individuals is measured in units of the number of infected hosts. Its density is consequently limited by the density of the host population. The disease spreads from infected to susceptible (uninfected) individuals by contact, by diffusion through air or water, or by some other mechanism.

The simplest assumption is that all infected individuals are equally contagious, regardless of how long they have been ill, that no host either recovers or dies, and that no new hosts are born. One then has an isolated population in a fixed habitat, which may be modelled by the Fisher model (4.1) or a recursion model of the form (4.2). An epidemic described by such a model is called a simple epidemic.

In most diseases an infected individual gradually becomes contagious (usually after an incubation period) and eventually either dies or recovers and becomes immune. A discrete-time model for a spatially homogeneous epidemic of such a disease in a fixed host population was introduced by Kermack and McKendrick [27]. Let s_n be the number of hosts which are not infected (the susceptibles) at time k. Then $s_{n+1} - s_n$ individuals become infected between the times n and $n+1$. Kermack and McKendrick assumed that this infection rate is equal to the product of the number of susceptibles s_n and an infectivity, which is a sum over all the infected individuals of a set of nonnegative infectivity factors which depend on how long the individual has been ill. Thus

$$(5.1) \qquad s_{n+1} - s_n = s_n \sum_{j=0}^{\infty} A_j [s_{n-j} - s_{n-1-j}] .$$

Kermack and McKendrick looked at the limiting case of (5.1) as the length of the time unit approaches zero, which leads to the integral equation

$$(5.2) \qquad \frac{ds}{dt} = s(t) \int_0^{\infty} A(\tau) \frac{ds}{dt} (t - \tau) \, d\tau .$$

It is assumed that the nonincreasing function $s(t)$ is known for $t \leq 0$, and that (5.2) holds for $t > 0$.

In the special case

$$A(t) = \varkappa e^{-\ell t}$$

where $\varkappa$ and ℓ are positive constants, this integro-differential equation is equivalent to the system of differential equations

$$(5.3) \qquad \begin{aligned} \frac{ds}{dt} &= -\varkappa si \\ \frac{di}{dt} &= \varkappa si - \ell i , \end{aligned}$$

where

$$(5.4) \qquad i(t) = -\int_0^\infty e^{-\varkappa\tau} \frac{ds}{dt}(t-\tau)\, d\tau \ .$$

This model is often called the general epidemic, although it is actually a rather special case. In this case the infectivity of an individual decays exponentially from the time of infection.

One obtains the same system if one assumes that all infected individuals have equal infectivity, but that they either die or become immune at a rate ℓ. This interpretation of the general epidemic model is frequently used.

Kermack and McKendrick noticed that the initial value problem $s(0) = s_o$, $i(0) = i_o$ for the system (5.3) can be solved explicity. It follows from this solution that if $s_o \le \ell/\varkappa$, i is monotone decreasing to zero while if $s_o > \ell/\varkappa$, i first increases and then decreases to zero. If one identifies the initial increase of i with an epidemic, the model predicts that an epidemic occurs if and only if s_o exceeds the threshhold value $\ell/\varkappa$.

Kendall [24] found another threshhold result of this kind. It is easily seen that for all positive values of s_o and i_o the function $s(t)$ decreases to a positive limit $s_\infty(s_o, i_o)$. Kendall showed that the limit $s_\infty(s_o, 0)$ as i_o approaches 0 satisfies the relation

$$(5.5) \qquad s_o - s_\infty(s_o, 0) \begin{cases} = 0 & \text{for } s_o \le \frac{\ell}{\varkappa} \\ > 0 & \text{for } s_o > \frac{\ell}{\varkappa} \end{cases}$$

Thus the limiting case of a small initial infection produces a positive number of cases if and only if $s_o > \frac{\ell}{\varkappa}$.

Kendall [25] modified the general epidemic model (5.3) to take account of the distribution of the population in space. The variable $s(x,t)$ now represents the density of susceptibles, ds/dt is replaced by $\partial s/\partial t$ in the definition (5.4) of $i(x,t)$, and the infectivity is assumed to be proportional to a spatial average $\bar{i}(x,t)$ of i. The equations (5.3) thus become

$$(5.6) \qquad \begin{aligned} \frac{\partial s}{\partial t} &= -\varkappa s \bar{i} \\ \frac{\partial i}{\partial t} &= \varkappa s \bar{i} - \ell s \ , \end{aligned}$$

where

$$\bar{i}(x,t) = \int V(x-y)\, i(y,t)\, dy$$

with $V \ge 0$ and $\int V\, dy = 1$.

Kendall [25] examined the initial value problem for the system (5.6) and observed that the variable

$$r(x,t) = s(x,0) + i(x,0) - s(x,t) - i(x,t)$$

satisfies the equation

$$(5.7)\qquad \frac{\partial r}{\partial t} = \ell i \quad .$$

Since this equation represents i as a derivative with respect to t, one can integrate the first equation in (5.6) to find that

$$(5.8)\qquad s(x,t) = s(x,0)\, e^{-\frac{\varkappa}{\ell}\int V(x-y)r(y,t)\,dy} \quad .$$

Kendall substituted this expression into the definition of r, solved the resulting equation for i, and substituted the result into (5.6) to find the single equation

$$(5.9)\qquad \frac{\partial r}{\partial t} = \ell\{ s(x,0)\,[1 - e^{-\frac{\varkappa}{\ell}\int V(x-y)r(y,t)dy}] + i(x,0) - r(x,t)\}$$

for r. We remark that if one interprets i as the density of diseased individuals all of whom are equally infective until they are removed by dying or recovering and becoming immune, then $r(x,t)$ represents the density of individuals who have been removed since $t=0$.

Kendall proved that if $s(x,0)+i(x,0) = s_o$, a constant, and if $i(x,0) = 0$ for $|x| > |x_o|$, then the limit

$$\lim_{|x|\to\infty}\ \lim_{t\to\infty} r(x,t) = s_o - \lim_{|x|\to\infty}\ \lim_{t\to\infty} s(x,t)$$

exists and is equal to the quantity $s_o - s_\infty(s_o, 0)$ which appears in the threshhold result (5.5). In particular, the set of victims of the epidemic has a uniformly positive density if and only if $s_o > \ell/\varkappa$. This striking result is known as Kendall's pandemic threshold theorem.

In [26], Kendall looked for travelling wave solutions of the form $s(x-ct)$, $i(x-ct)$ of the system of partial differential equations which results when $\bar{i}$ on the right of (5.6) is approximated by $i + k\partial^2 i/\partial x^2$ and $s(+\infty) \equiv s_o$ is prescribed. He showed that there are no such non-constant solutions when $s_o \leq \ell/\varkappa$, but that when $s_o > \ell/\varkappa$, there is a travelling wave of speed c with $s(\infty) = s_o$ and s strictly increasing if and only if

$$c \geq c^* \equiv 2[\varkappa s_o k(\varkappa s_o - \ell)]^{1/2} \quad .$$

The corresponding result for the system (5.6) was obtained for a special kernel $V(x)$ by Mollison [36,37], and for a wide class of kernels by Atkinson and Reuter [4] and Brown and Carr [7].

Kendall [26] conjectured that the minimal wave speed c^* gives a propagation speed for an epidemic. Aronson [1] studied the initial value problem for the system (5.6) by looking at the equivalent equation (5.9) with $r(x,0) = 0$. He proved that if $s(x,0) \equiv s_o$ and if $i(x,0) = 0$ for all sufficiently large

x , then if $s_o \leq \ell/\varkappa$ (sub-threshold), the function $r(x+ct,t)$ approaches zero as $t \to \infty$ for every positive c , while if $s_o > \ell/\varkappa$, there is a positive number c^* such that

$$(5.10) \qquad \lim_{t\to\infty} r(x+ct,t) = \begin{cases} 0 & \text{if } c > c^* \\ s_\infty(s_o, 0) & \text{if } c < c^* \end{cases}$$

Here $s_\infty(s_o, 0)$ is the quantity which appears in (5.5), and

$$(5.11) \qquad c^* = \inf_{\lambda > 0} \frac{1}{\lambda} \{ s_o \varkappa \int V(y)\, e^{\lambda y}\, dy - \ell \} .$$

We wish to show how this result is related to the theorems of Section 3. Define $Q[z]$ to be the value at $t=1$ of the initial value problem for the differential equation (5.9) with $r(x,0) = z(x)$. Then the sequence of functions

$$r_n(x) = r(x,n) ,$$

satisfies the recursion

$$r_{n+1} = Q[r_n] ,$$
$$r_o = 0 .$$

The operator Q is easily seen to be order preserving, but it is not translation invariant because of the presence of the function $i(x,0)$.

Of course, $Q \geq Q_-$ where $Q_-[z]$ is the value at $t=1$ of the solution of the problem

$$(5.12) \qquad \frac{\partial r}{\partial t} = \ell \{ s_o [1 - e^{-\frac{\varkappa}{\ell}\int V(x-y)\, r(y)\, dy}] - r \} , \quad r(x,0) = z(x) .$$

The operator Q_- is translation invariant, and its linearization L_- is obtained by linearizing the right-hand side of the differential equation. The second part of (5.10) and the formula (5.11) thus follow from Theorem 3.4 and Corollary 3.1.

On the other hand, if we assume that $i(x,0) = 0$ for $x \geq x_o$ and that $V(x) = 0$ for $|x| > A$, the right-hand side of (5.1) coincides with that of (5.12) for $x \geq x_o + A$. Because $1 - e^{-z} \leq z$ for $z \geq 0$, a computation shows that if c is defined by the formula

$$c = \frac{1}{\lambda} \{ s_o \int V(y)\, e^{\lambda y}\, dy - \ell \} ,$$

for some positive λ , if $x_1 \geq x_o + A$, and if the constant M satisfies $M \geq s_o + i(x,0)$, then the function

$$w = \min \{ M, Me^{-\lambda(x-ct-x_1)} \}$$

satisfies the inequality

$$\frac{\partial w}{\partial t} \geq \ell \{ s_o [1 - e^{-\frac{\varkappa}{\ell}\int V(x-y)w(y)dy}] + i(x,0) - w \}$$

both for $x < x_1 + ct$ and for $x > x_1 + ct$. It follows that the sequence $w_n(x) = w(x - nc)$ satisfies the inequality $w_{n+1} \geq Q[w_n]$, so that the first part of (5.10) is an immediate consequence of the Proposition of Section 3.

When $s_o \varkappa < \ell$, one can obtain arbitrarily small positive c by choosing λ, which implies the nonpropagation result.

This method of proof shows that the conclusions are still valid under the weaker hypotheses

$$s(x,0) = s_o \qquad \text{for} \quad x \geq x_o ,$$

$$i(x,0) = 0 \qquad \text{for} \quad x \geq x_o ,$$

$$s(x,0) \leq s_o \leq s(x,0) + i(x,0) \leq M .$$

A model which, like the Kermack-McKendrick model, takes into account the fact that an infected individual's infectivity depends upon how long he has been infected, was considered independently by Thieme [45] and Diekmann [8]. If $s(x,t)$ again represents the density of susceptible individuals, the rate of infection $-\partial s / \partial t$ is assumed to be determined by the equation

$$(5.13) \qquad \frac{\partial s}{\partial t}(x,t) = s(x,t) \int_0^\infty \int A(y,\tau) \frac{\partial s}{\partial t}(x-y,t-\tau)\, dy\, d\tau .$$

The integral on the right represents the sum of the infectivities at (x,t) of the individuals at $x - y$ who were infected at time $t - \tau$. The kernel $A(y,\tau)$ tells to what extent a person who has been ill for a time τ spreads the disease to a distance y. If there is an incubation period σ before an infected individual becomes infectuous, then $A(y,\tau) = 0$ for $\tau \leq \sigma$.

If there is a natural resistance to the disease, the integral on the right of (5.13) may represent a weighted time average of the infectivity over the past. A more general model of this kind was considered by Thieme [45].

Diekmann and Thieme noticed that the integral on the right of (5.13) is a t-derivative so that this equation may be integrated once. If one introduces the variable $u = \log [s_o / s(x,t)]$ where s_o is any positive constant, one finds that

$$(5.14) \qquad u(x,t) = \int_0^\infty \int A(y,\tau)\, s_o [1 - e^{-u(x-y,t-\tau)}]\, dy\, d\tau + q(x)$$

where $q(x)$ is an arbitrary function of x only. This is of the form (4.3).

If $\lim_{t \to -\infty} s(x,t) = s_o$, then $q \equiv 0$. When $A(y,\tau)$ is of the form $H(\tau)V(|y|)$ and $s_o \int V(y)\, dy \int H(\tau)\, d\tau > 1$, Diekmann [8] proved that there are nonconstant travelling wave solutions $u(x,t) = W_c(x - ct)$ of this equation of all speeds $c > c^*$ where c^* is defined by (4.4).

If $s(x,0) = s_o$, then (5.13) becomes

$$(5.15) \qquad u(x,t) = \int_0^t \int A(y,\tau)\, s_o [1 - e^{-u(x-y,t-\tau)}]\, dy\, d\tau + m(x,t) \ ,$$

where $m(x,t)$ may be thought of as the t - integral from 0 to t of the infectivity that is produced by those individuals who were already infected at $t = 0$. This equation is of the form (4.3) and if one assumes that m is bounded and vanishes for sufficiently large x , the results of Diekmann [9] and of Thieme [46] generalize the results (5.10) of Aronson.

If $s(x,0)$ is not constant, one obtains the integral equation

$$(5.16) \qquad u(x,t) = \int_0^t \int A(y,\tau)\, s(x-y,0)\, [1 - e^{-u(x-y,t-\tau)}]\, dy\, d\tau + m(x,t)$$

where m is defined as above but

$$u(x,t) = \log\, [s(x,0) / s(x,t)] \ .$$

Because $s(x,0)$ varies, the corresponding operator $Q[v,w]$ which was defined in the preceding section is not translation invariant. However, if $s(x,0) = s_o$ and $m(x,t) = 0$ for all sufficiently large x and $s(x,0) \geq s_1$ where $s_1 \iint A(y,\tau)\, dy\, d\tau > 1$, Diekmann [9] and Thieme [46] still prove (5.10).

A discrete version of the Diekmann-Thieme model (5.13) can be obtained by putting time averages in the sum on the right-hand side of the Kermack-McKendrick model (5.1). However, if one wants to obtain a first integral analogous to (5.14), one needs to replace this model by an equation of the form

$$s_{n+1}(\vec{x}) = s_n(\vec{x})\, e^{-\sum_{j=0}^{\infty} \sum_{\vec{y}} A_j(\vec{y})[s_{n-j-1}(\vec{x}-\vec{y}) - s_{n-j}(\vec{x}-\vec{y})]} \ .$$

6. Need for future work.

There are clearly some gaps left in the mathematical results we have presented.

For example, the asymptotic results such as Theorem 2.2 and 2.3 say that the solution eventually looks like a spreading wave, but do not say how long one has to wait to see this phenomenon. Since the models with which we deal are by no means exact, one may wonder whether the wave appears before the model no longer describes the phenomenon or before one is no longer interested in it. Thus it would be useful to have estimates on how quickly the limits in these theorems are attained.

The results of Fife and McLeod [14], Uchiyama [48,49], and Rothe [43] state that in one dimension the solution of an initial value problem for Fisher's equation converges "in shape" to a travelling wave. That is, the distance between points where p attains two given values approaches a constant. On the other hand, Theorems 2.2, and 2.3, and 2.4 allow this distance to grow linearly in time. It is natural to ask whether the distance between level curves remains bounded for solutions of Fisher's equation in two (or more) dimensions and, more generally, for solutions of a recursion of the form $p_{n+1} = Q[p_n]$.

A more important question is to what extent the results which were presented here can be extended to more realistic models.

It is clear that all the above models are greatly oversimplified. The oversimplification is not so much the result of replacing stochastic models by deterministic ones. It comes from neglecting nonnegligible interactions in order to produce models with order preserving properties.

The oversimplification is most easily seen in the population growth models. No population lives in complete isolation. In most cases one must discuss the interactions of several sexes, several age groups, and several species in a food chain. Thus instead of a single variable p_n , one needs to consider a vector-valued variable $\vec{p}_n$ whose components give the sizes of the various interacting populations at time n . A discrete-time model is again given by a recursion of the form $\vec{p}_{n+1} = Q[\vec{p}_n]$ where Q transforms vector valued functions into vector valued functions.

The techniques we have presented in Section 3 can be extended to this case if Q is order preserving. This means that an increase of one of the populations at one place at time n increases all the populations everywhere at time $n+1$. Such a condition can be described as <u>perfect symbiosis</u>.

This condition applies, for example, in Thieme's model (4.3) for an age structured population when $g(u)$ is an increasing function. While this model permits competition of the immature individuals of the same age in the same location with each other, it does not provide for any competition between juveniles or adults of different ages. The migration, reproduction, and death patterns of the adults are assumed to be independent of the presence or absence of other individuals. Such a model cannot contain any mechanism for regulating the total population size.

More generally, almost any population regulation mechanism which involves a competition for the same living space or resource between individuals of different kinds implies that the operator Q is not order preserving. For example if one has a two-sex model with a fixed carrying capacity for the total population and if the population is at the carrying capacity, then increasing the number of males will tend to decrease the number of females, unless the population regulation is accomplished exclusively through competition among the males. Thus realistic population models will usually involve a recursion with an operator which is not order preserving.

The above comments apply directly to population genetics models. Even the simplest of these models involves populations of three genotypes. The reduction from three variables to one involves assumptions which are not likely to be satisfied. A model for allelic variation at one or more gene loci in which the total population is controlled by a more or less realistic mechanism will usually involve a recursion of the form $\vec{p}_{n+1} = Q\vec{p}_n$ in which Q is not order preserving.

As we have already pointed out, an epidemic model is a population growth model in which the total population density of infected individuals is bounded by the population density of all individuals. Therefore one would not expect an order preserving recursion, except in the case of a simple epidemic, in which one is dealing with a single class consisting of all infected individuals, regardless of the time since infection. What saves the day here is the reduction of (5.6) to (5.9) and of (5.13) to (5.14) by means of an integration. This integration depends upon the fact that $s^{-1}\partial s/\partial t$ is equal to a derivative with respect to t. If the model (5.13) is altered slightly by replacing the integral on the right by a nonlinear function of this integral to take account of resistance to disease, this mathematical trick breaks down and it is not likely that one obtain a problem involving an order preserving operator.

We see from the above discussion that more realistic models are not likely to have the order preserving properties on which the results presented here depend. More realistic models can certainly be formulated with the help of experts in the field of application. In fact, some very detailed models of epidemics, which are presently only being studied by numerical computation, have been formulated. (See, e.g. [12])

If mathematics is to have a real impact in biology, it must produce mathematical tools which are capable of treating the models which actually arise, rather than confining its attention to those models which are amenable to presently known mathematical techniques. The question arises whether the methods which use order preserving properties can be stretched to deal with related but more exact models which do not quite have these properties. If not, the order-preserving methods may well occupy the position of the dinosaurs, having been well suited to a certain stage of development of the theory of biological spread but being unable to adapt to a more advanced stage of the subject. In any case, new mathematical ideas and tools are sorely needed.

References

1. D.G. Aronson. The asymptotic speed of propagation of a simple epidemic. Nonlinear Diffusion, ed. W.E. Fitzgibbon and H.F. Walker. Research Notes in Mathematics 14, Pitman, London, 1977, pp. 1-23.

2. D.G. Aronson and H.F. Weinberger. Nonlinear diffusion in population genetics, combustion, and nerve propagation. Partial Differential Equations and Related Topics, ed. J. Goldstein, Lecture Notes in Mathematics, vol. 446, Springer, 1975, pp. 5-49.

3. D.G Aronson and H.F. Weinberger. Multidimensional nonlinear diffusion arising in population genetics. Adv. in Math. 30 (1978), pp. 33-76.

4. C. Atkinson and G.E.H. Reuter. Deterministic epidemic waves. Cambridge Phil. Soc., Math Proc. 80 (1976), pp. 315-330.

5. N.T.J. Bailey. The Mathematical Theory of Epidemics. Griffin, London, 1957.

6. M. Bramson. Maximal displacement of branching Brownian motion. Comm. in Pure and Appl. Math. 31 (1978), pp. 531-581.

7. K.J. Brown and J. Carr. Deterministic epidemic waves of critical velocity. Math. Proc. of the Cambridge Phil. Soc. 81 (1977), pp. 431-435.

8. O. Diekmann. Thresholds and travelling waves for the geographical spread of infection. J. of Math. Biol. 6 (1978), pp. 109-130.

9. O. Diekmann. Run for your life. A note on the asymptotic speed of propagation of an epidemic. Mathematical Centre, Amsterdam, Report TW 176/78, 1978.

10. O. Diekmann and H.G. Kaper. On the bounded solutions of a nonlinear convolution equation. J. Nonlin. Analysis-Theory, Methods, and Applic. (in print).

11. O. Diekmann and N.M. Temme. Nonlinear Diffusion Problems. Mathematisch Centrum, Amsterdam, 1976.

12. L.R. Elveback, E. Ackerman, L. Gatewood, et. al. Stochastic two-agent epidemic simulation models for a community of families. Amer. J. of Epidem. 93 (1971), pp. 267-280.

13. P.C. Fife. Mathematical Aspects of Reacting and Diffusing Systems. Lecture Notes in Biomath. 28, Springer, New York, 1979.

14. P.C. Fife and J.B. McLeod. The approach of solutions of nonlinear diffusion equations to travelling wave solutions. A.M.S. Bull. 81 (1975), pp. 1076-1078 and Arch. for Rat. Mech. and Anal. 65 (1977), pp. 335-361.

15. R.A. Fisher. The advance of advantageous genes. Ann. of Eugenics 7 (1937), pp. 355-369.

16. A. Friedman. Partial Differential Equations of Parabolic Type. Prentice-Hall, Englewood Cliffs, N.J., 1964.

17. K.P. Hadeler and F. Rothe. Travelling fronts in nonlinear diffusion equations. J. Math. Biol. 2 (1975), pp. 251-263.

18. J.M. Hammersley. Postulates for subadditive processes. Annals of Prob. 2 (1974), pp. 652-680.

19. F. Hoppensteadt. Mathematical Theories of Populations: Demographics, Genetics, and Epidemics. Reg. Conf. Ser. in Appl. Math. 20, SIAM, Philadelphia, 1975.

20. Y. Kametaka. On the nonlinear diffusion equations of Kolmogorov-Petrovsky-Piskunov type. Osaka J. Math. 13 (1976), pp. 11-66.

21. Ja. I. Kanel'. Stabilization of solutions of the Cauchy problem for equations encountered in combustion theory. Mat. Sbornik (N.S.) 59 (101) (1962), supplement, pp. 245-288.

22. Ja. I. Kanel'. On the stability of solutions of the equations of combustion theory for finite initial functions. Mat. Sbornik (N.S.) 65 (107) (1964), pp. 398-413.

23. S. Karlin. Population subdivision and selection migration interaction. In Population Genetics and Ecology (ed. S. Karlin and E. Nevo), Academic Press, New York, 1976, pp. 617-657.

24. D.G. Kendall. Deterministic and stochastic epidemics in closed populations. Proc. of the Third Berkeley Symposium on Mathematical Statistics and Probability, ed. J. Neyman, IV (1956), pp. 149-165.

25. D.G. Kendall. Discussion of a paper of M.S. Bartlett. J. of the Royal Statistical Soc., Ser. A, 120 (1957), pp. 64-67.

26. D.G. Kendall. Mathematical models of the spread of infection. Mathematics and Computer Science in Biology and Medicine. H.M.S.O., London, 1965, pp. 213-225.

27. W.D. Kermack and A.G. McKendrick. A contribution to the mathematical theory of epidemics. Proc. Royal Soc. A 115 (1927), pp. 700-721.

28. H. Kesten. Random processes in random environments. In this volume.

29. M. Kimura and G. Weiss. Genetics 49 (1964), pp. 561-576.

30. A. Kolmogoroff, I. Petrovsky, and N. Piscounoff. Étude de l'équations de la diffusion avec croissance de la quantité de matière et son application a un problème biologique. Bull. Univ. Moscow, Ser. Internat., Sec. A, 1 (1937) #6, pp. 1-25.

31. T.G. Kurtz. Relationships between stochastic and deterministic population models. In this volume.

32. T.M. Liggett. Interacting Markov processes. In this volume.

33. G. Malécot. The Mathematics of Heredity. W.H. Freeman, San Francisco, 1969.

34. T.R. Malthus. An essay on the Principle of Population. Printed for J.Johnson in St. Paul's Churchyard, London. First edition 1798, third edition 1806.

35. A.G. McKendrick. Applications of mathematics to medical problems. Proc. of the Edinburgh Math. Soc. 44 (1925-26), pp. 98-130.

36. D. Mollison. Possible velocities for a simple epidemic. Adv. in Appl. Prob. 4 (1972), pp. 233-257.

37. D. Mollison. The rate of spatial propagation of simple epidemics. Proc. Sixth Berkeley Symp. on Math., Stat., and Prob. 3 (1972), pp. 579-614.

38. D. Mollison. Spatial contact models for ecological and epidemiological spread. J. Royal Stat. Soc. B 39 (1977), pp. 283-326.

39. T. Nagylaki. Selection in One- and Two-locus Systems. Lecture Notes in Biomathematics 15, Springer, New York, 1977.

40. M.H. Protter and H.F. Weinberger. Maximum Principles in Differential Equations. Prentice-Hall, Englewood Cliffs, N.J., 1967.

41. W.E. Ricker. Stock and recruitment. J. Fish. Res. Bd. Can. 11 (1954), pp. 559-623.

42. A. Robertson. Embryogenesis through cellular interactions. In this volume.

43. F. Rothe. Convergence to pushed fronts. In this volume.

44. F.R. Sharpe and A.J. Lotka. A problem in age distribution. Phil. Mag. 21 (1911), pp. 435-438.

45. H.R. Thieme. A model for the spatial spread of an epidemic. J. Math. Biol. 4 (1977), pp. 337-351.

46. H.R. Thieme. A symptotic estimates of the solutions of nonlinear integral equations and asymptotic speeds for the spread of populations. J. für die reine und angew. Math. 306 (1979), pp. 94-121.

47. H.R. Thieme. Density-dependent regulation of spatially distributed populations and their asymptotic speeds of spread. J. of Math. Biol. (in print).

48. K. Uchiyama. The behavior of solutions of the equation of Kolmogorov-Petrovsky-Piskunov. Proc. Japan. Acad. Ser. A, 53 (1977), pp. 225-228.

49. K. Uchiyama. The behavior of solutions of some non-linear diffusion equations for large time. J. of Math. of Kyoto Univ. 18 (1978), pp. 453-508.

50. P.F. Verhulst. Notice sur la loi que la population suit dans son accroissement. Correspondence Mathématique et Physique Publiée par A. Quételet 10 (1838), pp. 113-121. English translation in D. Smith and N. Keyfitz, Mathematical Demography, Springer, New York, 1977, pp. 333-337.

51. P. Waltman. Deterministic Threshold Models in the theory of Epidemics. Lecture Notes in Biomathematics 1, Springer, New York, 1974.

52. H.F. Weinberger. Asymptotic behavior of a model in population genetics. Nonlinear Partial Differential Equations and Applications, ed. J. Chadam. Lecture Notes in Math 648, Springer, New York, 1978, pp. 47-98.

53. H.F. Weinberger. Asymptotic behavior of a class of discrete-time models in population genetics. In Applied Nonlinear Analysis, ed. V. Lakshmikantham. Academic Press 1979, pp. 407-422.

54. H.F. Weinberger. Genetic wave propagation, convex sets, and semi-infinite programming in Constructive Approaches to Mathematical Models, ed. C.V. Coffman and G.J. Fix. Academic Press, New York, 1979, pp. 293-317.

This work was supported by the National Science Foundation through Grant MCS 7802182.

TOPIC IV

Models for cell motility

Discussion by the Editors

Motility in biological systems can be expressed in a number of ways and can also be investigated from different points of view. Three volumes on cell motility in the series "Cold Spring Harbor Conferences on Cell Proliferation"(1976), the volume edited by G. Hazelbauer, "Taxis and Behavior. Elementary Sensory Systems in Biology"(1978), and the volume edited by J.Nicholls, "The Role of Intercellular Signals: Navigation, Encounter, Outcome"(1979) contain the most recent but diverse information in this field. In his introduction to the book "Mathematical Models for Cell Rearrangement" (1975), R.D.Campbell noticed that cellular chemotactic movement has until recently not been treated mathematically. This fact is surprising for anyone who a priori considers that chemotaxis might be more appropriate for "mathematical reconstruction" than flagellar locomotion. One problem suggested by Campbell, has been the lack of experimental data concerning the statistical or individual behaviour of cells in defined chemical gradients.

Indeed, the mathematical investigation of cell motility and locomotion begins in 1970 with the publication of two papers - one by E.F.Keller and L.E.Segel on the slime mold (D. discoideum) aggregation, and the other by L.E.Blumenson on the spread of cancer cells. One year later, in 1971, M.H.Cohen and A.Robertson consider the effect of aggregative movement on organizing wave propagation in slime molds, and E.F.Keller and L.E.Segel investigate chemotaxis by a Brownian motion model. It is, thus, interesting to re-examine the Keller-Segel model(s) after nearly a decade. Evelyn Keller did so at our conference. In particular, she examines the correctness of the original assumptions about the three basic model parameters, namely motility, substrate consumption and chemotactic sensitivity. Her remarks lead us to the following comments. Firstly, it seems that the mathematical models for kinesis and taxis are very closely coupled with experimental data. Moreover, certain measurements may make indistinguishable two alternative models. This, however, is not unexpected in a domain where new knowledge is rapidly accumulated. For instance, with respect to the proteins involved, the motility of eukaryotic cells may be classified into two major groups: (i) actin-myosin based (amoeboid movement, cytoplasmic streaming, cytokinesis, etc.) and (ii) tubulin-dynein based (ciliary and flagellar movements, bending of axostyles, etc.). The introduction of biophysical and biochemical data continuously improves the mathematical models. This remark makes significant the motto in the paper by J.R. Lapidus and M.Levandowski and their last paragraph on experiments. Secondly, mathematics, in this field, has a descriptive function. Is thus the explanatory function of a mathematical model lost? Maybe not, as L.Segel suggested: "Later in the development of theory, however, more dignified mathematics may be involved"(1978). The third comment deals with the adequacy of a model for cell motility: "Which is a more appropriate representation of cellular motion - asks L. Segel - a biased random walk or a continuous diffusion process supplemented by an additional gradient-dependent flux?" The answer is still not clear. In their paper, J.R.Lapidus and M.Levandowsky distinguish three models for 'chemokinesis', namely, in order of simplicity, the diffusion model, the biased random walk, and a certain non-Markovian model. A precise model of kinesis in a chemotactic gradient is constructed by W.Alt by introducing a new biased random walk. The reader will find in the substantial survey lecture given by R.Nossal the basic information about the mathematical theories of topotaxis.

Cell adhesion is the final step in cell motility. J.G.Nicholls calls it "encounter": cells stop moving and interact with each other to form adhesive bonds and specialized junctions. In his paper, G.Bell adds to this "encounter" the cell adhesion to a (specific) substrate. The list of encounter processes is large, ranging from morphogenesis to metastasis. G.Bell refers here only to the bond formation. (In a Technical Report by Carla Wofsy, an encounter process for allergic reactions is studied.)

An attempt to incorporate chemotaxis, cell motility, phagocytosis and spatial distribution of cells into a mathematical model for tissue inflammation is presented by D.Lauffenburger. This is a valuable effort that shows, as Evelyn Keller said, a "healthy use" of some cell motility models.

ORIENTATION OF CELLS MIGRATING IN A CHEMOTACTIC GRADIENT

Wolfgang Alt
Institut für Angewandte Mathematik
Im Neuenheimer Feld 294
D-6900 Heidelberg / BRD

Swimming cells like flagellated bacteria explore the concentration profile of a chemical substance by sensing a temporal gradient during the motion of their cell body, and then react by increasing the length of path segments running up the gradient, whereas their turning behavior stays random. See BERG [4] and NOSSAL [24], this volume.

Contrary, cells migrating on a surface or in a tissue like amoebae or leukocytes, for example, use a biased turn-angle distribution in order to orient their motion with respect to a chemotactic gradient, cp. [15,24,25] . Obviously there has to be a mechanism by which even resting cells find out favorable directions leading them into regions of higher chemoattractant concentrations. RAMSEY [26], ZIGMOND [31], WILKINSON [30] and others presumed that a cell measures spatial differences of the chemoattractant over the length of its cell body (there is also an amplification model based on this assumption [21]). However, observations of cell surface activities of various cell types [1 ,6] suggest that pseudopods or other active protrusions of the cell membrane can contribute to a gradient sensing mechanism: for amoeba this possibility was first considered by GERISCH et al.[13] and for leukocytes by BOYARSKY, NOBLE [5]. The locomotion of neutrophils or dictyostelium cells on a two-dimensional surface is usually induced by the protrusion, stagnation and partial withdrawal of pseudopods (or lamellipods). Some of them constitute a "leading front", which for a certain time pulls the cell (or its nucleus) into a nearly straight direction, cp.[13,fig.3] [30,fig.9.2], [34,fig.2] . The stagnation of the anterior protrusion is preceded by the appearance of one or more lateral pseudopods which seem to compete with one another. Finally one of these will extend further and determine the direction of the subsequent cell (nucleus) displacement, compare [8 , 11] .

The resulting two-dimensional motion of such a cell (resp. the nucleus) can efficiently be modelled by a stochastic random walk assuming a piecewise straight movement with turning frequency β (Poisson distributed) and turn-angle distribution k_ϕ .We refer to [5 ,15,24],and to [2] for a more detailed model in arbitrary dimensions. If the medium is isotropic then the probability $k_\phi(\alpha)$ that a cell, which has been moving at an angle ϕ (with respect to the x-axis in the x,y-plane) and which changes direction, turns through an angle $\alpha \in [-\pi,\pi]$ is independent of ϕ and symmetric:

(1) $k_\phi(\alpha) = h(\alpha)$,

where h is a symmetric probability distribution on $[-\pi,\pi]$, for example

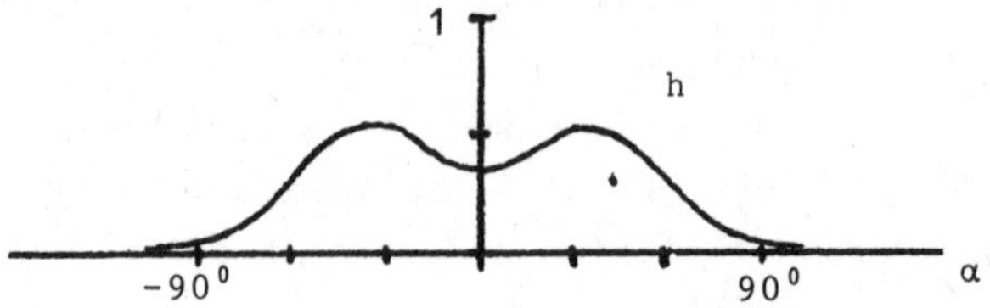

Figure 1. Turn-angle distribution for polymorphonuclear (PMN) leukocytes in an isotropic medium. Values are obtained by a realistic modification of the data in [25, fig.3] regarding the measurements and comments in [15].

These data indicate that in a homogeneous situation the activity of lateral surface extensions is symmetrically distributed around the moving angle ϕ . If, however, the cell migrates in a chemotactic gradient then the turn-angle distribution k_ϕ is no longer symmetric: the measured values in fig.2 show a clear preference for turn angles α having a sign opposite to that of ϕ :

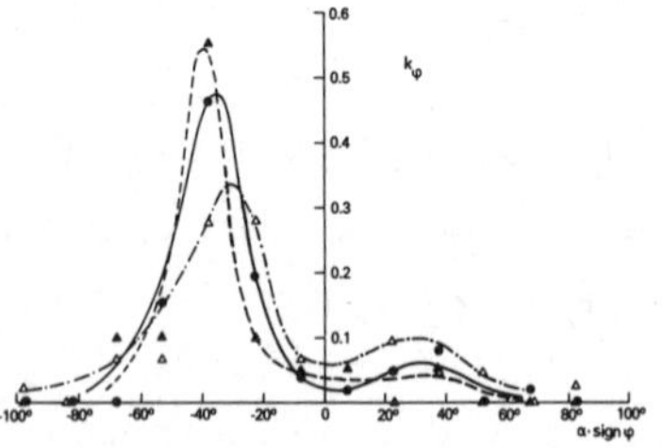

Figure 2. Turn-angle distribution for PMN leukocytes migrating in a chemotactic gradient parallel to the x-axis (corresponding to 0^0). Data with kind permission from [25, fig. 2b]. Drawn interpolation curves are slightly changed. Distributions over $\alpha \cdot \operatorname{sign}\varphi$, for different moving angles: $0 < |\varphi| \leq 15^0$ (Δ / -·-·-) $15^0 < |\varphi| \leq 30^0$ (• / ——) and $30^0 < |\varphi| \leq 45^0$ (▲/ ----) .

NOSSAL [24,25] approximated k_ϕ by a suitably shifted symmetric distribution p according to

(2) $$k_\phi(\alpha) = h + (1-h)\, p(\alpha+f(\phi)) ,$$

where h is a constant and f is some odd function. This description neglects the varying height (depending on ϕ) of the left maximum in fig. 2 as well as the appearance of the smaller maxima to the right. In the following we propose an alternative model which explains these phenomena and which is based on a more detailed biochemical consideration.

Receptor model for the bias of the turn-angle distribution

Assume that on the membrane of a migrating cell there are specific receptors, cp. [30] for a review, which are symmetrically distributed around the momentary displacement direction $\eta_\phi = (\cos\phi , \sin\phi)$ with a density $r = r(\alpha)$. Suppose that diffusing molecules of the chemoattractant S with a concentration ρ bind to a $(m-1)$-fold occupied receptor site R_{m-1} according to the reaction scheme

$$[S] + [R_{m-1}] \underset{k_{-m}}{\overset{k_m}{\rightleftharpoons}} [R_m] \xrightarrow{\tilde{k}_m} [R_{m-1}] + [P] , \quad m = 1,..,M$$

such that an intracellular substance P is produced, which might be degraded elsewhere. With the usual pseudo-steady-state assumption the resulting P-concentration density below the membrane equals $b(\rho)\, r(\alpha)$, where $b(\rho)$ represents the receptor kinetics depending on the extracellular concentration ρ : for $M = 1$ we have the Michaelis-Menten-kinetics and for $M = 2$

(3) $$b(\rho) = \frac{\tilde{k}_1 K_2 \rho + \tilde{k}_2 \rho^2}{K_1 K_2 + K_2 \rho + \rho^2}$$

where $K_m = (k_{-m} + \tilde{k}_m)/k_m$ denote the related dissociation constants.

According to the distribution in fig. 1 the extension of a pseudopod in direction $\eta_\phi(\alpha) = \big(\cos(\phi+\alpha), \sin(\phi+\alpha)\big)$ is initiated with probability $h(\alpha)$. If it elongates with a (mean) speed c_o then the receptors on it are moved into the surrounding medium. If the medium is isotropic, the P-concentration density inside the pseudopod remains constant, whereas in a chemotactic gradient after a (mean) protrusion time δ_o it undergoes a change of amount

(4) $$s_\alpha(\phi) = b'(\rho)\, \delta_o c_o\, r(\alpha)\, \eta_\phi(\alpha) \cdot \nabla\rho \quad .$$

provided the (mean) length $\delta_o c_o$ of such a sensing protrusion is small.

Finally suppose that $s_\alpha(\phi)$ serves as a signal pulse inducing the local increase of a transmitter substance (cytoplasmic Ca^{++} or cGMP for example [16,20,33]) proportional to $f(s_\alpha(\phi))$ with some nonnegative function f. If this "effector" density — by some comparison mechanism inside the cell, probably involving the diffusion of the transmitter and the activation of the locomotory system [9 ,12,28] — describes the relative advantage of the considered protrusion to extend further and determine the direction of the subsequent cell displacement, then the turn-angle probability can be expressed as

(5) $$k_\phi(\alpha) = h(\alpha)\,\frac{1}{F_\phi}\,f(s_\alpha(\phi)) \quad ,$$

where

$$F_\phi = \int_{-\pi}^{\pi} h(\alpha)\, f(s_\alpha(\phi))\, d\alpha$$

denotes the mean effect averaged over all possible protrusions during such a reorientation phase.

The simplest model for the function f would be linear

(6) $$f(s) = 1 + \kappa s$$

with some constant $\kappa \leq 1/s_o$, $s_o := \delta_o c_o\, \alpha_o\, b'(\rho)\, |\nabla\rho|\, \max r$.

The measured turn-angle distributions in fig.2, however, can be reproduced by this linear model only qualitatively. Furthermore the comparison between competing protrusions mentioned above probably is achieved by some threshold mechanism which can be reflected by a function f of the following type

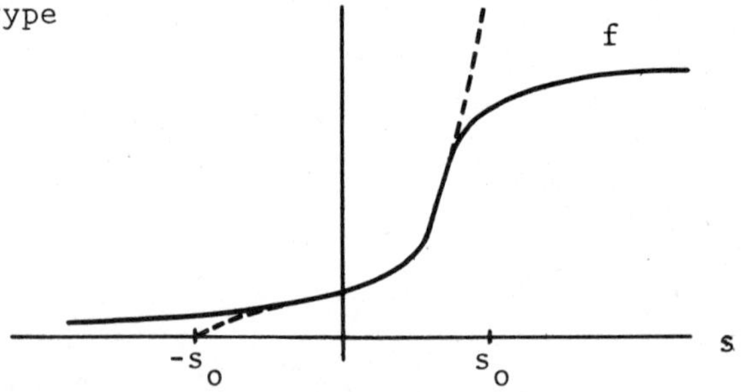

Figure 3. Hypothetical model for the signal - transmitter (- effector) dependence.

In order to fit the data in fig.2 we use a third order approximation to the function f in fig.3 (----):

(7) $$f(s) = 1 + \kappa_1 s + \kappa_2 s^2 + \kappa_3 s^3 + O(\varepsilon s^4) \quad , \quad |s| \leq s_o \leq 1/\kappa_1 \ ,$$

with suitable κ_i and small ε . Then the choice of the unbiased turn-angle distribution h in fig.1 and of the following receptor distribution r :

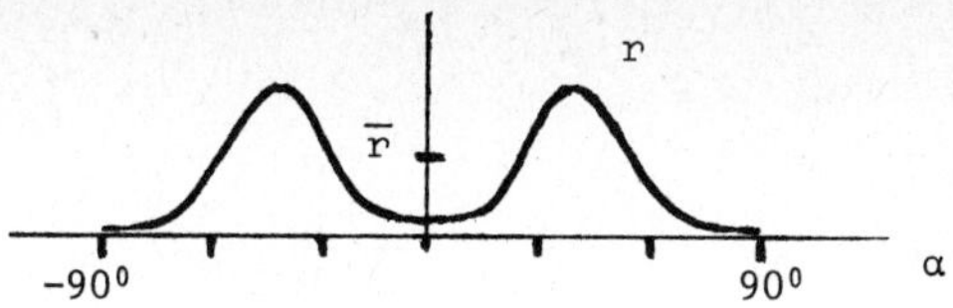

Figure 4. Distribution of receptors to both sides of a cell at the moment of reorientation (hypothetical model). The mean value $\bar{r}$ is defined in (11).

leads to a relatively good agreement of the theoretical turn-angle distribution k_ϕ in (5) with the experimental data from fig.2 :

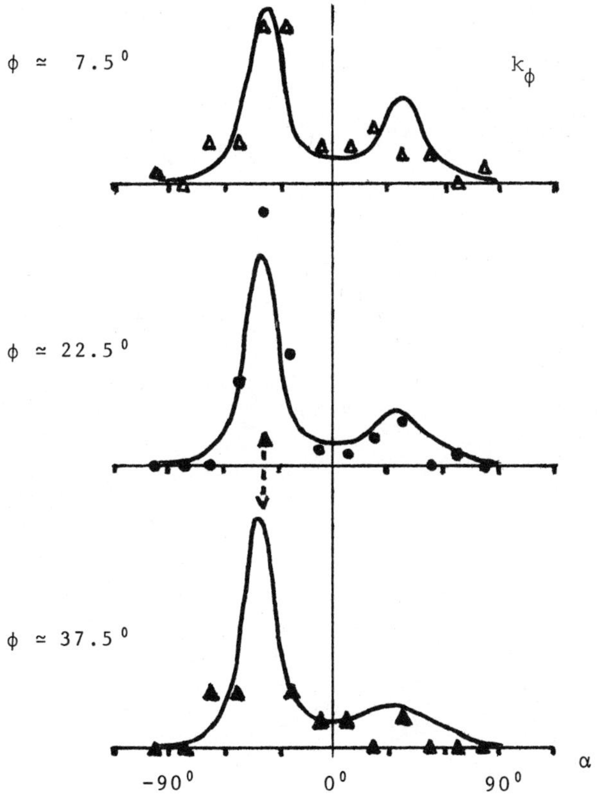

Figure 5. Turn-angle distribution calculated from equation (5) for different values of the moving angle φ , (△, ●, ▲) denoting the corresponding measured data from figure 2. Particular parameters: $\kappa_2 = \kappa_3 = 2.5\ \kappa_1$, $\varepsilon = o$ and $\delta = o.48$, δ defined in (10) . Note that the approximation (7) holds for $\delta \lesssim o.5$.

Like in NOSSAL's model (2) the theoretical curves in fig.5 show a "background" component due to the distribution h which is superimposed by the bias effect due to the receptor distribution r . Also the slight shift of both maxima to the left is fairly well reproduced by the turn-angle model (5) . The type of the r-distribution used here (fig.4) suggests the biologically reasonable conjecture that "new" receptors, which might be produced in course of the plasma streaming into the leading front of pseudopods, in the phase of reorientation mainly are placed at both sides of the front region, whereas the "used" receptors are moved backwards in order to undergo endocytosis [28] .

Modelling equations

Let $\sigma(t,z,\phi)$ denote the density of cells moving (independently of each other) at time t near a point z in direction η_ϕ . If we assume that the times between two reorientations are exponentially distributed with factor β independent of ϕ , see [15,25] , and that the (mean) speed c of cell displacement is constant then σ satisfies the following differential-integral equation

$$(8) \qquad \partial_t\sigma + c\,\eta_\phi\cdot\nabla_z\sigma + \beta(\mathrm{id} - T)\sigma = 0$$

where the turn-angle operator T has the kernel $k_{\phi-\alpha}(\alpha)$, namely

$$(9) \qquad T\sigma(\phi) = \int_{-\pi}^{\pi} \frac{\sigma(\phi-\alpha)}{F_{\phi-\alpha}}\, h(\alpha)\, f(\delta\, \hat{r}(\alpha) \cos\phi)\, d\alpha \quad .$$

Here we set

$$(10) \qquad \delta := \kappa_1\, \delta_o c_o \bar{r}\, b'(\rho)\, |\nabla\rho| \quad ,$$

$$\hat{r}(\alpha) = r(\alpha)/\bar{r} \quad \text{and}$$

$$(11) \qquad \bar{r} = \int_{-\pi}^{\pi} h(\alpha)\, r(\alpha)\, d\alpha$$

denoting the total amount of receptors placed on active protrusions. A derivation of (8) and some more refined models can be found in [2] . For later use we define the following Fourier coefficients

$$(12) \qquad \psi_{\mu\nu} = \int_{-\pi}^{\pi} h(\alpha)\, \hat{r}^{\mu}(\alpha) \cos(\nu\alpha)\, d\alpha \quad , \quad \psi_\nu = \psi_{o\nu} \quad .$$

Since a general treatment of transport equations like (8) is complicated, see [7], and asymptotic properties (for $t \to \infty$) have been analysed by NOSSAL [23] , we restrict the following considerations to special cases and approximative results.

The case of constant gradient

First let us consider experimental situations, where the gradient $\nabla\rho$ is fixed and constant, say in x-direction, and the coefficients β and δ are (nearly) constant. The measurements on the ZIGMOND-bridge [32], for example, are performed in a limited region inmidst the bridge, so that this assumption is justified. Indeed [32,fig.4] shows a steady state of oriented cells after a short time during which the gradient is established. A stationary solution $\sigma = \sigma(\phi)$ of (8) is just a fixed point

(13) $$\sigma = T\sigma$$

of the turn-angle operator T. Defining the operators

$$T_\mu\sigma(\phi) = \int_{-\pi}^{\pi} \sigma(\phi-\alpha)\, h(\alpha)\, \hat{r}^\mu(\alpha)\, d\alpha$$

and

$$M\sigma(\phi) = \cos\phi\, \sigma(\phi) \quad ,$$

and introducing the new function

(14) $$\sigma_*(\phi) = \sigma(\phi)\,/\,F_\phi$$

we see that (13) is equivalent to

(15) $$(\mathrm{id} - T_o)\,\sigma_* = \sum_{\mu\geq 1} \kappa_\mu \delta^\mu \left(M^\mu T_\mu\, \sigma_* - (T_\mu M^\mu 1)\cdot\sigma_* \right)$$

provided the function in fig.3 can be represented as

$$f(s) = 1 + \sum_{\mu\geq 1} \kappa_\mu s^\mu \quad .$$

In the special linear case (6), with $\kappa_\mu = o$ for $\mu \geq 2$, equation (15) has an exact solution given by

$$\sigma_*(\phi) = \frac{u_*}{2\pi}\left(1 + \delta\, \frac{1 - \psi_{11}}{1 - \psi_1}\cos\phi\right) \quad .$$

Together with (14) it follows that

$$\sigma(\phi) = \frac{u}{2\pi}\left(1 + 2\,\omega\cos\phi + 2\,\zeta_2\cos 2\phi\right)$$

is a solution of the fixed point equation (13), where u is the given constant cell density,

$$\omega = \frac{\delta}{2}\quad \frac{1 - \psi_1\psi_{11}}{1 - \psi_1}\left(1 + \frac{\delta^2}{4}\,\psi_{11}\,\frac{1 - \psi_{11}}{1 - \psi_1}\right)^{-1}$$

and $\zeta_2 = O(\delta^2)$ is a similar rational expression. Moreover, for sufficiently small δ the solution σ is unique (given u).

In the general case (7) this uniqueness statement analogously holds, and with the aid of a Fourier decomposition of equation (15) the solution σ can be determined as a Fourier series:

$$(16) \qquad \sigma(\phi) = \frac{u}{\pi}\left(\frac{1}{2} + \omega \cos\phi + \sum_{\nu \geq 2} \zeta_\nu \cos(\nu\phi)\right)$$

with

$$(17) \qquad \omega = \frac{\delta}{2}\,\frac{1 - \psi_1\psi_{11}}{1 - \psi_1} + \delta^3\tilde{\omega} + O(\delta^5)$$

$$(18) \qquad \zeta_2 = \frac{\delta^2}{4}\left(\psi_{11}\frac{1 - \psi_{11}}{1 - \psi_1} + \frac{\kappa_2}{\kappa_1^2}\,\frac{\psi_{20} - \psi_2\psi_{22}}{1 - \psi_2}\right) + O(\delta^4)$$

$$\zeta_3 = \delta^3\xi + O(\delta^5)$$

and

$$\zeta_\nu = O(\delta^\nu) \quad , \quad \nu \geq 4 \quad ,$$

where $\tilde{\omega}$ and ξ are explicit rational expressions in terms of the coefficients $\psi_{\mu\nu}$ and κ_ν .

Degree of orientation

We note that the so-called mean cosine [15,24] or forward index [2]

$$(19) \qquad \omega = \frac{1}{u}\int_{-\pi}^{\pi} \sigma(\phi)\cos\phi \; d\phi$$

is equivalent to the mean asymptotic chemotropism index describing the expected mean distance, which a cell travels in gradient direction, relative to the total path length, cp. [24] . With the particular model parameters fitting the data in fig.5 we obtain from (17)

$$\omega = 0.84\,\delta + 2.80\,\delta^3 + O(\delta^5)$$

which for $\delta = 0.48$ gives $\omega = 0.71 \pm 0.05$. This theoretical value of the chemotropism index fits the corresponding empirical data [31] .

Furthermore , the same parameters yield a rather good prediction by the theoretical model concerning another experimentally measured quantity namely the fraction of cells moving up the gradient [24,32] :

$$(20) \qquad f_o = \frac{1}{u}\int_{-\frac{\pi}{2}}^{\frac{\pi}{2}} \sigma(\phi)\, d\phi = \frac{1}{2} + \delta\frac{1}{\pi}\,\frac{1 - \psi_1\psi_{11}}{1 - \psi_1} + \delta^3\frac{2}{3\pi}(3\tilde{\omega} - \xi) + O(\delta^5)$$

For the particular parameters above we obtain

$$(21) \qquad f_o = 0.5 + 0.54\,\delta + 1.60\,\delta^3 + O(\delta^5) \; ,$$

where δ is again given by (10) .

In the experiments by ZIGMOND [32] an m-fold gradient corresponds to $|\nabla\rho| / \rho = 2(m-1)/(m+1)$, where the concentration in the middle region is

$$\rho = \frac{m+1}{2m} \rho_h$$

and where ρ_h denotes the higher concentration in one of the wells. Then from (10) we get

$$\delta = \kappa_1 \delta_o c_o \bar{r} \rho b'(\rho) 2\frac{m-1}{m+1} , \qquad (22)$$

so that we can rearrange the data of [32,fig.7] in the following $f_o = f_o(\delta)$ - diagram

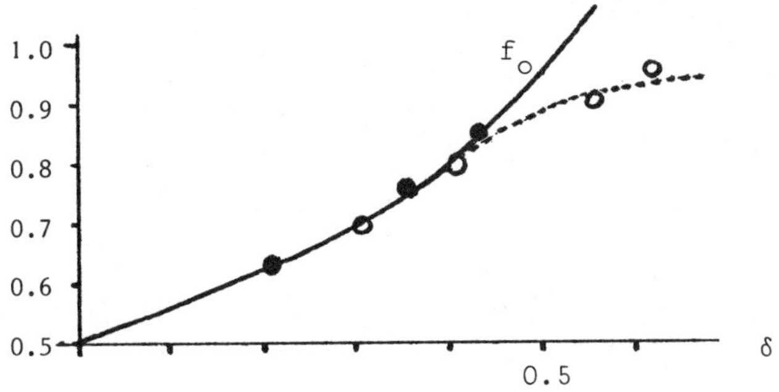

Figure 6. Fraction of cells moving in positive gradient direction, plotted in dependence on the value δ in (22). (● , ○) denote the measured data from [32,fig.7] for $\rho_h = 10^{-6}$M, 10^{-5}M, and for different values of m . The lined curve is calculated from the third order approximation of f_o in (21). The deviation of the empirical (dotted) curve for values $\delta \gtrsim 0.5$ is consistent with the deviation of (7) from f in figure 3 . Note that always $f_o \leq 1$.

Moreover, the whole density distribution measured by ZIGMOND[32,fig.8] can be reproduced by the theoretical solution σ in (16):

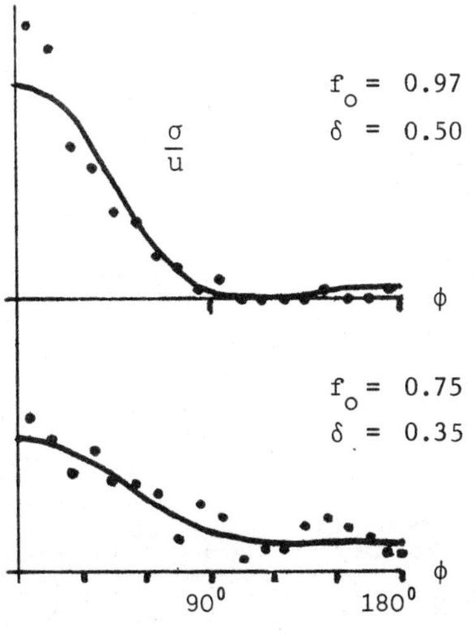

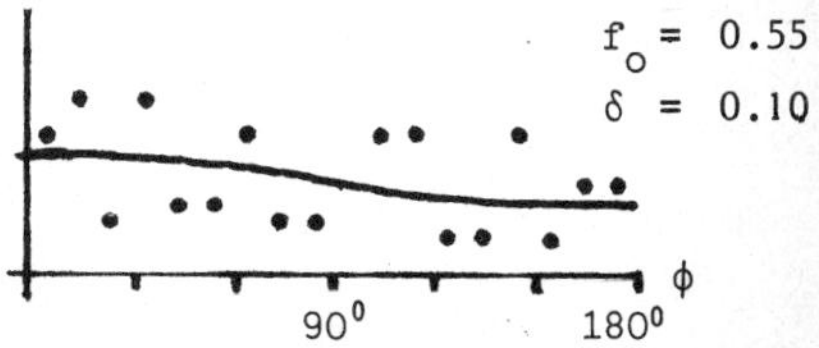

Figure 7. Distribution of cells migrating with an angle ϕ of deviation from the gradient direction (for different strength of the chemotactic response). Curves are calculated from the third order approximation of σ in (16). Dots denote corresponding measured values taken from [32,fig. 8] .

For a weak chemotactic response ($\delta \lesssim 0.3$, "shallow gradient" hypothesis) the data scatter around a $\cos\phi$ - distribution depending linear on $\delta \sim |\nabla\rho|$. However, for stronger responses the cells achieve a more accurate orientation with respect to the gradient [32]. In the theoretical solution this is mainly expressed by a $\cos 2\phi$ - superposition with amplitude ζ_2 proportional to $\delta^2 \sim |\nabla\rho|^2$, see equation (18).

Note that this δ^2- expression does not influence the chemotropism index (17)

$$\omega = \frac{\delta}{2}\left(1 + \psi_1 \frac{1-\psi_{11}}{1-\psi_1}\right) + O(\delta^3)$$

and the orientation factor (20)

$$f_o = \frac{1}{2} + \frac{2}{\pi}\,\omega + O(\delta^3) ,$$

both being linear in $\delta \sim |\nabla\rho|$ modulo $O(\delta^3)$. Therefore in experiments with temperate chemotactic response the "mean double cosine" ζ_2 becomes important for quantifying the accuracy of orientation, which might be measured in terms of the fraction of cells moving in a $(\pm 45^0)$ - sector with respect to the gradient:

$$(23) \qquad f_2 = \frac{1}{u}\int_{-\pi/4}^{\pi/4} \sigma(\phi)\, d\phi = \frac{1}{\pi}\left(\frac{\pi}{4} + \sqrt{2}\,\omega + \zeta_2\right) + O(\delta^3) \quad .$$

Optimal chemotactic response

In order to analyse how the strength of the chemotactic response depend on the receptor distribution r (fig.4) I have investigated the behavior of the two characteristics f_o and f_2 while r is a nearly Gaussian distribution concentrated around some angle α_*, assuming that the basal activity distribution h in fig.1 is given. It results that both values ω and ζ_2 take on a maximum for α_* between 40^0 and 50^0. Thus the receptor distribution r in fig.4 with a fairly concentrated maximum at $\alpha_o \simeq 40^0$ (representing the experimental findings) turns out to be nearly optimal in the sense described above.

A second question is how the chemotactic reponse depends on the extracellular ρ-concentration, while the relative gradient $|\nabla\rho| / \rho$ is held fixed. ZIGMOND [32, fig. 5] plotted the "% orientation" (= $100\, f_o$) of leukocytes in an m-fold gradient as a function of the higher concentration ρ_h. The two curves (m=3 and m=10) show an optimal response for ρ between 10^{-6} and 10^{-5} M. We used [32, fig. 5] together with (21) and (22) to find out suitable receptor kinetics $b = b(\rho)$ fitting these experimental data. Assuming a second order kinetics as in (3) we got the following dissociation constants:

$$K_1 = 1.2\ 10^{-7} \quad \text{and} \quad K_2 = 5.2\ 10^{-6} ,$$

where the transmission rates $\tilde{k}_m$ have to satisfy $\tilde{k}_1/\tilde{k}_2 = 0.3$. This is consistent with the binding kinetics for the tripeptide FMMM determined by SNYDERMAN, PIKE [27 , fig. 2 A] . Moreover, assuming that the length of a typical "sensing" protrusion is about 3 μm we are able to compute the local signal pulse (4) in terms of the mean increase of bounded receptors (producing the intracellular P-substance) on each pseudopod. It results that in cases of a weak chemotactic response ($f_o \approx 0.6$) , for instance, the relative increase is about 1% . This sensitivity is comparable with estimates based on a spatial gradient sensing mechanism [33] and ten times higher than the corresponding value derived by MATO et al.[19] in order to rule out the temporal gradient sensing mechanism by pseudopodal extension, cp. also [14] .

Diffusion approximation

In many experiments observing the chemotactic response of migrating cells only the density of the cell population is measured rather than the orientation of cells or even their locomotory behavior. Examples are the micropore filter and the agarose assays [18 ,22] . Moreover, the function of chemotaxis in vivo usually is cell accumulation [12 ,29]. Now the most common way to describe the time evolution of the cell density is to use diffusion equations [17 ,35]. Here we summarize conditions, which justify such a diffusion approximation, and we briefly show how this procedure preserves the statements about the orientation of cells in a gradient.

Let X be the size of the experimental region and T the duration of of a typical measurement. Then we suppose the following

(24) Parameter conditions:

(a) The mean speed c of cell (nucleus) displacement is relatively large and the mean protrusion speed c_o of "sensing" pseudopods is even larger, more precisely

$$c_o \gtrsim c \gg X/T \quad . \quad \text{Then} \quad \alpha := \frac{X}{cT} \ll 1 .$$

(b) The mean time $1/\beta$ of straight cell displacement is relatively small, namely

$$1/\beta \lesssim \alpha^2 \cdot T$$

implying that the mean dislacement length satisfies

$$c/\beta \lesssim \alpha \cdot X .$$

(c) The mean protrusion length is at most of the same size:

$$\delta_o c_o \;\lesssim\; \alpha \cdot X \ .$$

For simplicity assume that the chemotactic substance (and hence the chemotactic response, too) only depends on the spatial variable x . Then with the aid of singular perturbation theory the following result can be derived (see [3] for a general proof and [2] for some details):

Theorem. Let the parameter conditions (24) be fulfilled with sufficiently small α and with mean displacement time $1/\beta \sim \alpha^2 \cdot T$. Then, modulo an initial layer term, each solution of equation (8) can approximately (in the L^2-mean) be written as

$$\sigma(\cdot,\phi) = \frac{u^o}{\pi}\left(\frac{1}{2} + \alpha\,\omega^o \cos\phi + \alpha^2 \zeta^o \cos 2\phi\right) + \alpha^2 \frac{u^1}{2\pi} + O(\alpha^3) \tag{25}$$

where the approximate cell density u^o satisfies the well-known Patlak-Keller-Segel diffusion equation

$$\partial_t u^o = \partial_x\left(\frac{\mu}{c}\,\partial_x(c\,u^o) - \chi\,u^o\right) \tag{26}$$

with the motility coefficient and the chemotaxis coefficient

$$\mu = \frac{c^2}{2\beta(1-\psi_1)} \qquad\qquad \chi = \frac{\delta c}{2}\,\frac{1-\psi_1\psi_{11}}{1-\psi_1} \ .$$

(Note that $\delta = \kappa_1 \delta_o c_o \mp b'(\rho)\,\partial_x \rho \;\lesssim\; \alpha$.) The α-linear and the α-quadratic expression in (25) are determined by the dimensionless quantities

$$\omega^o = \frac{\delta}{2\alpha}\,\frac{1-\psi_1\psi_{11}}{1-\psi_1} - \frac{1}{2\alpha\beta(1-\psi_1)}\,\frac{1}{u^o}\,\partial_x(c\,u^o) \tag{27}$$

and

$$\zeta^o = \frac{\delta^2}{4\alpha^2}\left(\frac{\kappa_2}{\kappa_1^2}\,\frac{\psi_{20}-\psi_2\psi_{22}}{1-\psi_2} - \psi_{11}^2\right) + \frac{\delta}{2\alpha}\,\psi_{11}\,\omega^o - \frac{1}{2\alpha\beta(1-\psi_2)}\,\frac{1}{u^o}\,\partial_x(c\,u^o\,\omega^o) \ . \tag{28}$$

Furthermore, the "correcting" density u^1 has mean value zero and fulfils a diffusion equation (26) but with an additional term depending on u^o, ω^o, ζ^o and their derivatives.

This result means that experimental assays satisfying the conditions in (24) allow the application of the following mathematical tools:
(i) The diffusion approximation u^o describes the time evolution of the mean expected cell density modulo an error of order α^2 .
(ii) The mean cosine (19) quantifying the degree of orientation in the gradient is, modulo $O(\alpha^3)$, given by $\alpha\,\omega^o$, see (27), where the first term

proportional to the chemotaxis coefficient χ can be diminished in the the presence of a density gradient $\partial_x u^o$. This might be called "diffusive deorientation".

(iii) The mean double cosine related to the accuracy of orientation (23) is approximately given by $\alpha^2 \zeta^o$, see (28), where (with the particular parameters from fig. 5) the first term is positive. Thus ζ^o stays positive even if ω^o vanishes, which can occur if a weak chemotactic response is compensated by diffusive deorientation, meaning that slightly ($\sim\alpha^2$) more cells are migrating up and down the gradient than in the orthogonal direction.

In any case equation (25) implies that the mean expected probability for a cell, moving at time t near (x,y), to migrate with an angle ϕ is

$$2\pi \frac{\sigma(t,x,y,\phi)}{u(t,x,y)} = 1 + 2\alpha\,\omega^o(t,x,y)\cos\phi + 2\alpha^2\zeta^o(t,x,y)\cos 2\phi + O(\alpha^3)$$

(see also figure 7). Here ω^o by (27) and ζ^o by (28) are expressed in terms of the model parameters and of the first and second spatial derivatives of the approximating density $u^o(t,x,y)$ which is a solution of the diffusion equation (26).

References

1. ALBRECHT-BUEHLER, G. and R.M. LANCASTER: J. Cell Biol. 71, 370-382 (1976)
2. ALT, W.: J. Math. Biol. to appear (1980)
3. ALT, W.: Singular Perturbation of Diff.int.equations describing biased random walks (preprint)
4. BERG, H.C.: Ann. Rev. Biophys. Bioeng. 4, 119-136 (1975)
5. BOYARSKY, A. and P.B. NOBLE: Can. J. Physiol. Pharmacol. 55, 1 - 6 (1977)
6. DiPASQUALE, A.: Exp. Cell Res. 94, 191-215 (1975)
7. DUDERSTADT, J.J. and W.R. MARTIN: Transport Theory. Wiley + Sons. New York 1979
8. ENGEL, H.J.; R. SCHÜTZ and E. ZERBST: Die Blutzellen im Vitalpräparat (Film) IWF Göttingen C 851/1962
9. GALLIN, J.I.; E.K. GALLIN; H.L. MALECH and E.B. CRAMER: In Leukocyte Chemotaxis see [10], 123-141
10. GALLIN, J.I. and P.G. QUIE (eds.): Leukocyte Chemotaxis: Methods, Physiology and Clinical Implications. Raven Press. New York 1978
11. GERISCH, G.: Dictyostelium minutum. Aggregation (Film) IWF Göttingen E 673/1964
12. GERISCH, G.: Biol. Cellulaire 32, 61-68 (1978)
13. GERISCH, G. et al.: Phil. Trans. R. Soc. Lond. B 272, 181-192 (1975)
14. GERISCH, G. et al.: In Developments and Differentiation in the Cellular Slime Moulds (eds. CAPPUCCINELLI, ASHWORTH) Elsevier 1977. p.105
15. HALL, R.L. and S.C. PETERSON: Biophys. J. 25, 365-372 (1979)
16. HILL, H.R.: In Leukocyte Chemotaxis [10] , 179-193

17. LAUFFENBURGER, D.: This Proceedings

18. MADERAZO, E.G. and C.L. WORONICK: In Leukocyte Chemotaxis [10] , 43-55

19. MATO, J.M. et al.: Proc. Nat. Acad. Sci. USA 72, 4991-4993 (1975)

20. MATO, J.M. et al.: Proc. Nat. Acad. Sci. USA 74, 2348-2351 (1977)

21. MEINHARDT, H. and A. GIERER: J. Cell Sci. 15, 321-346 (1974)

22. NELSON, R.D. et al.: In Leukocyte Chemotaxis [10] , 25-42

23. NOSSAL, R.: Math. Biosci. 31, 121-129 (1976)

24. NOSSAL, R.: This Proceedings

25. NOSSAL, R. and S.H. ZIGMOND: Biophys. J. 16, 1171-1182 (1976)

26. RAMSEY, W.S.: Exp. Cell Res. 86, 184-187 (1974)

27. SNYDERMAN, R. and M.C. PIKE: In Leukocyte Chemotaxis [10] , 357-378

28. STOSSEL, T.P.: In Leukocyte Chemotaxis [10] , 143-160

29. TILL, G.: In Monogr. Allergy Vol. 12 (eds. KALLOS et al.) Karger. Basel 1977, 169-178

30. WILKINSON. P.C.: In Taxis and Behavior: Elementary Sensory Systems in Biology (ed. HAZELBAUER) Chapman and Hall. London 1978, 293-329

31. ZIGMOND, S.H.: Nature 249, 450-452 (1974)

32. ZIGMOND, S.H.: J. Cell Biol. 75, 606-616 (1977)

33. ZIGMOND, S.H.: J. Cell Biol. 77, 269-287 (1979)

34. ZIGMOND, S.H.: In Leukocyte Chemotaxis [10] , 57-66

35. ZIGMOND, S.H.: In Leukocyte Chemotaxis [10] , 87-96

Theoretical Models for the Specific Adhesion of Cells to Cells or to Surfaces

George I. Bell
Los Alamos Scientific Laboratory
University of California
Los Alamos, New Mexico 87545

Introduction

There are many examples in which the adhesion of cell to cell or of cell to substrate is mediated by bonds between specific molecules. In some cases, the molecules are well understood. Consider, for example, a solution containing cells such as red blood cells which do not ordinarily stick to one another. Suppose that molecules, such as antibody molecules, which have two or more sites for binding to specific determinants on the cell surface are introduced into the solution. A single molecule may be able to bind simultaneously to determinants on two cells and thus a sufficient concentration of such molecules may agglutinate the cells. Such agglutination of red cells by specific antibodies is commonly used for determining blood types. In other experiments the agglutination of various cells by protein molecules called lectins[1] has been analysed. These studies generated considerable excitement about a decade ago when it was found that tumor cells are more readily agglutinated[2] than normal cells.

In these examples, both the bridging molecules and the cell surface determinants are fairly well characterized, while in many systems *in vivo* the adhesive interactions appear to involve specific molecules but are less completely understood. For example, in various developing systems, such as aggregating cells of the cellular slime mold[3] or sponge cells[4] or in the developing retina[5], cell-cell adhesion seems to involve specific molecular interactions but the molecules are not completely characterized. In addition, it appears that the immune system involves cell-cell interactions mediated by antibody molecules, antigen molecules, antigen -antibody complexes, or by the interactions of mutually complementary cell surface receptors. Finally many studies have been made of the adhesion of cells to surfaces where the interaction appears to involve interaction between extracellular molecules such as fibronectin[6] and cell surface determinants. Again such studies have been motivated by findings that tumor cells have altered adhesive properties[6,7] which may account in some essential ways for the invasive and metastatic properties of the tumor cells.

During the past couple of years I have attempted to construct some parts of a theoretical framework for the analysis of cell adhesion mediated by specific molecular interactions. In this paper, I will review some components of this framework and describe some recent experimental and theoretical results.

The Conceptual Model

I have in mind the fluid mosaic model of the cell membrane[8] wherein the membrane is regarded as a phospholipid bilayer in which various integral membrane proteins are retained by virtue of their favorable free energy in the hydrophobic interior of the membrane. Proteins of interest for cell to cell binding have portions which extend beyond the lipid bilayer on the outside of the cell and most are glycoproteins, that is, they include covalently attached sugar moieties. The proteins are more or less free to move translationally in the plane of the membrane and

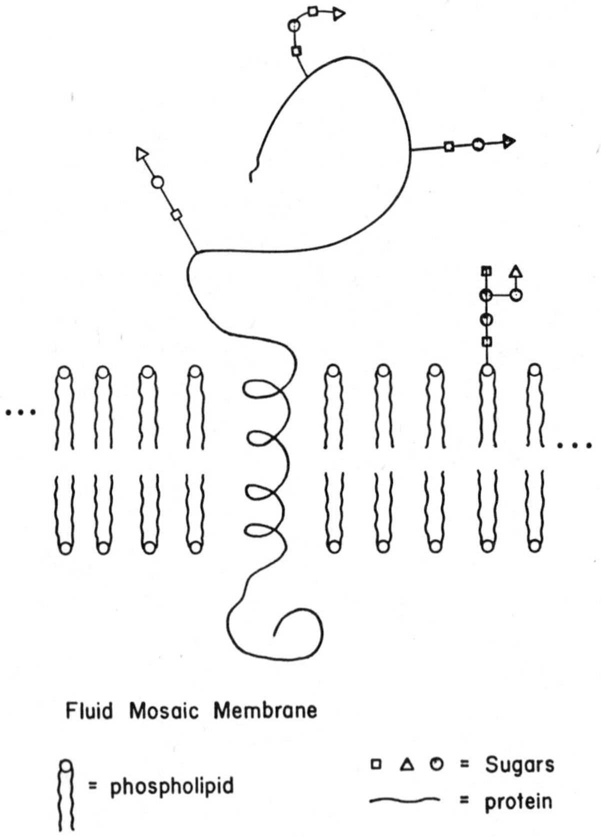

Figure 1. Fluid mosaic model of a cell membrane.

to rotate about an axis perpendicular to the membrane. The translational diffusion coefficients of several integral membrane proteins have been measured and values $\sim 10^{-10}$ cm^2/sec are regarded as typical[9]. However, some surface proteins are virtually immobile[10] while others may move more freely (see following sections). By contrast the diffusion coefficients for typical proteins of molecular weight $\sim 10^5$ in aqueous solutions are $\sim 5 \times 10^{-7}$ cm^2/sec.

For present purposes, let us designate as receptors those cell surface molecules which may mediate the adhesion of cell to cell or to substrate. Thus various membrane proteins or glycolipids might serve as receptors. For example, certain lymphocytes have antibody-like surface molecules which can serve as receptors in binding the cells to complementary soluble antigens or to antigenic determinants on the surfaces of other cells or on substrates. We shall regard such antibody-antigen interaction as a paradigm for the interaction between complementary receptors which may produce cell-cell adhesion.

In addition, soluble ligands including antibodies, antigens, lectins, and antigen-antibody complexes can mediate cell-cell adhesion. In all cases the ligand must have at least two binding sites, one of which interacts with each cell so as to form a molecular bridge between the cells.

Both receptor molecules and ligands may have multiple binding sites and any general theoretical treatment would be quite complex. In order to clarify the physical effects, I shall emphasize the simplest situation, in which the adhesion is produced by interaction between mutually complementary receptors, each with a single binding site. Adhesion produced by bivalent ligands interacting with monovalent receptors is only slightly more complex[11]. The main complication is that the ligands may crosslink receptors on each cell as well as form bridges between cells. Although I will emphasize cell-cell interactions, it should be recognized that most of the results would hold with minor modifications for cell-substrate interactions and that some would apply to the interaction of a cell with itself, in which, for example a cell protrusion sticks to some portion of the cell surface.

Although it is possible for cells with immobile receptors to stick to each other, this generally requires a high density of surface receptors on each cell. The problem has been treated by Chak and Hart[12].

Rate and Extent of Intercellular Bond Formation

Assume that two cells with complementary mobile receptors come into contact with each other. Unless there are very many receptors per unit area on say the second cell, it is unlikely that a particular receptor on the first cell will immediately find a complementary receptor on the second cell so positioned and oriented as to permit immediate binding. However insofar as the receptors are mobile they can diffuse about until by chance they achieve positions and orientations which do permit bond formation. A theory has been developed for predicting such rates of bond formation assuming that the receptors are diffusing on two dimensional surfaces with known diffusion coefficients[13,14]. In addition, something must be known about the intrinsic rate constants for reaction of the receptors. This information can be obtained from measured reaction rates in solution, if available.

If $N_{1f}(x,t)$ and $N_{2f}(x,t)$ are the numbers of free receptors per unit area on the two cells at position x and time t, $N_b(x,t)$ is the number of intercellular bonds, and k_+ and k_- are the forward and reverse rate constants for intercellular bond formation, we may write

$$\frac{dN_b}{dt} = k_+ N_{1f} N_{2f} - k_- N_b \tag{1}$$

According to the theory[13,14] the rate constants may be calculated from the receptor diffusion constants, D_1 and D_2, which are measurable, and solution reaction rates. In particular

$$k_+ = 2\pi E(D_1 + D_2) \quad , \tag{2}$$

where E is a factor (< 1) by which the reaction rate falls short of the diffusion limit. Moreover the equilibrium constant, $K^m \equiv k_+/k_-$ for membrane bound reactants may be related to the equilibrium constant K^s for reactants in solution by

$$K^m = \frac{K^s}{R} \tag{3}$$

where R is some distance ∿ 10 - 100 Å, relative to a lipid bilayer, within which the reactants may be localized.

From these equations we can deduce some important conclusions. First of all, the rate of bond formation can be very large. Consider for example, cells such as small lymphocytes having $\sim 10^5$ receptors on their surfaces, which is a representative value for the number of antibody like molecules. These cells have radii ∿4μm and hence area $\sim 200\mu m^2$ and thus ∿500 receptors per μm^2. When these cells first come into contact, $N_b = 0$ and $N_{if} \cong N_i$, the total number of receptors per unit area. If we take[9] $D_1 = D_2 = 10^{-10}$ cm^2/sec and $E = 0.1$ then $k_+ \cong 10^{-10} cm^2/sec = 10^{-2} \mu m^2/sec$ and hence from equation (1) $dN_b/dt \cong 2.5 \times 10^3/\mu m^2 sec$. Thus if these two cells are in contact over an area $\sim 1\mu^2$ for a few msec, they can establish ∿10 bonds, which it will be seen, may suffice to provide rather tight binding.

After the cells have been in contact for a while, $N_b(x,t)$ will increase at positions in the contact area. Bond formation will reduce N_{fi} locally but further free receptors will diffuse into the contact area. Indeed, if the contact area is only a suitably small fraction of the cell surface, and if the diffusion of free receptors is not impeded by the presence of bound receptors, then[15] at equilibrium $N_{fi} \cong N_i$ so that the number of bonds per unit area is, from equations (1) and (3)

$$N_b \cong K^m N_1 N_2 = \frac{K^s}{R} N_1 N_2 \qquad (4)$$

This means that the number of bonds per unit area can greatly excede the initial number of receptors per unit area. Indeed this can happen even for moderate values of K^s because of the very high local concentrations of receptors adjacent to the cell surface. Suppose, for example, that $K^s = 10^6\ M^{-1} = 1.7\times10^{-15} cm^3$/molecule and R = 20Å. Then $K^m \cong 10^{-8} cm^2 = 1 \mu m^2$ so that for $N_i = 500/\mu m^2$, $K^m N_i = 500$. This means that these parameters could lead to a five hundred fold concentration of receptors in the contact area, i.e. $N_b \cong 500\ N_j$, where i=1,2 and j=2,1.

Another way of interpreting these results is to observe that N_i/R can be interpreted as the local concentration of receptors adjacent to the surface. The above parameters give $N_i/R = 500\times10^8/2\times10^{-7} = 2.5\times10^{17}$ molecules/cm³ = 0.4×10^{-3} M, a remarkably high concentration of specific biological macromolecules.

Note that equation (4) gives a criterion for receptor redistribution. If $K^m N_1 \gg 1$, then $N_b \gg N_2$ so that the receptors on the second cell will accumulate in the contact area. Similarly if $K^m N_2 \gg 1$, then $N_b \gg N_1$ so that receptors will accumulate in the contact area on the first cell. The time for receptor redistribution is of the order of r^2/D where, r is the radius of the contact area[15], which is in the range of a few seconds to a few minutes for 0.1 μm < r < 1 μm and $D \cong 10^{-10}$ cm²/sec.

It should be noted that the foregoing model neglects biological complications which may be essential in many cases. Obviously, we have assumed that the receptors are mutually accessible to each other. This implies that the two cells must be able to get close enough together and that there must not be intervening macromolecules masking the receptors from each other. It is predicted[16,17] that non-specific electrical forces will permit the lipid bilayers to approach to within 50-100 Å of each other and indeed will favor such equilibrium separations. On the other hand it appears that at least in vivo cells are often separated by such molecules as collagen, fibronectin, or mucopolysaccharides[7]. In such cases, the adhesion of cells to molecules of the extracellular matrix is important, while opportunities for direct contact between integral membrane glycoproteins may be minimal unless the matrix is disrupted.

Another problem concerns the mobility of cell receptors. For one thing, in photobleaching experiments[9] a fraction of the cell surface molecules often appear to be immobile. There are various explanations for this apparently immobile fraction, but one possibility might be that there are local variations of receptor mobility over the cell surface such that, for example, receptors on cell protrusions such as microvilli are relatively immobile. This is believed to be the case for intestinal epithelial cells. In addition, there is evidence that the adhesion of cells to objects can reduce the mobility of receptors on the whole cell surface[18], presumably by modulating the linkage of receptors to the cytoskeleton, and that cytoskeletal connections accumulate in regions of cell-cell contact[19].

Thus under various circumstances cells can modulate the mobility of their receptors. In particular since adhesion per se may lead to mobility changes and indeed to receptor phagocytosis (eating) or excytosis (shedding), I suggest that the foregoing model may represent only the early stages of cell-cell binding and receptor redistribution. The sequelae may be complex and biologically important but are outside the scope of the model.

Strength of Specific Bonds

It is of interest to consider how effective specific bonds are in holding two cells together, or a cell to a substrate, in opposition to hydrodynamic or other forces. To this end I have estimated the force which is required in order to rapidly break a typical antigen antibody bond[13]. In general, bonding may be viewed as due to some free energy minimum, of depth $-E_o$, relative to well separated reactants, and of range r_o in some reaction coordinate. The force which is required in order to eliminate bonding is of the order of $F_o = E_o/r_o$, which for a typical antigen-antibody bond is around 1.2×10^{-5} dynes.

Of course each bond is spontaneously reversible and no force at all is required in order to break it if one is willing to wait long enough. But if two cells are stuck together by many bonds there is virtually no chance that they will all be broken at one time, unless a force is applied so as to stress the bonds (or unless the cells do something active to terminate binding). Under these circumstances, a bond lifetime τ, is expected to depend exponentially on the force per bond, f,

$$\tau \cong \tau_o \exp[(E_o - r_o f)/kT] \qquad (5)$$

where τ_o is some natural bond frequency. τ_o can be estimated from E_o and $\tau(f=0)$, which is the reciprocal of the reverse rate constant for bond formation. In equation 5, T is the absolute temperature and k is Boltzmann's constant.

Using this approach, I have estimated[13] that a force $\cong 4\times10^{-6}$ dynes/ bond ($\cong$1/3 F_o) will suffice for separating the cells. These estimates were made for particular bond parameters, E_o = 8.5 kcal/mole, r_o = 5Å, and for $K^m N_2 = 10^3$ and could vary by a factor two or more for other bond parameters of interest.

This force may be compared with other forces to which a cell may be subject. First of all there are non-specific electrical forces between cells because they are charged and polarizable objects. It has been predicted[16,17] that the long range (van der Waals) forces are attractive and that a force $\sim 10^{-5}$ dynes μm^{-2} is required to separate two cells from this attraction. Note that this is equivalent to about 2 of our specific bonds/μm^2. Since much larger receptor densities are expected, it is clear that specific bonds can be much stronger than the non specific attractive electrical forces.

The force required to hold a cell in a fluid stream can be estimated by Stokes law for laminar flow around a sphere. The result is[13], for a sphere of radius 4μm that ∿13v bonds will suffice to resist a flow of v cm/sec. If, for example we are considering the adhesion of a lymphocyte to an endothelial cell in the venule of a lymph node[20] where v ≅ 0.3 cm/sec, this tells us that around 4 bonds should suffice to attach the cell.

Such estimates assume that all of the bonds are equally stressed. This may be a reasonable approximation for attachment of a lymphocyte to an endothelial cell, where a single microvillus on the lymphocyte appears to stick to a local pit on the endothelial cell[21], but it greatly overestimates the strength of attachment in other cases. For example, experiments have been performed[22] with cells which have been permitted to adhere to the surface of a circular disc. The disc is then spun in a fluid at an angular velocity such that cells near the periphery are stripped off while those near the axis are unperturbed. At some intermediate radius the cells are just barely removed.

The fluid flow near the rotating disc is nearly laminar and analysis shows[23] that the drag force or stress per unit area on the disc is

$$F_D = 0.8\ \mu\ r\ \omega^{3/2}\ \nu^{-1/2} \qquad (6)$$

where μ is the fluid viscosity (in dyne-seconds per cm²), r is the radius under consideration, ω is the angular velocity and ν is μ/ρ with ρ the fluid density. In water μ = ν = 0.01 so that

$$F_D = 0.08\ r\omega^{3/2}\ \text{dynes/cm}^2. \qquad (7)$$

This force acts on the disc parallel to its surface and if there is a flattened cell on the surface this stress will tend to remove the cell from the surface. At some radius r there will be a critical sheer stress F_{DC} which is just sufficient to remove the cells. The force on a cell of area A, prior to removal is then $F_{DC}A$ which is nearly balanced by the stressing of bonds which hold the cell to the surface. In an example, a critical stress of ∿50 dynes/cm² was measured[22] and if the cell area were ∿200μm² = $2\text{x}10^{-6}$cm² the force per cell would be $\sim10^{-4}$ dynes. This could be sustained by ∿25 of our typical bonds.

In this example, the force is parallel to the surface and many of the cell-substrate bonds may be ineffective in opposing such motion. The only ones that can be clearly effective are those near the trailing edge of the cell which must break in order to permit cell motion. If further bonds were attached to the cell cyto-skeleton they might also impede cell motion but the facts are not clear. It is observed that cells subject to near the critical stress are strongly deformed but details of their attachment are unclear. Nevertheless this example illustrates that as in all theories of adhesion, it is not straight-forward to predict macroscopic yield stresses from theoretical microscopic bond strengths.

In our estimate of bond strength, it was assumed that as a receptor-receptor bond is stressed, this will be the weakest link in the chain. It is easy to see that covalent bonds in a receptor are much stronger for they have about tenfold larger values of E_o and threefold smaller values of r_o. However it is less clear that the receptor will not pull out of the lipid bilayer. For a particular integral membrane protein, glycophorin, I have estimated[13] that the force required to uproot this receptor is about $1.0x10^{-5}$ dynes. For a ganglioside (lipid molecule with attached sugars) the corresponding force was estimated to be near $5x10^{-6}$ dynes. It thus appears that the competition between receptor uprooting and bond breaking will depend on the precise nature of the receptor and strength of the bond.

Some Recent Experiments

The foregoing theoretical framework may be used to design and interpret a wide range of experiments. Consider, for example a surface which has been coated with some ligand to which certain cell receptors can bind. The cell can attach to the surface by means of receptor-ligand interactions and receptors on the top surface may, if still mobile, diffuse to the bottom surface and stick there. This motion has been exploited by S. Silverstein and colleagues[24] to measure the diffusion constant of Fc receptors on macrophages. These receptors are protein molecules which bind to the Fc or stem portion of certain classes of antibodies. The experiment is performed such that the cells are first allowed to spread on the surface which is then coated with antibodies in the cold. The cells are then warmed and a rapid disappearance of Fc receptors from the top surface is observed while other receptors on top do not disappear. From the observed rate of disappearance and assuming that receptors which diffuse off the top never return, it is possible to deduce for the receptors in the cell membrane, a diffusion coefficient $\geq 10^{-9} cm^2/sec$. This is a much larger value of D than expected from photobleaching experiments on other receptors and other cells. In the photobleaching experiments, fluorescent ligands are first bound to the receptors and then bleached by a pulse of laser light. The question is thus raised whether in such experiments the mobility of the receptor may be reduced by either ligand attachment or the light pulse. In this context it is of interest to note[25] that hybrid antibodies, one arm of which binds to a receptor and the other arm to ferritin or to a virus, are capable of inducing receptor clustering, a reaction normally associated with the crosslinking of receptors by bivalent ligands. These results raise the possiblity that fluorescein labeled ligands might cause unexpected receptor-receptor and/or receptor-cytoskeleton interactions and reduce receptor mobilities.

A quite different set of experiments[26] has concerned the fusion of intracellular granules with the cell membrane, which is the common mechanism whereby cells secrete hormones, neurotransmitters, and other specialized biochemicals. These granules are lipid bilayers, with imbedded proteins, inside of which are the molecules to be secreted. The secretory event seems to be triggered by a transient increase in the concentration of cations, usually Ca^{++}, in the vicinity of the granule which facilitates adhesion of the granule to the cell wall. The two lipid bilayers then break down in the vicinity of the adhesive contact allowing the contents of the granule to be released outside the cell. All of these events can take place in a fraction of a second.

New light may have been shed on these secretory processes by studying the adhesion between chromaffin granules from cells of the adrenal glands[26]. It was found that in the presence of sufficient cations, which are presumably required to neutralize charges on the granule surfaces, the granules will stick together on nearly every collision. If however, the proteins are removed, the sticking probabilities are reduced by around two orders of magnitude. The interpretation[27] is that protein-protein interactions facilitate efficient sticking.

After two granules are firmly stuck together, freeze-fracture electron microscopy suggests that proteins are systematically excluded from the contact area. It has been suggested[27] that this is because certain lipids become concentrated in the contact area. That is, the protein mediated contact facilitates a phase separation of lipids which in turn excludes proteins, save around the periphery of the contact area.

This example suggests that some of our model considerations may be applicable to rather different and important biological situations. It also points out that other phenomena such as, in this case, the requirement for cations and the phase separation of lipids may be involved.

I have sketched elsewhere[15] some of the applications which may be expected when adhesion involves two different kinds of interacting receptors on each cell. These considerations were developed as a possible model for the genetic restriction of immune responses, but more generally, I suggest that this example and those cited above show that diverse and complex interactions may be expected to result from cell-cell contact mediated by specific receptors and/or ligands.

References:

1. Nicolson, G.L., Int. Rev. Cytol. 39, 89 (1974).

2. Burger, M., Proc. Nat. Acad. Sci. USA 62, 994 (1969).

3. Barondes, S.H. and Rosen, S.D. in Neuronal Recognition, ed. Barondes, S.H., Plenum Press, NY, (1976).

4. Burger, M.M., Turner, R.S., Kuhns, W.J. and Weinbaum, G., Phil. Trans. R. Soc. Lond. B 271, 379 (1975).

5. Rutishauser, U., Thiery, J.-P., Brackenbury, R., Sela, B.-A., and Edelman, G., Proc. Nat. Acad. Sci. USA 73, 577 (1976).

6. Grinnell, F., Int. Rev. of Cytology, 53, 65 (1978).

7. Surfaces of Normal and Malignant Cells, ed. R. Hynes, Wyley Int. in press.

8. Singer, S.J. and Nicolson, G.L. Science 175, 720 (1972).

9. Webb, W.W., Frontiers of Biol. Energetics 2, 1333 (1978).

10. Schlessinger, J., Barak, L.S., Hammer, G.G., Yamada, K.M., Pastan, I., Webb, W.W. and Elson, E.L. Proc. Nat. Acad. Sci. USA 74, 2909 (1977).

11. Bell, G.I., Cell Biophysics, 1, (1979).

12. Chak, K.C., and Hart, H., Bull. Math. Biol., in press.

13. Bell, G.I., Science 200, 618 (1978).

14. Dembo, M., Goldstein, B., Sobotka, A.K. and Lichtenstein, L.M., J. Immunol. 122, 518 (1979).

15. Bell, G.I. in Physical Chemical Aspects of Cell Surface Events in Cellular Recognition, eds. DeLisi, C. and Blumenthal, R., Elsevier/North Holland, NY (1979).

16. Parsegian, V.A., Ann. Rev. Biophys. Bioeng. 2, 221 (1973).

17. Nir, S. and Anderson, M., J. Memb. Biol. 31, 1 (1977).

18. Edelman, G.M., Science 192, 218 (1976).

19. Bourguignon, L.Y.W., Hyman, R., Trowbridge, I. and Singer, S.J., Proc. Nat. Acad Sci. USA 75, 2406 (1978).

20. de Sousa, M. in Receptors and Recognition, A2, Cuatracasas, P. and Greaves, M.F. eds., Chapman and Hall, London, pp 105-163 (1976).

21. Anderson, A.O. and Anderson, N.D., Immunology 31, 731 (1976).

22. Smith, L., Univ. Utah, Bioeng. Dept., private comm.

23. Levich, V.G., Physiochemical Hydrodynamics, Prentice Hall, Inc. (1962).

24. Silverstein, S., Rockefeller Univ., Manuscript in prep.

25. Stackpole, C.W., DeMilo, L.T., Hammerling, U., Jacobson, J.B. and Lardis, M.P., Proc. Nat. Acad. Sci. USA 71, 932 (1974).

26. Morris, S.J., Chiu, V.C.K., and Haynes, D.H., Memb. Biochem. 2, 163 (1979).

27. Haynes, D.H., Kolber, M.A. and Morris, S.J., J. Theor. Biol., in press.

CHEMOTAXIS IN BACTERIA: A BEGINNER'S GUIDE TO THE LITERATURE

Howard C. Berg

Department of Molecular, Cellular and Developmental Biology
University of Colorado, Boulder, Colorado 80309, USA

Chemotaxis is a behavioral process that enables an organism to move toward regions in its environment that are chemically favorable or to avoid those that are not. The process involves data acquisition (chemoreception), data reduction (signalling, adaptation), and output (changes in locomotion). How are these steps carried out? The organism of choice for studies at the molecular level has proved to be the bacterium *Escherichia coli* (or its close relative *Salmonella typhimurium*). These are unicellular organisms of microscopic size (rods about 10^{-4} cm in diameter by 2 x 10^{-4} cm long). They possess specific receptors for a variety of sugars and amino acids. They swim by rotating thin helical filaments (flagella) that project into the external medium. The filaments spin alternately clockwise and counterclockwise. The direction of rotation depends, in part, on the way in which the occupancy of the receptors changes with time. Relatively little is known about the coupling between the receptors and the flagella. Here is a beginner's guide to the literature:

For the lay reader:

Adler, J. (1976) "The sensing of chemicals by bacteria," Sci. Am. 234(4), 40-47.
Berg, H. C. (1975) "How bacteria swim," Sci. Am. 233(2), 36-44.

Reviews with an emphasis on chemoreception:

Adler, J. (1975) "Chemotaxis in bacteria," Ann. Rev. Biochem. 44, 341-356.
Berg, H. C. (1975) "Chemotaxis in bacteria," Ann. Rev. Biophys. Bioeng. 4, 119-136.
Hazelbauer, G. L. & Parkinson, J. S. (1977) "Bacterial chemotaxis," in Receptors and Recognition, Ser. B, Vol. 3, Microbial Interactions, ed. by Reissig, J. L. (Chapman & Hall, London) pp. 59-98.
Koshland, D. E. Jr. (1977) "A response regulator model in a simple sensory system," Science 196, 1055-1063.
Parkinson, J. S. (1977) "Behavioral genetics in bacteria," Ann. Rev. Genet. 11, 397-414.

Reviews with an emphasis on flagellar function:

Berg, H. C. (1975) "Bacterial behavior," Nature 254, 389-392.
Iino, T. (1977) "Genetics of structure and function of bacterial flagella," Ann. Rev. Genet. 11, 161-182.
Macnab, R. M. (1978) "Bacterial motility and chemotaxis: The molecular biology of a behavioral system," CRC Crit. Rev. Biochem. 5, 291-341.

Silverman, M. & Simon, M. (1977) "Bacterial flagella," Ann. Rev. Microbiol. 31, 397-419.

Papers that deal with the physics of chemoreception:

Berg, H. C. & Purcell, E. M. (1977) "Physics of chemoreception," Biophys. J. 20, 193-219.

Papers that deal with the physics of flagellar function:

Brennen, C. & Winet, H. (1977) "Fluid mechanics of propulsion by cilia and flagella," Ann. Rev. Fluid Mech. 9, 339-398.

Purcell, E. M. (1977) "Life at low Reynolds number," Am. J. Phys. 45, 3-11.

ASSESSING THE KELLER-SEGEL MODEL: HOW HAS IT FARED?

E.F. Keller
Visiting Fellow, Program in Science, Technology and Society
Massachusetts Institute of Technology
Cambridge, Massachusetts 02139 USA
(On leave from State University of New York at Purchase
Purchase, New York 10577)

Nine years ago, Lee Segel and I (1970) introduced an equation to describe the macroscopic motion of motile, chemotactic organisms, namely:

$$b_t = \nabla \cdot (\mu(s)\nabla b) - \nabla \cdot (\chi(s)b\nabla s)$$

where b(x,t) is the density of organisms, s(x,t) of substrate, $\mu(s)$ represents the motility and $\chi(s)$ the chemotactic sensitivity of the organisms. Wishing to apply this equation to the migrating bands of chemotactic bacteria which had been studied by Adler (1966), we added a second equation to describe the diffusion and consumption of substrate (Keller and Segel, 1971):

$$s_t = D\nabla^2 s - k(s)b$$

and looked for a solution to these equations which would mimic the behavior which Adler had observed. Accordingly, we sought one dimensional traveling waves, reducing the equation to a pair of ordinary differential equations:

$$cs' = k(s)b - Ds'', \text{ and } cb' = (\chi(s)bs')' - (\mu(s)b')'$$

where the prime denotes differentiation with respect to $\xi = x - ct$, the wave variable. To solve these, a number of important assumptions were made regarding the form of the functions $\mu(s)$, $\chi(s)$, and k(s). Although a certain amount of biological support was available, these assumptions were made primarily on the grounds of simplicity. They worked well, and together with the very rough estimates we made for the relevant parameters, again on the basis of somewhat limited biological data, they gave a reasonably good fit with the shape of the bands that Adler had observed. Our hope was that, with better experimental data, our rather rough guesses could be replaced by more informed choices, and that the model could thereby be improved. Ultimately, we hoped that such a macroscopic picture, supported by sufficiently careful measurements, could be used to shed light on the kinds of microscopic properties in which biologists are generally interested. Since that time, numerous elaborations of this model—both experimental and theoretical—have appeared. These elaborations make it possible to look back to try to assess the success of the model and the legitimacy of its simplifying assumptions. They also make it possible, and even appropriate, to raise some questions about the nature of model building. In the history of this very simple model, I think it is possible to see some important lessons of more general import, and it is for this reason that I welcomed the opportunity to review its progress.

The basic assumptions which we made in 1971 were threefold:

$$\mu(s) = \mu_o, \tag{i}$$
$$k(s) = k_o, \text{ and} \tag{ii}$$
$$\chi(s) = \delta s^{-1}. \tag{iii}$$

The first assumption was made in the virtual absence of information. The value of the constant μ_o was taken from some rough early measurements of Adler and Dahl (1967), indicating an order of magnitude greater than the diffusion constants for substrates used. On this basis, the diffusion of the substrate was then ignored. We pointed out that the dependence of motility on substrate concentration is of considerable importance not only to the model itself, but also to any attempt to model the chemotactic mechanism. Unfortunately, this property remains relatively unexplored. Subsequent estimates of the magnitude of the motility coefficient have, however, been revised downward (Lovely and Dahlquist, 1975; Holz and Chen, 1978 and 1979), thus calling into question the legitimacy of neglecting substrate diffusion.

Our second vital assumption was that k(s) is constant. Again, this assumption was made largely on the grounds of simplicity, although Adler had published some very limited data supporting the expectation that, at least for large values of substrate, the rate of consumption would not be substrate-limited. However, it is clear that this assumption cannot hold as s goes to zero (as it appears to do in the rear of the band), and that furthermore, the shape of the function k(s) for small s would critically affect the equations. An obvious guess might be $k(s) \propto \frac{s}{K + s}$, but since the measurement of k(s) seemed to be relatively straightforward, we had hoped someone would do it. Unfortunately, the available data relating to this question is essentially the same now as it was then, causing, as I will discuss later, a serious insufficiency in the theory.

Finally, the chemotactic sensitivity, perhaps the most crucial function of the model, was assumed to be of the form $\chi(s) = \delta s^{-1}$, where the constant δ was to be inferred from the shape of the band. This choice was dictated by several considerations. First, and perhaps most important, was the observation that, given our other assumptions, in particular the assumption that μ = constant, it was necessary that $\chi(s)$ have a singularity at least of order one if the equations were to generate a traveling wave. Simplicity and the tradition of the Weber-Fechner law dictated that we take this singularity to be of order one. With these assumptions, we were able to obtain a reasonably good fit to the data. Subsequently, the experiments of Dahlquist, Lovely and Koshland (1972) appeared to provide some confirmation for this assumption (see Segel and Jackson, 1973), and more recently, Lauffenbarger and Keller (1979) have concluded that data on the swarming of myxobacteria is also consistent with this assumption. Of course, such support can hardly be regarded as definitive, and the more direct data of Brown and Berg (1974) tends to support the much more biochemically plausible form:

$$\chi(s) = \frac{\delta k}{(k + s)^2}.$$

As has been frequently pointed out, this form can be approximated by $\chi \sim s^{-1}$ in the range where $s \sim k$, but, as $s \to 0$, $\chi(s)$ approaches a constant. Here we have a dilemma. The mathematical analysis unambiguously states that, in order to have a traveling wave solution, at least given the other assumptions of the model, $\chi(s)$ must be singular. Well, suppose we relax the demand for exact traveling waves? Thus far, attempts to produce numerical solutions which approximate Adler's bands, using the more realistic form of χ, have failed.* This has led some researchers, e.g., Lapidus and Schiller (1976) to include a growth term in the equations, others to ignore the difficulty, and still others to explore the implications of modifying the other assumptions.

Among those who have pursued our original model, in spite of the uncertainties and difficulties, are Holz and Chen (1978 and 1979). They have provided the most impressive attempt to date to test this model. With the use of laser light scattering they have been able to record the profile of the migrating bands as a function of time, with considerably higher accuracy than had been previously available. On the whole, their results indicate an agreement between theory and experiment which is quite gratifying, but, as should be evident by now, a number of questions arise in that comparison. The results from their

---

*The numerical calculations of Scribner, Segel and Rogers (1974) are not here considered a counter instance in view of the fact that in those calculations $\chi(s)$ was modified only for extremely low values of s, a range of s in which the uncertainties in the computational procedures were considerable.

first paper are summarized in their Figure 5, where the profiles of the band at a sequence of times (ranging from one hour to approximately five hours after band formation) are superimposed and compared with a set of band shapes predicted by our model with varying values of $\bar{\delta} = \frac{\delta}{\mu}$. The first three profiles are closely clustered, and seem to correspond quite well to the theoretical curve for $\bar{\delta} = 1.33$. The fourth profile indicates a deterioration of the band, and indeed, by the time of the fifth profile, there is hardly a band left. The reasons for this deterioration are unclear; they might reflect damage to the bacteria inflicted by the technology used. Holz and Chen concentrate on the first three profiles to obtain numerical estimates of the relevant parameters. They find that fitting our model to their data requires a reassessment of the value of μ, closer to that obtained by Lovely and Dahlquist. However, they note that the theoretical band speed remains too high by a factor of 50%—a discrepancy, they conclude, worthy of some concern. We should note, though, that the calculation of band speed, c, depends critically on the value of k, (c is proportional to k), and that k is itself in need of more careful measurement. Furthermore, in their subsequent paper, in which the technology was considerably refined to permit a much more rapid scanning, and hence much faster measurements of the position, velocity and shape of the band, they find that although the experimental measurements of c remain locally constant, they vary quite widely over the course of the experiment. In their second paper, their analysis includes the effects of diffusion, now necessary because of the reduced value of μ (at least when oxygen is the substrate). The authors conclude that the agreement with experiment is much improved by inclusion of D in the equations. The disparity between experimental and theoretical values of c is, however, not resolved, nor is the variation of c over the course of the experiment. Since neither the possibility of initial growth nor of subsequent damage, even death, of the bacteria has been precluded, a serious difficulty with the entire experiment is suggested. Not only would a variation in μ over time explain the variation in c, but it would also undermine the applicability of our model in the first place.

The questions which arise at this point are twofold. We have here a situation in which some very sophisticated technology is put to work to produce measurements of extremely high accuracy in order to test a model which has built into it a number of very qualitative features. The first kind of question we need to ask ourselves pertains to the disparity between high- and low-precision data used in the same equations. If I suggest that, given that disparity, the precision of these experiments may be inappropriate to the model, then we are left with the following quite general question: How can we estimate the uncertainty implicit in the biological unknowns of a system in order to determine the level of precision with which it is appropriate to assess a model? Or, in other words, how can we judge how literally it is appropriate to interpret the predictions of our models?

My second set of questions has to do with a more fundamental issue. Even if the experimental data were of uniformly high precision, and provided an ideal fit with the theoretical curves, in what sense, if any, could we then say the model had been validated? To what extent could such a fit provide confirmation of the basic assumptions of the model? To answer this question, it is instructive to look at the more general formulation of the model which Odell and I analyzed in 1975 (Keller and Odell, 1975). In that paper, we solved the original pair of equations without the restrictive assumptions which Segel and I had made on the forms of the functions μ, χ, and $k(s)$. We then derived the necessary and sufficient conditions for the existence of a traveling wave solution, continuing, however, to neglect diffusion. Taking $k(s) = k_0 s^{\alpha}$, $\mu(s) = \mu_0 s^r$, and $\chi(s) = \delta s^{-p}$, (for values of s approaching zero), we found the necessary and sufficient conditions to be:

$$\alpha < 1, \qquad \text{(i)}$$

$$\alpha < p, \text{ and} \qquad \text{(ii)}$$

$$p + r > 1. \qquad \text{(iii)}$$

Our hope was that, with such a general formulation, we might be able, on the one hand, to relax the unphysical demand for a singularity in χ, and, on the other hand, to infer the properties of these functions from experimental data such as that of Holz and Chen.

With regard to the first point, we can see from the necessary and sufficient conditions cited above that, in principle, we might approach a traveling wave with a non-singular $\chi(s)$ provided that we could take $\mu(s) = \mu_0 s$. Under these conditions, $k(s)$ would have to be constant. The point here is only a suggestive one. It suggests that, if the model is basically correct, the bacteria might approach a traveling band with a strong (probably temporary) dependence of μ on s, in lieu of a singularity in χ, I suggest a temporary dependence because the existing evidence indicates that $\mu(s)$ does <u>not</u> go to zero under conditions where adaptation has occurred. Clearly, this suggestion needs to be pursued both experimentally and numerically. Unfortunately, in the search for numerical solutions, there still remains a large parameter space to play with and such solutions remain to be exhibited.

I can, however, report some recent calculations of Odell's which bear directly on the second point. These calculations not only relate to the methodological questions I raise, they also make some theoretical points in their own right. What Odell has done has been to exhibit the solutions to the initial value problem, as a function of time, as well as the time-independent traveling wave solutions, for a variety of choices of parameters α, r, and p. For simplicity, the equations were first rewritten in dimensionless form, thereby necessitating a rescaling before any quantitative comparison can be made with the data. Such rescaling would generate the absolute values of the parameters that best fit the data. In all cases, the value of the band speed in the traveling wave solution is dictated by the solution to the initial value problem, and Odell's initial conditions were chosen to expedite the calculations.

Two such calculations are illustrative of the point I wish to make. In both of these, alternative sets of assumptions (i.e., alternative to the ones originally made by Segel and myself) were found to yield traveling waves which could be fit to the data of Holz and Chen at least as well as the solutions to the original equations. In the first of these two cases, the functions were chosen to be:

$$k(s) = \sqrt{s}\,, \tag{i}$$

$$\mu(s) = \sqrt{s}\,, \tag{ii}$$

$$\chi(s) = s^{-1}, \tag{iii}$$

(yielding $\bar{\delta} = \frac{\delta}{\mu_0} = 1$). The absolute values of the parameters, as determined by comparison with the data, are in the process of being computed. Two features of this solution which are not dependent on scaling are worth noting. They are:

(1) The band is now symmetric for $\bar{\delta} = 1$, whereas for $k(s) = k$, $\mu(s) = 1$, and $\chi(s) \propto s^{-1}$, symmetry is achieved only for $\bar{\delta} = 2$; and

(2) The band takes longer to reach its steady state form (roughly three hours in real time), during which initial period it steadily attenuates.

Comparison with data of Holz and Chen suggests, therefore, another possible interpretation of the attenuation they observed. But the central moral of this comparison lies elsewhere. It would appear that, given the wide range of alternatives available in the form and value of the functions $\mu(s)$, $k(s)$, and $\chi(s)$, that measurements of band shape, no matter now precise, may be insufficient to distinguish between these alternatives. Independent determination of at least two of the functions appears to be necessary.

Similar calculations of Odell's make a different point, and although it is a point parenthetical to the main issues of this paper, it addresses a problem that has been raised in the literature and therefore deserves to be answered. In these calculations, μ and k were taken to be constant, but $\overline{\delta}$ was permitted to assume values less than one, thereby illustrating the rather "kinky" solutions which Odell and I had called attention to in 1976. These solutions had been overlooked in the original Keller-Segel analysis, and, although they are qualitatively different from the solutions for $\overline{\delta} > 1$, they appear nevertheless to be real, despite a suggestion of Rosen's (1976) to the contrary. Where, in the original solutions, b and s approach zero asymptotically for $\xi \to -\infty$, in these solutions b and s reach zero at finite values of ξ. For values of $\overline{\delta} \leq 1/2$, b' is no longer continuous at $\xi = 0$, although the net bacterial flux still is. In our note on this point, Odell and I conjectured that solutions for $\overline{\delta} \leq 1/2$ would not be stable, and Odell's recent calculations bear this conjecture out. A stable band is observed for $\overline{\delta} = .75$, but is seen to be unable to form when $\overline{\delta} = .4$. In the latter case, a band attempts to form, but attenuates rapidly. The point made by these calculations may be academic in the sense that the empirical evidence suggests that for migrating bands of bacteria, at least in the context of the original model, the appropriate value of $\overline{\delta}$ is indeed greater than 1. It is, however, a point worth making if we are either to reexamine that model or try to extend the analysis beyond bacteria—in short, in the larger theoretical context—and it is to this larger context that I would like to address my concluding remarks.

The gist of my remarks so far is that while attempts to subject the predictions fo the Keller-Segel model to fine-scale empirical test have resulted in an agreement which can only be gratifying, there are serious questions about the significance of such agreement, and perhaps even about the appropriateness of the effort. In my view, these questions have import for the larger problem of the role of mathematics in biology. In retrospect, we can see more clearly than ever the priority of independent determination of the biological functions k(s), μ, and χ. To date, even though such measurements are incomplete, they remain the most reliable information we have. Our early hope that macroscopic measurements of band shape could by themselves lead to microscopic insight appear now to have been unrealistic—there seems to be, even in principle, too much latitude in the theoretical description to make this possible. Worse yet, there is at least the possibility of a contradiction between what the model does imply about the shape of the relevant functions and what biology tells us, though as yet this remains unclear.

What, then, can be said for the value of our model? Has it been useful in any ways? I think it has. Even though I now think that we, as well as subsequent investigators, may have gone seriously awry in taking our model too literally, it is clear that that model has provoked a number of quite interesting and useful studies. Its primary value, I would argue, has come from what is probably its mathematically most trivial aspect, and that was in providing a general mathematical description of chemotaxis* which could yield a qualitative picture of a variety of phenomena, of which traveling bands of migrating bacteria is only one. In hindsight, the emphasis placed on traveling bands in bacteria appears to have been misplaced—a large variety of interesting chemotactic phenomena may be found in many other organisms. These may also be described mathematically be appropriate modifications of our equations, and this is being done by a number of investigators. The extent to which these descriptions turn out to be useful seems to be directly proportional to the extent to which attention is focused on the biological rather than the mathematical phenomena. Mathematics here finds itself in the rather unusual posture of a descriptive science, one which may point but cannot lead the way. It can identify important parameters to measure, but can rarely determine them. In this connection, I would like to cite Lauffenburger's paper, in this conference, as a prime instance of what I would regard as a healthy use of these models. Other efforts could also be cited. Applications are beginning to be made not only to other microorganisms, but even to predator-prey interactions of higher organisms. Chemotaxis appears to be a very general property of biological organisms, playing a role not only in the organization of complex organisms, but also in the organization of ecological communities. Insofar as it can lead to better feeding, to aggregation, to niche formation, it is clearly an important factor in evolution. If our model could be of any help in studying these larger questions, then I think it would have served a very real and considerable value.

---

*In fairness, it needs to be pointed out that even here we cannot claim too much credit, for it turned out that a very similar equation to ours had been written down long ago by Patlak (1953).

REFERENCES

Adler, J. (1966), Science, 153, 708.

Adler, J. and M. Dahl. (1967), J. Gen. Microbiol., 46, 161.

Brown, D.A. and H.C. Berg. (1974), Proc. Nat. Acad. Sci., USA, 71, 1388.

Dahlquist, F.W., P. Lovely, and D.E. Koshland, Jr. (1972), Nature New Biol., (London), 236, 120.

Holz, M. and S. Chen. (1978), Biophysical J., 23, 15.

Holz, M. and S. Chen. (1979), Biophysical J., in press.

Keller, E.F. and G.M. Odell. (1975), Mathematical Biosciences, 27, 309.

Keller, E.F. and L.A. Segel. (1970), J. Theoret. Biol., 26, 399.

Keller, E.F. and L.A. Segel. (1971), J. Theoret. Biol., 30, 225.

Lapidus, I. and R. Schiller. (1976), Biophysical J., 16, 779.

Lauffenburger, D. and K.H. Keller. (1979), J. Theoret. Biol., 81, 475.

Lovely, P.S. and F.W. Dahlquist. (1975), J. Theoret Biol., 50, 477.

Odell, G.M. and E.F. Keller. (1976), J. Theoret. Biol., 56, 243.

Patlak, C.S. (1953), Bull. Math. Biophysics, 15, 311.

Rosen, G. (1976), J. Theoret. Biol., 59, 243.

Scribner, T.L., L.A. Segel, and E.H. Rogers. (1974), J. Theoret. Biol., 46, 189.

Segel, L.A. and J.L. Jackson. (1973), J. Mechanochem., "Cell Motility," 2, 25.

MODELING CHEMOSENSORY RESPONSES OF SWIMMING EUKARYOTES

I. Richard Lapidus
Department of Physics/Engineering Physics
Stevens Institute of Technology
Hoboken, New Jersey 07030

M. Levandowsky
Haskins Laboratories of Pace University
41 Park Row
New York, New York 10038

ABSTRACT

The application of biased random walk and diffusion types of models to chemosensory responses of populations of swimming eukaryotes is described. Three general types of models and their physiological bases are discussed. The distinction between steady-state and transient solutions, and the importance of various time scales in the experiments to be modeled are emphasized. Experiments with *Paramecium*,*Tetrahymena* and *Crypthecodinium* are discussed in this context.

INTRODUCTION

"It is a capital mistake to theorize before one has data. Insensibly one begins to twist facts to suit theories, instead of theories to suit facts." Sherlock Holmes.

Models of aggregation or dispersion of microorganisms in response to chemical gradients have been based largely on experiments with swimming prokaryotes (bacteria), or with non-swimming, ameboid eukaryotes (slime molds, leukocytes). However, recent work with swimming eukaryotes, particularly ciliate protozoa, is beginning to provide data which may be suitable for modeling.

Not surprisingly, the swimming behavior of these larger, structurally more complex organisms appears to be much richer than that reported for the bacteria. Nevertheless, in the well-studied case of the large ciliate *Paramecium*, it appears that models of a type similar to those used for bacteria and slime molds may provide at least a reasonable starting point.

MODELS

An important kind of behavior in *Paramecium* consists of straight "runs" (actually, spiral swimming along a straight axis), punctuated by "avoiding reactions" (ARs) in which the cell stops, then swims backward briefly before starting a new run in another direction. Although the anatomical and physiological bases are quite different, this behavior is similar to that of the much smaller swimming bacteria, and may be approximated by a random walk type of model. For a large population of cells, over sufficiently large distances and long time periods, compared to characteristic distances and times associated with individual cell motion, this stochastic model may be approximated by different equations.

Let $N(\vec{r},t)$ be the cell number density in an experimental container. If there are a large number of cells in a volume which is small compared to the size of the container, we may treat N as continuous. In this case the number of cells passing through a unit area per unit time (the flux, or current-density), $\vec{J}$, is also a continuous variable and we can write an equation of continuity,

$$\frac{\partial N}{\partial t} = - \vec{\nabla} \cdot \vec{J} , \tag{1}$$

which expresses the conservation of total cell number. In the time-scale of most experiments with swimming protozoa this is a reasonable assumption. For long time periods, however, when either growth or death of cells is appreciable, additional terms would be required, as in some models of long-term bacterial behavior. In this paper, population growth and death are ignored.

If cells are not responding to a special signal, but merely swimming randomly, without significant interference with each other, then the total current is merely the diffusion current, i.e. $\vec{J} = \vec{J}_r$, where

$$\vec{J}_r = -\mu \nabla N , \tag{2}$$

and μ, the "random motility" or diffusion coefficient is determined by the average swimming speed, v, and the average turning frequency, f.

$$\mu = v^2/f . \tag{3}$$

Here a major difference between prokaryotic and eukaryotic swimming appears. In bacteria, such as *Escherichia coli* and *Salmonella typhimurium*, v is a constant and therefore changes in μ are due to changes in f, the turning frequency. In eukaryotes such as *Paramecium* and the smaller ciliate *Tetrahymena*, however, both v and f can vary.

If the motility μ is constant over space and an isotropic random walk is assumed, then substitution of Equation (2) in Equation (1) yields the familiar diffusion equation,

$$\partial N/\partial t = \mu \nabla^2 N . \tag{4}$$

(Note: in this case we are dealing with biological diffusion based on swimming behavior, and μ is therefore orders of magnitude greater than in the more familiar case of molecular diffusion, or the diffusion of particles due to Brownian motion).

In some species, when the cell population is exposed to a non-uniform external stimulus, such as a chemical gradient, a biased random walk is observed experimentally. Before discussing this, however, some general remarks are in order.

We shall be concerned here only with models of situations where the cells respond to a signal by modifying a non-directed swimming rather than by an oriented motion toward or away from the stimulus (such as the source of a diffusing chemical cue). In the very useful terminology of Fraenkel and Gunn (1961; see also Diehn et al 1977), such non-oriented responses are termed kineses, and oriented responses are taxes. It is unfortunate that bacteriologists are accustomed to calling bacterial chemokinesis a taxis. Actually, a true taxis has not been reported in bacteria. In fact, one may question whether it is physiologically possible in these small organisms. In eukaryotes, however, both phenomena are observed, and the distinction is important. We therefore follow the very useful terminology of Fraenkel and Gunn here.

We distinguish 3 types of models for chemokinesis. For each of these there is some evidence in various species.

First, and perhaps the simplest way to introduce a chemokinetic effect, is to let μ, the random motility, be a function of the concentration, S, of the chemical signal: $\mu = \mu(s)$, we call this type I. While this type of response may produce apparent chemokinetic movement of a population of cells, the phenomenon should be called pseudochemokinesis because the long-term steady-state distribution of organisms must be uniform, regardless of the particular functional form of $\mu(s)$ (see below).

A second, somewhat more familiar approach is to treat the basic motion as a biased random walk. We call this type II. Successive excursions ("walks") are independent and random in orientation, but the mean free path is a function of orientation with respect to a perceived signal. The usual way to model this bias, or "drift," is to add a chemokinetic flux term, $\vec{J}_c$, to the random flux $\vec{J}_r$:

$$\vec{J} = \vec{J}_r + \vec{J}_c$$

$$= -\mu\nabla N + N\delta\nabla F \; . \tag{5}$$

Here δ is a constant with the units of a motility or diffusion constant, and F, the sensitivity function, models the cell's perception and response to the stimulus, S.

The functional form of F in Equation (5) is of considerable physiological interest. First of all, if ∇F is simply a constant, we have the Kolmogorov forward equation (the Fokker-Planck equation) of biased diffusion. Keller and Segel (1971), modeling slime mold aggregation and bacterial swimming, suggested a logarithmic sensitivity

$$F(S) = \ln S, \tag{6}$$

in analogy with the empirical Weber-Fechner law of human physiology. In this case, for example, if S is an exponential function of x in one dimension, then F = constant. Lapidus and Schiller (1976) later suggested a hyperbolic function,

$$F(S) = S/(S + K), \tag{7}$$

where K is a constant. This arises from a model in which the cell's perception of the chemical cue is mediated by reversible binding to receptor molecules (presumably, proteins on the cell surface), with first-order (mass-action) kinetics as in the well known Langmuir adsorption, or Michaelis-Menten enzyme kinetics. Experiments by Berg and Brown (1972) support the latter expression in the case of bacterial chemokinesis. Over a sufficiently restricted range, equations (6) and (7) are equivalent.

Alternatively, second order kinetics may be appropriate for some sensory responses, e.g.

$$F(S) = S^2/(S^2 + K^2) \, , \tag{8}$$

indicating a cooperative binding effect involving two or more receptor units, as in the well-known case of hemoglobin-oxygen binding. Such a sigmoid sensitivity function might offer the organism an advantage in the form of a controlled sensitivity threshold, lacking in the hyperbolic expression. Other possible sensitivity functions can be hypothesized, but the above would probably appeal to the physiologist's intuition.

A third, mathematically more difficult sort of model, for which there is also some evidence in *Paramecium*, explicitly incorporates a "memory," and is therefore non-Markovian. We call this type III. Here we suppose that at any given time behavior responds to previous as well as to current stimuli. While this is the observed phenomenon in some species, mathematical representation on a macroscopic, populational level is difficult, and would not take the form of a simple partial differential equation. To date there appears to have been little directly applicable theoretical work using this approach.

Nonetheless, there is a close relation between the last two kinds of models, which is worth dwelling upon. Consider the basic physiological question: by what mechanism could a single cell detect a chemical gradient? This appears to be largely a matter of scale. Some very large cells may detect a sharp enough chemical gradient by comparing concentrations at opposite ends of the cell itself. In this case the possibility of actual orientation, or "aiming" of the cell exists and we might expect such cells to exhibit a true taxis rather than a non-oriented kinesis. Anecdotal accounts suggest this may be the case with some long, slender ciliates, such as psammophiles from tide-pools (Faure-Fremiet, 1967), or certain large histophagic species (Parducz, 1964). As noted earlier, however, we shall not deal with such behavior in this paper, and, in fact, there appears to be little mathematical work on true taxes (for a review, see Okubo, 1979).

In other, possibly more common cases, "a clock acts as a ruler," and a kind of "memory" is required to detect a spatial gradient. While swimming in the gradient, the cell responds to a perceived temporal change in concentration, and thereby also to the spatial gradient. Experiments by several investigators (Macnab and Koshland, 1972; Berg and Brown, 1972) indicate that this is the method used by bacteria to detect a gradient. Van Houten (1976, 1977, 1978, 1979) has shown that *Paramecium* responds in this manner to some chemical cues: cells transferred abruptly from a low to a high concentration, exhibit increased motility at first, but return to their pre-transfer motility - "adapt" - after 10 minutes. Such a long adaptation period, coupled with the fast swimming speed of *Paramecium* suggests the applicability of the third type of model above.

Now, imagine a series of hypothetical organisms (they may turn out to be real!) with shorter and shorter adaptation times, and slower and slower swimming speeds. Ultimately we approach bacterial behavior in which adaptation is so rapid and swimming so slow that the behavioral response tracks the gradient of the perceived signal closely. Thus, the third model goes smoothly into a biased random walk type of model. The apparent difference between the two types may actually be only a matter of scale.

SOLUTIONS

Of the three types of model referred to above, the first two (modulated diffusion and diffusion plus drift) have been most studied. We shall first note some important qualitative features of their steady-state solutions, which could be predictive for experiments lasting long enough for a steady-state to be achieved. In the following section we shall contrast these with some distinctive short-term, transient properties of solutions, which are more relevant to some other experiments.

1. Steady-state solutions. In our basic model,

$$\partial N/\partial t = -\vec{\nabla} \cdot \vec{J} \ , \tag{9}$$

$$\begin{aligned} \vec{J} &= \vec{J}_r + \vec{J}_c \\ &= -\mu \nabla N + N\delta\nabla F \ . \end{aligned} \tag{10}$$

A steady-state solution is given by $\partial N/\partial t = 0$. This implies

$$\vec{\nabla} \cdot \vec{J} = 0 \tag{11}$$

or

$$\vec{J} = \text{constant} \tag{12}$$

Since the flux vanishes at the boundaries of an experimental container, this implies that

$$\vec{J} \equiv 0 \ . \tag{13}$$

Now, in case there is no drift term, i.e., $\delta = 0$, giving a model of the first type, we have from the above that

$$\nabla N = 0 \tag{14}$$

$$N = \text{constant} \tag{15}$$

Regardless of spatial variation of μ, the steady-state concentration of cells throughout the container is uniform.

Thus, if a model of the first type holds - i.e., no drift term, but μ varies spatially due to a non-uniform concentration of a chemical cue - then experiments which last long enough to give a steady state distribution of organisms will produce a uniform distribution.

An intuitive feeling for this phenomenon may be obtained using a physical model suggested by Patlak (1953). Consider the temperature distribution in a well-insulated object of variable heat conductivity. If the object is truly adiabatic, then regardless of the variations in conductivity, an ultimate steady state will be reached in which temperature is uniform throughout.

If, however, there is a non-zero drift term ($\delta \neq 0$ in our model) then we have a model of the second type. The steady state solution is now given by

$$\mu \nabla N = N\delta \nabla F \ , \tag{16}$$

$$N/N_0 = \exp[\int (\delta/\mu)\nabla F \cdot d\vec{r} \ . \tag{17}$$

In this case the steady state solution is not constant but varies spatially in a way determined by the function F (and μ, if μ is not constant).

We see that the steady state solutions are quite different in models of types I and II, and this fact may be used to distinguish them experimentally.

2. Transient states. Many experiments never reach a steady state. Hence, the transient (non-steady state) solution are of interest in analyzing such experiments.

In models of type I, for example, although the steady state solution is a uniform distribution, transient states are not uniform. For a spatially variable μ we have

$$\begin{aligned} \partial N/\partial t &= -\vec{\nabla} \cdot \vec{J} \\ &= \nabla \cdot (-\mu(\vec{r}) \ \nabla N) \\ &= \mu \nabla^2 N + \nabla \mu \cdot \nabla N \ . \end{aligned} \tag{18}$$

The first term on the right is a standard diffusion term (but with spatially variable motility), but the second term introduces a new effect. If $\nabla\mu$ is appreciable, this term can be significant, leading to transient aggregation or dispersion effects which in many cases may be what is seen in experiments. A convenient term for transient, non-steady state aggregation by this mechanism is pseudochemokinesis as noted earlier.

EXPERIMENTS

1. *Paramecium*. Most of the experimental data on behavior of ciliates has been obtained with this large eucaryotic microorganism. Van Houten et al (1975) developed a convenient assay method using a 3-way glass stopcock. In this apparatus, a suspension of organisms is introduced into one arm, consisting of a glass tube with an inner diameter of 1 cm. This can then be connected to two others containing test solutions by turning the stopcock, creating a "T-maze" on a protozoan scale. At the end of an

experiment the stopcock is closed and the organisms which accumulated in each arm are counted. In this way the response of *Paramecium* to a variety of chemicals has been assayed. Accumulation ("attraction") occured in solutions of folic acid and acetate, whereas quinidine solutions were avoided ("repulsion"). Van Houten has sought explanations for these results in terms of the effects of the chemicals on both swimming speed, V, and the frequency of the avoiding response, F_{AR} (as noted earlier, the avoiding response is the typical response of *Paramecium* and consists of turning after a straight "run", corresponding roughly to the "tumble" in bacterial chemokinesis) (Van Houten, 1977).

In normal cells, F_{AR} and V are simultaneously affected by events at the cell membrane, particularly the membrane potential E_m (Eckert, 1972) and thus one cannot vary these two parameters independently. However, a number of behavioral mutants have been obtained that form a powerful set of tools for dissection of the two parameters. Results of this work to date are intriguing, and indicate something of the complexity of interactions between the two variables. According to Van Houten one can distinguish between

> "two groups of chemicals that cause chemokinesis by two different mechanisms, I and II. In mechanism I, attraction and repulsion are correlated with decreased and increased F_{AR}. The associated increase and decrease in velocity seem to be unimportant in determining net effect. Moreover in 'pawns' (mutants with no avoiding reaction) agents of group I do not cause appreciable attraction and repulsion. In mechanisms II, it is the response swimming velocity that predominates. Repulsion is associated with increased V and a decrease of F_{AR} to zero, and attraction is associated with decreased V due to slow swimming and time spent in frequent turning in the avoiding reaction. Pawns are attracted and repelled by agents of group II. Attractants I and repellents II cause the same qualitative changes in behavior (decreased F_{AR} and decreased V) but have opposite chemokinesis results." (Van Houten, 1979).

Van Houten found also a nonlinear relation of "repulsion" and "attraction" (as assayed in the stopcock apparatus) to effects of chemicals on membrane potential, E_m: attraction is associated with weak hyperpolarization or strong hypopolarization, whereas repulsion is associated with strong hyperpolarization or weak hypopolarization

An explanation of these results, in terms of the models discussed earlier, has not yet been developed. Given the swimming speed of *Paramecium*, the size of the apparatus, and the duration time of experiments, a model of type I seems unlikely. If the mechanism of response were simply a change in the motility, μ, due to the changes of V and F_{AR}, a uniform distribution would be expected, since there seems ample time for a steady-state to be achieved. Thus, relevant experimental measurements are probably the response to gradients (type II models) or adaptation period ("memory") to perceived chemical change (type III models). As noted, Van Houten has also provided some evidence of the latter. It will be of interest to see how those observations are related to the patterns related above.

2. *Tetrahymena*. Because it is easily grown in pure culture on defined medium, this smaller ciliate has been the object of an enormous amount of biochemical research, but its behavioral physiology has been explored only recently. In our laboratory and elsewhere (Almagor et al, 1977; A. Ron, personal communication), a number of chemosensory responses have been studied using capillary assays similar to those used by bacteriologists. We have observed that aggregation in capillaries filled with attractive substances, e.g. casein hydrolysate, is associated with slower swimming speeds and also lower turning frequency, in such a way that the mean free path remains almost unchanged but the motility decreases. Thus, in one experiment, cells in a salt solution had an average speed of 1.5 mm/sec and an average turning frequency of 1.3/sec; in .1% casein hydrolysate (a strong attractant) cells from the same preparation had an average speed of 0.31 mm/sec and an average turning frequency of 0.28/sec. From these data it appears that the average mean free path is little changed, whereas the motility, μ, decreases from 1.6 mm^2/sec to 0.33 mm^2/sec. On the basis of such preliminary results, a model of type I seems quite plausible, but further work will be needed to test this possibility. The aggregation we observed during a typical experiment with capillaries are probably not steady-state situations, because of the much slower swimming speed of these small ciliates.

The detailed swimming behavior of *Tetrahymena* is also rather different from that seen in either bacteria or *Paramecium*. Straight runs are not separated by sharp events corresponding to the avoiding reaction or the tumble, but usually by relatively smooth turns. However it is possible to measure both swimming speed and turning frequency and to compare experimental data with predictions of random walk-derived models. Further progress in this area is to be expected.

3. *Other species*. A number of other swimming eukaryotes exhibit chemosensory behavior (Levandowsky and Hauser, 1978), but very little data exists yet that could be related to the theories discussed above. A recent observation with a marine dinoflagellate, *Crypthecodinium*, indicates a biased random walk response to a carbon dioxide gradient (Hauser et al, 1978), and this type of behavior might be described by a model of the Fokker-Planck type, though this has not been done yet.

SUMMARY

Most of the relevant experiments have been done with *Paramecium* but preliminary work with other species look very promising. In no case has a kinetic model been successfully fitted to experimental data, but this is partly due to the fact that experiments have not been designed to test theoretical predictions. One purpose of this paper is to present several plausible theoretical frameworks in such a way that future experiments will be designed to test and distinguish between the various possible mechanisms of chemokinetic aggregation and dispersion.

ACKNOWLEDGEMENTS

We thank Andrew Kehr and Nicholas Sauter for their technical assistance in obtaining some of the experimental results on *Tetrahymena* in our laboratory. I.R.L. was a Visiting Senior Scientist at Haskins Laboratories while some of this work was carried out. Some of this work was supported by USPHS grant #FR05596 to A. M. Liberman of Haskins Laboratories.

REFERENCES

Almagor, M., Bar-Tana, J. and Ron, A. 1977. J. Cell Biol. 75, 78a.

Berg, H. C. and Brown, D. A. 1972. Nature 239, 500.

Diehn, B., Feinlieb, M., Haupt, W., Hildebrand, E., Lenci, F. and Nultsch, W. 1977. Photochem. Photobiol. 26, 559.

Eckert, R. 1972. Science 176, 473.

Faure-Fremiet, E. 1967. In "Chemical Zoology" (G. W. Kidder, Ed.), Vol. I, 21. Academic Press, New York.

Fraenkel, G. S. and Gunn, D. L. 1961. "The Orientation of Animals". Dover Publications, Inc. New York.

Hauser, D. C. R., Petrylak, D., Singer, G., and Levandowsky, M. 1978. Nature 273, 230.

Keller, E. and Segel, L. A. 1971. J. Theoret. Biol. 30, 225.

Lapidus, I. R. and Schiller, R., 1976. Biophys. J. 16, 779.

Levandowsky, M. and Hauser, D. C. R. 1978. Int. Rev. Cytrol. 53, 145.

Macnab, R. and Koshland, D. E., Jr. 1972. Proc. Nat. Acad. Sci. USA 69, 2509.

Okubo, A. 1979. "Diffusion and Ecological Problems". Springer Verlag, Heidelberg.

Parducz, B. 1964. Acta. Protozool. 2, 367.

Patlak, C. S. 1953. Bull. Math. Biophys. 15, 311.

Van Houten, J. 1976. Thesis, University of California.

Van Houten, J. 1977. Science 198, 746.

Van Houten, J. 1978. J. Comp. Physiol. 127, 167.

Van Houten, J. 1979. Science, 204, 1100.

Van Houten, J. Hansma, H. and Kung, C. 1975. J. Comp. Physiol. 104, 427.

Mathematical Model for Tissue Inflammation: Effects of Spatial Distribution, Cell Motility, and Chemotaxis

Douglas Lauffenburger, Department of Chemical Engineering and Materials Science, University of Minnesota, Minneapolis, Minnesota, U.S.A. 55455 *

* address after July 1979: Department of Chemical and Biochemical Engineering
University of Pennsylvania
Philadelphia, Pennsylvania, U.S.A. 19104

Abstract

Clinical evidence has accumulated in the past decade correlating defective host phagocyte function with increased susceptibility to and severity of infection. In particular, deficient leukocyte random motility and chemotaxis are associated with decreased resistance to infection.

In an attempt to increase understanding of the roles of the component processes of the inflammatory response, we have developed a mathematical model for tissue inflammation. The aim is to relate the dynamics of bacterial growth to the kinetic and transport parameters of bacteria and leukocytes in tissue. The model considers a local tissue region in the vicinity of a venule and applies continuum unsteady-state species conservation equations to the bacterial and leukocyte population densities and to the concentration of a chemotactically active chemical mediator produced by the bacteria. The analysis quantifies the effects of key parameters, such as leukocyte random motility and chemotactic coefficients, bacterial growth and destruction rate constants, and leukocyte vessel emigration coefficient, upon host ability to eliminate the bacteria.

Introduction

Inflammation is the primary reaction by a host to tissue injury or invasion. It is generally a helpful defense against infection (Metchnikoff, 1968) although sometimes it may become a pathological condition itself (see, for example, Sell (1975)). The elements principally responsible for elimination of offending agents from tissue are the phagocytic white blood cells. In particular, the polymorphonuclear leukocytes (PMN) are the "first line" of defense against infection (Stossel, 1972), being mobilized to enter tissue from the microcirculation bloodstream within a few hours after the appearance of microbes, chemical antigens, or products of thermal or radiative damage in that local tissue region. This defense response is effective if the injurious agents are eliminated before they can cause major disruption of local or systemic function.

The general steps in the inflammation process can be summarized as follows (see, for example, Hersh & Bodey, 1970). After tissue damage or invasion by a foreign agent, vasodilation and increased vascular permeability begin within minutes. The number of leukocytes marginating from the bloodstream and adhering to the vascular endothelial walls increases, especially in the post-capillary venules. Adherent leukocytes migrate between the endothelial cells of the venules, into the tissue space. Once in the tissue, the leukocytes move about and phagocytize the harmful agents. Chemical mediators released by the foreign agent or damaged tissue can influence this movement. The major effect is to cause directed movement, or chemotaxis, by the leukocytes toward increasing concentrations of the mediators, called chemotactic agents (Wilkinson, 1974).

Inadequate performance of a process involved in any of these steps may affect the host's ability to mount an effective defense against infection. In fact, there is a rapidly increasing body of literature describing the correlation between abnormalities in the inflammatory response and pathological conditions (see, for example, Rytel (1976) and von Graevenitz (1977)). Of special interest to us is the recognition that deficient leukocyte motility and chemotaxis are strongly correlated with increased susceptibility to and severity of infection. The deficiencies may be due to a number of causes; among them are genetic cell abnormalities, the presence of motility or chemotaxis inhibitors in the blood or tissue, and inadequate production of attractants in the tissue. Examples of situations in which deficient PMN motility or chemotaxis are present in patients with recurrent infection are discussed by Miller (1975), Senn & Jungi (1975), and Quie & Cates (1977), among others.

Therefore, leukocyte motility assays might be useful for diagnosis or prediction of host defense response (Keller et al, 1975). Further, a number of researchers have proposed therapeutic alteration or manipulation of leukocyte motility properties in order to provide increased protection against infection (Isturiz et al, 1978; Goetzl et al, 1974; Hogan & Hill, 1978) or to diminish some of the pathological effects of inflammation (Mizushima et al, 1977). Hence, it is important to be able to understand and to predict in a quantitative fashion the relation between transport phenomena such as motility and chemotaxis and the susceptibility to infection. Not only would this allow a better understanding of the dependence of host defense effectiveness upon these parameters, but it would provide a more informed basis for using chemotaxis and motility assays in clinical diagnosis and therapy. In this paper we discuss some of our initial efforts in this direction, developing a continuum, spatially-distributed model of bacterial and leukocyte population dynamics in a vascularized tissue.

Mathematical model

We consider a region of tissue surrounding a venule. Bacteria introduced into this region multiply at a rate allowed by available nutrient levels. Leukocytes can enter the tissue from the venule bloodstream, and can phagocytize and kill the bacteria. The leukocytes have a finite life-span within the tissue. Both cell types may possess random motility, and the leukocytes may be chemotactically attracted to the bacteria through the action of chemotactic agents. See Figure 1.

We are primarily interested in the tissue concentrations of three species; bacteria, leukocytes, and chemotactic attractant, as functions of space and time. In this work, since our aim is to develop the simplest mathematical model for inflammation dynamics, we will assume the attractant concentration distribution to be parallel with the bacterial density distribution. This will be reasonable if the attractant concentration is everywhere proportional to the bacterial density, as discussed in more detail elsewhere (Lauffenburger & Keller, 1979). Thus, the driving

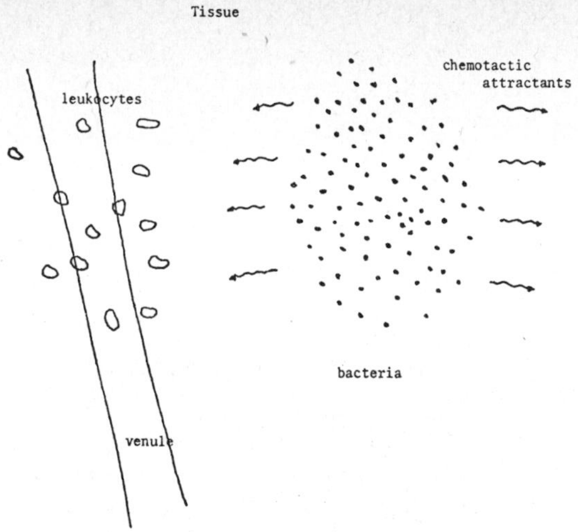

Figure 1. Illustration of inflammatory response to bacterial invasion of tissue.

force for chemotactic movement in this model will be the bacterial density gradient. Therefore, we write conservation equations in the region of interest, for the bacterial density b and leukocyte density c:

$$\frac{\partial b}{\partial t} = - \nabla \cdot \underset{\sim}{J}_b + G_b \tag{1a}$$

$$\frac{\partial c}{\partial t} = - \nabla \cdot \underset{\sim}{J}_c + G_c \tag{1b}$$

$\underset{\sim}{J}_b$ and $\underset{\sim}{J}_c$ are the local fluxes of the species at any point in the tissue space. G_b and G_c are the local net generation rates of the species.

We consider that the tissue is perfused by parallel blood vessels, so that it is composed of cylindrical regions each fed by a single venule and effectively isolated from adjacent regions by a cylindrical surface of symmetry. This geometry is illustrated in Figure 2a. The geometry of the model is axial symmetry appropriately

(a)

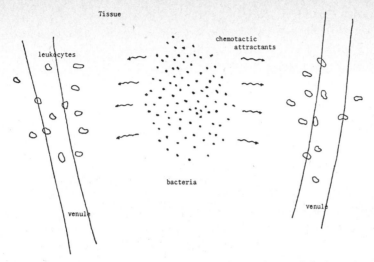

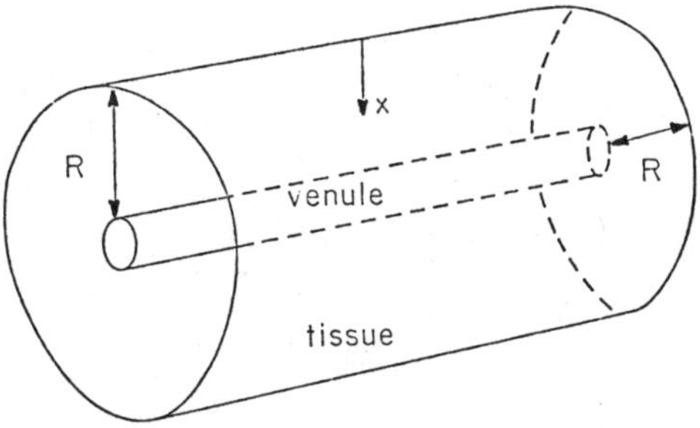

Figure 2a. Tissue cylinder model.

treated in cylindrical coordinates; however, no serious loss of generality is incurred by treating it as one-dimensional, of thickness 2R, where R is the radius of the tissue cylinder, as illustrated in Figure 2b.

(b)

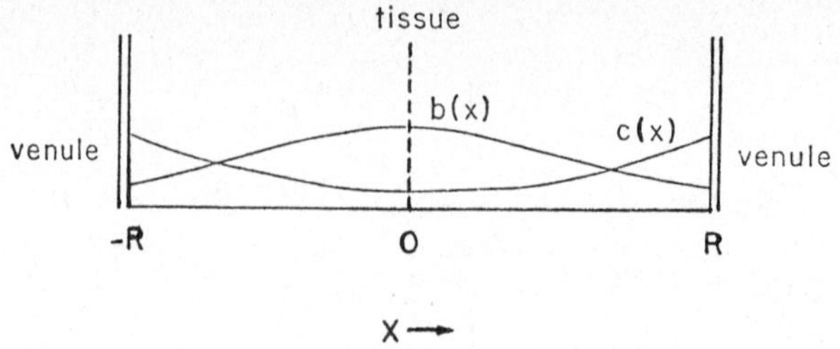

Figure 2b. One-dimensional radially-distributed model geometry.

Bacterial and leukocyte motility are modeled by the phenomenological expression originated by Keller and Segel (1971), and applied to leukocytes by Rosen (1976). The bacterial random motility coefficient is D_b, and the leukocyte random motility and chemotactic coefficients are D_c and χ, respectively. In this initial modeling work we will assume them to be constants in order to establish the fundamental behavior. Thus, with convective flow in the tissue space neglected, we have

$$\underset{\sim}{J}_b = - D_b \nabla b \qquad (2a)$$

$$\underset{\sim}{J}_c = - D_c \nabla c + \chi c \nabla b \qquad (2b)$$

Remember that the attractant concentration is assumed proportional to the bacterial density, with the proportionality constant implicit in χ.

Growth of bacteria in body tissue is assumed to be exponential since required nutrients should generally be present in the interstitial fluid. Therefore,

$$\text{bacterial growth rate} = k_g b \qquad (3)$$

where k_g is the growth rate constant.

Experiments by Stossel (1973) indicate that the appropriate form for the phagocytic uptake of particles by leukocytes is

$$\text{phagocytosis rate} = \left[\frac{k_{dm} b}{K_b + b}\right] c \qquad (4)$$

where k_{dm} is the maximum phagocytosis rate constant and K_b is a saturation density constant. We assume that all ingested bacteria are killed.

For this simple model, we also assume that K_b is very large compared to the bacterial density, so that phagocytosis is simply proportional to bacteria and leukocyte densities:

$$\text{phagocytosis rate} = k_d \; b \; c \qquad (5)$$

with $k_d = k_{dm}/K_b$.

Because leukocytes have a finite total life-span (Craddock, 1972) and since emigration from the blood is a random event, we assume that leukocyte death in the tissues is also an exponential process, in that

$$\text{leukocyte death rate} = g_c c \qquad (6)$$

where g is the death rate constant. g_c may be a function of b, but here we will assume g_c to be constant.

Using equations (3), (5), and (6) in equations (1a,b), and assuming the one-dimensional geometry transverse to the venule axis, we obtain

$$\frac{\partial b}{\partial t} = D_b \frac{\partial^2 b}{\partial x^2} + k_g b - k_d bc \tag{7a}$$

$$\frac{\partial c}{\partial t} = D_c \frac{\partial^2 c}{\partial x^2} - \chi \frac{\partial}{\partial x} (c \frac{\partial b}{\partial x}) - g_c c \tag{7b}$$

in $-R \leq x \leq R$. If the one-dimensional tissue region is symmetric, we need consider only half of the region, for example, $0 \leq x \leq R$. At $x = 0$, symmetry requires the derivatives of cell densities and attractant concentration to vanish, so that the boundary conditions there are

$$\frac{\partial b}{\partial x} = \frac{\partial c}{\partial x} = 0 \tag{8a,b}$$

At the vessel wall, bacteria are assumed to be able to move from bloodstream to tissue, and vice versa, across the venule wall. The rate of net movement across the wall is postulated to be proportional to the difference between the densities at the two sides of the wall. The constant of proportionality is h_b, called the bacterial emigration coefficient, and represents the ease of movement across the wall. If there is no net accumulation at the interface between tissue and bloodstream, the flux from tissue to bloodstream must equal the flux within the tissue to the interface. Assuming the bacterial density in the local venule bloodstream to be zero, the attractant boundary condition at $x = R$ is

$$D_b \frac{\partial b}{\partial x} = - h_b b \tag{9a}$$

The leukocyte boundary condition is a crucial element of the model, since one of the key events in inflammation is the increased leukocyte emigration from the venules. First of all, the cell flux from the bloodstream across the endothelial layer must be equal to the cell flux away from the endothelium into the tissue interior. Movement of cells from the tissue into the bloodstream is not observed *in vivo* (Athens et al, 1961). *In vitro* experiments have indicated that leukocyte migration into the tissue is proportional to the local bloodstream density initially, but is diminished as leukocytes accumulate on the epithelial cells of the vessel wall (Beesley et al, 1978). Such a diminution could be due to physical hindrance as the leukocytes squeeze through the junctions between epithelial cells. Whether this effect actually occurs *in vivo* needs to be verified. If the emigration rate is diminished by local accumulation of leukocytes at the vessel wall, the leukocyte boundary condition is analogous to that for the bacteria:

$$D_c \frac{\partial c}{\partial x} - \chi c \frac{\partial b}{\partial x} = h_c \left[c_b - c\right] \tag{9b}$$

h_c is the leukocyte emigration coefficient, and c_b is the venule density. This equation appears to allow movement of cells from the tissue to the bloodstream when $c(R)$ is greater than c_b. Therefore, we require that $h_c = 0$ whenever $c(R)$ is greater than

c_b, which can occur when a large peak of leukocyte density in the tissue disperses.

h_c accounts for the ability of the circulating leukocytes to marginate and adhere to the wall. There is evidence that both margination and adherence increase during inflammation, perhaps due to slowing of blood flow after vasodilation (Beesley et al, 1978) and to chemical "stickiness" factors (Fehr and Jacob, 1977). Thus, h_c may not be constant, but may be a function of b. This possibility will not be explored in this paper.

Analysis

The equations can be nondimensionalized by introducing the following normalized variables and dimensionless parameters:

$$\tau = k_g t \qquad \xi = \frac{x}{R} \qquad v = \frac{b}{b_i} \qquad u = \frac{c}{c_b} \qquad \mu_b = \frac{D_b}{k_g R^2} \qquad \mu_c = \frac{D_c}{k_g R^2} \qquad \theta = \frac{g_c}{k_g}$$

$$\gamma = \frac{k_d c_b}{k_g} \qquad \beta_b = \frac{h_b}{k_g R} \qquad \beta_c = \frac{h_c}{k_g R} \qquad \delta = \frac{\chi b_i}{k_g R^2}$$

b_i is a characteristic initial density of the invading bacterial population. The resulting equations are then

$$v_\tau = \mu_b v_{\xi\xi} + (1 - \gamma u)v \tag{10a}$$

$$u_\tau = \mu_c u_{\xi\xi} - \delta(uv_\xi)_\xi - \theta u \tag{10b}$$

with boundary conditions

$$\xi = 0 \qquad v_\xi = 0 \qquad u_\xi = 0 \tag{11a,b}$$

$$\xi = 1 \qquad \mu_b v_\xi = -\beta_b v \qquad \mu_c u_\xi - \delta u v_\xi = \beta_c(1 - u) \tag{12a,b}$$

Since we are primarily concerned with the effects of the motility properties, we will first study the case $\delta = 0$, then proceed to consider the influence of chemotaxis by allowing $\delta > 0$.

I. $\delta = 0$. In this situation, the leukocytes possess only random motility. The leukocyte equation (10b) can be solved independently, and the solution of (10a) follows.

Consider the steady-state solutions, with $v_\tau = u_\tau = 0$. The steady-state leukocyte distribution, $u_{ss}(\xi)$, is

$$u_{ss}(\xi) = A\left(\frac{\cosh \alpha\xi}{\cosh \alpha}\right) \tag{13}$$

where $A = (1 + \frac{\alpha}{\phi}\tanh \alpha)^{-1}$, $\alpha^2 = \theta/\mu_c$, and $\phi = \beta_c/\mu_c$. The effect of μ_c on the leukocyte density profile is shown in Figure 3. The maximum value of $u_{ss}(\xi)$ occurs at the venule wall, $\xi = 1$, and is $u_{ss}(1) = A$. The minimum value occurs at the edge of the tissue region farthest from the venule wall, $\xi = 0$, and is $u_{ss}(0) = A/\cosh \alpha$. As μ_c gets very large, the ability of the leukocytes to traverse the tissue region relative to their death rate becomes great, so that in the limit $\mu_c \to \infty$, the leukocyte distribution is uniform. As μ_c gets very small, on the other hand, the cells are likely to die before they can move deep into the tissue, so the leukocyte density is

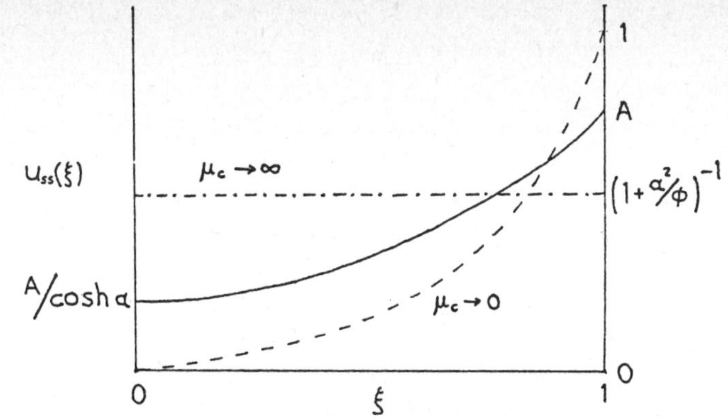

Figure 3. Leukocyte density profile in tissue, for $\delta = 0$ (no chemotaxis).

very low far from the venule. These possible variations in tissue leukocyte distribution may have crucial implications for the effectiveness of the defense response, since the bacteria will multiply locally when $u < \gamma^{-1}$, and will be depleted locally when $u > \gamma^{-1}$.

Equation (13) implies that a steady-state distribution of leukocytes is always present in the tissue, which is consistent with experimental observations. It may be reasonable to imagine that β_c is very small under normal conditions, yielding a very small tissue leukocyte cell density, and that it increases at the beginning of inflammation.

With the steady-state phagocyte distribution determined, the steady-state bacteria equation becomes linear. There are three possible results for this equation: 1) $v_{ss} \equiv 0$ and the bacteria are eliminated; 2) $v \to \infty$, so the bacteria grow without bound, and there is no steady state; 3) $v_{ss} = v_{ss}(\xi)$ is finite and positive. The first result represents effective host defense, the second represents establishment of acute infection, and the third a tolerant compromise state.

The existence and stability properties of these three states can be determined conveniently in terms of a single parameter,

$$\psi = \gamma A - 1 \tag{14}$$

and are summarized in Figure 4, for the simplest case $\beta_b = 0$. The elimination steady state is globally stable for $\psi > \cosh\alpha - 1$, and unstable for $\psi < 0$. The tolerant state can only exist for $0 < \psi < \cosh\alpha - 1$, and therefore the acute infection result will be obtained for $\psi < 0$. Since equation (10a) and its boundary conditions are linear and homogeneous for $\delta = 0$, the tolerant state does not exist in general in the system described by these model equations for this case.

Consider now the behavior in the region $0 < \psi < \cosh\alpha - 1$. The tissue will be divided into two sections by $\xi = \omega$, where $\omega = \frac{1}{\alpha}\cosh^{-1}\left|\frac{\cosh\alpha}{\gamma A}\right|$ is the value of ξ at which $\gamma u_{ss}(\xi) = 1$. For $\xi < \omega$, there will be net bacteria growth, while for $\xi > \omega$, there will be net bacteria death. Thus the elimination state will be approached if the net growth is less than the net death, and the infection state will result if the

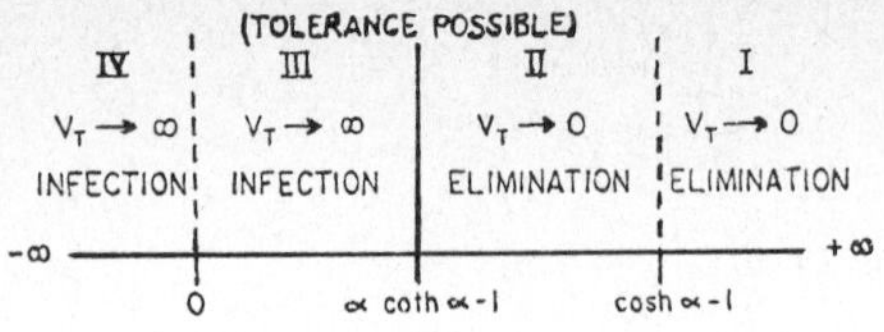

Figure 4. Dependence of inflammation defense effectiveness upon ψ, for $\delta = 0$ (no chemotaxis).

reverse is true. It can be shown (Lauffenburger & Keller, 1979) that the region can be divided between the elimination and infection states, approximately by the value

$$\psi = \alpha \coth \alpha - 1 \tag{15}$$

For $\beta_b > 0$, we can still prove uniqueness and stability for the elimination steady state for $\psi > \cosh \alpha - 1$. Since the equation remains a homogeneous problem, the tolerance steady state will still not generally exist. However, now the boundary between the acute infection state and the possibility of tolerance is $\psi = -\mu_b \pi^2$ rather than $\psi = 0$. The sufficient condition for stability of the elimination state in the region $-\mu_b \pi^2 < \psi < \cosh \alpha - 1$ will remain the same to a first approximation.

II. $\delta > 0$. Chemotaxis will not affect the linear stability results for the elimination steady state. Thus, the effect of chemotaxis will probably be seen in the transient system behavior, rather than in the steady-state behavior. There is now, however, the possibility of a tolerant steady state.

Therefore we turn to numerical solution of the transient equations. The initial bacterial distribution is assumed to be uniform with density b_i, and the initial leukocyte distribution to be uniform with density $0.1 \times c_b$. This initial condition represents, essentially, the "surveillant" tissue leukocyte population usually present due to the normally small value of h_c when there is no tissue invasion. At $\tau = 0$, the bacteria arrive in the tissue, and h_c increases to its inflammatory level. Since we are interested in the possible role for chemotaxis of providing effective bacteria elimination when the defense is inadequate in the absence of chemotaxis, we have chosen parameter values such that ψ falls in region IV of Figure 4. That is, if $\delta = 0$, then the bacterial population would grow indefinitely.

The powerful potential effect of chemotaxis is demonstrated by Figure 5, which shows the transient behavior of the cell population densities in the tissue, V_T and U_T, given the initial conditions discussed above. V_T and U_T are defined by the equations

$$V_T = \int_0^1 v(\xi)\, d\xi \tag{16a}$$

$$U_T = \int_0^1 u(\xi)\, d\xi \tag{16b}$$

We see that for $\delta = 0$, the bacteria grow rapidly. As δ is increased, the

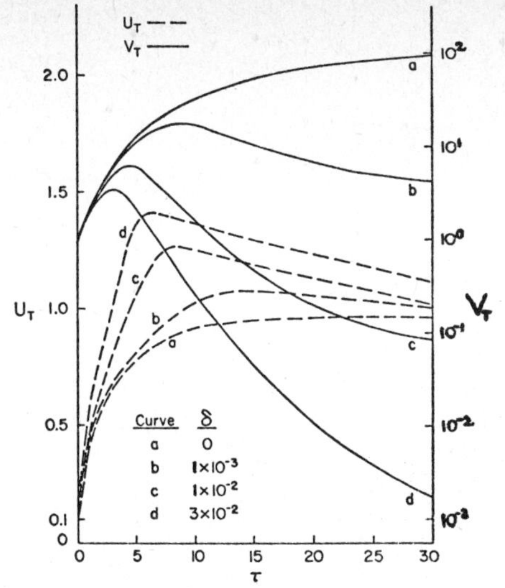

Figure 5. Transient behavior of total cell population densities in tissue, for varying values of δ. Parameter values used: $\mu_b = 10^{-3}$, $\mu_c = 10^{-1}$, $\gamma = 1$, $\theta = 10^{-2}$, $\beta_b = 4$, $\beta_c = 40$, $v_T(0) = 1$, $u_T(0) = 10^{-1}$.

transient behavior changes from rapid bacterial growth to rapid elimination. Thus chemotaxis can allow effective elimination in situations where pure random motility cannot. In fact, a threshold behavior is apparent for the effect of chemotactic coefficient on the host defense results, if the computations are terminated after a physiologically appropriate time. This is an important consideration because it has been observed that the adequacy of the cellular inflammatory response may be determined within 12 to 24 hours (Miles, 1956). For this reason, long time transient computations and steady state considerations are less important than the initial dynamics.

Remembering our observation that chemotaxis cannot actually stabilize the elimination state for small ψ, it is apparent that there can come a point in time at which the bacteria are reduced to a level so low that the chemotactic stimulus would be insufficient for continued leukocyte attraction. Then the leukocyte density would slowly diminish and when it fell below the density necessary for net reduction of bacterial population, the bacteria could proliferate again. If the bacteria have been reduced to such a low level as to be practically eliminated, however, no renewel of growth can occur, and the invasion will be effectively repulsed.

Since the time required for elimination is an important consideration for defense effectiveness, it should be pointed out that increasing δ can greatly improve host defense even when the parameter values are such that ψ falls in regions I or II. That is, even if the elimination state is stable for δ = 0, the bacterial population may increase initially (because the "surveillant" leukocyte population present at τ=0 is inadequate), and the time necessary for reduction of bacterial density to low levels may be unacceptably large. Increasing δ may decrease the time required for satisfactory reduction of the bacterial population, thus allowing adequate defense.

Discussion

Figure 4 summarizes the parameter dependencies of the outcome of the inflammatory defense response to a microbial invasion, in the case where there is no leukocyte chemotactic movement. Understanding of the results depends upon realization that a sufficient condition for bacterial elimination is $u(0) > \gamma^{-1}$, while a sufficient condition for proliferation is $u(1) < \gamma^{-1}$. These conditions yield the borders between regions I and II, and regions III and IV, respectively. There is an attractive simplicity in the "success/failure" dependence upon the value of the parameter ψ. Through this quantity the sensitivity of the defense to each individual parameter can be traced. The elimination criterion equation (15), translated into the original dimensional quantities, is

$$\frac{k_d c_b}{k_g} > \cosh \frac{g_c^{1/2} R}{D_c^{1/2}} \left(1 + \frac{g_c^{1/2} D_c^{1/2}}{h_c} \tanh \frac{g_c^{1/2} R}{D_c^{1/2}}\right) \equiv \rho \tag{17}$$

Establishment of infection can be due either to intrinsic defects in the host system, or to extraordinary features of the invading microbes. Examples of such cases include:

a) low c_b - low numbers of circulating leukocytes; this is a pathological condition known as neutropenia

b) low k_d - defective phagocytosis, possibly due to inadequate opsonization or to resistant bacteria

c) high g_c - short-lived leukocytes, which may be caused by toxins produced by the bacteria

d) large R - tissue region with inadequate access to venules.

The properties with which we are primarily concerned in this work are h_c, D_c, and χ. The emigration coefficient h_c incorporates several steps involved in the movement of white cells from the bloodstream into the tissue: 1) venule dilation and margination of cells to the venule wall; 2) adherence of the cells to the wall; 3) migration across the wall cell layer. Inequality (17) shows that low values of h_c may allow bacterial proliferation; therefore, defects in any of these steps could lead to infection.

The elimination criterion also favors large values of the random motility coeffecient. However, as D_c increases, ρ decreases monotonically toward an asymptotic value, so that there is a point after which an increase in the random motility yields little benefit. In fact, a highly suggestive result obtains when $D_c >> g_c R^2$ and $h_c >> g_c R$. These may be expected to be valid for a normally functioning defense, since the leukocyte death rate should be small relative to the movement into and across the tissue region. When they hold true, inequality (17) becomes

$$\frac{k_d c_b}{k_g} > 1 \tag{18}$$

That is, the elimination criterion becomes <u>independent</u> of the motility properties.

This result suggests the need for chemotaxis -- there would otherwise be no possibility of preventing infection by a bacterial species possessing properties violating inequality (18).

Regarding the mechanisms by which chemotaxis can help provide adequate defense, one evident function is to dramatically alter the leukocyte profiles within the tissue. For $\delta = 0$, the maximum leukocyte density must be found at $\xi = 1$, and is bounded by the value c_b. For $\delta > 0$, the leukocyte density can exceed c_b in the tissue. This is because the maximum principle for differential equations does not hold for equation (13b) when $\delta > 0$. This means that peaks of leukocyte density may form within the tissue. These peaks can localize the leukocytes at the points of greatest bacterial density once within the tissue. According to the transient computations, the leukocytes will seek the region of large bacterial density by following the bacterial density gradient. This allows efficient destruction of the harmful agent. An example is illustrated in Figures 6-8.

A less obvious consequence of chemotaxis is the possibility for a tremendous increase of the leukocyte infiltration into the tissue, as is apparent in Figure 4. This increase in recruitment is effected by the modification of the leukocyte boundary condition at $\xi = 1$ (compare equation (14b) for $\delta = 0$ and $\delta > 0$). For $\delta > 0$, it is possible for the slope $u_\xi(1)$ to be negative; hence, $u(1)$ may decrease. Since the leukocyte influx is diminished as $u(1)$ increases approaching 1, decreasing $u(1)$ allows greater infiltration. Thus, chemotaxis can promote more rapid movement from the vessel wall into the deep tissue regions, permitting more cells to migrate across the wall. This is independent of any change in h_c. Notice, however, that this effect will only be important when the limiting step for infiltration is movement from the wall into the tissue; that is, when leukocyte accumulation at the vessel wall inhibits further immigration as $c(R)$ approaches c_b. The effect of h_c on infiltration may be more important.

Another function of chemotaxis that is possible but is not considered in this model could be to promote the movement of cells from the venules into the tissue by actually increasing the value of h_c. This would be possible if a chemotactic mediator enhances any of the three steps involved in emigration mentioned earlier. Chemotaxis might also help increase the value of c_b; if chemotactic factor diffuses into the bloodstream, it may serve to attract white cells from their bone marrow sources (Kass & de Bruyn, 1976; Miller et al, 1971; Rother, 1971).

Conclusions

A mathematical model of the inflammatory response to bacterial invasion of tissue has been developed in order to better understand the relations between cellular kinetic processes and transport phenomena, and host susceptibility to infection. The use of _in vitro_ quantitative measurements of host leukocyte properties such as random motility and chemotactic coefficients in models that predict _in vivo_ inflammatory response behavior could be a useful technique in clinical diagnosis and

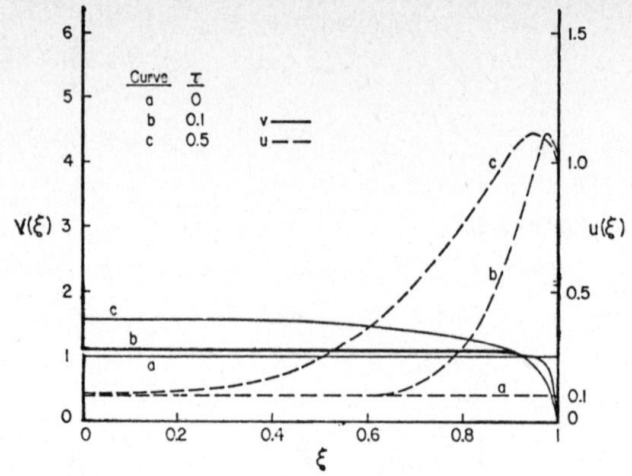

Figure 6. Examples of transient cell density profiles in tissue, as τ increases. Parameter values used are the same as for Figure 5, with $\delta = 3 \times 10^{-2}$.

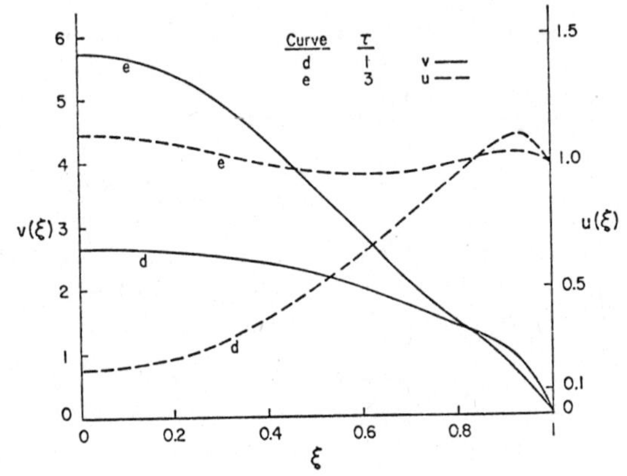

Figure 7. Examples of transient cell density profiles in tissue, as τ increases. Parameter values used are the same as for Figure 6.

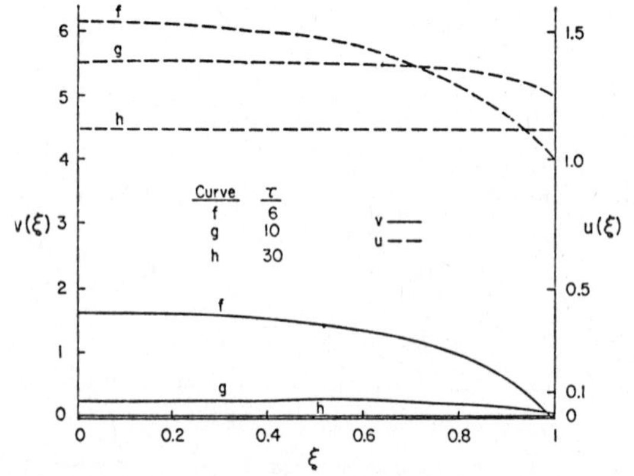

Figure 8. Examples of transient cell density profiles in tissue, as τ increases. Parameter values used are the same as for Figure 6.

treatment. *In vitro* assays of leukocyte motility and chemotaxis are already being used commonly, although there is incomplete understanding of the connection between *in vitro* measurements and *in vivo* defense performance. We hope that our initial modeling efforts, some of which were presented here, will stimulate further investigation of the dynamics of the inflammatory response. The other important task is the development of methods for quantitative analysis of the various *in vitro* assays, in order for the assays to yield useful predictive information. We have recently presented our first results in this area (Lauffenburger and Aris, 1979), and the potential for further useful work appears to be great.

Finally, our model highlights the complicated nature of leukocyte infiltration. Therefore, the mechanism and kinetics of the emigration process need further investigation, especially in regard to their possible dependence upon chemotaxis.

Acknowledgements

This work was undertaken as part of a doctoral dissertation under the direction of Professors Rutherford Aris and Kenneth H. Keller, Department of Chemical Engineering and Materials Science, University of Minnesota. Also gratefully acknowledged is the help provided by Dr. Sally Zigmond, Department of Biology, University of Pennsylvania.

References

Athens, J., O. Haab, S. Raab, A. Mauer, H. Ashenbrucker, G. Cartwright, and M. Wintrobe (1961): J. Clin Invest. 40, 989-995.
Beesley, J., J. Pearson, J. Carleton, A. Hutchings, and J. Gordon (1978): J. Cell Sci. 33, 85-101.
Craddock, C. (1972). In Hematology (W. Williams, E. Beutler, A. Erslav, and R. Rundles, eds.), McGraw-Hill, New York, pp. 607-617.
Eccles, S. and P. Alexander (1974): Nature 250, 667-669.
Fehr, J. and H. Jacob (1977): J. Exp. Med. 146, 641-652.
Goetzl, E., S. Wasserman, I. Gigli, and K. Austen (1974): J. Clin Invest. 53, 813-818.
Hersh, E. and G. Bodey (1970): Ann. Rev. Med. 21, 105-132.
Hogan, N. and H. Hill (1978): J. Inf. Dis. 138, 437-444.
Isturiz, M., A. Sandberg, E. Schiffman, S. Wahl, and A. Notkins (1978): Science 200, 554-556.
Kass, L. and P. de Bruyn (1976): Anat. Rec. 159, 115-123.
Keller, E. and L. Segel (1971): J. Theor. Biol. 30, 225-234.
Keller, H., M. Hess, and H. Cottier (1975): Sem. Hemat. 12, 47-57.
Lauffenburger, D. and R. Aris (1979): Math. Biosci. 44, 121-138.
Lauffenburger, D. and K. Keller (1979): J. Theor. Biol. 81, 475-503.
Metchnikoff, E. (1968): Lectures on the Comparative Pathology of Inflammation, Dover Press, New York.
Miles, A. (1956): Ann. NY Acad. Sci. 66, 356-369.
Miller, M. (1975): Sem. Hemat. 23, 59-82.
Miller, M., F. Oski, and M. Harris (1971): Lancet i, 665-669.
Mizushima, Y., N. Matsumara, M. Mori, T. Shimizu, B. Fukushima, and Y. Mimura (1977): Lancet ii, 1037.
Quie, P. and K. Cates (1977): Am. J. Path. 88, 711-725.
Rosen, G. (1976): J. Theor. Biol. 59, 371-380.
Rother, K. (1971): J. Imm. 107, 316.
Rytel, M. (1973): Inf. Dis. Rev. 3, 185-196.
Sell, S. (1975): Immunology, Immunopathology, and Immunity, Harper & Row, Hagerstown,MD.
Senn, H. and W. Jungi (1975): Sem. Hemat. 12, 27-45.
Stossel, T. (1972): NE J. Med. 286, 776-777.
Stossel, T. (1973): J. Cell Biol. 58, 345-356.
Von Graevenitz, A. (1977): Ann. Rev. Microb. 31, 447-471.
Wilkinson, P.C. (1974): Chemotaxis and Inflammation, Churchill-Livingstone, Edinburgh and London.

MATHEMATICAL THEORIES OF TOPOTAXIS

Ralph Nossal

National Institutes of Health, Bethesda, Maryland 20205

I. Introduction

The attraction of microorganisms by external stimuli has fascinated biologists and other natural scientists for many years. Classical papers on the behavior of plant sperm and bacteria moving in response to chemical gradients appeared almost 100 years ago (Pfeffer, 1884; Engelmann, 1881). Other stimuli which are known to affect locomotory behavior of unicellular organisms are light, heat, gravity, mechanical perturbation, and magnetic fields. However, in recent years movement of cells in chemical fields has been the most thoroughly studied. Coincidentally, most mathematical modelling of cell migration phenomena has concerned response to chemical signals.

Not only are phenomena of chemotaxis and chemokinesis (see Sec.II for terminology) of intrinsic biological interest, but they are important elements in the way higher organisms fight disease. The inflammatory response of neutrophils and other blood cells now is known to be strongly correlated with those cells' ability to respond to chemotactic stimuli (Wilkinson, 1974; Gallin and Quie, 1978). Although blood cells are the only normal mammalian cells which show chemotactic response, certain neoplastic cells, too, are attracted by tumour specific mediators (Wilkinson, 1974). Also, there is evidence that the detection of chemical gradients is important in the development of polarity in the vascularization of wounds and tumors (Folkman and Cotran, 1976), and one can conceive of ecological phenomena where organisms respond to gradients of environmental stimuli. A reader who is interested in detailed information about basic biological phenomena and current research might wish to refer to the references just cited, or to recent reviews by Becker(1977), Zigmond(1978), and MacNab(1978).

II. Terminology and Definitions

The terminology used to describe locomotory responses of microorganisms oftentimes has been somewhat ambiguous. Consequently, several investigators who contribute to this field (Diehn, et al.,(1977); Keller, et al.,(1977)) recently have attempted to clarify various pertinent terms. Their nomenclature generally follows that given in the classical treatise of Fraenkel and Gunn (1940) but, because it may be unfamiliar, we now provide a brief review.

If the steady-state activity of the motor apparatus depends upon the absolute magnitude of a stimulus intensity, the response associated with a changing stimulus is termed kinesis. An example is the increased swimming speed which occurs when certain blue-green algae are illuminated by light of increased intensity (Nultsch,1962). The term orthokinesis indicates that the speed of a microorganism is affected. The term klinokinesis is used to describe a change in the frequency of directional change. (For cells locomoting on a surface by essentially straight-line paths, klinokinesis implies that the time between turns is affected (see Fig. 1).) *Paramecia* show klinokinesis and orthokinesis in response to spatial variations of cation concentrations (Van Houten, 1978).

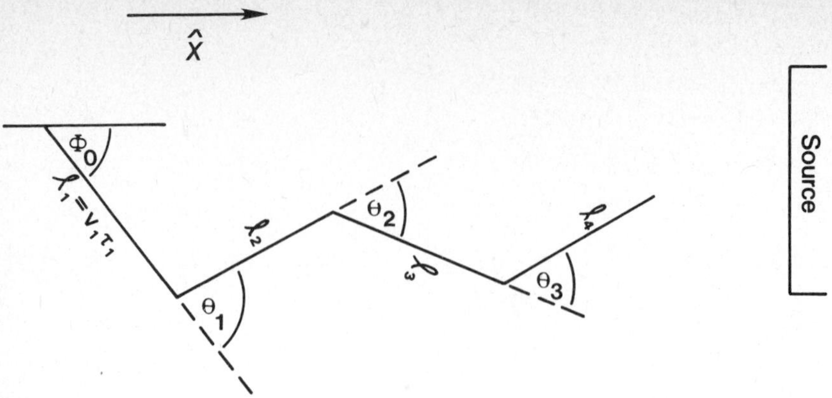

Figure 1: Trajectory of cell locomotion

If the movement of the microorganism bears no relationship to any fixed direction, its locomotion is said to be randomly directed. Most locomotion of microorganisms contains some stochastic content, and the migration of individual cells has the appearance of a generalized random walk. In the case of a spatially isotropic, randomly directed, movement, such motion oftentimes can be characterized by a mobility parameter analogous to the diffusion coefficient of molecular migration. In order that this representation be appropriate, observations must be performed only after a delay which is long compared with times of microscopic motion such as the average duration between turns. Gail and Boone (1970) showed that fibroblast locomotion can be described by such a parameter. In Secs.IV and V. we explicitly indicate how the mobility parameter depends on the speeds and the magnitude or frequency of turns.

A stimulus gradient can lead to spatial accumulation of microorganisms merely as a result of klinokinesis. However, accumulation oftentimes is more efficiently elicited by tactic responses. The term taxis signifies that a stimulus causes an orientational change in the locomotion of the microorganism. An animal might move directly and unswervingly to a stimulus source, but more generally some stochastic parameter of the random walk is affected, resulting in a spatially biased displacement. In the case of leukocytes (neutrophils) which move on a surface, the turn-angle probability of the cells is affected by a chemotactic field (Zigmond, 1974) and the cells tend to zig-zag towards the attractant (Nossal and Zigmond, 1976). Although average orientation is towards the source, the cells still execute a random walk as they locomote.

Topotaxis is a term which was used in early work, but it has the same meaning as the term 'taxis' alone (Fraenkel and Gunn, 1940) and now is only rarely used. Positive taxis signifies that the microorganism has a tendency to move towards a stimulus, whereas negative taxis implies movement away from a stimulus; similar definitions hold for kineses. If a change in stimulus intensity results in but a temporary change in locomotion, the response sometimes is referred to as phobic.

Many different types of stimulus are known, usually designated by a prefix such as chemo (signifying movement in a chemical gradient), photo (responding to light), geo (influenced by a gravitational field), etc. Substances which elicit chemotactic response oftentimes are referred to as chemotaxins.

Flagellated bacteria exhibit a biased motion in gradients of chemical attractants which is quite different than that shown by leukocytes. The distribution of direction angles remains spatially isotropic when these organisms move, but the frequency of turns is decreased when a cell moves from a lower to higher concentration (Berg and Brown, 1972). However, the same behavior is noted if the medium is spatially isotropic and the concentration of chemoattractant is increased uniformly by enzymatic reactions (Brown and Berg, 1974). Thus, the frequency of turns is not dependent on the steady state stimulus but, rather, depends on a temporal change in stimulus. A distinction must be made between this type of behavior and simple klinokinesis. Although such response does not fit the narrow definition of taxis given above, we nonetheless shall refer to it in the following as chemotaxis. In Sec.IV we show explicitly how such a mechanism leads to spatial accumulation, and we obtain mathematical expressions which relate microscopic responses resulting from temporal detection of chemoattractant to parameters describing gradient response on a macroscopic scale.

III. Kinetic Equations for Macroscopic Motions

"Behavioral" studies of locomotion frequently involve experimental timescales which are long compared with times characterizing intrinsic microscopic cell response. Examples of the latter are the intervals between the turns of a stochastic trajectory and the times spent in turning, both of which usually are of the order of seconds or minutes. Macroscopic timescales, in contrast, are minutes or hours, examples being those pertaining to clinical assays. We now review equations used for describing the movements of cell populations on macroscopic scales of space and time.

Let $n(\mathbf{r},t)d^3r$ designate the number density of microorganisms located in a differential volume element d^3r about $\mathbf{r}$ at time t. Conservation of particles implies that the time evolution of $n(\mathbf{r},t)$ is given by the continuity equation as

$$\frac{\partial n(\mathbf{r},t)}{\partial t} = -\nabla\cdot\mathbf{J}(\mathbf{r},t) + G(\mathbf{r},t) \qquad (3.1)$$

where $\mathbf{J}(\mathbf{r},t)$ is a current representing a flow of cells into or out of the volume, and $G(\underset{\sim}{r},t)$ is a source term which accounts for cell growth and/or destruction. If the macroscopic time scale is sufficiently long that many fundamental stochastic decisions have been made before a system is observed (e.g., many steps on a random walk have occurred), the current $\underset{\sim}{J}(\underset{\sim}{r},t)$ can be written as (Keller and Segel, 1971)

$$\underset{\sim}{J}(\underset{\sim}{r},t) = -\mu(\underset{\sim}{r},t)\underset{\sim}{\nabla}n(\underset{\sim}{r},t) + \underset{\sim}{v}_d(\underset{\sim}{r},t)n(\underset{\sim}{r},t) \qquad (3.2)$$

where $\mu(\underset{\sim}{r},t)$ is the "mobility coefficient", and $\underset{\sim}{v}_d(\underset{\sim}{r},t)$ is a drift velocity engendered by a spatially anisotropic stimulus.

The mobility coefficient is the analog of the diffusion coefficient; for a stationary, spatially isotropic system, the moments of the distribution $n(\underset{\sim}{r},t)$ evolve according to

$$\overline{(\underset{\sim}{r}-\overline{r_t})^2} = [\overline{r_t^2} - (\overline{\underset{\sim}{r}_t}\cdot\overline{\underset{\sim}{r}_t})] = 2\delta\mu t \qquad (3.3)$$

where δ is the dimensionality, and $\overline{r_t^2}$ and $\overline{r_t}$ are given as

$$\overline{r_t^2} \equiv \int \underset{\sim}{r}\cdot\underset{\sim}{r}n(r,t)d^3r; \quad \overline{\underset{\sim}{r}}_t \equiv \int \underset{\sim}{r}n(r,t)d^3r \quad . \qquad (3.4)$$

When μ is calculated from a stochastic description of the motion of individual cells, the moments $\overline{r_t^2}$ and $\overline{\underset{\sim}{r}_t}$ are replaced by expected values; $2\delta\mu$ is the proportionality coefficient between time and the mean-square displacement of a representative cell.

If the system is not spatially isotropic, the mobility coefficients may differ according to the direction in which the cells are moving. In this case Eqs.(3.1)-(3.2) would appear as (Crank, 1956)

$$\frac{\partial n(\underset{\sim}{r},t)}{\partial t} = \frac{\partial}{\partial x}\mu_x(\underset{\sim}{r},t)\frac{\partial n(\underset{\sim}{r},t)}{\partial x} + \frac{\partial}{\partial y}\mu_y(\underset{\sim}{r},t)\frac{\partial n(\underset{\sim}{r},t)}{\partial y}$$

$$+ \frac{\partial}{\partial z}\mu_z(\underset{\sim}{r},t)\frac{\partial n(\underset{\sim}{r},t)}{\partial z} - \underset{\sim}{\nabla}\cdot[\underset{\sim}{v}_d(\underset{\sim}{r},t)n(\underset{\sim}{r},t)] + G(\underset{\sim}{r},t) \quad , \qquad (3.5)$$

when the distribution of angles through which the cells turn is symmetric about zero (Nossal and Weiss, 1974a; Lovely, 1974). Expressions for anisotropic mobility coefficients μ_x, μ_y, and μ_z have been derived by Nossal and Weiss (1974a) and by Lovely (1974). (Lovely also considered the more general situation of asymmetric turn-angle distributions, in which case Eq.(3.5) contains additional terms because the mobility tensor no longer necessarily is diagonal and symmetric.) If a cell orients strongly in an external stimulus field and moves on a stochastic trajectory towards the source, one might expect that the dispersion $\langle r_i^2 - \langle r_i \rangle^2 \rangle$ in the direction $\hat{x}_i$ of the source would increase whereas diffusion in a perpendicular direction would decrease.

When the external stimulus is fixed in space and time, Eqs.(3.1) and (3.2) are linear equations. Such equations form the basis of several assays for bacterial or leukocyte chemotaxis which are reviewed below. The fundamental constitutive relationship for the response of flagellated bacteria to a gradient of chemoattractant is (Brown and Berg, 1974)

$$\underset{\sim}{v}_d \sim \frac{K_D C}{(K_D + C)^2} \underset{\sim}{\nabla} \ln C \qquad (3.6)$$

where C is the concentration of the attractant and K_D is a constant. If the chemoattractant gradient is created by the metabolic activity of the bacteria an additional relationship is needed to relate the time change of C to the cell density $n(\underset{\sim}{r},t)$. A typical expression is (Keller and Segel, 1971)

$$\frac{\partial C(\underset{\sim}{r},t)}{\partial t} = \underset{\sim}{\nabla} \cdot D \underset{\sim}{\nabla} C(\underset{\sim}{r},t) - kn(\underset{\sim}{r},t)C^{\alpha} \qquad (3.7)$$

where D is the diffusion coefficient of the chemoattractant, k is a rate constant related to the metabolism of the bacteria, $0 \leq \alpha \leq 1$.

If the concentration of chemoattractant is within a certain range (specifically, if $C \approx K_D$), the expression given by Eq.(3.6) becomes

$$\underset{\sim}{v}_d \sim \frac{1}{C} \underset{\sim}{\nabla} C \quad . \qquad (3.8)$$

This dependence on relative stimulus intensity has been observed in many sensory transduction processes, and is known to physiologists as the "Weber-Fechner law". Although it is tempting to use a similar constitutive relation when formulating mathematical models for polymorphonuclear leukocytes (neutrophils), experimental evidence suggests that the response may be somewhat different (see Sec.VI).

Note that even if the signal which elicits a true tactic response is absent the microorganisms can redistribute themselves and exhibit localized accumulation. If the mobility coefficient varies with position, Eqs.(3.1) and (3.2) yield

$$\frac{\partial n(\underset{\sim}{r},t)}{\partial t} = \mu(\underset{\sim}{r})\nabla^2 n(\underset{\sim}{r},t) + \underset{\sim}{\nabla}\mu(\underset{\sim}{r}) \cdot \underset{\sim}{\nabla} n(\underset{\sim}{r},t) \quad . \qquad (3.9)$$

The second term on the right has the same functional form as vectorial drift arising from tactic stimulation. However, in this instance the cells experience kinesis. For example, if the mobility depends on the concentration of a chemical, the cells experience chemokinesis; yet, no angle bias will be observed, in contrast with certain chemotactic responses where an average orientation towards the source occurs.

Equations such as those discussed here have been used to analyze assays for leukocyte chemotaxis (Zigmond and Hirsch, 1973; see, also references given in Sec.VI), and to describe band formation in populations of chemotactic bacteria (Keller and Segel, 1971; see, also, references given in Sec.VII). Somewhat more elaborate equations have been used to model the aggregation of slime mold amoeba (Keller and Segel, 1970; see Sec.VII). Also, a similar set of equations which includes a time varying mobility coefficient, but which does not contain drift induced by an external stimulus, has been used to model an in vitro immune assay for cellular sensitivity (Nossal, 1977,1978).

Until now we have implicitly assumed that all motile particles have the same properties. However, in certain cases an assembly might contain different species, e.g., a "fast" and a "slow" population, and appropriate modifications would be required.

IV. Stochastic Models (Bacteria)

In principle, if the detailed movements of individual cells were known the locomotion of a population of such cells could be described. Moreover, the microscopic behavior of the cells must be studied in order to fully understand the basic biological mechanisms underlying locomotion. Thus, information of motion on a 'microscopic' scale in a sense is more fundamental than characterization of macroscopic behavioral response. Unfortunately, the efforts involved in analyzing trajectories frequently are quite tedious (e.g., the tracing of cells from cinemicrographs) or require an unusually high level of technical expertise (an example being the observation of bacteria with a computer controlled tracking microscope (Berg, 1971); also, when motion picture film is used there is a considerable delay before data can be analyzed because of time needed for processing. Consequently, quick macroscopic assays oftentimes are preferred, particularly for clinical applications. In order to fully appreciate the significance of parameters measured by macroscopic means, their relationship to microscopic motions must be understood.

Several investigators recently have examined the trajectories of cells in order to characterize the stochastic behavior of representative individual cells within the population. For example, Berg and Brown (1972) investigated the movements of chemotactic E. Coli bacteria; leukocytes have been studied by Peterson and Noble(1972), Zigmond(1974), Nossal and Zigmond(1976), and Hall and Peterson(1979); slime mold amoebae have have been investigated by Hall(1977); 3T3 mouse fibroblasts have been observed by Gail and Boone(1970) and Albrecht-Buehler(1979); and mycoplasma were investigated by Bredt et al(1970). Other investigators also have analyzed trajectories of individual cells, but the abovementioned studies particularly emphasize the stochastic character of motion.

To a good approximation the trajectories of these cells can be characterized as straight-line paths, separated by discrete turns (see Fig.1). In the case of bacteria, the cells actually stop and "twiddle" while seeming to hunt for a new direction of motion (Berg and Brown, 1972). The execution of turns by amoeboid-like cells is less dramatic, but the cells frequently experience a slowing down while changing directions. In both cases, the stochastic parameters of interest are the speeds and directions of movement, the run lengths (either time or distance travelled between turns), the angles between turns, times spent in turning, and the bias by an external field. Not all of these parameters are independent, and different investigators choose different subsets in order to describe their observations; in this paper trajectories usually are specified in terms of the average speeds along each path segment $\{v_i\}$, the time durations of such motion $\{\tau_i\}$, the turn-angles between segments $\{\theta_i\}$, and the delays while cells are resting between turns (Nossal and Weiss, 1974a,b).

The parameters μ and $\underset{\sim}{v}_d$ which appear in the formulation given by Eqs.(3.1)-(3.3) depend upon the characteristics of the cell system being studied. These are related to the expected values of the distances which the cells travel as (Chandrasekhar, 1943)

$$\mu \sim \lim_{t\to\infty} \frac{1}{t} [\langle r_t^2\rangle - \langle r_t\rangle\cdot\langle r_t\rangle] \qquad (4.1)$$

$$\underset{\sim}{v}_d = \lim_{t\to\infty} \frac{1}{t} \langle \underset{\sim}{r}_t \rangle \quad . \tag{4.2}$$

The proportionality coefficient in Eq.(4.1) depends on the dimensionality of the system and, for locomotion in two dimensions, is 1/4. The parameter $\underset{\sim}{v}_d$ measures the directional response of the cells; if there is no net drift due to external fields, one has $\langle \underset{\sim}{x}_t \rangle = 0$ and $\underset{\sim}{v}_d = 0$, $\mu = \frac{1}{4}\lim_{t\to\infty} t^{-1}\langle \overline{r_t^{\,2}} \rangle$.

Information obtained recently on the motion of bacteria and leukocytes helps us focus on particular aspects of the mathematical modelling needed to compute these quantities. However, the influence of external fields on the locomotion of these cells is somewhat different, and we first discuss the migration of bacteria. As has already been mentioned, when bacteria move in a gradient of chemoattractant they do not seem to orient in the field, that is, the instantaneous directions of the cells' trajectories are spatially isotropic. All angles $\{\theta_i\}$ are equally probable (Figure 1), and the probability of turning through an angle θ does not depend upon the direction of motion prior to the turn. Net movements of cells towards a source occurs because the probability of stopping and changing directions is dependent on orientation; in other words, the run-times $\{\tau_i\}$ are biased by the external field so that motion towards the source occurs for longer periods than does oppositely directed motion.

Actually, it has been shown that the bacterial flagella move in a coordinated fashion for longer periods when the bacteria sense an increase in attractant concentration with time (Brown and Berg, 1974). We now examine how the latter behavior leads to accumulation in a spatially varying field. For simplicity, we here consider motion in two dimensions (Nossal and Weiss, 1974a,b); some related results have been derived elsewhere for motion in three dimensions (Lovely and Dahlquist, 1975; Hall, 1977).

Because the cells do not orient in the field, the chemotactic velocity can be determined in a rather simple manner. When a cell moves for discrete time intervals $\{\tau_i\}$, the distance travelled in a given direction $\hat{x}$ (taken, e.g., as the direction towards the stimulus) is

$$x_t = \sum_i \int_0^{\tau_i} x_i(s)ds \tag{4.3}$$

where $\sum_{i=1}^{n} \tau_i = t$. If the angle between the $\hat{x}$ axis and the instantaneous velocity $\underset{\sim}{v}(s)$ is designated as $\phi_i(s)$, it then follows that the expected value $\langle x \rangle$ at time t is

$$\langle x_t \rangle = \langle \sum_i \int_0^{\tau_i} v_i(s)\cos\phi_i(s)ds \rangle \quad . \tag{4.4}$$

If we assume that the velocities are constant between turns, we then have

$$\langle x_t \rangle = \langle \sum_i v_i \tau_i \cos\phi_i \rangle$$

$$= \sum_{i=1}^{n} \int v_i \tau_i \cos\phi_i \mathbb{P}(\{v_i\};\{\tau_i\};\{\phi_i\}) d\{v_i\} d\{\tau_i\} d\{\phi_i\} \tag{4.5}$$

where $\mathbb{P}(\ldots)$ is the joint probability that a cell is moving with speed v_i for time τ_i at angle ϕ_i on the ith link, etc. Other assumptions which seem to be permissible are: i) the locomotion of a cell is Markoffian, i.e., the probability of responding to a stimulus at any given time is a function only of the 'state' (position, speed, direction) of a cell at that time, and not of its prior history; ii) the speeds exhibited by a cell on successive links are uncorrelated (this assumption can be relaxed to account for 'fast' and 'slow' cells) and are not correlated with directions; iii) run length distributions, which do depend on directions, are exponentially distributed. The latter assumption is based on experimental observation of bacterial motion, and shall be relaxed when we discuss leukocyte migration. Of course, we also assume a stationary random process.

It now follows from Eq.(4.5) that $\langle x_t \rangle$ is given as

$$\langle x_t \rangle = n \int_0^\infty P(v)dv \int_{-\pi}^{\pi} p(\phi)d\phi \int_0^\infty \cos\phi \cdot v \cdot \tau \cdot P(\tau|\phi,v)d\tau \qquad (4.6)$$

$$= (2\pi)^{-1} n \int_0^\infty P(v)vdv \int_{-\pi}^{\pi} \cos\phi d\phi \int_0^\infty \tau\lambda(\phi,v)e^{-\lambda(\phi,v)\tau}d\tau \qquad (4.7)$$

where, to get Eq.(4.7) from (4.6) we used the facts that the distribution of directions is spatially isotropic, i.e., $p(\phi)=1/2\pi$, and that the distribution of run lengths is exponential, $P(\tau|\phi,v) = \lambda(\phi,v)\exp(-\lambda(\phi,v)\tau)$. The speed distribution P(v) can be any well behaved function.

The experiments which provided a quantitative measure of the temporal dependence of chemotactic response in flagellated bacteria were performed by Brown and Berg (1974). Bacteria were suspended in a homogeneous medium and the concentration of attractant (L-glutamate) was changed enzymatically. By changing the relative amounts of substrates and products, the amount of attractant either could be increased or decreased. When attractant was generated the frequency of directional changes exhibited by the bacteria was decreased, but destruction of chemoattractant had no effect on the motion. The mean run length was found to be proportional to the time rate of change of bound chemoattractant. Data were fit to the expression

$$\langle \ell n\lambda \rangle = \begin{cases} \ell n\lambda_0 - \alpha dP_b/dt & \text{if } dP_b/dt > 0 \\ \ell n\lambda_0 & \text{otherwise} \end{cases} . \qquad (4.8)$$

where dP_b/dt is the time rate of change of the fractional amount of occupied chemoreceptors (Mesibov, Ordal, and Adler, 1973)

$$dP_b/dt = [K_D/(K_D+C)^2]dC/dt \quad . \qquad (4.9)$$

K_D here is the dissociation constant for binding of attractant to receptors.

How does the information contained in Eqs.(4.8) and (4.9) relate to the manner in which the drift velocity $\underset{\sim}{v}_d$ depends on a spatial gradient of chemoattractant? Consider a uniform gradient in the $\hat{x}$ direction, in which case as a bacterium swims it will sense a temporal change in chemoattractant concentration given as $dC/dt = v \cos\phi \, dC/dx$, where $v \cos \phi$ is the component of motion along the gradient. Let us assume that α is small, in which case we have from Eqs.(4.8)-(4.9)

$$\langle\lambda\rangle^{-1} \cong \begin{cases} \lambda_0^{-1}\left(1 + \dfrac{\alpha K_D}{(K_D+C)^2} v\cos\phi\dfrac{dC}{dx}\right) & \text{if } |\phi|<\pi/2 \\ \lambda_0^{-1} & \text{if } |\phi|\geq\pi/2\ . \end{cases} \quad (4.10)$$

$$\equiv \begin{cases} \lambda_0^{-1} + \varepsilon v\cos\phi & \text{if } |\phi|<\pi/2 \\ \lambda_0^{-1} & \text{if } |\phi|\geq\pi/2 \end{cases} \quad (4.11)$$

where ε is given as

$$\varepsilon \equiv \frac{\alpha K_D C}{\lambda_0 (K_D+C)^2} \frac{d\ell nC}{dx} \quad . \quad (4.12)$$

Thus, from Eq.(4.7) we find

$$\langle x_t\rangle = (2\pi)^{-1} n \int_0^\infty vP(v)dv \int_{-\pi}^{\pi} d\phi\cos\phi\lambda^{-1}(\phi,v) \quad (4.13).$$

$$= \frac{\varepsilon}{4} n\langle v^2\rangle \quad , \quad (4.14)$$

where $\langle v^2\rangle = \int_0^\infty v^2 P(v)dv$.

The chemotactic velocity is defined as in Eq.(4.2). If the time between turns is negligible, the total time is related to the average run-length as

$$t = n\langle\tau\rangle = n(2\pi)^{-1}\int_0^\infty P(v)dv \int_{-\pi}^{\pi} d\phi\lambda^{-1}(\phi,v) \quad (4.15)$$

$$= n\lambda_0^{-1} + \mathcal{O}(\varepsilon) \quad (4.16)$$

where, to get Eq.(4.16) from Eq.(4.15) we again used Eq.(4.11). Thus, from Eqs.(4.2) and (4.12) we finally have

$$v_d \approx \frac{\alpha}{4} \langle v^2\rangle \frac{K_D C}{(K_D+C)^2} \frac{d\ln C}{dx} \quad (4.17)$$

which, over a limited range of chemoattractant concentration ($C = K$), has the form of the "Weber-Fechner law" $\underset{\sim}{v}_d \sim \underset{\sim}{\nabla} \ln C$ (Mesibov, Ordal, and Adler, 1973; Brown and Berg, 1974). The concentration dependence given by the function $K_D C/(K +C)^2$ has been noted in studies of bacteria swimming in exponential gradients (Dahlquist, Lovely, and Koshland, 1972). The $\langle v^2 \rangle$ dependence seen in Eq.(4.17) also has been derived theoretically by Segel (1976), who analyzed a detailed model of enzymatically mediated attractant-receptor interactions. However, such speed dependence has not yet been verified experimentally.

The expression given in Eq.(4.16) should be modified if the bacteria hesitate for a significant time between turns. The total time would be given as

$$t = n(\lambda_0^{-1} + \tau_w) \quad (4.16')$$

where τ_w is the average time spent in "twiddling", and $\lambda_0\tau_w$ is the ratio of the time spend in the 'rest' phase to that spent in active locomotion. Instead of Eq.(4.17) one then obtains

$$v_d = \frac{\alpha\langle v^2\rangle}{4(1+\lambda_0\tau_w)} \frac{K_D C}{(K_D+C)^2} \frac{d\ln C}{dx} \quad . \quad (4.18)$$

This analytic relationship between temporal stimulus and gradient response seems to be a new result, but the expressions which we have just derived are closely related to others obtained by a different scheme in earlier analyses of cell migration in two dimensions (Nossal and Weiss, 1974a,b). Lovely and Dahlquist (1975) derived a general expression for directed motion in three dimensions whose mathematical form essentially is identical to that obtained by Nossal and Weiss (1974a,b) for surface locomotion, and we thus expect that the result given by Eq.(4.18) is appropriate for three-dimensional swimming motion as well as two-dimensional locomotion.

Nossal and Weiss (1974a,b) also derived expressions for anisotropic diffusion in two dimensions when the anisotropy arises from taxis (general relationships were obtained by Lovely(1974)). The derivations and resulting expressions are somewhat complicated and the interested reader should consult the original sources for details. For isotropic two-dimensional motion (in the absence of a field), one has (Nossal and Weiss, 1974a; Lovely and Dahlquist, 1975; Hall, 1977)

$$\mu = \frac{1}{2\lambda_0}\left[[\langle v^2\rangle - \langle v\rangle^2] + \frac{\langle v\rangle^2}{1-\langle\cos\phi\rangle} \right] \qquad (4.19)$$

where λ_0^{-1} is the mean duration of a trajectory and $\langle\cos\phi\rangle$ is averaged over turn-angle distributions. When motion occurs in three dimensions the multiplicative factor in Eq.(4.19) is 1/3, rather than 1/2.

For the special case of chemotactic movement of bacteria treated above, one finds (Nossal and Weiss, 1974a) that the correction resulting from drift is of the order $\varepsilon^2 \sim (\nabla C)^2$ (cf. Eq.(4.14)). Thus, for weakly biasing fields μ can be considered not to depend on the chemotactic stimulus, even though it might vary with absolute chemoattractant concentration through its dependence on $\langle v^2\rangle$ and $\langle v\rangle$. Of course, for sufficiently strong fields, the expression for mobility will show a non-linear dependence on stimulus strength, as will the drift velocity.

V. Stochastic Motions (Neutrophils)

The ease with which an expression for v_d was derived in the previous section results from the fact that the probability of turning through an angle θ is not dependent upon the cell's orientation. When the turn-angle distribution depends on the path angle prior to a turn, it appears that the mathematics needed to relate macroscopic parameters of locomotion to the stochastic movements of individual organisms is a bit more complicated.

Studies of leukocyte locomotion (Zigmond, 1974; Nossal and Zigmond, 1976) show that the conditional turn angle distribution function $p(\theta|\phi)$ can be well represented as $p(\theta|\phi)=p(\theta+f(\phi))$. An expression for v_d for this case was derived (Nossal, 1976) by obtaining an expression for the Laplace time transform of $\langle x_t \rangle$ and then examining the asymptotic behavior $\lim_{t\to\infty} \langle x_t \rangle$ by invoking a Tauberian theorem. Assumptions which were invoked are that: i) successive path segments are uncorrelated; ii) velocities are not correlated either with run lengths, turn-angle probabilities, or orientation in the field*; iii) the shape of the turn-angle probability density function is not affected by cell orientation; and, iv) the function $f(\phi)$ is odd. The latter implies that when a cell moves obliquely with respect to a field vector it has a preference to turn back towards the field direction, the consequence of which is that a cell zigzags towards the attractant.

The manipulations which in this instance are needed to obtain an expression for v_d from Eq.(4.5) are somewhat tedious (Nossal, 1976). They involve successive averaging with respect to velocities $\{v\}$, run-lengths $\{\tau\}$, and turn-angle distributions, and a reordering of a series expansion to provide an expression for the Laplace transform of

---

* A modification to account for orientation dependent velocity distributions $\mathcal{S}(v|\phi)$ has been given in Nossal and Zigmond (1976).

$\langle x_t \rangle$ as an integral over a function which itself is the solution of a Fredholm integral equation. Asymptotic analysis of the Fourier coefficients of that function then provides an expression for v_d in terms of the solution of an infinite set of coupled equations which, fortunately, has been found to be amenable to truncation (Nossal and Zigmond, 1976).

The resulting expression for v_d is

$$v_d = \langle v \rangle \, a_0 \tag{5.1}$$

where a_0 is given as the solution of the infinite set of equations

$$\frac{1}{2}\left[\beta_{m+1}+\beta_{m-1}\right] - a_0\beta_m = \sum_{\ell=-\infty}^{\infty} b_\ell \{\delta^{r}_{\ell,m} - p_\ell I(m,0,\ell) \tag{5.2}$$

$$m = \ldots -2,-1,0,1,2\ldots\ldots$$

and the $\{b_i\}$ also must be obtained from these equations. The coefficients $\{\beta_m\}$, $\{p_m\}$ and $\{I(m,0,\ell)\}$ are related to the probability distributions of turn angles as

$$\beta_m = \beta_{-m} = (2\pi)^{-1} \int_{-\pi}^{\pi} e^{-im\phi} d\phi \int_0^{\infty} \tau P(\tau|\phi) d\tau \tag{5.3}$$

$$p_m = p_{-m} = \int_{-\pi}^{\pi} e^{-im\theta} \, p(\theta) d\theta \tag{5.4}$$

and

$$I(m,0,\ell) = (2\pi)^{-1} \int_{-\pi}^{\pi} e^{-im\phi} \, e^{i\ell[\phi - f(\phi)]} d\phi \quad . \tag{5.5}$$

In the case of leukocyte chemotaxis, run-lengths are not correlated with orientation (Zigmond, 1974), in which case $\beta_i = \beta_0\delta^{kr}_{i0}$ and some simplification in Eq.(5.2) occurs. When the offset bias is taken to be constant in magnitude, so that $f(\phi)$ is

$$f(\phi) = \begin{cases} \bar{\Phi} & \text{if } \phi > 0 \\ -\bar{\Phi} & \text{if } \phi < 0 \end{cases} \tag{5.6}$$

,

the analysis of Eq.(5.2) yields

$$a_0 = \frac{2}{\pi} \sum_{\ell=0}^{\infty} (\underset{\approx}{A}^{-1})_{2\ell+1,1} / (2\ell+1)^2 \tag{5.7}$$

where $\underset{\approx}{A}^{-1}$ is the inverse of the matrix $\underset{\approx}{A}$ whose elements are

$$A_{m,m} = (1-p_m \cos m\bar{\Phi})/(mp_m \sin m\bar{\Phi})$$

$$A_{m,\ell} = \begin{cases} 4/\pi(\ell^2-m^2) & \text{if } \ell+m \text{ is odd} \\ 0 & \text{if } \ell+m \text{ is even}, \quad m \neq \ell. \end{cases} \tag{5.8}$$

Numerical evaluation indicates that convergence occurs even when $\underset{\approx}{A}$ is approximated by only an 8x8 matrix (Nossal and Zigmond, 1976).

The parameters in this theoretical model can be varied to examine their effect on the chemotactic response velocity. When the turn angle distribution function $p(\theta)$ is represented by a simple saw-tooth function, viz.,

$$p(\theta) = \begin{cases} (\theta^*)^{-2}[\theta^* - |\theta|], & |\theta| < \theta^* \\ 0 \end{cases} \tag{5.9}$$

the curves shown in Figure 2 are generated. Several interesting observations may be made from the figure. We expect that the offset bias $\bar{\Phi}$ somehow is related to the strength of a chemotactic gradient and, when leukocytes encounter a chemotactic field, directed response occurs because $\bar{\Phi}$ changes from 0 to some value $\bar{\Phi} \neq 0$. It is seen from Figure 2 that the transition from random motion to strong chemotactic reponse is enhanced if the width of the turn-angle distribution is narrow; not only is the rate at which v_d changes with

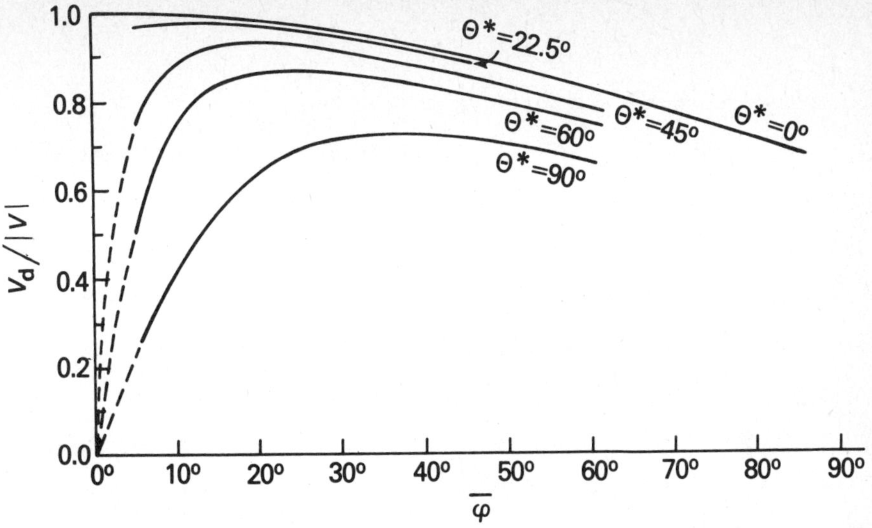

Figure 2. Dependence of v_d on turn-angle bias

$\overline{\phi}$ increased when θ* is small, but the maximal response also is greater.

Furthermore, we learn from this calculation that when the stimulus becomes so strong that offset bias exceeds 30-40°, the macroscopic responsiveness of the cells actually decreases because they tend to overcompensate in turning towards the source axis. Although Figure 2 was calculated for the functions given by Eqs.(5.6) and (5.9), other reasonable descriptions of cell motion lead to similar behavior (e.g., if the absolute value of offset f(ϕ) increases with $|\phi|$ - see Nossal and Zigmond (1976)). It thus is apparent that considerable caution must be used when deducing mechanisms of chemoreception from macroscopic observations; in particular, a leveling off in v_d or similar measure of responsiveness when a stimulus increases is insufficient evidence that chemoreceptors are being saturated. It could be that the cells are being overstimulated, so that they zigzag too avidly and thus move inefficiently towards the source.

Two other measures of macroscopic response which have been used to investigate leukocyte chemotaxis are the "chemotropism index",CI, and the fraction of cells oriented towards the source f_0. The chemotropism index (McCutcheon, 1946) is defined as

$CI = \Sigma v_i \tau_i \cos\phi_i / \Sigma v_i \tau_i$, i.e., as the net distance moved by a cell in the direction of a source, divided by the total distance travelled. In order to determine a value of CI one must keep account of the identities of individual cells as they move and, if a sufficiently long time period is involved, one effectively measures the chemotactic velocity since $\lim_{t\to\infty} \langle CI \rangle \doteq v_d / \langle v \rangle = a_0$ (cf. Eq.(5.1)).

An orientation assay (Zigmond, 1977) can be used if cells have morphological markers which indicate the directions of their locomotion. In the case of leukocytes, the assay involves placing cells on a cover slip which then is mounted on a specially fabricated microscope slide into which two wells have been cut, one of which contains chemoattractant. A bridge of fluid is set up between the wells and a chemoattractant gradient is established to which the leukocytes respond. This test has great virtue in that it involves observation at only one time, that the technology needed for obtaining primary data is relatively simple, and that information is obtained with minimal delay (e.g., it is not necessary to process movie film.)

Since the speeds of cells moving in a chemoattractant gradient are only weakly dependent upon direction (Zigmond, 1974), the chemotactic drift velocity is related to the probability density $n(\phi)$ that a cell be moving in direction ϕ (see Figure 1) as

$$v_d = \langle v \rangle \int_{-\pi}^{\pi} n(\phi) \cos\phi \, d\phi \quad . \qquad (5.10)$$

This expression is valid only if the cell locomotion is a stationary process and if $n(\phi)$ is measured at a time long enough after the start of an experiment that a stationary state has been achieved.

The fraction of cells f_0 moving towards the source is the integral $\int_{-\pi/2}^{\pi/2} n(\phi)d\phi$ and it is seen from Eq.(5.10) that f_0 bears no quantitative relationship to v_d except for rather special forms of $n(\phi)$. The total fraction of oriented cells is an even less precise measure of basic chemotactic responsiveness than is the chemotropism index or v_d, and it is subject to the same sort of "smoothing" artifacts as we noted when discussing the saturation of response seen in Figure 2. However, one important advantage of the orientation assay is that, with only slightly greater effort at data reduction, one can deduce the complete distribution $n(\phi)$. Such data could provide a better indication of the influence of chemotactic stimuli on essential response than does v_d. For example, even if increased offset bias $\bar{\Phi}$ gave saturating response in v_d as indicated in Figure 2, discernible changes in $n(\phi)$ might occur. Knowledge of $n(\phi)$ also would be helpful in establishing the proper macroscopic force-response relation for leukocytes (see Sec.VI), because v_d then could be calculated from Eq.(5.10).

Finally, we note that throughout this section it implicitly has been assumed that cells are at sufficiently low density that collisions can be ignored. However, crowding and cell-cell interactions affect locomotion so that, for example, leukocyte mobility increases at high cell concentrations (Patten, Gallin, Clark, and Kimball, 1973). If the crowding has the effect of narrowing the turn-angle distributions, it seems that both μ and v_d would increase (see Eq.(4.19) and the discussion following Figure 2).

VI. Macroscopic Assays for Leukocyte Chemotaxis

The locomotory behavior of leukocytes and other blood cells is an important factor in the way higher organisms respond to immunological challenge (Gallin and Quie, 1978). Thus, there is strong incentive to develop reliable and well understood clinical assays for the response of these cells.

The most frequently used in vitro assay for leukocyte chemotaxis is some version of the barrier assay originally developed by Boyden(1962). In this case cells are layered upon a micropore filter which separates two compartments, one of which (usually the lower) contains the putative chemoattractant. A concentration gradient is set up across the filter and movement of responsive cells into the barrier is enhanced. Unfortunately, the random mobility of the cells often also is affected by the concentration of chemotaxin, in which case chemokinesis occurs simultaneously with chemotaxis and, in certain instances, interpretation of the response becomes somewhat complicated.

One difficulty in designing mathematical models for such assays is that the basic stimulus-response relationship for neutrophil chemotaxis has not yet been unequivically determined. Using a Boyden chamber assay, Keller et al. (1977b) examined responses to a chemotaxin which first was shown not to affect the random locomotion of leukocytes. The measure of chemotactic response in this case was the number of cells which moved into the filter in a given time. For most of the concentrations which were tested, the response was found to increase with increasing concentration of chemotaxin C_f placed in the test chamber. Since, at any particular time and position in the filter, the ratio $\nabla C/C$ of chemoattractant is independent of C_f, the latter result implies that the Weber-Fechner law generally does not hold. However, for certain concentration ranges the response was found to be independent of C_f, in which case the Weber-Fechner relation is a good approximation.

Using an orientation assay (see Sec.V), Zigmond (1977) also found deviations from the Weber-Fechner law, but the deviations seem to be somewhat less pronounced (e.g., the number of cells orienting in the field when a 3-fold gradient was impressed across the sample was 55% if the maximal concentration was 10^{-7} M chemotaxin (an N-formyl methionyl peptide) and approximately 60% when the concentration was 10^{-6} M.) Also, several years ago Hu and Barnes (1970) studied neutrophils moving on a planar surface in response to materials released by damaged red blood cells. They found that the chemotactic response velocity v_d seemed to depend on the gradient of chemoattractant as $v_d \sim (\nabla C/C)^{0.25}$. Subsequently, Grimes and Barnes (1973) used well defined gradients of cAMP (cyclic 3',5'-adenosine monophosphate) and observed the behavior similar to that shown in Figure 2. No functional relationship between v_d and the ratio of the gradient to the concentration was discovered, although a gradient threshold and saturation effect were observed.

Thus, the basic responsiveness of leukocytes to chemotactic gradients is not yet clearly characterized. For many applications it might be possible to write $v_d \sim g(C)f(\nabla C/C)$, with $f(z) \simeq z$ and $g(c) \simeq$ const. The Weber-Fechner description would be appropriate, but only over a restricted concentration range (probably one which is close to the value of the binding constant for association of chemotaxin with cell bound receptors.) In other circumstances, when responsiveness varies linearly with concentration (Keller _et al_, 1977) the data probably can be fit as $g \sim C$, in which case $v_d \sim \nabla C$ would suffice for purposes of mathematical modelling. Clearly, however, because of this uncertainty about basic phenomenology, mathematical analyses of leukocyte chemotaxis must be performed with some caution.

The mathematical modelling of the barrier assay is based on Eqs.(3.1) and (3.2). Although seemingly quite straightforward, complications arise even when one assumes that a steady-state gradient of chemotaxin is established, in which case C is given as $C = C_i + (C_f - C_i)x/L$ where L is the width of the filter and C_i and C_f are the chemotaxin concentrations in the upper and lower compartments.

The difficulty is that μ and v_d generally depend on C, and therefore one must solve a partial differential equation, obtained from Eq.(3.9), having x-dependent coefficients.

Rosen (1976) has obtained a solution to these equations when μ and v_d are given as $\mu(C) = \mu_0 + \mu_1 f(C)$ and $v_d = \chi_0 \nabla C$, μ_0, μ_1, and χ_0 being positive constant parameters. (With slight modification the solution would hold for any response where v_d depends only on the gradient of C.) Infinite boundaries are assumed and an approximate solution is found to be

$$n(x,t) \approx (4\pi\bar{\mu}t)^{-1/2} N e^{-(x-vt)^2/4\bar{\mu}t} \qquad (6.1)$$

where N is the total number of neutrophils per unit area of filter, and $\bar{\mu}$ and v are defined as

$$\bar{\mu} = \mu_0 + \mu_1 f(\bar{C}), \quad v = L^{-1}(C_f - C_i)[\chi_0 + \mu_1 f'(\bar{C})] \qquad (6.2)$$

with $\bar{C}$ being the average concentration of chemattractant within the filter, viz., $\bar{C} = (C_i + C_f)/2$. The concentration dependence of μ can be determined by incubating the cells with equal concentrations of chemoattractant above and below the filter, i.e., $C_i = C_f$. (One finds, for example, that μ first increases as C_i increases, but it then decreases.) However, the assumption of concentration independent force law (i.e. $\chi(C) = \chi_0$ =const.) only can be verified *a posteriori* by comparing computations with experimental measurements. Comparison was made between theory and experimental results of the response of neutrophils to gradients of casein. Data of the positions of the "leading front" of the experimental distributions were compared with values x_f for which $n(x_f,t)$ in Eq.(6.1) was $n(x_f,t)/\max[n(x,t)] = e^{-1/2}$. Although the relation between x_f and the observed leading front is not entirely clear, fair agreement between experimental and calculated values were noted. However, the effects of the several approximations which were needed in order to obtain Eq.(6.1) are not fully understood. For example, it would be interesting to know how $\bar{\mu}$ and v depend on different analytical forms of f(C) and v_d, and how they relate to values obtained from corresponding exact mathematical solutions.

Another assay which is used in clinical research is the so called "migration under agarose" assay (Nelson, Quie, and Simmons, 1975). In this case cells move between an agarose gel and the bottom of the plate in which the gel is formed. The cells are placed in a well cut midway between two other wells, one of which contains tissue culture medium plus putative chemoattractant and the other of which contains medium alone as a control. If the test substance is active, the distribution of cells moving from the center well will be spatially asymmetric. Recently, Lauffenberger and Aris (1979) provided a mathematical analysis of the assay. Equation (3.9) was taken as the basic model and, for simplicity, a solution was sought in one dimension assuming constant mobility μ and constant chemotactic coefficient (so that $v = \chi_0 \underset{\sim}{\nabla} C$). However,the situation is complicated because C is a time-varying function, given by solution of the diffusion equation with the well acting as a source of constant concentration C_0. An analytic solution was obtained by assuming that random motility is small compared with chemotactic response. The distribution n(x,t) is expressed as a perturbation series, the first term of which is the exact solution obtained when $\mu=0$, higher order terms being proportional to successive powers of $\mu/\chi_0 C_0$. The mathematical manipulations include solving equations by the method of characteristics and using a singular perturbation technique to calculate the dispersion of the moving front of the cells.

A somewhat similar migration assay, the "MIF assay", has been used extensively as an in vitro test for cellular immune sensitivity (David, 1973; Morely, 1974) In a typical version of the assay, capillaries are filled with white cells from peripheral blood, so that the cell population consists of monocytes and lymphocytes in addition to neutrophils. The capillaries are placed in a tissue culture medium containing a test antigen. If the patient has been sensitized to the antigen, his lymphocytes then will produce soluble factors known as lymphokines, certain of which inhibit the migration of neutrophils and other motile cells of the population. Consequently, the fan of cells migrating from the capillary will have a smaller radius than will the fan of cells moving from another capillary into medium which lacks the antigen.

This assay has been analyzed using a theory based on Eqs.(1) and (2), where $G(\underset{\sim}{r},t)$ is taken to be a point source of cells escaping from the mouth of the capillary (Nossal, 1977, 1978). The mobility coefficient varies as the lymphokine concentration changes with time. Expressions were derived for the mean-square radius of the distribution, and corresponding measures of the response (the decrease of migration area arising from the action of the lymphokines) were computed. Possible chemotactic attraction of cells by materials released by other cells (Zigmond and Hirsch, 1973) and spatial distribution of lymphokine concentration both were ignored, and the theory perhaps should be modified to account for these features.

Before concluding, a conceptual difficulty concerning these leukocyte migration models must be mentioned. The solution of the diffusion equation is a continuous distribution $n(\underset{\sim}{r},t)$ which, for large values of $|r|$, decreases <u>asymptotically</u> to zero. However, in contrast with ideal gas particles, there exists some maximal speed which locomoting cells cannot exceed. Thus, the actual cell distribution must be <u>absolutely</u> zero beyond a certain value of $|r|$ at any given time. Perhaps, therefore, a modified equation such as the telegraphist's equation would constitute a more faithful mathematical model, particularly for response at early times.

VII. Pattern formation induced by cellular metabolism

Some of the most interesting mathematical problems relating to chemotaxis concern the way spatial structure arises in cell distributions when chemotactic gradients are created by the cells themselves. The cell systems which have been most studied in this regard are flagellated bacteria (Keller and Segel, 1971) and cellular slime molds (Keller and Segel, 1970; Cohen and Robertson, 1971). Our review of these topics will be very brief, principally because they have been discussed by other speakers at this Conference.

When bacteria move in a medium containing chemoattractants which can be metabolized, a spatially inhomogeneous population of cells will create gradients capable of eliciting chemotactic response. The original work on mathematical modelling of the resulting spatial bands was performed by Keller and Segel (1971) and was based, essentially, on Eqs.(3.1)-(3.2) and (3.7)-(3.8). Motion in only one dimension was considered, in which case bands can propagate without bacterial growth (in contrast, because of geometrical constraints, ring formation requires a source term (Nossal, 1972,a,b).) Experimental observations on band shapes and speeds were reproduced by a theory in which chemoattractant diffusion was neglected and only one chemoattractant was presumed to be present. Recently, Holz and Chen (1979) reexamined this problem in relation to laser densitometry measurements of the response of E. Coli to oxygen gradients. These studies indicate that substrate diffusion must be taken into account. Holz and Chen also discerned fine structure in the band profiles when more than one chemoattractant was present, for which a theoretical analysis has yet to be provided.

Other recent work on bacterial chemotaxis includes a study by Scribner, Segel, and Rogers (1974) on the mechanism by which a band first forms from an initial inoculum (some of the inoculum does not enter the band and remains behind). In this study kinetic equations in which the mobility coefficient and chemotatic response terms were dependent upon chemoattractant concentration were employed, and numerical methods were used to obtain solutions. Conditions on μ and v_d which are necessary for the existence of unique solutions recently have been examined by Keller and Odell(1976) and Rascle(1978).

Yet more complicated and fascinating phenomenology is exhibited by populations of the cellular slime mold Dicteostylium discoideum. In this case the cells show tactic response to gradients of cAMP. The situation is made mathematically challenging by the fact that, upon stimulation with exogenous cAMP, the cells release cAMP; to add complexity, the chemoattractant is degraded by an enzyme (phosphodiesterase) which also is secreted by the cells. Delays and refractory periods, during which the cells cannot secrete cAMP, result in pulsatile production of chemoattractant. Investigations of equations for these phenomena lead to interesting stability analyses. A recent report of a computer simulation of pulsatile cell aggregation (Parnas and Segel, 1978) contains an up-to-date list of references.

REFERENCES

Albrecht-Buehler, G. (1979) J. Cell Biol. **80**, 53-60.

Becker, E. L. (1977) Arch. Pathol. Lab. Med. **101**, 509-513.

Berg, H. C. (1971) Rev. Sci. Instr. **42**, 868-871.

Berg, H. C. and D. A. Brown. (1972) Nature **239**, 500-504.

Boyden, S. (1962) J. Exptl. Med. **115**, 453-466.

Bredt, W., K. H. Hofling, H. H. Heunert, and B. Milthaler. (1970) Z. med. Mikrobiol. u. Immunol. **156**, 39-43.

Brown, D. A. and H. C. Berg. (1974) Proc. Nat. Acad. Sci. USA **71**, 1388-1392.

Chandrasekhar, S. (1943) Rev. Mod. Phys. **15**, 1-89.

Cohen, M. and A. Robertson. (1971) J. Theoret. Biol. **31**, 101-118.

Crank, J. (1956) The Mathematics of Diffusion. Clarendon Press, Oxford.

Dahlquist, F. W., P. Lovely, and D. E. Koshland, Jr.(1972) Nature New Biol. (London) **236**, 120-123.

David, J. R. (1973) N. Engl. J. Med. **288**, 143-149.

Diehn, B., M. Feinleib, W. Haupt, E. Hildebrand, F. Lenci, and W. Nultsch. (1977) Photochem. Photobiol. **26**, 559-560.

Engelmann, T. W. (1881) Botanishe Zeit. **39**, 441- .

Folkman, J. and R. Cotran. (1976) Int. Rev. Exp. Pathol. **16**, 207-248.

Fraenkel, G. S. and D. L. Gunn. (1940) The Orientation of Animals, Kineses, Taxes, and Compass Reactions. Oxford University Press. (Reprinted by Dover Publications, New York, 1961).

Gail, M. and C. W. Boone. (1970) Biophys. J. **10**, 980-993.

Gallin, J. A. and P. G. Quie, eds. (1978) Leukocyte Chemotaxis: Methods, Physiology, and Clinical Implications. Raven Press, New York.
Grimes, G. J. and F. S. Barnes. (1973) Exptl Cell Res 79, 375-385.
Hall, R. L. (1977) J. Math. Biol. 4, 327-335.
Hall, R. L. and S. C. Peterson. (1979) Biophys. J. 25, 365-372.
Holz, M. and S-H. Chen. (1979) Biophys. J. 26, 243-261.
Hu, C. L. and F. S. Barnes. (1970) Biophys. J. 10, 958-969.
Keller, E. F., and L. A. Segel. (1970) J. Theoret. Biol. 26, 399-415.
Keller, E. F. and L. A. Segel. (1971) J. Theoret. Biol. 30, 225-234.
Keller, E. F. and G. M. Odell. (1976) J. Theoret. Biol. 56, 243-247.
Keller, H. U., P. C. Wilkinson, M. Abercrombie, E. L. Becker, J. G. Hirsch, M. E. Miller, W. S. Ramsey, and S. H. Zigmond. (1977a) Clin. Exp. Immunol. 27, 377-380.
Keller, H. U., J. H. Wissler, M. W. Hess and H. Cottier. (1977b) Experientia 33, 534-536.
Lauffenburger, D and R. Aris. (1979) Math. Biosci. 44, 121-138.
Lovely, P. S. (1974) Taxis, Kinesis, and Anisotropic Diffusion, Ph.D. Dissertation, University of Oregon.
Lovely, P. S. and F. W. Dahlquist. (1975) J. Theoret. Biol. 50, 477-496.
MacNab, R. (1978) CRC Crit. Rev. Biochem. 5, 291-342.
McCutcheon, M. (1946) Physiol Rev. 26, 319-336.
Mesibov, R., G. W. Ordal, and J. Adler. (1973) J. Gen. Physiol. 62, 203-223.
Morley, J. (1974) Acta Allergol. 29, 185-208.
Nelson, R.D., P. G. Quie,, and R. L. Simmons. (1975) J. Immunol. 118, 1650-1656.
Nossal, R. (1972a) Math. Biosci. 13, 397-406.
Nossal, R. (1972b) Exptl. Cell Res. 75, 138-142.
Nossal, R. (1976) Math. Biosci. 31, 121-129.
Nossal, R. (1977) J. Theoret. Biol. 64, 703-722.
Nossal, R. (1978) Chapt. 5 in Theoretical Immunology, ed. G. I. Bell, A. S. Perelson, and G. H. Pimbley, Jr. Marcel Dekker, Inc., New York.
Nossal, R. and G. H. Weiss. (1974a) J. Theoret. Biol. 47, 103-113.
Nossal, R. and G. H. Weiss. (1974b) J. Stat. Phys. 10, 245-253.
Nossal, R. and S. Zigmond. (1976) Biophys. J. 16, 1171-1182.
Nultsch, W. (1962) Planta 57, 613- .
Parnas, H. and L. A. Segel. (1978) J. Theoret. Biol. 71, 185-207.
Patten, E., J. I. Gallin, R. A. Clark, and H. R. Kimball. (1973) Blood 41, 711-719.
Peterson, S. C. and P. B. Noble. (1972) Biophys. J. 12, 1048-1055.
Pfeffer, W. (1884) Unter. bot. Inst. Tubingen 1, 363- .
Rascle, M. (1978) Comptes Rendues Acad. Sci. (Paris) 286A, 555-558; "Prepublication," Universite de Saint-Etienne (1978), No. 1; J. Diff. Eqs. (to appear).
Rosen, G. (1976) J. Theoret. Biol. 59, 371-380.
Scribner, T. L., L. A. Segel, and E. H. Rogers. (1974) J. Theoret. Biol. 46, 189-219.
Segel, L. A. (1976) J. Theoret. Biol. 57, 23-42.
Van Houten, J. (1978) J. Comp. Physiol. A127, 167-174.
Wilkinson, P. C. (1974) Chemotaxis and Inflammation. Churchill Livingstone, Edinburg.
Zigmond, S. H. (1974) Nature (Lond.) 249, 450-452.
Zigmond, S. (1977) J. Cell Biol. 75, 606-616.
Zigmond, S. (1978) J. Cell Biol. 77, 269-287.
Zigmond, S. H. and J. G. Hirsch. (1973) J. Exp. Med. 137, 387-410.

TOPIC V

Stochastic versus deterministic approaches

Discussion by the Editors

The three papers selected for this section may be thought of as the first epilogue of the conference. Our special purposes have been not only to improve communication between biologists and mathematicians but also between mathematicians using mainly deterministic or stochastic theories. We thought that both approaches are not only compatible but also formally related. The differences lie in the natural laws considered: as Alexander Pope wrote, "The Universal Cause/Acts to one end, but acts by various laws". The reader surely noticed the same preoccupation in the paper given by Hans Weinberger (Topic III).

The main theme in the work of T.G.Kurtz is the way in which deterministic models arise as the limiting cases of stochastic models. His exemplary investigation offers some seeds of research for the foundations of a general mathematical theory of population processes; the meta-theorems he introduced as obvious indicate the source of randomness and its weight:

I. If the population is large, the internal fluctuations are negligible.

II. If the environmental (spatial of temporal) fluctuations are much faster than the population dynamics, they average out.

T.G.Kurtz gives arguments for discrete Markov population (jump) processes as well as for non-Markovian processes where there is an age- or location-dependent growth. His spatial competition model on a lattice will find various, direct or refined, applications in ecology, genetics or cell proliferation.

The other two papers investigate the connection between deterministic and stochastic models, utilizing tools from functional analysis. After his 1977 paper on the links between stochastic and ordinary differential equations, H.Doss analyses now the connections between stochastic and ordinary integral equations. The analytical tools used by G.Papanicolaou widen subtle differences in examining the asymptotics for partial differential equations with random coefficients.

For some readers the material presented in this section might appear too abstract or too technical. We do not intend to argue that mathematics today is "an enormous, powerful, complex enterprise largely beyond the language or intuition of the nonspecialist" (L.A. Steen, 1978). Contrarily, we are going to pinpoint the rigorous inquiry mathematics pursues, apparently for its own sake (or enjoyment), which actually confers it the reasonable or "unreasonable" effectiveness. Logically speaking, the method used in these "too abstract" papers is the construction of a strongly connected chain of reasoning of the type "If this, then that, and then it follows still further that ... ". Of course, "this" and "that" are certain mathematical expressions. Perhaps this is the root of the fact that the problems mathematics generates and the structures required to solve them share a common logical basis; they differ primarily in the form in which they are expressed. A.N.Whitehead has understood the mathematically powerful research as a paradox:"The utmost abstractions are the true weapons with which to control our thoughts of concrete fact".

CONNECTIONS BETWEEN STOCHASTIC AND ORDINARY INTEGRAL EQUATIONS

by

Halim DOSS

Let $\sigma : \mathbb{R}^n \to \mathcal{M}_{m,n}$ (the space of matrices with m columns and n rows) and $b : \mathbb{R}^n \to \mathbb{R}^n$ be two Lipschitz continuous maps. Suppose that σ is of class C^2. Let $B = (B_t)_{t \geq 0}$ $(B_0=0)$ be a standard $\mathbb{R}^m$ valued Brownian motion defined on a probability space $(\Omega, \mathcal{F}, \mathcal{F}_t, P)$. Consider the solution X^x of the following equation :

$$(1) \qquad X_t^x = x + s.\int_0^t \sigma(X_s^x)dB_s + \int_0^t b(X_s^x)ds, \quad x \in \mathbb{R}^n.$$

The letter S in the equality (1) designates the Stratonovich integral. By using the Itô integral we have :

$$(1') \qquad d\,X_t^{i,x} = \sum_{j=1}^{m} \sigma_{i,j}(X_s^x)dB_j + \left[\frac{1}{2} \sum_{k=1}^{n} \sum_{\ell=1}^{m} \sigma_{k,\ell} \frac{\partial}{\partial x_k} \sigma_{i,\ell}(X_s^x) + b_i(X_s^x)\right] ds$$

if $X^x = (X^{i,x})_{i=1,\dots,n}$.

Let us make the following hypothesis (I) about the coefficient σ : there exists a map $h : \mathbb{R}^n \times \mathbb{R}^m \to \mathbb{R}^n$ satisfying the total differential equation :

$$(2) \qquad \begin{cases} \dfrac{\partial h_i}{\partial \beta_j}(\alpha,\beta) = \sigma_{i,j}(h(\alpha,\beta)) \quad 1 \leq i \leq n, \; 1 \leq j \leq m \\ h(\alpha,0) = \alpha \qquad (\alpha,\beta) \in \mathbb{R}^n \times \mathbb{R}^m \end{cases}$$

We note $C^o_{\mathbb{R}_+}(\mathbb{R}^m)$ the space of continuous maps from $\mathbb{R}_+$ to $\mathbb{R}^m$ which vanish at zero.

THEOREM 1 :

There exists a unique map $\phi : \mathbb{R}^n \times C^o_{\mathbb{R}_+}(\mathbb{R}^m) \to C_{\mathbb{R}_+}(\mathbb{R}^n)$ satisfying the following conditions :

i) If $u \in C^o_{\mathbb{R}_+}(\mathbb{R}^m)$ is of bounded variation on each compact interval then $\phi(x,u)$ is the solution of :

$$\phi(x,u)(t) = x + \int_0^t \sigma(\phi(x,u)(s))du_s + \int_0^t b(\phi(x,u)(s))ds \tag{3}$$

ii) ϕ is continuous when we put on the spaces $C^o_{\mathbb{R}_+}(\mathbb{R}^m)$ and $C_{\mathbb{R}_+}(\mathbb{R}^n)$ the topology of uniform convergence on compact intervals.

Furthermore one has the following explicit formulae :

(4) $\phi(x,u)(t) = h(D(x,u)(t),u_t)$ where $D=D(x,u)$ is determined by the differential equation:

$$\frac{\partial D}{\partial t} = \frac{\partial h}{\partial \alpha}(h(D_t,u_t), - u_t)\ \{b(h(D_t,u_t))\}\ ;\ D(o) = x. \tag{5}$$

One can then represent the solution X^x of equation (1) by $X^x = \phi(x,B)$ a.s.

Proof : Let $u \in C^o_{\mathbb{R}_+}(\mathbb{R}^m)$ be of bounded variation on each compact interval and let $D=D(x,u)$ be the solution of equation (5). Consider $\bar{\phi}(t) = h(D_t,u_t)$. We easily see, thanks to the ordinary change of variable formula, that $\bar{\phi}$ is a solution of equation (3) and then that $\bar{\phi}(t)=\phi(x,u)(t)$. If ω is an element of the probability space Ω consider $D(\omega)(t) = D(x,B(\omega))(t)$ where $B(\omega) = (B_t(\omega))_{t \geq o} \in C^o_{\mathbb{R}_+}(\mathbb{R}^m)$ is a trajectory of the Brownian motion B.

By using Itô formula we show that the process $(\tilde{\phi}(t))_{t \geq o}$ defined by $\tilde{\phi}(t,\omega) = h(D(\omega)(t), B_t(\omega))$ is a solution of the stochastic differential equation (1) and then, since this solution is unique, it follows that $X_t^x = \tilde{\phi}(t)$ a.s. The fact that the trajectory $D=D(x,u)$ depends continuously on the point $(x,u) \in \mathbb{R}^n \times C^o_{\mathbb{R}_+}(\mathbb{R}^m)$ is a consequence of a classical theorem about solutions of ordinary differential equations depending on a parameter varying in a Banach space. The properties of the functional ϕ stated in Theorem 1 are then easily verified, see [3].

REMARK 2 :

Consider the vector fields $A_j = \sum_{i=1}^{n} \sigma_{i,j}(x) \frac{\partial}{\partial x_i}$ $(j=1,\dots,m)$; the hypothesis (I) means that, for each (j,j'), the Lie bracket $[A_j, A_{j'}] = A_j A_{j'} - A_{j'} A_j$ is equal to zero. This condition is always satisfied if $m=1$.

We can see that this condition is also necessary for the validity of Theorem 1 ; it suffices to consider the diffusion process $X(t) = (X_1(t), X_2(t))$ defined by : $X_1(t) = B_1(t)$; $X_2(t) = \int_0^t B_1(s)dB_2(s)$ where B_1 and B_2 are two independent one dimensional Brownian motions. The process X is in fact a measurable functional ϕ of the couple (B_1,B_2) but ϕ is not continuous for the topology of uniform convergence on compact intervals. Y. Yamato [8] has recently generalized the preceding theorem when the coefficients σ and b are of class C^∞ under the less rectrictive hypothesis that the Lie algebra generated by the vector fields A_j, $j=1,\dots,m$ is nilpotent of step p $(p < \infty)$. He gives then a representation of the diffusion process X^x in terms of multiple Wiener integrals.

Some consequences of Theorem 1 :

1) Under hypothesis (I) it is possible to study "trajectory by trajectory" and not only in probability or in L^2 the dependence of the diffusion process $(X_t^x)_{t \geq 0}$ with respect to a parameter or to the initial data.

2) Let $u \in C^o_{\mathbb{R}_+}(\mathbb{R}^m)$ be of bounded variation on compact intervals and $\phi = \phi(x,u)$ the solution of equation (3) ; fix a number $0 < T < \infty$, then for each $\varepsilon > 0$ there exists $\alpha > 0$ such that

$$\{||B-u||_T < \alpha\} \underset{\text{a.s.}}{\subseteq} \{||X^x - \phi||_T < \varepsilon\}$$

where $||.||_T$ is the sup norm on the interval $[0,T]$.

Let $S_x = \{\phi(x,u),\ u \in C^o_{\mathbb{R}_+}(\mathbb{R}^m)$ is of bounded variation on compact intervals$\}$.

If P_x is the law of the diffusion process X^x on $C_{\mathbb{R}_+}(\mathbb{R}^n)$, the preceding property shows thats the topological support of the law P_x is equal to the closure $\overline{S}_x$ of the set $S_x \subseteq C_{\mathbb{R}_+}(\mathbb{R}^n)$; it's a theorem of Stroock and Varadhan.

3) Ventsel and Freidlin estimates :

Fix $0 < T < \infty$ and $x \in \mathbb{R}^n$. Consider the map ϕ defined in Theorem 1, restricted to the interval $[0,T]$, as a continuous map from the Banach space $\Theta = C^o_{[0,T]}(\mathbb{R}^m)$ to the Banach space $C_{[0,T]}(\mathbb{R}^n)$. Consider also the Brownian motion $B = (B_t)_{t \in [0,T]}$ as a Θ valued Gaussian random variable.

Let A be a Borel set of Θ. We know then that, [2],

$$(6) \qquad -\Lambda(\overset{o}{A}) \leq \underline{\lim}_{\varepsilon \searrow 0} \varepsilon^2 \operatorname{Log} P[\varepsilon B \in A] \leq \overline{\lim}_{\varepsilon \searrow 0} \varepsilon^2 \operatorname{Log} P[\varepsilon B \in A] \leq -\Lambda(\overline{A})$$

where $\Lambda(E) = \inf_{\omega \in E} \mu(\omega)$ if $E \subseteq \Theta$ and $\mu(\omega) = \frac{1}{2}\int_0^T |\dot{\omega}_s|^2\, ds$

if $\omega \in C^o_{[0,T]}(\mathbb{R}^m)$ admits a derivative $\dot{\omega}$ in the Lebesgue sense and $\dot{\omega} \in L^2_{[0,T]}$, $\mu(\omega) = +\infty$ if not.

Let $(X^\varepsilon_t)_{t \in [0,T]}$ be the diffusion process solution of

$$(7) \qquad X^\varepsilon_t = x + \varepsilon.\text{s.} \int_0^t \sigma(X^\varepsilon_s)\, dB_s + \int_0^t b(X^\varepsilon_s)\, ds.$$

Consider the process X^ε as a $C_{[0,T]}(\mathbb{R}^n)$ valued random variable. By Theorem 1 we know that $X^\varepsilon = \phi(\varepsilon B)$ a.s. Let D be a Borel set of $C_{[0,T]}(\mathbb{R}^n)$. We have : $(X^\varepsilon \in D) = (\varepsilon B \in \phi^{-1}(D))$ and by (6):

$$(8) \qquad -\Lambda(\phi^{-1}(\overset{o}{D})) \leq -\Lambda(\overset{o}{\overbrace{\phi^{-1}(D)}}) \leq \overline{\lim}_{\varepsilon \searrow 0} \varepsilon^2 \operatorname{Log} P(X^\varepsilon \in D) \leq -\Lambda(\overline{\phi^{-1}(D)}) \leq -\Lambda(\phi^{-1}(\overline{D}))$$

because ϕ is continuous so that

$$\phi^{-1}(\overline{D}) \supseteq \overline{\phi^{-1}(D)} \quad \text{and} \quad \phi^{-1}(\overset{o}{D}) \subseteq \overset{o}{\overbrace{\phi^{-1}(D)}}$$

The inequalities (8) are the Ventsel and Freidlin estimates [7].

4) Asymptotic expansions :

Under hypothesis (I), when the coefficient σ is of class C^{k+1} and the drift b is of class C^k it is easy to see that the map ϕ from the Banach space $\Theta = C^0_{[0,T]}(\mathbb{R}^m)$ to the Banach space $\Omega' = C_{[0,T]}(\mathbb{R}^n)$ is differentiable of class C^k $(k \geq 1)$.

Let χ and δ be two measurable maps from Ω' to $\mathbb{R}$ and $\alpha_\delta = \sup_{\omega \in \Theta} (\delta(\phi(\omega)) - \mu(\omega))$.

Suppose that the maps χ and δ satisfie the following conditions (II) :

i) The supremum α_δ is reached at a unique point $\omega_0 \in \Theta$.

ii) $|\chi(\phi(\omega))| \leq K_1 \exp(K_2|\omega|^2)$ a.s. $K_1, K_2 \in \mathbb{R}_+$ and $|\omega| = ||\omega||_T$

iii) $\delta(\phi(\omega)) \leq L_1 + L_2\, |\omega|^2$ a.s. and $L_2 < \frac{1}{4T}$

iv) $\delta\circ\phi$ is upper semi-continuous and uniformly continuous in a neighbourhood of the origin containing the ball of radius $2\left|\frac{L_1+1-\alpha_\delta}{2L_2 - 1/2T}\right|^{1/2}$

v) $\delta\circ\phi$ is of class C^2 in a neighbourhood of the point ω_0 and $\sup_{\substack{|\eta| < \alpha \\ \eta \in \Theta}} \mathbb{E}\left[\exp((1+\varepsilon)\frac{D^2}{2}(\delta\circ\phi)(\omega_0+\eta)(B,B))\right] < \infty$ for at least one couple (α,ε), $\alpha > 0$ and $\varepsilon > 0$.

THEOREM 2

Under hypothesis (I) and (II) and if, in a neighbourhood of the point ω_0, $\delta\circ\phi$ is of class C^k and $\chi\circ\phi$ is of class C^{k-2} $(k \geq 3)$, one has, X^ε being the solution of equation (7), the following asymptotic expansion :

$$(9) \qquad \exp(-\alpha_\delta\, \varepsilon^{-2})\, E[\chi(X^\varepsilon)\, \exp(\varepsilon^{-2}\, \delta(X^\varepsilon))]$$

$$= \Gamma_0 + \Gamma_1\,\varepsilon + \ldots + \Gamma_{k-3}\,\varepsilon^{k-3} + O(\varepsilon^{k-2}),\ \varepsilon \to 0$$

where $\Gamma_0 = \chi(\phi(\omega_0))\ E\left[\exp(\frac{D^2}{2}\ (\delta\circ\phi)(B,B))\right]$ and where the constants Γ_i are determined by the differentials of $\delta\circ\phi$ and $\chi\circ\phi$ at the point ω_0.

Proof : With slightly different hypotheses Schilder [5] proved the asymptotic expansion (9) in the one dimensional Brownian motion case. In fact one can see that (9) is true in the m-dimensional Brownian motion case under conditions (I) and (II), see [4]. The diffusion process X^ε being, under condition (I), equal a.s. to the process $\phi(\varepsilon B)$ we see that the proof of Theorem 2 is immediate because we are reduced to the study of the brownian motion B thanks to the functional ϕ which is quite regular.

REFERENCES

[1] R. AZENCOTT : Cours à Saint-Flour, Lecture Notes in Maths. Springer-Verlag, Eté 1978 (to appear).

[2] M.D. DONSKER & S.R.S. VARADHAN : Asymptotic evaluation of certain Markov Process Expectations for large time III. Comm. on pure and applied maths. Vol. XXIX, 389-461 (1976).

[3] H. DOSS : Liens entre equations differentielles stochastiques et ordinaires. Annales Institut H. Poincaré, Vol. XIII, n° 2, (1977) p. 99-125.

[4] H. DOSS : Quelques formules asymptotiques pour les petites perturbations de systèmes dynamiques. A paraître aux Annales de l'Institut H. Poincaré.

[5] M. SCHILDER : Some asymptotic formulae for Wiener integrals. Trans. Amer. Math. Soc. Vol. 125, 1966, p. 63-85.

[6] D.W. STROOCK & S.R.S. VARADHAN : On the support of diffusion processes with application to the strong maximum principle. 6th Berkeley Symposium Vol. III, 1972.

[7] A.D. VENTSEL & M.I. FREIDLIN : On small random perturbations of dynamical systems. Russian Math. surveys, Vol. XXV, 1970, p. 1-55.

[8] Y. YAMATO : Stochastic differential Equations and Nilpotent Lie Algebras. Z. Wahrscheinlich. Verw Gebiete 47, 213-229 (1979).

N.B.- I wish to mention that H.J. Sussmann has independently obtained some results similar to Theorem 1 : "On the gap between deterministic and stochastic ordinary differential equations" by Hector H. Sussmann. The Annals of Probability, 1978, Vol. 6, n° 1, 19-41.

Relationships between stochastic and deterministic population models

Thomas G. Kurtz
Department of Mathematics
University of Wisconsin-Madison
Madison, Wisconsin 53706
USA

Abstract

The infinitesimal parameters of a variety of Markov population models can be written as $q^{(N)}_{k,k+\ell} = N\beta_\ell(N^{-1}k)$ where $k,\ell \in \mathbb{Z}^d$ and N is a parameter which is of the same order of magnitude as the population size. Under appropriate conditions, a family of Markov processes $\{X_N\}$ with these parameters satisfies $\lim_{N\to\infty} N^{-1}X_N(t) = X(t)$, in probability, where $X(t)$ is a solution of the differential equation $\dot{X} = F(X) \equiv \sum_\ell \ell\beta_\ell(X)$.

Several approaches to studying the error $N^{-1}X_N(t) - X(t)$ have been considered by a number of authors, including limit theorems with error bounds for the sequence $\sqrt{N}(N^{-1}X_N(t) - X(t))$.

These results can be generalized in a variety of ways. We will consider examples that are not Markovian (i.e. epidemic models in which the infectious period is not exponentially distributed) and examples that take into account the spatial distribution of the population. In particular we will obtain the solution of Fisher's Equation $u_t = au_{xx} + \alpha u(1-u)$ as the deterministic limit of Markov process models for two competing plant species.

0. Introduction

The primary purpose of this paper is to explore the interrelationships among some of the various ways of modeling the same biological or physical phenomenon. The models we will consider are best described as population models although we have a broad definition of the term population in mind. Typically these models will involve collections of "particles," each particle being one of a finite number of "types." Particles of a given type may have other "attributes" (e.g. age or location) which play a role in the formulation of the model. We do not intend to place heavy emphasis (or any emphasis at all) on the distinction between "type" and "attribute" but the distinction will usually be clear.

The "relationship" which will attract most of our attention is the relationship between stochastic and deterministic versions of the same model. Typically in

---

AMS 1979 Subject Classification - Primary: 60J25, 60J70, 60K30, 92A15, 92A10.

Key words and phrases: Population processes, approximation, reaction-diffusion, epidemic models, Markov processes, jump processes.

population models randomness arises in three ways. Fluctuations due to "sampling" mechanisms, random motion, and fluctuations due to randomness in the "environment." Typically (or at least frequently) the first two are called internal fluctuations, and the latter, external fluctuations. Typically the first two types are modeled by objects probabilists like—Markov chains, Brownian motion, etc., while the latter type is modeled by assuming the parameters of the models for the first two (sometimes called models of the dynamics) are themselves random variables or stochastic processes.

Our primary theme will be the ways in which deterministic models arise as the "limiting cases" of stochastic models. There are two meta-theorems which will come as no surprise to anyone.

I. If the population is large the internal fluctuations are negligible.

II. If the environmental fluctuations are much faster than the population dynamics they average out.

The second meta-theorem refers to spatial as well as temporal fluctuations (see the papers of Kesten and Papanicolaou in this volume). Of course from the point of view of a moving particle spatial fluctuations become temporal. We will concentrate on the first meta-theorem, looking first at models without special attributes and then at models with age and location dependence.

Section 1 includes general comments on the modeling process, and Section 2 is devoted to the basic results for families of stochastic models given by discrete Markov processes. Finally examples of extensions of the results in Section 2 to models involving age and location dependence are given in Sections 3 and 4.

1. Construction of stochastic models

We will be considering models involving a finite number of different types of particles each of which may possess some additional attributes. A description of the state of such a process will include the numbers of particles of each type and possibly a list of attributes (e.g. a list of ages). Clearly, in a well behaved model, creation or destruction of particles and changes of type will occur at discrete times while attributes may change at discrete times (e.g. particles jump) or continuously in time (e.g. particles diffuse or age).

Attribute changes can be modeled in a great variety of ways. "Aging" is obvious while changes of location might be modeled by specifying that each particle independently "undergoes Brownian motion."

Discrete events can be modeled in two ways. First, one can simply specify the distribution for the length of time between two particular events, for example, between the birth of a particular particle and its death. Alternatively, one can specify the intensity for the occurrence of an event. That is, if $\mathcal{F}_t$ represents the history of the process up to time t (i.e. the σ-algebra generated by the process up to time t), then $\lambda(t)$, a non-negative random variable depending on the history of the process up to time t, is the intensity of the event (at least formally) if

$$(1.1) \qquad P\{\text{occurrence in } (t,t+\Delta t] \mid \mathcal{F}_t\} = \lambda(t)\Delta t + o(\Delta t) .$$

Modeling of this sort typically involves specifying the intensity $\lambda(t)$ as a function of the state at time t and at earlier times.

One may want to specify intensities for a single, specific event, such as the death of a particular particle. For example if τ_d is the time of death of the particle and $\alpha(t)$ is the age of the particle at time t then $\lambda(t)$ might be of the form

$$(1.2) \qquad \lambda(t) = \begin{cases} \mu(\alpha(t)) & t < \tau_d \\ 0 & t \geq \tau_d \end{cases}$$

or referring to (1.1)

$$(1.3) \qquad P\{\tau_d \in (t,t+\Delta t) \mid \mathcal{F}_t\} = \begin{cases} \mu(\alpha(t))\Delta t + o(\Delta t), & t < \tau_d \\ 0 & t \geq \tau_d . \end{cases}$$

On the other hand one may want to specify the intensity of a category of events, such as the times at which a new particle "immigrates" into the system. For example $\lambda(t)$ might be a constant λ (non random) independent of t. In this case, if $M(t)$ is the counting process giving the number of immigrants arriving up to time t, then $M(t)$ is a Poisson process with parameter λ i.e.

$$(1.4) \qquad P\{M(t) = k\} = e^{-\lambda t}\frac{(\lambda t)^k}{k!} , \qquad k = 0,1,2,\cdots .$$

We have not given a precise formulation of what we mean by an intensity (see Aalen (1978) for a recent rigorous discussion and additional references) but the intuitive formulation in (1.1) should suffice for our purposes. We note that the exponential distribution and the Poisson process are intimately related to the specification of discrete events using intensities. We leave it as an exercise for the reader to show (at least formally) that if τ_d (defined above) is finite a.s. then $\int_0^{\tau_d} \mu(\alpha(t))\,dt$ is exponentially distributed. Similarly if $\lambda(t)$ is an intensity for a category of events, $M(t)$ is the counting process giving the number of occurrences up to time t, $\int_0^{\infty} \lambda(t)\,dt = \infty$ a.s. and $\tau(u)$ is defined as the solution of

$$(1.5) \qquad \int_0^{\tau(u)} \lambda(t)\,dt = u ,$$

then $Y(u) = M(\tau(u))$ is a Poisson process with parameter 1. (See, for example, Meyer (1971).) It follows, of course, that

$$M(t) = Y(\tau^{-1}(t)) = Y(\int_0^t \lambda(s)\,ds) \tag{1.6}$$

that is $M(t)$ is a random time change of a Poisson process. We will exploit this fact in the next section.

A final observation concerning intensities is that they add, provided the events involved cannot occur simultaneously. Suppose, for example, we are interested in two categories of events with intensities $\lambda_1(t)$ and $\lambda_2(t)$. If (with probability one) no event in category 1 occurs at the same time as an event in category 2, then the occurrence of some event in one of the two categories has intensity $\lambda_1(t) + \lambda_2(t)$.

We illustrate the use of intensities to specify models with three simple models given by jump Markov processes.

Logistic Population Growth. We consider a model for a population with a single type. Because it will be important in the next section we introduce a parameter N which can be interpreted as the area of the region occupied by the population. If the population size is X then the population density is X/N. There are two categories of events we will consider, births and deaths, and we assume only one individual is born or dies at a time. We will assume the intensity for births is proportional to the population size and hence the birth intensity is

$$\lambda X = N\lambda\, X/N \tag{1.7}$$

when the population size is X. Note that on the right we have written the intensity as N times a function of the density. We assume the death rate increases as the population density increases and take the death intensity to be

$$(\mu + \gamma\, X/N)X = N(\mu + \gamma\, X/N)\, X/N\ . \tag{1.8}$$

Again we note that on the right we have written the intensity as N times a function of the density.

Epidemic Model. Here we assume that a population of constant size N is divided into three types of individuals, individuals that are infected by a certain disease, individuals that are not infected and are susceptible, and individuals that are immune. Since the total population is constant, we can calculate the number of immune individuals if we know the number of susceptible individuals S and the number of infected individuals I. Consequently we can describe the state as (S, I). There are two categories of events, $(S,I) \to (S-1, I+1)$, a susceptible individual becomes infected, and $(S,I) \to (S, I-1)$, an infected individual becomes immune. We assume that the intensity for the event that a particular susceptible becomes infected is proportional to the fraction of the population that is infected (i.e. the intensity is $\lambda I/N$ up to the time the individual is infected). Since intensities add the intensity for the category of events $(S, I) \to (S-1, I+1)$ is

$$\lambda S I/N = N\lambda (S/N)(I/N) \quad . \tag{1.9}$$

We assume the intensity for $(S, I) \to (S, I-1)$ is

$$\mu I = N\mu I/N \quad . \tag{1.10}$$

Again note that both intensities can be written as N times a function of the numbers of each type divided by N.

Once the intensities of the events are specified one must use these intensities to characterize a process. For the Markov models described above, the usual approach is to derive a system of differential equations for the transition probabilities of the process. However, by exploiting the relationship in (1.6) we can obtain these processes as solutions of systems of stochastic equations involving independent Poisson processes. We illustrate this approach with the epidemic model.

Let Y_1 and Y_2 be independent Poisson processes with intensity 1, and consider the following system of equations:

$$M_1(t) = Y_1\left(\int_0^t \lambda I(u)\, S(u)/N \, du\right), \tag{1.11}$$

$$M_2(t) = Y_2\left(\int_0^t \mu I(u)\, du\right) , \tag{1.12}$$

$$I(t) = I(0) + M_1(t) - M_2(t) \tag{1.13}$$

and

$$S(t) = S(0) - M_1(t) \quad . \tag{1.14}$$

It is not difficult to see that this system of equations has a unique solution for any choice of $I(0)$ and $S(0)$, that M_1 counts the number of transitions of the form $(S, I) \to (S-1, I+1)$ and M_2 counts the number of transitions of the form $(S, I) \to (S, I-1)$. In addition, it is at least intuitively clear that M_1 and M_2 have the correct intensities. For example

$$P\{M_1(t+\Delta t) - M_1(t) > 0 \mid \mathfrak{F}_t\} = P\{Y_1\left(\int_0^{t+\Delta t} \lambda I(u)\, S(u)/N \, du\right) - Y_1\left(\int_0^t \lambda I(u)\, S(u)/N \, du\right) > 0 \mid \mathfrak{F}_t\} \quad . \tag{1.15}$$

The right side of (1.15) is just the conditional probability that a Poisson process jumps in an interval of length $\int_t^{t+\Delta t} \lambda I(u)\, S(u)/N \, du$ which should be (and is)

$$E\left[\int_t^{t+\Delta t} \lambda I(u) S(u)/N \, du \mid \mathfrak{F}_t\right] + o(\Delta t) = (\lambda I(t) S(t)/N)\Delta t + o(\Delta t) \quad . \tag{1.16}$$

Consequently the solution of the system is a process corresponding to the specified intensities. In fact it is a Markov process with the transition probabilities obtained by the usual approach.

Finally we note that M_1 and M_2 can be eliminated from the system to give a single vector equation

$$(1.17)\qquad (S(t), I(t)) = (S(0), I(0)) + (-1,1) Y_1\left(\int_0^t \lambda I(u) S(u)/N \, du\right) + (0,-1) Y_2\left(\int_0^t \mu I(u)\, du\right) .$$

The analogous equation for the logistic population growth model is

$$(1.18)\qquad X(t) = X(0) + Y_1\left(\int_0^t \lambda X(u)\, du\right) - Y_2\left(\int_0^t (\mu + \gamma X(u)/N) X(u)\, du\right) .$$

These equations will be used in the next section to study the relationship between these stochastic models and their deterministic analogs.

Time change representations of this type are discussed in more detail in Kurtz (1978, preprint a).

2. Deterministic limits of density dependent families of jump Markov processes

Motivated by the population growth and epidemic models discussed in the previous section, we will consider families of jump Markov processes with values in $\mathbb{Z}^d$, the d-dimensional integer lattice, such that the intensity of a jump of the form $k \to k+\ell$, $k, \ell \in \mathbb{Z}^d$, is $N\beta_\ell(N^{-1}k)$. We note, for example, that for the epidemic model we can take $\beta_{(-1,1)}(x_1, x_2) = \lambda x_1 x_2$ and $\beta_{(0,-1)}(x_1, x_2) = \mu x_2$, for $x_1, x_2 \geq 0$, zero if $x_1 < 0$ or $x_2 < 0$, and $\beta_\ell \equiv 0$ for all other ℓ.

Since the intensities depend on the normalized quantities $N^{-1}k$, rather than consider the $\mathbb{Z}^d$-valued processes we will consider processes X_N with values in $\{N^{-1}k : k \in \mathbb{Z}^d\}$ where X_N is obtained by dividing each component of the original process by N. Referring to (1.17) we will characterize X_N as the solution of

$$(2.1)\qquad X_N(t) = X_N(0) + \sum_\ell N^{-1} \ell\, Y_\ell\left(N \int_0^t \beta_\ell(X_N(s))\, ds\right)$$

where the Y_ℓ are independent Poisson processes with $E[Y_\ell(t)] = \mathrm{Var}[Y_\ell(t)] = t$, provided a solution exists and is unique. This equation can be solved uniquely, at least on some time interval, simply by observing that $X_N(t)$ is constant except at times when one of the terms on the right jumps. The solution is given by $X_N(0)$ until the first jump, then is determined by the first jump until the second jump, etc. We have existence and uniqueness if we have conditions which ensure that there are only finitely many jumps in a finite amount of time. We state two theorems that give conditions for this. We assume $X_N(0)$ is not random.

Theorem (2.2). Suppose $\sup_{x \in K} \sum_{\ell} \beta_{\ell}(x) < \infty$ for every bounded set $K \subset \mathbb{R}^d$, and there exist constants $K_1, K_2 \geq 0$ such that

$$\sum_{\ell} N\beta_{\ell}(x)\,(|x + N^{-1}\ell| - |x|) \leq K_1 + K_2|x| \tag{2.3}$$

for all $x \in \mathbb{R}^d$. Then (2.1) has a unique solution and for $K_2 > 0$

$$P\{\sup_{s \leq t} |X_N(s)| \geq \lambda\} \leq \lambda^{-1}(|X_N(0)|e^{K_2 t} + \frac{K_1}{K_2}(e^{K_2 t} - 1)) . \tag{2.4}$$

If $K_2 = 0$ then the right side of (2.4) is replaced by $\lambda^{-1}(|X_N(0)| + K_1 t)$.

Remark. Essentially, (2.4) follows from the fact that

$$|X_N(t)| - \int_0^t \sum_{\ell} N\beta_{\ell}(X_N(s))\,(|X_N(s) + N^{-1}\ell| - |X_N(s)|)\,ds \tag{2.5}$$

is a martingale. The fact that $\sum_{\ell} \beta_{\ell}(x)$ is bounded on bounded sets and (2.4) then imply that X_N can have only finitely many jumps in a finite time interval.

For the population growth model

$$\sum_{\ell} N\beta_{\ell}(x)\,(|x + N^{-1}\ell| - |x|) = \lambda x - (\mu + \gamma x)x , \qquad x \geq N^{-1} \tag{2.6}$$

is bounded from above.

Theorem (2.7). For $y = \{y_{\ell}\}$, $y_{\ell} \geq 0$, $\ell \in \mathbb{Z}^d$, such that $\sum |\ell| y_{\ell} < \infty$, define

$$My \equiv \sum_{\ell} \ell y_{\ell} . \tag{2.8}$$

If there exist $K_1(x)$ and $K_2(x) \geq 0$ such that for $y_{\ell} \geq 0$ satisfying $\sum_{\ell} |\ell| y_{\ell} < \infty$

$$\sum_{\ell} \beta_{\ell}(x + My) \leq K_1(x) + K_2(x) \sum_{\ell} y_{\ell} , \tag{2.9}$$

then (2.1) has a unique solution and for $K_2(X_N(0)) > 0$

$$E[\sum Y_{\ell}(N \int_0^t \beta_{\ell}(X_N(s))\,ds)] \leq \frac{NK_1(X_N(0))}{K_2(X_N(0))}(e^{K_2(X_N(0))t} - 1) . \tag{2.10}$$

If $K_2(X_N(0)) = 0$ then the right side of (2.10) is replaced by $NK_1(X_N(0))t$.

Remark. It can be shown that (Kurtz (1978))

$$E[Y_{\ell}(N \int_0^t \beta_{\ell}(X_N(s))\,ds] = E[N \int_0^t \beta_{\ell}(X_N(s))\,ds] . \tag{2.11}$$

This fact and (2.9) give (2.10). For the epidemic model we can take $K_1(x_1, x_2) = \lambda(|x_1| + |x_2|)^2 + \mu(|x_1| + |x_2|)$ and $K_2 = 0$.

We want to use properties of the Poisson processes Y_ℓ to study X_N. In particular if we define $\hat{Y}_\ell(u) \equiv Y_\ell(u) - u$, then $\hat{Y}_\ell$ has mean zero and the law of large numbers gives

(2.12)
$$\lim_{N\to\infty} N^{-1} \sup_{s\le u} |\hat{Y}_\ell(Ns)| = 0 .$$

Renormalizing, it follows from the central limit theorem that the process $W_\ell^N(u) \equiv N^{-1/2}\hat{Y}_\ell(Nu)$ converges weakly as $N \to \infty$ to a Brownian motion $W_\ell(u)$. (See Billingsley (1968) for a discussion of weak convergence.) We rewrite (2.1) in terms of $\hat{Y}_\ell$. Defining

(2.13)
$$F(x) \equiv \sum_\ell \ell \beta_\ell(x)$$

we have

(2.14)
$$X_N(t) = X_N(0) + \sum_\ell N^{-1} \ell \hat{Y}_\ell\left(N \int_0^t \beta_\ell(X_N(s))\, ds\right) + \int_0^t F(X_N(s))\, ds .$$

Given (2.12) if $X_N(0) \to X(0)$ it is natural to expect X_N to converge to the solution of

(2.15)
$$X(t) = X(0) + \int_0^t F(X(s))\, ds .$$

<u>Theorem</u> (2.16). Suppose for each bounded set K, $\sum_\ell |\ell| \sup_{x\in K} \beta_\ell(x) < \infty$ and there exists $M_K > 0$ such that

(2.17)
$$|F(x) - F(y)| \le M_K |x - y| .$$

If the solution of (2.15) exists for all $t \ge 0$ and $X_N(0) \to X(0)$ then

(2.18)
$$\lim_{N\to\infty} \sup_{s\le t} |X_N(s) - X(s)| = 0 .$$

<u>Proof</u>. For simplicity we assume the conditions on the β_ℓ and F hold with K replaced by $\mathbb{R}^d$. The more general result in the statement of the theorem follows by replacing $\beta_\ell(x)$ by $\rho(nx)\beta_\ell(x)$ where ρ is smooth, $\rho(x) = 1$ for $|x| \le 1$, and $\rho(x) = 0$ for $|x| \ge 2$, and applying the result with K replaced by $\mathbb{R}^d$ for each $n > 0$.

Setting $\bar{\beta}_\ell = \sup_x \beta_\ell(x)$, we note that

$$\varepsilon_N(t) \equiv \sup_{s\le t} \left|\sum_\ell N^{-1}\ell\,\hat{Y}_\ell\left(N\int_0^s \beta_\ell(X_N(u))\,du\right)\right| \le \sum_\ell \sup_{s\le t} N^{-1}|\ell|\,|\hat{Y}_\ell(N\bar{\beta}_\ell s)| . \tag{2.19}$$

The terms in the sum on the right side of (2.19) go to zero by the law of large numbers and each term is dominated by $N^{-1}|\ell|\,(Y_\ell(N\bar{\beta}_\ell t) + N\bar{\beta}_\ell t)$. The law of large numbers applied to the process with independent increments

$$Q(u) = \sum_\ell |\ell|\,(Y_\ell(\bar{\beta}_\ell u) + \bar{\beta}_\ell u) \tag{2.20}$$

implies

$$\lim_{N\to\infty} N^{-1}Q(Nt) = \lim_{N\to\infty} \sum_\ell N^{-1}|\ell|\,(Y_\ell(N\bar{\beta}_\ell t) + N\bar{\beta}_\ell t) = \sum_\ell 2|\ell|\,\bar{\beta}_\ell t = \sum_\ell \lim_{N\to\infty} N^{-1}|\ell|\,(Y_\ell(N\bar{\beta}_\ell t) + N\bar{\beta}_\ell t) . \tag{2.21}$$

The fact that the limit and summation in (2.21) can be interchanged implies the same for (2.19) and we conclude $\lim_{N\to\infty} \varepsilon_N(t) = 0$ a.s.

Finally assuming $|F(x) - F(y)| \le M|x - y|$, Gronwall's Inequality gives

$$\sup_{s\le t} |X_N(s) - X(s)| \le (|X_N(0) - X(0)| + \varepsilon_N(t))e^{Mt} \tag{2.22}$$

and (2.18) follows. □

Without giving details we consider a central limit theorem for the deviation of X_N from X, the solution of (2.15). Here we need to assume $\sum_\ell |\ell|^2\bar{\beta}_\ell < \infty$. With W_ℓ^N defined as above and setting $V_N(t) \equiv \sqrt{N}\,(X_N(t) - X(t))$, we note that

$$V_N(t) = V_N(0) + \sum_\ell \ell W_\ell^N\left(\int_0^t \beta_\ell(X_N(s))\,ds\right) + \int_0^t N^{\frac{1}{2}}(F(X_N(s)) - F(X(s)))\,ds . \tag{2.23}$$

From the fact that $W_\ell^N \Longrightarrow W_\ell$,

$$\int_0^t \beta_\ell(X_N(s))\,ds \to \int_0^t \beta_\ell(X(s))\,ds$$

(assuming β_ℓ is continuous), and

$$N^{\frac{1}{2}}(F(X_N(s)) - F(X(s))) \approx \partial F(X(s))\,V_N(s)$$

(assuming F is smooth), we can conclude that $V_N \Longrightarrow V$ where V satisfies

$$V(t) = V(0) + \sum_{\ell} \ell W_{\ell}\left(\int_0^t \beta_{\ell}(X(s))\,ds\right) + \int_0^t \partial F(X(s)) \cdot V(s)\,ds \quad . \tag{2.24}$$

Since (2.24) is a linear equation in V and the second term on the right is a Gaussian process, V must also be Gaussian.

For details on the central limit theorem as outlined here see Kurtz (preprint b). Other versions of this and closely related theorems can be found in Rosen (1967), Norman (1972, 1974), Barbour (1974) and Kurtz (1971, 1976). Error estimates for approximation in distribution are given in Barbour (1974), Allain (1976 a, b) and Alm (1978), and for pathwise approximations in Kurtz (1978). Note that the theorems described here apply uniformly on bounded time intervals. Norman (1974) gives one result which applies uniformly on the infinite time interval (see also Alm (1978) and Barbour (1976)), but a good deal of work remains to be done in this case.

Returning briefly to the examples of Section 1, we note that for the epidemic model $F(x_1, x_2) = (-\lambda x_1 x_2, \lambda x_1 x_2 - \mu x_2)$, so that the limiting deterministic model in Theorem (2.16) is the familiar system

$$\begin{aligned} \dot{S} &= -\lambda S I \\ \dot{I} &= \lambda S I - \mu I \end{aligned} \quad . \tag{2.25}$$

For the population growth model

$$F(x) = \lambda x - (\mu + \gamma x)x = \gamma x \left(\frac{\lambda - \mu}{\gamma} - x\right)$$

so that in the case $\lambda > \mu$ the deterministic model is the logistic model.

3. An epidemic with general infection time

An underlying assumption in the epidemic model discussed above is that the length of time that an infected individual remains infected is exponentially distributed. In this section we describe briefly results of Wang (1975, 1977a) extending the theorems in the previous section to epidemic models in which the infection time has an arbitrary distribution.

As before, let N be the total population size, let $S_N(t)$ be the fraction of the total population that is susceptible and let $I_N(t)$ be the fraction that is infected. We assume

(i) the intensity for a particular susceptible to become infected is $\alpha(I_N(t))$ and hence the intensity for some susceptible to become infected is $N\alpha(I_N(t))\,S_N(t)$;

(ii) the time Δ an infected individual remains infected satisfies $P\{\Delta > t\} = F(t)$;

(iii) recovered individuals are immune.

We construct the model in a manner analogous to that used above. Let $\eta_1, \eta_2, \cdots, \Delta(1), \Delta(2), \cdots$ be independent random variables and Y be an independent Poisson process. The η_i will be the lengths of time the initially infected remain infected, $\Delta(k)$ will be the length of time the kth newly infected individual remains infected (hence $P\{\Delta(k) > t\} = F(t)$). If the number of infected individuals initially in the population is k_N, then $\sum_{i=1}^{k_N} \chi_{[0,\eta_i)}(t)$ is the number of these individuals that are still infected at time t.

Setting

$$(3.1) \qquad I_N^0(t) = N^{-1} \sum_{i=1}^{k_N} \chi_{[0,\eta_i)}(t)$$

the model is given by the solution of the following equations:

$$(3.2) \qquad S_N(t) = S_N(0) - N^{-1} Y(N \int_0^t \alpha(I_N(s)) S_N(s)\, ds);$$

$$(3.3) \qquad Z_N(t) = Y(N \int_0^t \alpha(I_N(s)) S_N(s)\, ds)$$

$$(3.4) \qquad I_N(t) = I_N^0(t) + N^{-1} \int_0^t \chi_{[s,\, s+\Delta(Z_N(s)))}(t)\, dZ_N(s) .$$

To justify (3.4) note that Z_N jumps by one each time a susceptible is infected. If the infection occurs at time s, the individual remains infected until time $s + \Delta(Z_N(s))$ and hence $\chi_{[s,\, s+\Delta(Z_N(s)))}(t)$ is one if the individual is infected at time t and zero otherwise.

For simplicity we assume that all those initially infected have just become infected at time zero and hence $P\{\eta_i > t\} = F(t)$.

<u>Theorem</u> (3.5). Suppose $|\alpha(x) - \alpha(y)| \le \mu |x-y|$. If $\lim_{N\to\infty} I_N(0) = I(0)$ and $\lim_{N\to\infty} S_N(0) = S(0)$ then

$$(3.6) \qquad \lim_{N\to\infty} E[|I_N(t) - I(t)|^2 + |S_N(t) - S(t)|^2] = 0$$

where

$$(3.7) \qquad S(t) = S(0) - \int_0^t \alpha(I(s)) S(s)\, ds$$

and

$$(3.8) \qquad I(t) = I(0) F(t) + \int_0^t F(t-s) \alpha(I(s)) S(s)\, ds .$$

<u>Sketch of Proof</u>. Rewrite (3.2) and (3.4) as

(3.9) $$S_N(t) = -N^{-1}\hat{Y}(N\int_0^t \alpha(I_N(s))\,S_N(s)\,ds) + S_N(0) - \int_0^t \alpha(I_N(s))\,S_N(s)\,ds$$

and

(3.10) $$I_N(t) = N^{-1}\int_0^t (\chi_{[s,\,s+\Delta(Z_N(s)))}(t) - F(t-s))\,dZ_N(s) + N^{-1}\int_0^t F(t-s)\,d\hat{Z}_N(s) + I_N^0(t) + \int_0^t F(t-s)\,\alpha(I_N(s))\,S_N(s)\,ds$$

where $\hat{Y}$ is the centered Poisson process and
$\hat{Z}_N = \hat{Y}(N\int_0^t \alpha(I_N(s))\,S_N(s)\,ds)$.

Let a_N denote the first term on the right of (3.9) and b_N and c_N the first and second terms on the right of (3.10). Then

(3.11) $$E[a_N^2] = N^{-1}E[\int_0^t \alpha(I_N(s))\,S_N(s)\,ds] ,$$

(3.12) $$E[b_N^2] = N^{-2}E[\int_0^t F(t-s)\,(1-F(t-s))\,dZ_N(s)] = N^{-1}E[\int_0^t F(t-s)\,(1-F(t-s))\,\alpha(I_N(s))\,S_N(s)\,ds] ,$$

and

(3.13) $$E[c_N^2] = N^{-1}E[\int_0^t F^2(t-s)\,\alpha(I_N(s))\,S_N(s)\,ds] .$$

Gronwall's inequality then gives (3.6) . □

Wang (1977a) gives Gaussian limit theorems for $N^{\frac{1}{2}}(S_N(t) - S(t))$ and $N^{\frac{1}{2}}(I_N(t) - I(t))$. Wang (1977b) contains related results for age-dependent population models.

4. A spatial competition model

We now give an example of a deterministic, large-population approximation for a model involving location dependence. We formulate the model as a model for competition between two annual plant species, but those familiar with models in population genetics will recognize its connection with models of selective advantage.

We assume that there are two plant species which occupy sites on the lattice $L_N = \{N^{-1}k : k \in \mathbb{Z}^2\}$. Each site will be occupied by exactly one plant. The model is based on the following assumptions.

(i) Species A produces $(1+s)$ times as many seeds as species B.

(ii) A fraction $q(x,y)$ of the seeds produced at site y are distributed to site x.

(iii) Exactly one seed at each site ("randomly" selected from the available seeds) grows.

We assume that the number of seeds produced by each plant is large and ignore random fluctuations in the number of seeds at a site. In particular the probability that the plant at a site next year is type A is just the fraction of seeds that are type A distributed to the site this year.

Let X^n be the function on L_N such that $X^n(x) = 1$ if the plant at site x in year n is of species A and $X^n(x) = 0$ if species B. Then

$$(4.1) \qquad P\{X^{n+1}(x) = 1 | X^n\} \equiv \frac{\sum_y q(x,y)(1+s)X^n(y)}{\sum q(x,y)(1+s)X^n(y) + \sum_y q(x,y)(1-X^n(y))} \equiv \Pi^{n+1}_{(x)} .$$

We assume that X^n is a Markov chain and that the random variables $X^{n+1}(x)$, $x \in L_N$ are conditionally independent given X^n. This assumption along with (4.1) determines the transition mechanism for the chain. We now consider a sequence of such processes X_N parameterized by N with $s = \alpha/N$ for some fixed $\alpha > 0$, and $q(x,y)$ replaced by $q_N(x,y)$. We define

$$(4.2) \qquad T_N f(x) \equiv \sum_y q_N(x,y) f(y) / \sum_y q_N(x,y)$$

and note that

$$(4.3) \qquad P\{X_N^{n+1}(x) = 1 | X_N^n\} = \frac{(1+N^{-1}\alpha) T_N X_N^n(x)}{1 + N^{-1}\alpha T_N X_N^n(x)} = \Pi_N^{n+1}(x) .$$

Setting $\xi_N^{n+1} = X_N^{n+1} - \Pi_N^{n+1}$ (i.e. center X_N^{n+1} at its conditional expectation), we have

$$(4.4) \qquad X_N^{n+1} = X_N^{n+1} - \Pi_N^{n+1} + \frac{(1+N^{-1}\alpha) T_N X_N^n}{1 + N^{-1}\alpha T_N X_N^n} - T_N X_N^n + T_N X_N^n$$

$$= \xi_N^{n+1} + \frac{N^{-1}\alpha T_N X_N^n (1 - T_N X_N^n)}{1 + N^{-1}\alpha T_N X_N^n} + T_N X_N^n$$

and defining $Y_N^n \equiv T_N X_N^n$ this gives

$$(4.5) \qquad Y_N^{n+1} = T_N \xi_N^{n+1} + T_N Y_N^n + N^{-1} T_N \frac{\alpha Y_N^n (1 - Y_N^n)}{1 + N^{-1}\alpha Y_N^n} .$$

Iteration gives

$$Y_N^n = T_N^n Y_N^0 + N^{-1} \sum_{k=0}^{n-1} T_N^{n-k} \Gamma_N(Y_N^k) + \sum_{k=0}^{n-1} T_N^{n-k} \xi_N^{k+1} \tag{4.6}$$

where $\Gamma_N(y) = \alpha y(1-y)/(1+N^{-1}\alpha y)$. Note that (4.6) is essentially a deterministic equation for Y_N with a random forcing term. If the random term (the third on the right) in (4.6) is small then we would expect Y_N to be close to the solution of the deterministic equation. We give an example of conditions under which this is valid.

Theorem (4.7). Suppose seeds are distributed uniformly over all sites within a distance $N^{-1/2}$ of the parent plant, and that

$$\lim_{N\to\infty} \sup_x |Y_N^0(x) - u_0(x)| = 0 \tag{4.8}$$

where u_0 is uniformly continuous.

Let $T(t)$ denote the semigroup generated by $\frac{1}{6}\Delta$ (i.e. the semigroup corresponding to a Brownian motion). Let u be the solution of

$$u(t) = T(t)u_0 + \int_0^t T(t-s)\alpha u(s)(1-u(s))\,ds \tag{4.9}$$

i.e. the solution of

$$u_t = \frac{1}{6}\Delta u + \alpha u(1-u) \tag{4.10}$$

with $u(0,x) = u_0(x)$. Then

$$\lim_{N\to\infty} \sup_x E[|Y_N^{[Nt]}(x) - u(t,x)|] = 0 . \tag{4.11}$$

Proof. We leave it to the reader to verify that if f has two bounded uniformly continuous derivatives, then

$$\lim_{N\to\infty} \sup_x |N(T_N f(x) - f(x)) - \frac{1}{6}\Delta f(x)| = 0 . \tag{4.12}$$

This convergence implies

$$\lim_{N\to\infty} \sup_x \sup_{s\le t} |T_N^{[Ns]} f(x) - T(s)f(x)| = 0 \tag{4.13}$$

for all bounded uniformly continuous f (see Trotter (1958) or Kurtz (1969)).

First we compare Y_N^n to Z_N^n given by

$$Z_N^n = T_N^n Y_N^0 + N^{-1} \sum_{k=0}^{n-1} T_N^{n-k} \Gamma_N(Z_N^k) . \tag{4.14}$$

We define the norm

$$(4.15) \qquad \|\eta\| = \sup_{x \in L_N} E[|\eta(x)|^2]^{1/2}$$

for random processes indexed by L_N and observe that

$$(4.16) \qquad \|T_N \eta\| \le \|\eta\|$$

and

$$(4.17) \qquad \|\Gamma_N(\eta_1) - \Gamma_N(\eta_2)\| \le \alpha \|\eta_1 - \eta_2\| \quad \text{for non-negative } \eta_1, \eta_2 .$$

Consequently

$$(4.18) \qquad \|Y_N^{[Nt]} - Z_N^{[Nt]}\| \le \|N^{-1} \sum_{k=0}^{[Nt]-1} T_N^{[Nt]-k} (\Gamma_N(Y_N^k) - \Gamma_N(Z_N^k))\|$$

$$+ \|\sum_{k=0}^{[Nt]-1} T_N^{[Nt]-k} \xi_N^{k+1}\|$$

$$\le N^{-1} \sum_{k=0}^{[Nt]-1} \alpha \|Y_N^k - Z_N^k\| + \|\sum_{k=0}^{[Nt]-1} T_N^{[Nt]-k} \xi_N^{k+1}\|$$

$$\le \int_0^t \alpha \|Y_N^{[Ns]} - Z_N^{[Ns]}\| \, ds + \|\sum_{k=0}^{[Nt]-1} T_N^{[Nt]-k} \xi_N^{k+1}\| .$$

Gronwall's inequality gives

$$(4.19) \qquad \|Y_N^{[Nt]} - Z_N^{[Nt]}\| \le \|\sum_{k=0}^{[Nt]-1} T_N^{[Nt]-k} \xi_N^{k+1}\| e^{\alpha t} .$$

To estimate the right side of (4.19) define $p_N(k, x, y)$ so that

$$(4.20) \qquad T_N^k f(x) = \sum_y p_N(k, x, y) f(y)$$

and note that

$$(4.21) \qquad E[\xi_N^k(x) \xi_N^\ell(y)] = 0$$

as long as $x \neq y$ or $k \neq \ell$. Therefore

$$(4.22) \qquad E[(\sum_{k=0}^{[Nt]-1} T_N^{[Nt]-k} \xi_N^{k+1}(x))^2] = \sum_{k=0}^{[Nt]-1} \sum_y p_N^2([Nt]-k, x, y) E[(\xi_N^{k+1}(y))^2]$$

$$\le \sum_{\ell=1}^{[Nt]} \sum_y p_N^2(\ell, x, y) .$$

Let $C_N = \#\{k \in \mathbb{Z}^2 : |k| < N^{\frac{1}{2}}\}$. Then

$$(4.23) \qquad p_N(1,x,y) = \begin{cases} C_N^{-1} & |x-y| < N^{-\frac{1}{2}} \\ 0 & \text{otherwise.} \end{cases}$$

Let $U_N(\ell)$, $\ell = 0,1,2\cdots$ be a random walk with $U_N(0) = 0$ and transition function p_N. By symmetry $p_N(\ell,x,y) = p_N(\ell,y,x)$ and the right side of (4.22) becomes

$$(4.24) \qquad \sum_{\ell=1}^{[Nt]} \sum_y p_N(\ell,x,y)\, p_N(\ell,y,x) = \sum_{\ell=1}^{[Nt]} p_N(2\ell,x,x) = \sum_{\ell=1}^{[Nt]} p_N(2\ell,0,0)$$

$$= \sum_{\ell=1}^{[Nt]} C_N^{-1} P\{|U_N(2\ell-1)| < N^{-\frac{1}{2}}\}$$

$$= \sum_{\ell=1}^{[Nt]} C_N^{-1} P\{|N^{\frac{1}{2}} U_N(2\ell-1)| < 1\} .$$

We observe that

$$(4.25) \qquad \lim_{N\to\infty} P\{|N^{\frac{1}{2}} U_N(m)| < 1\} = P\{|U(m)| < 1\}$$

where U is a random walk in the plane with jumps uniformly distributed over the unit disk. Since $C_N = 0(N)$ and $\lim_{m\to\infty} P\{|U(m)| < 1\} = 0$ we see that the right side of (4.24) goes to zero as $N \to \infty$.

To complete the proof of the theorem we need to show

$$(4.26) \qquad \lim_{N\to\infty} \sup_{s \le t} \|Z_N^{[Ns]} - u(s)\| = 0 .$$

We rewrite (4.14) as

$$(4.27) \qquad Z_N^{[Nt]} = T_N^{[Nt]} Y_N^0 + \int_0^{[Nt]N^{-1}} T_N^{[Nt]-[Ns]} \Gamma_N(Z_N^{[Ns]})\, ds .$$

Setting $\Gamma(y) = \alpha y(1-y)$, we have

$$(4.28) \qquad \|Z_N^{[Nt]} - u(t)\| \le \|T_N^{[Nt]}(Y_N^0 - u_0)\| + \|(T_N^{[Nt]} - T(t))\,u_0\|$$

$$+ \left\|\int_0^{[Nt]N^{-1}} T_N^{[Nt]-[Ns]}(\Gamma_N(Z_N^{[Ns]}) - \Gamma_N(u(s)))\,ds\right\|$$

$$+ \left\|\int_0^{[Nt]N^{-1}} T_N^{[Nt]-[Ns]}(\Gamma_N(u(s)) - \Gamma(u(s)))\,ds\right\|$$

$$+ \left\|\int_0^{[Nt]N^{-1}} (T_N^{[Nt]-[Ns]} - T(t-s))\,\Gamma(u(s))\,ds\right\|$$

$$+ \left\|\int_{[Nt]N^{-1}}^{t} T(t-s)\,\Gamma(u(s))\,ds\right\| .$$

It is easy to see that the first, second, fourth, fifth and sixth terms on the right of (4.28) tend to zero so we lump them together as ε_N. Referring to (4.17) we then have

$$(4.29) \qquad \|Z_N^{[Nt]} - u(t)\| \le \varepsilon_N(t) + \int_0^t \|\Gamma_N(Z_N^{[Ns]}) - \Gamma_N(u(s))\|\,ds$$

$$\le \varepsilon_N(t) + \int_0^t \alpha\,\|Z_N^{[Ns]} - u(s)\|\,ds ,$$

and hence

$$(4.30) \qquad \|Z_N^{[Nt]} - u(t)\| \le \varepsilon_N(t)\,e^{\alpha t}$$

which, since $\sup_{s \le t} \varepsilon_N(s) \to 0$, implies (4.26). □

<u>Remark</u>. We would, of course, like to relate the convergence of $Y_N^{[Nt]}$ to the original process $X_N^{[Nt]}$. Let φ be a continuous function with compact support. Then

$$(4.31) \qquad \lim_{N\to\infty} N^{-2} \sum_{x \in L_N} \varphi(x)\,Y_N^{[Nt]}(x) = \lim_{N\to\infty} N^{-2} \sum_{x \in L_N} \varphi(x)\,T_N X_N^{[Nt]}(x)$$

$$= \lim_{N\to\infty} N^{-2} \sum_{x \in L_N} T_N\,\varphi(x)\,X_N^{[Nt]}(x)$$

$$= \int \varphi(x)\,u(t,x)\,dx .$$

It is easy to see $\lim_{N\to\infty} \sup_x |T_N\,\varphi(x) - \varphi(x)| = 0$, and hence (4.31) implies

$$(4.32) \qquad \lim_{N\to\infty} N^{-2} \sum_{x \in L_N} \varphi(x)\,X_N^{[Nt]}(x)\,dx = \int \varphi(x)\,u(t,x)\,dx .$$

Let G be any bounded open set in $\mathbb{R}^2$ whose boundary has Lebesgue measure zero. Then (4.32) implies that

$$(4.33) \qquad \lim_{N \to \infty} N^{-2} \, \# \{x \in G \cap L_N : X_N^{[Nt]}(x) = 1\} = \int_G u(t,x)\, dx \quad .$$

Arnold and Theodosopulu (preprint) consider location dependent chemical reaction models and obtain results similar to Theorem (4.7). Lang and Xanh (preprint) obtain similar results for Smoluchowski's model of coagulation in colloids.

References

1. Aalen, Odd (1978). Nonparametric inference for a family of counting processes. Ann. Statistics 6, 701-726

2. Allain, Marie-France (1976a). Approximation par un processus de diffusion des oscillations, autour d'une valeur moyenne, d'une suite de processus de Markov de saut pur. C. R. Acad. Sc. Paris, 282, 891-894.

3. Allain, Marie-France (1976b). Étude de la vitesse de convergence d'une suite de processus de Markov de saut pur. C. R. Acad. Sc. Paris, 282, 1015-1018.

4. Alm, Sven Erick (1978). On the rate of convergence in diffusion approximation of jump Markov processes. Dissertation, Uppsala University, Uppsala, Sweden.

5. Arnold, Ludwig and M. Theodosopulu (preprint). Deterministic limit of the stochastic model of chemical reactions with diffusion.

6. Barbour, Andrew D. (1974). On a functional central limit theorem for Markov population processes. Adv. Appl. Prob. 6, 21-39.

7. Barbour, Andrew D. (1976). Quasi-stationary distributions in Markov population processes. Adv. Appl. Prob. 8, 296-314.

8. Billingsley, Patrick (1968). Convergence of Probability Measures. Wiley, New York.

9. Kurtz, Thomas G. (1969). Extensions of Trotter's operator semigroup approximation theorems. J. Functional Analysis 3, 111-132.

10. Kurtz, Thomas G. (1971). Limit theorems for sequences of jump Markov processes approximating ordinary differential processes. J. Appl. Prob. 8, 344-356.

11. Kurtz, Thomas G. (1976). Limit theorems and diffusion approximations for density dependent Markov chains. Mathematical Programming Study 5, 67-78, North-Holland, Amsterdam, New York, Oxford.

12. Kurtz, Thomas G. (1978). Strong approximation theorems for density dependent Markov chains. Stoch. Proc. Appl. 6, 223-240.

13. Kurtz, Thomas G. (preprint a). Representations of Markov processes as multiparameter time changes. (to appear, Ann. Probability).

14. Kurtz, Thomas G. (preprint b). Approximation of population processes. Lecture notes for NSF Regional Conference, Missoula, Montana, June 25-29, 1979 (to appear).

15. Lang, Reinhard, and Nguyen Xuan Xanh (preprint). Smoluchowski's theory of coagulation in colloids holds rigorously in the Boltzmann-Grad limit.

16. Meyer, P. A. (1971). Demonstration simplifée d'un théorème de Knight. Séminaire de Probabilités V, Université de Strasbourg, Lect. Notes in Math. 191, 191-195.

17. Norman, M. Frank (1972). Markov Processes and Learning Models. Academic Press, New York.

18. Norman, M. Frank (1974). A central limit theorem for Markov processes that move by small steps. Ann. Probability 2, 1065-1074.

19. Rosén, B. (1967). On the central limit theorem for sums of dependent random variables. Z. Wahrscheinlichkeits theorie und Verw. Gehrete 7, 48-82.

20. Trotter, H. F. (1958). Approximation of semigroups of operators. Pacific J. Math. 8, 887-919.

21. Wang, Frank J. S. (1975). Limit theorems for age and density dependent stochastic population models. J. Math. Biosci. 2, 373-400.

22. Wang, Frank J. S. (1977a). Gaussian approximation of some closed stochastic epidemic models. J. Appl. Prob. 14, 221-231.

23. Wang, Frank J. S. (1977b). A central limit theorem for age and density dependent population processes. Stoch. Proc. Appl. 5, 173-193.

Remarks on Limit Theorems for Partial Differential Equations with Random Coefficients

George C. Papanicolaou*
Courant Institute, New York University

We shall give some examples of asymptotics for partial differential equations with coefficients that are random functions. Roughly, the asymptotic analysis deals with cases in which the coefficients vary much more rapidly than the solution but there are many ways such a statement can be interpreted. We shall not give detailed statements or proofs; some essential points are presented in [1]. For boundary value problems of ordinary differential equations with random coefficients some of the examples are applications of the theorems of White and Franklin [2].

We shall consider first the equation

$$(1) \qquad (-\Delta + v^{\varepsilon})\phi^{\varepsilon}(x) = f(x) , \qquad x \in \mathbb{R}^d$$

where $v^{\varepsilon}(x) = v^{\varepsilon}(x,\omega) \geq 0$ is a given random process ($\omega \in \Omega$, (Ω, F, P) is some probability space) depending on a parameter $\varepsilon > 0$. The solution of (1) $\phi^{\varepsilon}(x) = \phi^{\varepsilon}(x,\omega)$ is also a random process while $f(x)$ is a given deterministic, for simplicity, function which let us say vanishes outside a compact set. In the simplest case the random potential $v^{\varepsilon}(x,\omega) = v(x/\varepsilon,\omega)$, where $v(x,\omega)$ is a given stationary random process with mean $\bar{v} = E\{v(x)\} \geq 0$. We expect that $\phi^{\varepsilon}(x)$ will be close, in some sense, to the solution of

$$(2) \qquad (-\Delta + \bar{v})\bar{\phi}(x) = f(x) , \qquad x \in \mathbb{R}^d ,$$

when ε is small.

Such a result is of course an ergodic theorem which relies on stationarity (or approximate stationarity) and we will discuss it further in a more involved, nonlinear, context below.

The simple association of (1) with (2) as some approximation when the potential v^{ε} is characterized completely by one spatial scale $(v^{\varepsilon}(x,\omega) = v(x/\varepsilon,\omega))$ allows for a simple analysis similar to homogenization [3] but brings up many other questions. For example, can one determine the behavior of $\phi^{\varepsilon}-\Phi$ (fluctuation theory), what about the spectrum of $-\Delta + v^{\varepsilon}$, in particular the behavior of the bottom of the

*Research supported by the U.S. Army Office of Scientific Research under Grant No. DAAG29-78-G-0177.

spectrum?

We introduce a class of potentials v^ε of the following type. Let $\{x_1,x_2,...\}$ be a Poisson system of points with mean density $\rho(x)$ and $w(x) \geq 0$ be a fixed function with compact support. Let represent the realization of points $\{x_1,x_2,...\}$ of the Poisson process and put

$$(3) \qquad v^\varepsilon(x,\omega) = \sum w\left(\frac{x-x_i}{\varepsilon}\right) .$$

Suppose in addition that the density ρ depends on ε, say ρ^ε and goes to infinity as $\varepsilon \to 0$. The question is again: how does the solution ϕ^2 of (1) behave for ε small but now with a difference, namely can one obtain the approximation uniformly in the height of w (i.e. w could be infinite in some region in its support)?

If w is bounded and $\rho^\varepsilon(x) = \varepsilon^{-d}\,\tilde\rho(x)$ then we are effectively in the situation of (2), namely ϕ^ε behaves like $\bar\phi$ which satisfies (2) with $\bar v = \tilde\rho(x)\bar w$, $\bar w = \int w(x)dx$. But the situation is very different when w can be unbounded because one must solve the single potential problem exactly. In fact the results are dimension dependent. For $d = 3$, $\rho^\varepsilon = \varepsilon^{-1}\tilde\rho$ then $\bar v$ turns out to be given by $\tilde\rho(x)\bar w$ where now $\bar w$ is obtained from an exact solution of the one-potential (rather than sum as in (3)) problem. When for example $w = +\infty$ inside some set D and zero outside, then $\bar w$ is the capacitance of the set D (recall $d = 3$ or $d \geq 3$ here). One can also ask, in this context, what happens when $\tilde\rho$ is large. (Can one obtain effective potentials $\bar v$ that in some sense are uniform in $\tilde\rho$? This leads to the theory of the coherent potential [4] that is not yet developed mathematically.

In the nonlinear case, consider the problem

$$(4) \qquad -\Delta\phi^\varepsilon + F\left(\phi^\varepsilon,x,\frac{x}{\varepsilon}\right) = 0 , \qquad x \in \mathbb{R}^d$$

where $F(\phi,x,y)$ is for each ϕ and x fixed a random field, stationary in y, with mean $\bar F(\phi,x) = E\big(F(\phi,x,y)\big)$. We may also consider non-linear random terms of the form (3) where the scaling parameter enters a little differently than in (4) but asymptotically has the same effect. Let $\bar\phi$ be the solution of

$$(5) \qquad -\Delta\bar\phi + \bar F(\bar\phi,x) = 0 , \qquad x \in \mathbb{R}^d$$

that is, suppose that (5) has a nice solution. Suppose further that the variational problem

$$(6) \qquad -\Delta z + \frac{\partial \bar{\delta}}{\partial \phi}(\bar{\phi},x)z = 0 \qquad x \in \mathbb{R}^d$$

has only the $z \equiv 0$ solution and $(-\Delta + \bar{F}_\phi)^{-1}$ is well defined. Then, as in [1] one can show that there is a set of realizations of the random field F for which (4) has a suitable solution and that this solution converges in probability to the one of (5). If condition (6) is not satisfied then we are in the bifurcation situation and this is discussed in [1].

Once the relation between (4) and (5) has been established one can ask about the behavior of $\varepsilon^{-\gamma}(\phi^\varepsilon - \bar{\phi}) = \psi^\varepsilon$ as $\varepsilon \to 0$ for a suitable constant γ (that depends on the dimension d). Under hypotheses of the familiar central limit theorem type on $F(\phi,x,y)$ one can show that $\psi^\varepsilon(x)$ converges as $\varepsilon \to 0$ to a gaussian process with mean zero and covariance that is easily computed using the inverse of $-\Delta + \bar{F}_\phi$. One interesting thing about the proof of this theorem is that it can be divided into two essentially separate steps. The first one involves only PDE estimates and relatively little probability while the second involves the usual spatial central limit theorem (for integrals) and no PDE work.

Are there interesting limit theorems in which the limit is neither deterministic (of the ergodic theory form) or gaussian (of the fluctuation form)? In the nonlinear case the situation appears to be very difficult because it is very hard to characterize processes that are not gaussian. Perhaps the spatial Markov property is relevant here. In the linear case the situation is manageable, at least formally, and a number of interesting processes emerge as in the context of waves in random media [5]. One should also mention, in the linear case, the qualitative and quantitative properties of the spectrum of (1). There are several interesting conjectures [4] in this context.

References

1 G.C. Papanicolaou, Stochastically perturbed bifurcation, Proceedings of IRIA Congress, Dec. 1979, J. L. Lions and R. Glewinsky, editors, Springer lecture notes in Mathematics, to appear.

2 B. White and J. Franklin, A limit theorem for stochastic two-point boundary-valve problems for ordinary differential equations, Comm. Pure Appl. Math. 32 (1979), pp. 253-276.

3 A. Bensoussan, J.L. Lions and G. Papanicolaou, Asymptotic analysis of periodic structures, North Holland, Amsterdam, 1978.

4 M. Lax, Wave propagation and conductivity in random media, in SIAM-AMS Proceedings, Vol. VI, Amer. Math. Soc., Providence, R.I., 1973.

5. A. Ishimaru, Wave propagation and scattering in random media, Vols. I, II, Academic Press, New York, 1978.

TOPIC VI

Further mathematical approaches and techniques

Discussion by the Editors

In this last section four papers are picked out which deal with some of the most discussed theories during the conference: the non-Markovian percolation theory (and the theory of subadditive processes) as well as the deterministic theory of reaction-diffusion processes. Accordingly, the material presented here can be thought of as the second epilogue of the conference, handling with future research lines. J.Hammersley was the designated lecturer for giving vistas in percolation theory. In his paper, the titles of the last paragraphs are research programmes, viz. spatially correlated percolation and percolation on random graphs. Co-author of a recent book on first-passage percolation, R.T.Smythe suggests in his paper future asymptotic statements for front velocities and the shape of the configuration. X.X.Nguyen introduces a class of stationary spatially homogeneous (strongly) subadditive processes and derives a result on the decomposability of such processes into an additive part with the same constant and a subadditive part with constant zero. Here, the decomposability is a consequence of a certain ergodic theorem. The biological significance of the problems treated here cannot slip to the reader's notice; we mention only the first directly suggested applications: the influence of the environment on a given configuration (Nguyen), the stochastic domination of a process by another (Smythe) or the relationship between the agent's pathogenicity and the susceptibility of its host (Hammersley).

The principal idea of the paper written by C.Conley and J.Smoller is to consider the basic reaction-diffusion equation as defining a semi-flow in an infinite-dimensional space and to find particular solutions with the aid of topological techniques. The authors actually examine the stability problem for steady state solutions of reaction-diffusion equations with homogeneous boundary conditions. Because these stable solutions may be viewed as models of dissipative stuctures, the consequences of this study for the investigation of self-organizing phenomena is easily recognizable.

Our epilogue to this discussion might be the following sentence of A.N.Whitehead:"Mathematics is the most powerful technique for the understanding of pattern, and for the analysis of the relation of patterns."

TOPOLOGICAL TECHNIQUES IN REACTION-DIFFUSION EQUATIONS

by Charles Conley and Joel Smoller*

§1. Introduction. In this note, we shall illustrate how some topological ideas can be used to obtain rather precise information about solutions of reaction-diffusion equations. The equations are of the form

(1) $$u_t = u_{xx} + f(u), \quad -L < x < L,$$

in a single space variable, with either homogeneous Dirichlet or Neumann boundary conditions. The function $f(u)$ is a cubic polynomial having three distinct real roots, with $f(-\infty) > 0 > f(\infty)$. The solutions of interest to us are the so-called "steady-state" solutions; i.e. solutions which are independent of t, and therefore satisfy the equation

(2) $$u'' + f(u) = 0, \quad -L < x < L,$$

with the same boundary conditions. Our first goal is to determine the dimension of the unstable manifold of solutions of (2); the solution being considered as a particular solution of (1). Our second goal is to show how to "continue" solutions when the boundary manifolds serve as the "continuation parameter."

There are three main ingredients in our program. The basic one is the computation of the "index" of the solution, which uses the Morse decomposition of the isolated invariant set, [1], in the Neumann problem, and the invariance of the index under continuation in the Dirichlet case. The second is the non-degeneracy of the solution, [3], and the third is the global bifurcation picture for the steady-state solutions, [4]. These ideas are discussed elsewhere in detail, and we shall not dwell on them here. As a biproduct of our techniques, we shall show how to continue solutions of the Neumann problem to solutions of the Dirichlet problem and vice-versa.

The picture that we have in mind is that we can consider the partial differential equation (1) as defining a semi-flow** in an infinite dimensional space. Here the solutions of (2) correspond precisely to the "rest-points" of (1). Our main goal is to find all of the rest points, and to determine the dimensions of the unstable

---

*Research of Conley supported by N.S.F., that of Smoller by A.F.O.S.R.

**In the appendix we show how to embed the semi-flow in a locally compact flow.

manifolds of the non-degenerate solutions. We will illustrate our methods by examples which contain all the general techniques.

§2. Non-Degenerate Solutions of the Dirichlet Problem.

In this section and the next, we take the function $f(u)$ to be a cubic polynomial having one positive and two negative roots, of the form $f(u) = -u^3 + \sigma u^2 - \tau u + \pi$.

As discussed in [2], solutions of the Dirichlet problem for (2) correspond to orbits of the vector field

$$\dot{u} = v, \qquad \dot{v} = -f(u) \tag{3}$$

which satisfy the boundary conditions

$$u(\pm L) = 0. \tag{4}$$

We shall restrict our attention to positive solutions; these correspond to solution curves of (3), which begin and end on the line $u = 0$, lie in $u \geq 0$, and take "time" $2L$ to make the journey; see Figure 1.

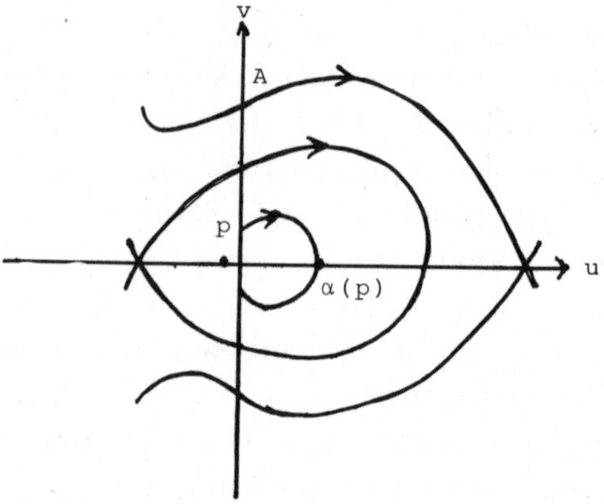

Figure 1

It is shown in [4], that if $\sigma^3 > (\frac{27}{4})^2 \pi$, then the global bifurcation picture for the positive solutions takes the form of Figure 2. Thus, for sufficiently short or long intervals; (i.e. $L < L_1$ or $L > L_2$) there is exactly one solution, while for $L_1 < L < L_2$ there are three solutions. These results are obtained by analysing the "time-map",

$$T(p) = \int_0^{\alpha(p)} (2F(\alpha(p)) - 2F(u))^{-1/2}\, du, \quad \text{where} \quad F' = f,$$

and showing that T has exactly two critical points.

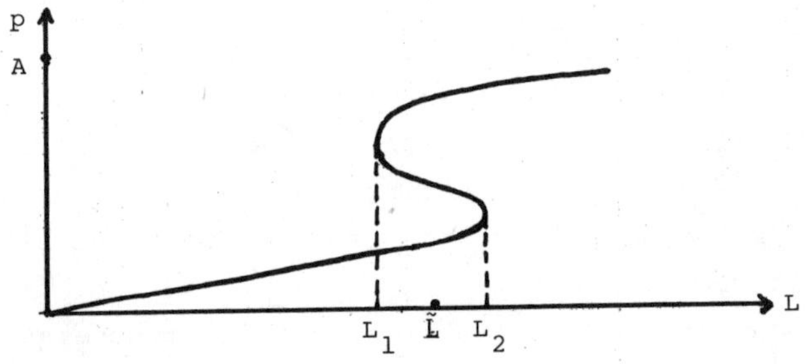

Figure 2

Having the global picture in mind, we next turn to the problem of non-degeneracy of these solutions. We recall that a solution $u_L(x)$ of (2), (4), is called non-degenerate if zero is not in the spectrum of the linearized operator $d^2/dx^2 + f'(u_L(x))$, (with homogeneous Dirichlet boundary conditions at $\pm L$). Now it was proved in [3] that whenever $T'(p) \neq 0$, then the corresponding solution is non-degenerate. Since T has exactly two critical points, it follows that except for the two distinguished solutions corresponding to the two turning points in Figure 2, all solutions under consideration are non-degenerate.

We shall now show how one can compute the index of the steady-state solutions. We begin by assuming $L = \tilde{L}$, $L_1 < \tilde{L} < L_2$ (see Figure 2). Then (2) has precisely three solutions on $-\tilde{L} < x < \tilde{L}$, satisfying homogeneous Dirichlet boundary conditions. Let us call these u_1, u_2 and u_3 corresponding to p_1, p_2 and p_3 respectively, with $p_1 < p_2 < p_3$. By using the fact that the index is invariant under continuation, we can proceed as in [2] to determine that u_1 and u_3 have index* Σ^0, and u_2 has index Σ^1. However since the u_i's are non-degenerate, it follows that u_1 and u_3 are attractors; i.e., they are stable solutions of (1), and that u_2 has a one-dimensional unstable manifold. The argument goes as follows.

---

* Σ^k denotes the homotopy-type of the pointed k-sphere.

Consider for example u_2; since $f'(u_2(x))$ is bounded, the operator $A = d^2/dx^2 + f'(u_2(x))$, with homogeneous Dirichlet boundary conditions, has at most a finite number, say k, of positive eigenvalues. Since 0 is not in the spectrum of A, it follows that u_2 has a k-dimensional unstable manifold, so that the index of u_2 is a pointed k-sphere; whence $k = 1$. The argument for u_1 and u_3 is similar.

Now using the invariance of the index under continuation, and taking L as a parameter, we see that all solutions corresponding to points on the curve in Figure 2 having (finite) positive slope are attractors, while the solutions with (finite) negative slope have one-dimensional unstable manifolds.

Finally, we consider the two distinguished degenerate solutions corresponding to $L = L_1$ and $L = L_2$ in Figure 2. Since they can be continued to the empty set, they have index $\overline{0}$, the homotopy type of the one point (pointed) space. No general statement can be made about their unstable manifold, except that it must be finite dimensional.

§3. Solutions of the Neumann Problem.

We consider here equation (2) with homogeneous Neumann boundary conditions

$$(5) \qquad u'(\pm L) = 0.$$

Thus solutions of (2), (5) are orbits of (3) which begin and end on the line $v = 0$, and take "time" $2L$ to make the journey. Observe that the three zeros of f are always solutions. In addition, it is easy to see that non-constant solutions must occur in pairs; this follows from the symmetry of the flow with respect to the line $v = 0$.

In order to study the global bifurcation picture, we again consider the map $T(p) = \int_p^{\beta(p)} (2F(\beta(p) - 2F(u))^{-1/2}\, du$ where $\beta(p)$ is that point on the line $v = 0$ satisfying $F(p) = F(\beta(p))$. It is proved in [4], that $T'(p) \neq 0$ for all p. Thus no non-constant solution of (2), (5) can bifurcate, ([3]), and in addition, all non-constant solutions are non-degenerate. The global bifurcation diagram takes the form of Figure 3, where each component corresponds to solutions having a given number of extrema. The points L_n, $n = 0,1,2,\ldots$, correspond to those L for which the spectrum of the operator* $d^2/dx^2 + f'(b)$ contains 0; i.e., $L_n = (2n+1)\pi/2\sqrt{f'(b)}$, $n = 0,1,2,\ldots$.

* b is that root of f with positive derivative.

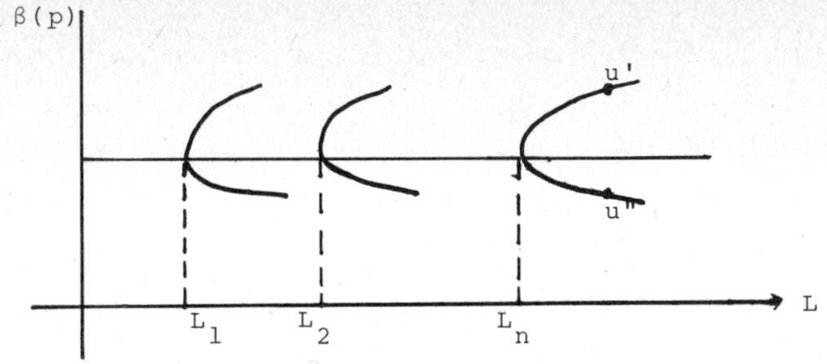

Figure 3

It is shown in [2], using the index discussed above, that no non-constant solution of (2) - (5) can be stable, where we again consider these solutions as rest points of (1). We will show in fact, that the dimension of the unstable manifold of the solution is precisely k, where k is the number of (internal) extrema of the corresponding solution. To do this, first observe that the constant solution $u \equiv b$ has an n-dimensional unstable manifold, if $L_n < L < L_{n+1}$; this follows easily from the remarks above. Next, we let S_n denote the isolated invariant set containing as its only rest points, the solution $u_n \equiv b$ and the two non-constant solutions which bifurcate out of zero, u_n' and u_n'' (see Figure 3). Moreover, symmetry considerations show that the index of u_n', $h(u_n')$, equals $h(u_n'')$, the index of u_n''. We shall inductively assume that $h(S_n) = \Sigma^n$, and that $h(u_n) = \Sigma^{n+1}$. (This is true for $n = 1$ where $L_1 < L < L_2$ since in this case, $h(S_1) = \Sigma^1$ and $h(b) = \Sigma^2$.)

Thus we have the general situation where we have an isolated invariant set containing exactly three critical points, the index of one of which is known, the total index of the isolated invariant set is known, and the index of the two other rest points is unknown. For the rest points u_n', u_n, u_n'', we denote by M_n', M_n, M_n'', the corresponding Morse sets (see [1]), having indices h_n', h_n and h_n', respectively. Our problem is to compute h_n'.

Now first we observe that there cannot exist an orbit connecting M_n' to M_n'' ; if so then there would exist, by symmetry, an orbit connecting these two sets, but running in the opposite direction. This however, is not possible for Morse sets. Hence, M_n' is not connected to M_n''. Next, if M_n' is not connected to M_n by an orbit, then we would have $h(S) = h_n' \vee h_n \vee h_n'$, or $\Sigma^n = h_n' \vee \Sigma^{n+1} \vee h_n'$. Now as in §2, h_n' must be a Σ^k for some k. But then this last equation is topologically impossible. It follows then, that there must be an orbit connecting M_n' to M_n, and by symmetry, one connecting M_n'' to M_n.

Now there are two cases to consider; namely
i) the orbits run from M_n to M_n' and M_n'', or
ii) the orbits run from M_n' and M_n'' and M_n.
We will consider case i. Before doing this however, we need some notation. We let H_k, H_k', and H_k'' denote the k-th homology groups of M, M' and M" respectively, and we denote by $\bar{H}_k$, the k-th homology group of the invariant set S. Since we are in case i), we have the boundary maps $\partial_k: H_{k+1} \to H_k' + H_k''$.

Now by hypothesis, we have

$$H_k = \begin{cases} 0, & k \neq n+1 \\ \mathbb{Z}, & k = n+1 \end{cases}, \text{ and } \bar{H}_k = \begin{cases} 0, & k \neq n \\ \mathbb{Z}, & k = n. \end{cases}$$

Moreover, we have the following (long) exact sequence:

$$\longrightarrow \bar{H}_{j+1} \longrightarrow H_{j+1} \longrightarrow H_j' + H_j'' \longrightarrow \bar{H}_j \longrightarrow H_j .$$

We shall show how this enables us to compute each H_j', $j = 1,2,\ldots$. First, if $j \neq n$, we have the exact sequence $0 \to H_j' + H_j'' \to 0$, which implies $H_j' = H_j'' = 0$. Now if $j = n$, we have the exact sequence

$$0 \longrightarrow \mathbb{Z} \longrightarrow H_n' + H_n'' \longrightarrow \mathbb{Z} \longrightarrow 0 .$$

Observe that by exactness, $H_n' + H_n''$ cannot map into 0, and thus the image of $H_n' + H_n''$ is $n\mathbb{Z}$, for some n. Then since $\mathbb{Z}$ maps onto we see that $H_n' + H_n''$ maps onto $\mathbb{Z}$. Thus $H_n' + H_n'' =$ $\mathbb{Z} + j(\mathbb{Z})$, where j is the map $\mathbb{Z} \to H_n' + H_n''$. Now $j(\mathbb{Z}) = \mathbb{Z}/n\mathbb{Z}$ or $j(Z) = \mathbb{Z}$. It follows that $H_n' + H_n'' = \mathbb{Z} + n\mathbb{Z}$, or $H_n' + H_n'' =$ $\mathbb{Z} + \mathbb{Z}$. Now the first equation implies $2(\text{rank } H_n') = 1$, which is impossible. Thus $H_n' + H_n'' = \mathbb{Z} + \mathbb{Z}$, so since H_n' has no elements of finite order and H_n' is finitely generated, H_n' must be the sum of a finite number of copies of $\mathbb{Z}$ and so $H_n' = \mathbb{Z}$.

Therefore, we have shown that

$$H_j' = \begin{cases} 0, & j \neq n \\ \mathbb{Z}, & j = n \end{cases} . \tag{6}$$

Now the solutions u_n' and u_n'' are non-degenerate, and have finite dimensional unstable manifolds, so $h_n' = \Sigma^k$ for some k. But in view of (6), $k = n$. Thus the two bifurcating solutions have n-dimensional unstable manifolds.

Observe now that for each n, there exist solutions $u_i^n(x,t)$, $i = 1,2$, of (2),(5) with $u_i^n(x,t) \longrightarrow b$ as $t \longrightarrow -\infty$, $i = 1,2$, and $u_1^n(x,t) \longrightarrow u_n'(x)$, $u_2^n(x,t) \longrightarrow u_n''(x)$, as $t \longrightarrow +\infty$.

Next, suppose that case ii) occurs; i.e. that there are orbits running from M_n' and M_n'' to M_n. Then (with the same notation as above), we have the (long) exact sequence

$$\longrightarrow H_j \longrightarrow H_{j-1}' + H_{j-1}'' \longrightarrow \overline{H}_{j-1} \longrightarrow H_{j-1} \longrightarrow$$

If we put $j \neq n + 1$, we have $0 \longrightarrow H_{j-1}' + H_{j-1}'' \longrightarrow 0$, while if $j = n + 1$, $\mathbb{Z} \longrightarrow H_n' + H_n'' \longrightarrow \mathbb{Z}$, exactly as in the previous case. Thus as before the two bifurcating solutions have n-dimensional unstable manifolds; this time however, the connecting orbits run in directions opposite to those of the previous case.

4. Continuation of Solutions by Changing the Boundary Data.

In this section, we shall take $f(u)$ to be the function $f(u) = -u(u - b)(u - 1)$, $0 < b < 1$. Our goal here is to show how to "continue" steady-state solutions of the Dirichlet problem to steady-state solutions of the Neumann problem, and vice-versa, by taking the boundary manifolds as bifurcation parameter. Of course we assume here that L is fixed. Thus we consider equation (2) together with the boundary conditions*

$$(7)_\alpha \qquad \alpha u(-L) + (\alpha - 1)u'(-L) = \alpha u(L) + (1 - \alpha)u'(L) = 0 .$$

Observe that for $\alpha = 1$ $(7)_\alpha$ yields Dirichlet data, while for $\alpha = 0$ $(7)_\alpha$ becomes Neumann data.

It was proved in [4], that for homogeneous Dirichlet data there are precisely three solutions of (2), if L is sufficiently large; say $L > L_0$. We depict the phase portrait in Figure 4. The constant $u \equiv 0$ is clearly a solution of the Dirichlet problem as are the solutions u_1^1 and u_1^2 as depicted in Figure 4. Now if $\alpha < 1$, these solutions continue to solutions $u = 0$, $u = u_\alpha^1, u_\alpha^2$, respectively, as depicted in Figure 5. That is, 0, u_α^1 and u_α^2 are solutions

---

* The flow defined by (1), $(7)_\alpha$ is gradient-like with respect to the functional $\Phi(u) = \int_{-L}^{L} (\frac{1}{2} uu_{xx} + F(u))dx$.

of (2), $(7)_\alpha$, which have the same index as the corresponding solutions 0, u_1^1, u_1^2, respectively. Since it was shown in [2] that $h(0) = \Sigma^0$,

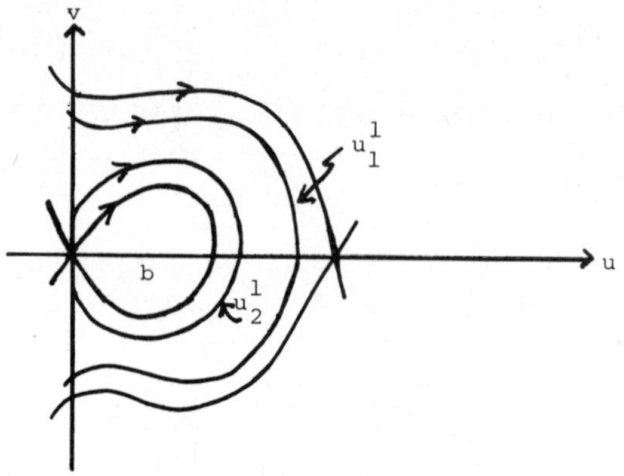

Figure 4

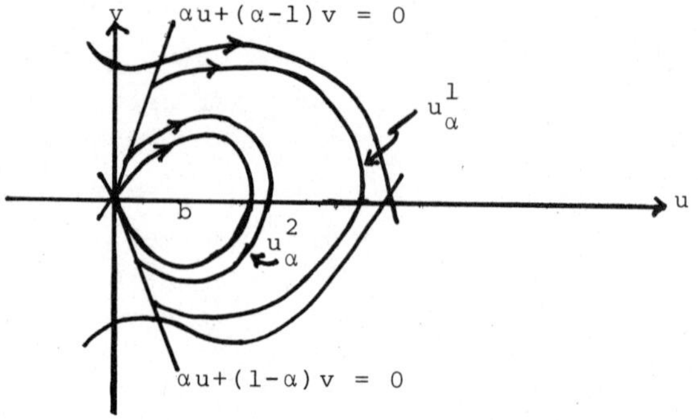

Figure 5

$h(u_1^1) = \Sigma^0$ and $h(u_1^2) = \Sigma^1$, we have that $h(0) = \Sigma^0$, $h(u_\alpha^1) = \Sigma^0$ and $h(u_\alpha^2) = \Sigma^1$. Now as $\alpha \longrightarrow 0$, it is easy to see that 0, u_α^1 and u_α^2 continue to the three rest points 0, 1 and b respectively, which are solutions of the Neumann problem.* This gives an alternate way of computing the indices of the three rest points, and of course agrees with the results in [2].

Suppose now that we start with solutions of the Neumann problem, and assume that L is so large that there exist non-constant solu-

---

* u_α^2 would bifurcate into several solutions as $\alpha \longrightarrow 0$, if L is large, but this new isolated invariant set continues to b; see below.

tions of the Neumann problem. These solutions must continue to solutions of (2), $(7)_\alpha$ for small $\alpha > 0$, but they must also continue to the unstable solution of the Dirichlet problem, see [2]. Our aim here is to describe this "cancellation" phenomenon.

Let α_1 be chosen such that the slope of the line $\alpha_1 u + (\alpha_1 - 1)v = 0$ equals the slope of the tangent line to the orbit homoclinic to the origin, at the origin. We assert that there is an α_2, $0 < \alpha_2 < \alpha_1$ with the property that there is a solution of (2), $(7)_{\alpha_2}$ which takes "time" 2L, and cuts each boundary line precisely once. This follows by a continuity argument; namely, for α near α_1, all such orbits take "time" greater-than 2L, while for α near 0, all such orbits take "time" less-than 2L. From the existence of α_2 it follows that there is an α_3, $\alpha_2 < \alpha_3 < \alpha_1$, with the property that the orbit running from the line $\alpha_3 u + (\alpha_3 - 1)v = 0$ to the line $\alpha_3 u + (1 - \alpha_3)v = 0$, which is tangent to both of these lines, takes "time" 2L. This again follows by continuity. To see this, consider the above solution of (2), $(7)_{\alpha_2}$. It lies on a periodic orbit which is tangent to some line $\tilde{\alpha}u + (\tilde{\alpha} - 1)v = 0$, with $\tilde{\alpha} > \alpha_2$. Hence the orbit which runs from $\tilde{\alpha}u + (\tilde{\alpha} - 1)v = 0$ to $\tilde{\alpha}u + (1 - \tilde{\alpha})v = 0$ obviously takes time great than 2L. For α near zero, all such tangency orbits take "time" near the time of the equation obtained by linearizing about b and by hypothesis, this "linearized time" is less-than 2L; thus α_3 exists. We let $\bar{\alpha} < \alpha_1$ be the largest such α_3. Now if $\alpha > \bar{\alpha}$, the problem (2), $(7)_\alpha$ has no non-constant solutions different from the "tangency" orbit which defines $\bar{\alpha}$. This follows since any other non-constant solution lies on an orbit tangent to two boundary lines having slope exceeding $\bar{\alpha}(1 - \bar{\alpha})^{-1}$. Also, observe that the "tangency" orbit which defines $\bar{\alpha}$ obviously continues to the unstable solution of the Dirichlet problem.

Finally, it may be of interest to see how the constant solution b continues as α becomes positive. Thus, first suppose that L is such that there are exactly two non-constant solutions of the Neumann problem, i.e. $h(b) = \Sigma^2$. Then it is easy to see that b continues to the solution E'F' in Figure 6; to see that this solution has index Σ^2, merely let L increase. Now suppose that L is such that there are 4 non-constant solutions of the Neumann problem, i.e., $h(b) = \Sigma^3$. Then b bifurcates into 3 solutions as depicted in Figure 7; two of these have index Σ^3 and the third has index Σ^4; of course, as before, there must exist solutions of (1), $(7)_\alpha$ which "connect" these rest points.

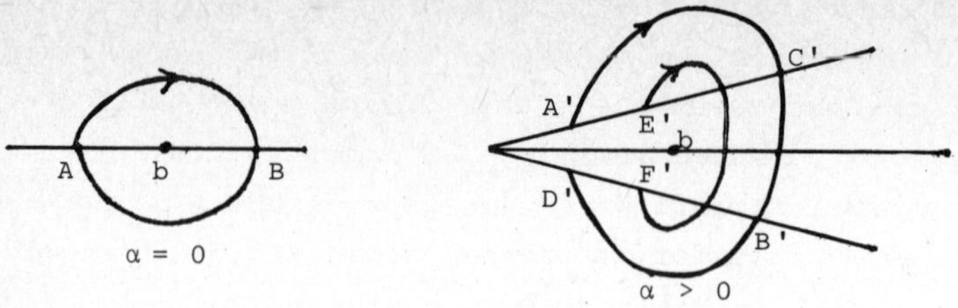

Figure 6. $L_1 < L < L_2$; AB continues to A'B', BA continues to C'D' ; b continues to E'F'.

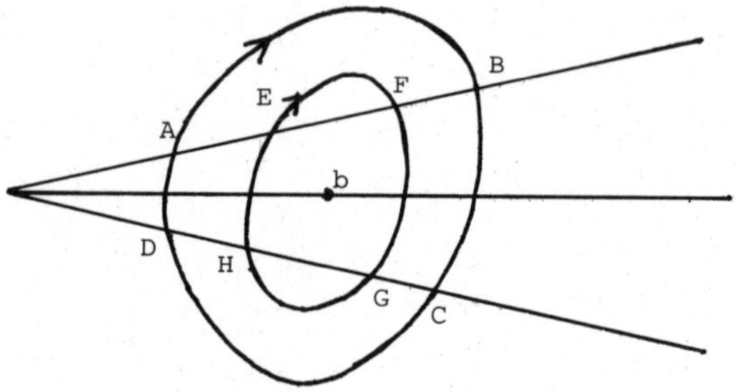

Figure 7. $L_2 < L < L_3$; b bifurcates into the three solutions A B C D A B C , B C D A B C D and E F G H E F G H .

Appendix. In order that the general abstract results in [1] can be applied to our situation, we must show how the semi-flow generated by equation (1), together with the corresponding homogeneous boundary conditions, can be embedded in a locally compact flow. Thus consider first the (abstract) semi flow $f: X \times \mathbb{R}^+ \longrightarrow X$, where X is compact, and assume that for the restriction $X \times [0,1] \longrightarrow X$, the "curves" $f(x,t)$, $t \geq 0$, through each $x \in X$, satisfy a uniform modulus of continuity θ. Let Φ be the set of curves in X having θ as a modulus of continuity which satisfy, for $t \geq 0$, $\gamma(t) = f(\gamma(0), t)$. Finally, let P denote those curves mapping $\mathbb{R}^-$ into X, having a modulus of continuity θ near $t = 0$. Now we have a natural map $\phi: P \longrightarrow \Gamma(X)$ (where $\Gamma(X)$ is the set of regular curves on X; see [1]) which takes $p \in P$ into the curve $\sigma(t)$ defined by

$$\sigma(t) = \begin{cases} f(p(0), t), & t \geq 0 \\ p(t), & t < 0 . \end{cases}$$

Now it is easy to check that the set $\phi(P) \cap \Phi$ is a local flow; and in fact that it is a locally compact local flow, in the compact-open topology. This is the general situation. Now for the equation (1), we note that $f(-\infty) > 0 > f(\infty)$ implies that the equation admits arbitrarily large bounded invariant regions (see [0]). The equation (1) always defines a semi-flow on smooth functions. Also if we start with a compact set of initial data, the above smoothness assumptions are satisfied. Thus as in the abstract case, we can append the past histories to determine a locally compact local flow.

References

0. Chueh, K.N., Conley, C., and J.A. Smoller, Positively invariant regions for systems of nonlinear diffusion equations, Ind. U. Math. J., 26, (1977), pp. 373-392.

1. Conley, C., Isolated Invariant Sets and the Morse Index, C.B.M.S. notes, #38, Amer. Math. Soc., Providence, 1978.

2. Conley, C., and J.A. Smoller, Remarks on the stability of steady-state solutions of reaction-diffusion equations, Proc. NATO Conf. on Bifurcation Theory, Cargese , 1979, to appear.

3. Smoller, J., Tromba, A., and A. Wasserman, Nondegenerate solutions of boundary-value problems, J. Nonlinear Anal., to appear.

4. Smoller, J., and A. Wasserman, Global bifurcation of steady-state solutions, to appear.

University of Wisconsin
Madison, Wisconsin 53706
U.S.A

University of Michigan
Ann Arbor, Michigan 48109
U.S.A.

PERCOLATION

by J.M. Hammersley

Trinity College, Oxford.

Introduction and terminology

This paper deals with situations where a fluid spreads randomly through a medium. Here fluid and medium are abstract terms, to be interpreted according to any particular context - liquid penetrating porous material, electrons migrating through semiconductors, cancer infecting biological tissue, etc. etc. Such a process is called either diffusion or percolation, depending upon whether it is natural to ascribe the underlying random mechanism to the fluid or the medium respectively. However, this distinction is not clear cut: some problems can be described in either way, and some require both descriptions.

A trivial example will exhibit the distinction. Suppose the medium is a straight line, on which a typical point has co-ordinate x. Let the fluid be a particle, which performs a succession of random jumps on the line. For the diffusion model, we furnish the particle with a sequence of independent identically distributed (=i.i.d.) random variables $\{d_0, d_1, \ldots\}$ from a given distribution D. The particle starts at x=0, and jumps a distance d_{n-1} at its nth jump. For the analogous percolation process, we ascribe a random variable d_x to the point x of the medium, where the collection $\{d_x : -\infty < x < \infty\}$ is an uncountable collection of i.i.d. random variables from D. The particle starts from x=0 and initially jumps a distance d_0; thereafter, whenever it lands on a point x, its next jump is of distance d_x. Now, if D is a continuous distribution, there will be no essential difference between the diffusion and percolation models, since the particle will (almost surely) never revisit any point. However, if D is a discrete distribution, say the one with just two values $d=\pm 1$ with equal probabilities $\frac{1}{2}$, then the diffusion and percolation models are quite unlike: in the diffusion model, the particle performs a Pólya walk and therefore visits every integer point sooner or later; whereas in the percolation model the particle moves steadily in one direction until it encounters a pair of points x and x+1 with $d_x \neq d_{x+1}$, whereafter it oscillates between these two points for ever. To get an example of a process requiring a mixture of diffusion and percolation, we might suppose that diffusion governs the even-numbered jumps and percolation the odd-numbered jumps.

Next we introduce some terminology for dealing with percolation processes in rather general terms. Subsequently we shall provide concrete examples by interpreting this terminology in ways that suit particular contexts. The medium will be a linear graph G, that is to say a collection of objects called vertices and edges (equivalent terms often used in the literature are vertex = point = node = atom, and edge = line = loop = bond). The edges are connections between certain selected pairs of vertices: we write E_{ij} for an edge which connects a vertex V_i to a vertex V_j. An edge may be directed, in which case it goes from V_i to V_j but not vice versa; or it may be undirected, in which case it may be traversed in either direction. Not every pair of vertices needs to be connected by an edge. Hence the specification of the structure of the graph consists in specifying exactly which pairs are to have edge connections. Two vertices are called neighbours if they are connected by an edge. A path is a sequence of vertices $V_1, V_2, \ldots, V_n$ together with a sequence of edges $E_{12}, E_{23}, \ldots, E_{n-1,n}$ such that we can travel from V_1 to V_n along these edges, with due regard to the direction of any edge that happens to be directed. A cluster of vertices is a set of vertices such that some path goes from any vertex of the set to any other vertex of the set. Similarly a cluster of edges is a set of edges such that every ordered pair of edges in the cluster belongs to some path on the graph. From time to time we employ various adjectives (e.g. wet, dry, immune, susceptible, etc.) to describe certain properties possessed by the vertices and/or edges of the graph; and these adjectives then carry over in a natural way to paths and clusters: e.g. a path is a susceptible path if all the vertices and edges on it are susceptible; a cluster of vertices is a susceptible cluster if all its vertices are connected by susceptible paths; and so on.

In a percolation process we have to ascribe some random mechanism to the medium; and we do this by attaching a random variable d_i to each vertex V_i, and a random variable d_{ij} to each edge E_{ij}. The complete set of all such d_i and d_{ij} is a random vector denoted by $\underset{\sim}{d}$: we call the d_i the <u>vertex-co-ordinates</u> of $\underset{\sim}{d}$, and the d_{ij} are the <u>edge-co-ordinates</u> of $\underset{\sim}{d}$. The random vector $\underset{\sim}{d}$ is distributed with some distribution D; and the specification of the random mechanism is achieved by specifying D to suit the context of any particular application.

The fluid is an entity which flows along the paths of the graph, and its motion is governed entirely by the structure of the graph G and the specification of the distribution D. We illustrate the possibilities by some particular cases.

<u>Bernoulli percolation</u>

Bernoulli percolation is percolation in its oldest and simplest form. In this version all co-ordinates of $\underset{\sim}{d}$ are mutually independent random variables taking only the values 0 and 1. The vertex co-ordinates have the distribution $d_i=1$ with probability p_V (and $d_i=0$ with probability $q_V=1-p_V$), while the edge co-ordinates have $d_{ij}=1$ with probability p_E (and $d_{ij}=0$ with probability $q_E=1-p_E$). Vertices and edges with co-ordinates 1 are said to be <u>open</u>, those with co-ordinates 0 are <u>closed</u>. We suppose that the graph G has infinitely many vertices and edges, and that we are given a specified subset of vertices S called <u>source vertices</u>. We regard G as a network of pipes (its edges) and junctions (its vertices), provided with taps that may be open or closed (according to the values of the respective co-ordinates d_i and d_{ij}). Fluid is supplied to all source vertices S, and thence flows along all open paths, thus wetting edges and vertices, and creating <u>wet clusters</u> and <u>dry clusters</u> in G. We define the <u>percolation probability</u> $P=P(p_V,p_E) = P(p_V,p_E,S,G)$ to be the probability that G contains at least one infinite wet cluster; and we seek to make numerical or qualitative statements about P.

For fixed S, G, the function $P(p_V, p_E)$ is defined on the square $0 \leq p_V \leq 1$, $0 \leq p_E \leq 1$; and it is clearly a non-decreasing function of its two arguments p_V and p_E. It takes the values $P(0,0) = 0$ and $P(1,1) = 1$. There is an important threshold theorem which asserts that $P>0$ only in the upper right hand corner of the square, the shaded region XYZ in Figure 1; and $P=0$ elsewhere in the unshaded region, provided that S is a finite set.

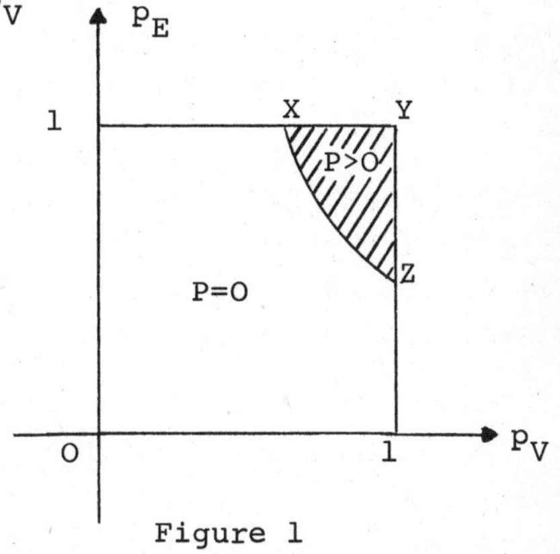

Figure 1

A second important theorem is due to McDiarmid [1], and it asserts that

$$\left[P(p_V, p_E) - P(p_E, p_V)\right] / (p_V - p_E) \geq 0.$$

(McDiarmid's paper has not yet been published; and I am greatly indebted to him for permission to quote his result).

To exemplify these two theorems, consider the spread of blight in an orchard. Let us represent the orchard by a square lattice G: the vertices of G are the points in the Euclidean plane having integer co-ordinates, and the edges of G are undirected and connect together vertices that are unit distance apart. We suppose that a fruit tree is planted at each vertex. We also suppose that the fruit trees are of two mutually exclusive kinds, <u>immune</u> and <u>susceptible</u>. An immune tree can never catch the blight, while a susceptible tree may become infected. Suppose that p_V is the proportion of susceptible trees, and that susceptible trees are independently and uniformly distributed throughout the orchard. We also suppose that p_E is the probability that a blighted tree will directly infect a neighbouring susceptible tree. As the source S of the blight, we suppose that the tree at the origin is initially blighted; and that thereafter the blight spreads randomly according to Bernoulli percolation. Then the percolation probability P represents the chance that the single source of infection will lead to large scale blight throughout the whole orchard, while the complementary probability $Q=1-P$ is the chance that the blight will be confined locally near S.

The threshold theorem tells us that the blight will be local provided that p_E and p_V are sufficiently small. When G is a square lattice, the point X has co-ordinates (0.59,1); and the point Z has co-ordinates (1,0.50).

Quite generally, whatever G is, the two parameters p_E and p_V are measures of <u>infectability</u> and <u>susceptibility</u> respectively. McDiarmid's theorem implies that $P(p-\delta,p) \leqslant P(p,p-\delta)$ when $\delta>0$; hence, if susceptibility and infectability are at equal levels p in (say) the spread of blight through an orchard or of cancer through biological tissue, then a reduction δ in susceptibility will more effectively inhibit extensive spread than an equal reduction δ in infectability. It would be of considerable medical significance to know how far McDiarmid's theorem can be generalized: thus, for what values of p and p' (not necessarily equal) will it still remain true that $P(p-\delta,p') \leqslant P(p,p'-\delta)$ when $\delta>0$?

The numerical evaluation of percolation probabilities is usually beyond the resources of analytical theory (except for rather special kinds of graph), so then recourse has to be made to computer simulations. It is however worth looking briefly at what is involved in a theoretical attack. To avoid complications, suppose that we consider the case where S is a single vertex and $p_E=1$ and $p_V=p$ (in the literature on physics, this is known as <u>atom percolation</u>). We can then write $P(p,1)=P(p)=1-Q(p)$, and $Q(p)$ will be the probability that S belongs to a finite cluster of vertices. Such a cluster will have a <u>boundary</u>, namely the set of closed vertices not in the cluster but having neighbours in the cluster. Suppose that there are n vertices (other than S) in the cluster and b vertices in the boundary. The probability of this is $p^n q^b$; and hence

$$Q(p) = \sum_{n,b} f_{nb} p^n q^b , \tag{1}$$

where f_{nb} is the number of clusters that contain S and n other vertices and have b vertices in their boundaries. Thus the calculation of Q reduces to determining f_{nb}. Even in the case when G is a regular lattice, the latter determination is a difficult combinatorial problem, known as the <u>polyomino problem</u>.

Sometimes however it is possible to write (1) in the form

$$Q(p) = f_1(p,q) f\left[f_2(p,q)\right] , \tag{2}$$

where f is an unknown invertible function, and f_1 and f_2 are known functions of p and q=1-p, and where f_2 has the following special properties: $f_2(0,1)=f_2(1,0)=0$ and $f_2(p,q)$ has a unique maximum in $[0,1]$ at (say) $p=p_0$. In this case the equation

$$f_2(p,q) = f_2(p',q'), \qquad 0\leq p\leq p_0\leq p'=1-q' \tag{3}$$

generates an involution $p \leftrightarrow p'$ on $[0,1]$, and $Q(p)/f_1(p,q)$ is fixed under this involution. From the threshold theorem we know that $Q(p)=1$ for sufficiently small p. Hence $Q(p')=f_1(p',q')/f_1(p,q)$ for p' sufficiently near 1. It also seems to be true (though I have no rigorous justification for this belief) that p_0 is actually the <u>critical probability</u>, namely the supremum of all p such that $Q(p)=1$; in which case the involution completely determines Q for all p. (The underlying heuristic reason behind this belief is that Q is likely to be a function with just one branch point, which can only be at $p=p_0$). The statement is certainly correct in those easy cases where it can be independently checked by other means: for example, it is true whenever G is a regular Bethe lattice.

First-passage percolation

In Bernoulli percolation, the random variables are 0 or 1. In first-passage percolation the co-ordinates of $\underset{\sim}{d}$ are mutually independent non-negative random variables; the vertex co-ordinates have a common distribution function U_V, and the edge co-ordinates a common distribution function U_E. The fluid does not percolate instantaneously along any path, but instead takes a random time t_{path} to do so, where t_{path} is equal to the sum of all edge and vertex co-ordinates on that path. If S is a given source vertex, and V is any other vertex on G, we define the first-passage time from S to V to be the infimum of t_{path}, taken over all paths from S to V. Thus the first-passage time is the time taken by the fluid to reach V from S.

If G is a lattice embedded in Euclidean space, we may divide the first passage time from S to V by the Euclidean distance between S and V, and study the limit of this quotient as V recedes to a great distance from S. This limit, denoted by $\mu(U_V,U_E)$ exists with probability 1, and is called the time constant of the process. In general, the time constant is not isotropic, that is to say it depends upon the direction of the straight line from S to V.

First-passage percolation gives information about the rate at which the fluid will spread through the medium. We can ask a number of questions about the shape and size of the wet cluster after a given lapse of time since the release of the fluid from S. We can also enquire whether there is an analogue of McDiarmid's theorem for $\mu(U_V,U_E)$ in place of $P(p_V,p_E)$: to the best of my knowledge, this question has not yet been tackled in the literature.

The overall features of first-passage percolation are typified by what happens when G is the square lattice in the plane; and most of the theoretical work has been devoted to this special case, on which the monograph [2] by Smythe and Wierman provides an excellent survey. Mention should also be made of a recent paper by Kesten [3], that resolves an important conjecture in this area.

Resistive percolation

In this model the vertex co-ordinates d_i are all zero, while the edge co-ordinates d_{ij} are independent and identically distributed non-negative random variables. We regard the graph G as an electrical network, and d_{ij} as the resistance of the wire from V_i to V_j. We then ask for the effective resistance been a specified pair of vertices in the network, or more generally between two such sets of vertices (a source set and a sink set). If we allow some of the resistors to have (at random) infinite resistance (which is equivalent to cutting a connection), we encounter typical threshold theorems as in Bernoulli percolation. The theory has been developed with particular reference to the study of disordered alloys; but perhaps, under a suitable reformulation, it may have a few biological applications. For an account of resistive percolation, see the article by Kirkpatrick [4].

High-density percolation

Let C be a cluster of open vertices on a graph G under Bernoulli percolation. Let m be a specified positive integer. Call a vertex within C a special vertex if it has at least m neighbours in C. These special vertices form subclusters C_k (k=1,2,...) within C: that is to say V_i and V_j belong to the same subcluster C_k if and only if there is a path from V_i to V_j which passes exclusively through special vertices. These subclusters are called clusters of density m. We can then ask the usual questions about the shape and size of clusters of density m, and in particular look for the m-density percolation probability $P_m(p_V,p_E)$ that the source vertex S belongs to an infinite subcluster of density m. For a discussion of this model, which has applications in the theory of liquid-glass transitions, see [5].

Bootstrap percolation

Bootstrap percolation, which arises in the study of magnetic spin-glasses, is somewhat similar to high-density percolation; but it also includes a relaxation procedure which reduces the sizes of the subclusters. We start with an ordinary vertex cluster in Bernoulli percolation. We then delete vertices from this cluster which do not have at least m neighbours in the cluster and we continue this process of deletion until a stable situation is reached in which every surviving vertex has at least m surviving neighbours. Again we ask the usual questions about the cluster of survivors. See [6] for a discussion of this model.

Spatially correlated percolation

In the foregoing discussion it has been assumed that the co-ordinates of the random vector $\underset{\sim}{d}$ are mutually independent. However, this is unnecessary, and is not even valid in some applications. The co-ordinates of $\underset{\sim}{d}$ may have any joint distribution D whatever. A certain amount of homogeneity in D is advisable if reasonable conclusions are to be obtained. The simplest way of achieving correlation while preserving adequate homogeneity is to take the distribution D to be that of a Markov field on G itself. This automatically links the percolation neighbourhood structure of G with the clan structure of G. For details of the way that a Markov field on a linear graph depends upon its underlying clan structure, consult Moussouris [7]. (A clan is a subset of vertices C such that every (if any) pair of vertices in C is a pair of neighbouring vertices; and, in a Gibbsian ensemble D, the logarithmic likelihood is a linear functional on the space of all open clans. A clan may contain only one vertex, or may be the empty set).

Percolation on random graphs

Most of the applications in the literature have been directed towards the physical sciences (see [8] for a review), particularly solid state physics; and accordingly G has usually been taken as a fixed crystalline lattice. Such an assumption is much less suitable in the biological sciences, where media are likely to be amorphous or randomly heterogeneous. There is however no reason in principle why G should not itself be a random graph, D being imposed on G after the realization of

G from a postulated population of graphs. For example, we might take the vertices of G to be the points of a Poisson process in three-dimensional space, and define two points to be neighbours if and only if the distance between them is less than a prescribed quantity. First-passage percolation properties would then remain valid, because the subadditivity of the process is preserved.

Concluding remarks

This paper has given a brief elementary outline sketch of the principal models employed for percolation. How far these models may be suitable for biological work, especially cancer or growth problems, is a matter which I must leave in the more knowledgeable hands of the biologists and medical men at this conference.

Postscript

The correct generalization of McDiarmid's theorem is that $P(p_V\delta, p_E) \leqslant P(p_V, p_E\delta)$ whenever $0 \leqslant \delta \leqslant 1$. The contours in the (p_V, p_E)-plane, on which P is constant, go in a south-easterly direction as p_V increases, and they cross (from above to below) the hyperbolae on which the product $p_V p_E$ is constant. I hope to publish a proof of this assertion elsewhere.

References

[1] C. McDiarmid (1979), "Dependent random variables, random graphs and percolation". (Private communication).

[2] R.T. Smythe and J.C. Wierman (1978), First-passage Percolation on the Square Lattice. Lecture Notes in Mathematics, No. 671 Springer-Verlag, Berlin.

[3] H. Kesten (1979), "On the time constant and path length of first-passage percolation". (To appear).

[4] S. Kirkpatrick (1973), "Percolation and conduction", Rev. Mod. Phys. **45**, 574-587.

[5] G.R. Reich and P.L. Leath (1978), "High-density percolation: exact solution on a Bethe lattice". J. Statist. Phys. **6**, 611-622.

[6] J. Chalupa, P.L. Leath and G.R. Reich (1979), "Bootstrap percolation on a Bethe lattice", J. Phys. C. Solid State Phys. **12**, L31-35.

[7] J. Moussouris (1974), "Gibbs and Markov random systems with constraints". J. Statist. Phys. **10**, 11-33.

[8] V.K.S. Shante and S. Kirkpatrick (1971), "An introduction to percolation processes", Adv. Phys. **20**, 325-357.

SOME ERGODIC THEOREMS FOR SPATIAL PROCESSES

Nguyen Xuan Xanh

Universität Bielefeld, Fakultät f. Math.
Universitätsstr., D-48 Bielefeld 1, Fed Rep Germany

§1. Definition. Let $(\Omega, \mathcal{F}, P)$ be some probability space and $\mathcal{C}$ the collection of bounded Borel subsets of the d-dimensional Euclidean space $\mathbb{R}^d$. All considerations carry over unchanged if $\mathbb{R}^d$ is replaced by the integer lattice $\mathbb{Z}^d$.

For any given family $X = (X_G)_{G \in \mathcal{C}}$ of random variables the couple $\mathfrak{X} = (X, P)$ is called a spatial process whenever the expectation $\mathbb{E}X_G$ is finite for any $G \in \mathcal{C}$.

In applications Ω is often the configuration space, or phase space, equal to the product space $\Gamma^{\mathbb{Z}^d}$ for some abstract set Γ in the discret case, or the space $\mathcal{M}$ of countable sequences of points of $\mathbb{R}^d$ having no limit point in bounded domains, in the continuous case. The points of Ω represent the microscopic states of some physical system, distributed according to the probability law P. X_G represents some physical quantity restricted to the domain G, like energy, entropy etc., depending on the state ω of the system.

We assume that there exists an Abelian group $(T_u, u \in \mathbb{Z}^d)$ of isomorphisms (shift-transformations) on Ω and that $\mathfrak{X}$ has the following two properties

(1) Spatial homogeneity

$$X_{G+u}(T_u) = X_G \quad \text{for any } G \in \mathcal{C} \text{ and } u \in \mathbb{Z}^d$$

almost surely;

(2) Stationarity

$$T_u P = P \quad \text{for any } u \in \mathbb{Z}^d .$$

Roughly speaking, property (1) means that if we translate the domain G and the configuration ω at the same time then the quantity X_G does not change its value. Any spatial process has this property in its canonical representation.

In view of (1), stationarity is equivalent to the usual one, i.e.

(2') $$P\left\{(X_{G_1+u},\ldots,X_{G_n+u})\in A\right\} = P\left\{(X_{G_1},\ldots,X_{G_n})\in A\right\}$$

for any $A\in\mathcal{B}(\mathbb{R}^n)$, $G_1,\ldots,G_n\in\mathcal{C}$, $n=1,2,\ldots$ and $u\in\mathbb{Z}^d$.

Processes which have properties (1) and (2) will be called <u>stationary spatial processes</u>.

§2. <u>Problem</u>. Once a spatial process $\mathfrak{X}$ is given we ask: do the averages $(1/|G|).X_G$ exhibit an asymptotical behaviour as G becomes large in some sense ?($|\cdot|$ denotes the Lebesgue measure on $\mathbb{R}^d$).

§3. <u>A generalized ergodic theorem</u>. One answer is the following individual ergodic theorem.

For $n=o,1,2,\ldots$ let

(3) $$F_n = \left\{(x^1,\ldots,x^d)\in\mathbb{R}^d : -\frac{2n+1}{2} \leq x^i < \frac{2n+1}{2} , \; i=1,\ldots,d\right\}$$

and F_+ be recursively defined as follows

$$d=1 \qquad F_+^1 = (-\infty,-1/2)$$

$$d=2 \qquad F_+^2 = \left[F_+^1\times(-\infty,+\infty)\right]\cup\left[F_o^1\times(-\infty,-1/2)\right]$$

$$\vdots$$

$$d\geq 2 \qquad F_+ = F_+^d = \left[F_+^{d-1}\times(-\infty,+\infty)\right]\cup\left[F_o^{d-1}\times(-\infty,-1/2)\right]$$

where

$$F_o^k = \left\{(x^1,\ldots,x^k)\in\mathbb{R}^k : -\tfrac{1}{2} \leq x^i < \tfrac{1}{2} , \; i=1,\ldots,k\right\}$$

$$F_o^d = F_o \qquad .$$

We set

(4) $$X_{G,\Lambda} = X_{G\cup\Lambda} - X_\Lambda \quad \text{for any } G,\Lambda\in\mathcal{C} \text{ with } G\cap\Lambda=\emptyset ,$$
$$X_\emptyset \equiv o$$

<u>Theorem 1</u>([9]). Let $\mathfrak{X}$ be an arbitrary stationary spatial process. Assume that there exist two random variables $Y,Z\in L^1(P)$ with $Y\geq o$

such that

(5) $\left|X_{F_o,\Lambda}\right| \leq Y$ a.s. for any $\Lambda \in \mathcal{F}_+ \cap \mathcal{C}$,

(6) $\lim_{\Lambda \nearrow F_+} X_{F_o,\Lambda} = Z$ a.s. $(\Lambda \in \mathcal{C})$.

Then

$$\lim \frac{1}{|G_n|} \cdot X_{G_n} = \xi := \frac{1}{|F_o|} \cdot \mathbb{E}(Z|\mathcal{J}) \quad \text{a.s.}$$

for any regular sequence $\{G_n\}$, $G_n \in \mathcal{C}$; $\mathcal{J}$ denoting the σ-field of the invariant subsets $A \in \mathcal{F}$.

For the definition of a regular sequence of subsets see [9]. Accordingly a sequence $\{G_n\}$ of convex subsets is regular if for example

$$G_n \subset G_{n+1} \quad \text{for any } n ,$$

and

$$d(G_n) := \sup\left\{ \varrho \geq 0 : \text{the sphere } S(x,\varrho) \subset G_n ,\ x \in \mathbb{R}^d \right\} \longrightarrow \infty$$

when $n \to \infty$.

For the sake of simplicity, the sets G_n in Theorem 1 as well as later in Theorem 2 are restricted to be parallelepipeds whose sides are parallel to the coordinate axes of a fixed basis in $\mathbb{R}^d$. The result can be proved for the more general convex sets. See [9], [10] .

In general, the random variable $X_{G,\Lambda}$ has the interpretation as the value of the process in G given the configuration in the environment Λ outside. It measures the influence of Λ on G . In this sense the conditions of Theorem 1 say that if the environment Λ becomes larger and larger, its influence on F_o remains "moderate" and "well-behaved". They remind us of conditions in Lebesgue's convergence theorem.

In dimension one and in the discrete case Theorem 1 reduces to a modified version of the classical individual ergodic theorem used by Breiman [2] to prove the so-called Breiman's version in information theory. This modified version is already contained in a paper by Maker [8] seventeen years earlier.

There is a L^1-version of Theorem 1 which can easily be proved. See

[9].

§4. Subadditive processes.

An important class of spatial processes is that of subadditive processes. A process $\mathfrak{X}$ is called subadditive if for almost all ω

(7) $$X_{G\cup\Lambda}(\omega) \leq X_G(\omega) + X_\Lambda(\omega) \quad \text{for any } G,\Lambda \in \mathcal{C} \text{ with } G\cap\Lambda = \emptyset ,$$

and strongly subadditive if for almost all ω

(8) $$X_{G\cup\Lambda}(\omega) + X_{G\cap\Lambda}(\omega) \leq X_G(\omega) + X_\Lambda(\omega) \quad \text{for any arbitrary } G,\Lambda \in \mathcal{C} .$$

Hammersley and Welsh [4] formulated for the first time the axioms of subadditive processes (in dimension one) and conjectured ergodic theorems. Such theorems would naturally generalize the classical ergodic theorems by Birkhoff and von Neumann.

About three years later Kingman [6] solved these problems completely. Many very nice applications followed. See[7].

We give the analogous version for the case of dimension $d \geq 2$. Let $\gamma = \gamma(\mathfrak{X})$ be the constant of the process, defined as follows

$$\gamma = \inf_n \mathbb{E}X_{F_n} / |F_n|$$

Theorem 2.

(I) Let $\mathfrak{X}$ be a subadditive stationary spatial process with finite constant γ.

Then there exists an invariant random variable ξ with expectation γ such that

(9) $$\lim_{d(G_n)\to\infty} \frac{1}{|G_n|}\cdot X_{G_n} = \xi \qquad \text{in } L^1(P)$$

and

(1o) $$\limsup_{n\to\infty} \frac{1}{|G_n|}\cdot X_{G_n} = \xi \qquad \text{a.s.}$$

for any regular sequence $\{G_n\}$.

(II) If in addition the process is strongly subadditive, then we have the almost sure convergence

$$(11) \qquad \lim \frac{1}{|G_n|} \cdot X_{G_n} = \xi \qquad \text{a.s.}$$

for any regular sequence $\{G_n\}$.

We give an idea of the proof of Theorem 2.

Part (I) is the "easy part". It can be proved analogously to Kingman, by using the classical ergodic theorems in the most general form elaborated by Tempel'man [12] and by exploiting the spatial homogeneity. See [1o].

Part (II), the almost sure convergence, turns out to be a special case of Theorem 1. How can this be seen ?

Observe first the following nice property of the strong subadditivity.

Lemma 1. $\mathfrak{X}$ is strongly subadditive if and only if for any subset $G \in \mathcal{C}$ the set function

$$\Lambda \longrightarrow X_{G,\Lambda}(\omega) \qquad (\Lambda \in \mathcal{C} \, , \Lambda \cap G = \emptyset)$$

is monotonically decreasing for almost all ω .

This means for stongly subadditive processes that the influence of the environment Λ on any domain G decreases for increasing Λ . Hence the limit

$$(12) \qquad Z = \lim_{\Lambda \to F_+} X_{F_o,\Lambda} \qquad (\Lambda \in \mathcal{C})$$

exists almost surely and defines a random variable.

On the other hand, from Lemma 1 again, we have for any $\Lambda \in F_+ \cap \mathcal{C}$ the inequalities

$$Z \le X_{F_o,\Lambda} \le X_{F_o,\emptyset} = X_{F_o} \qquad \text{a.s.}$$

Thus

$$(13) \qquad \left| X_{F_o,\Lambda} \right| \le \max \left(|Z| \, , |X_{F_o}| \right) \qquad \text{a.s.}$$

for any $\Lambda \in F_+ \cap \mathcal{C}$.

In view of (12) and (13) the conditions of Theorem 1 will be satisfied if the random variable Z is integrable. And this is the case, because

<u>Lemma 2</u>. Z is integrable if and only if γ is finite.

Hence Theorem 1 applies and Part (II) of Theorem 2 is proved.

The decomposition of the process into an essential additive part with the same constant and a subadditive part with constant zero is a consequence of Theorem 1.

In dimension one, Kingman arrived at a more subtle result by using the linear order structure of $\mathbb{Z}^1$. He showed it to be sufficient that the sequence

$$(X_{F_o,\Lambda})_\Lambda = (X_{o,i})_i \quad ,$$

where

$$X_{o,i} = x_{oi} - x_{1i} \quad ,$$

has a limit point in Cesàro sense, i.e. there exists a subsequence $\{i_n\}$ such that

$$\frac{1}{i_n}\cdot(X_{o,1}+\dots+X_{o,i_n}) \longrightarrow f \qquad \text{a.s.}$$

for some f . As a consequence the almost sure convergence holds for subadditive processes.

However in the case $d \geq 2$ the situation is more complicated.

For some applications see [9] and [1o].

§5. Let us conclude with some remarks concerning another type of subadditive processes.

Consider a process $\mathfrak{X}$ with the following property

(14) There exists a non-negative function $\theta(.,.,.)$ of ω, G, Λ such that almost surely

$$X_{G\cup\Lambda}(\omega) \leq X_G(\omega) + X_\Lambda(\omega) + \theta(\omega,G,\Lambda) \quad \text{for all disjoint sets } G,\Lambda\in\mathcal{C}.$$

Such a process is still called subadditive. The term θ gives an upper bound of the interaction between G and Λ which can depend on ω.

If the function θ is reasonably choosen we can expect ergodic theorems for the process $\mathfrak{X}$. New techniques are neccessary. In one dimension Kesten [5] and Hammersley [3] provided some results which have already nice applications (see e.g. [1], [11]).

Let us consider in higher dimension the simplest case $\theta \equiv o$. Then the process is subadditive in the sense of §4. The mean ergodic theorem has been proved above. Does the almost sure convergence hold without the requirement of strong subadditivity ?

Denote by

$$Q(m) = \left\{(x^1,\ldots,x^d)\in\mathbb{R}^d : o \leq x^i \leq m \ , \ 1\leq i\leq d\right\} \qquad \text{for } m\in\mathbb{N}$$

and

$$X_m := X_{Q(m)} \qquad .$$

Theorem 3. Let $\mathfrak{X}$ be a stationary subadditive process with finite constant γ and for which the variance $V(G) = V(X_G)$ exists for any $G\in\mathcal{C}$. Suppose the following.

i) There exists a constant δ with $o\leq\delta<1$ such that the correlation coefficient of X_G and X_Λ does not exceed δ for any disjoint $G,\Lambda\in\mathcal{C}$;

ii) $\mathbb{E}(X_m^-)^2 \leq A_m^2$ for $m\in\mathbb{N}$ where $o\leq A_1\leq A_2\leq \ldots$ and $\sum_{m=1}^{\infty} \frac{A_m}{m^{2d}} < \infty$

Then

$$(15) \qquad |Q(2^n)|^{-1}.X_{2^n} \longrightarrow \gamma \qquad \text{a.s.} \qquad \text{as} \quad n\rightarrow\infty \quad .$$

If in addition $(X_G)_{G\in\mathcal{C}}$ is an almost surely monotone family we can substitute in the convergence (15) the subsequence there by the whole sequence $|Q(m)|^{-1}.X_m$.

The main idea of the proof is the same as in [3]. With the help of some technical calculation we arrive at

$$\sum_{n\geq o} |Q(2^n)|^{-2}.V(X_{2^n}) < \infty$$

and by the Chebyshev inequality

$$\sum_{n\geq o} P\left(|Q(2^n)|^{-1} \cdot |X_{2^n} - \mathbb{E}(X_{2^n})| > \epsilon \right) < \infty$$

for any $\epsilon > 0$.

This together with the Borel-Cantelli lemma implies (15).

We hope we could return to this problem in the case $\theta \neq o$ and when P is not stationary at another opportunity.

References

1 M.Bramson, D.Griffeath: On the Williams-Bjerknes Tumour Growth Model II. Preprint 1979.

2 L.Breiman: The Individual Ergodic Theorem of Information Theory. Ann. Math. Stat. 28 (1958), 8o9-811.

3 J.Hammersley: Postulates for Subadditive Processes. Ann. Prob. 2 (1974), 652-68o.

4 J.H.Hammersley, D.J.A.Welsh: First-Passage Percolation, Subadditive Processes, Stochastic Networks, and Generalized Renewal Theory. Bernoulli-Bayes-Laplace Anniversary Volume, Berlin. Springer, 1965, pp.61-11o.

5 H.Kesten: Contribution to J.F.C.Kingman Subadditive Ergodic Theor' In [7] .

6 J.F.C.Kingman: The Ergodic Theory of Subadditive Processes. J. Roy. Statist. Soc. B2o (1968), 499-51o.

7 ——————: Subadditive Ergodic Theory. Ann.Prob. 1 (1973),883-9o9.

8 P.Maker: The Ergodic Theorem for a Sequence of Functions. Duke Math J. 6 (194o), 27-3o.

9 Nguyen X.X., H.Zessin: Ergodic Theorems for Spatial Processes. Z.f. Wahrscheinlichkeitstheorie 48 (1979), 133-158.

1o Nguyen X.X.:Ergodic Theorems for Subadditive Spatial Processes. Z.f.Wahrsch. 48 (1979), 159-176.

11 K.Schürger: On the Asymptotic Geometrical Behaviour of a Class of Contact Interaction Processes with a Monotone Infection Rate. Z.Wahrsch. 48 (1979), 35-48.

A.A.Tempel'man: Ergodic Theorems for General Dynamic Systems. Trans. Moscow Math. Soc. 26 (1972), 94-132.

PERCOLATION MODELS IN TWO AND THREE DIMENSIONS

R. T. Smythe
University of Oregon
Eugene, OR 97403

I. Introduction of the model.

Let Z^d, $d = 2,3$, denote the d-dimensional integer lattice. Call $x,y \in Z^d$ neighbors if $\|x - y\| = 1$ (here and in what follows $\|x - y\| = |x_1 - y_1| + |x_2 - y_2|$ for $d = 2$, with the corresponding definition for $d = 3$). Every pair of neighbors is joined by a bond. In the unoriented model, travel between neighbors can proceed in either direction along the bond; in oriented models some (or all) bonds permit travel in only one direction.

In the original percolation model each bond (independently of all other bonds) is open with probability p and closed with probability $1 - p$; travel proceeds in this model only along open bonds. An open cluster is defined as a maximal connected subgraph of the lattice with all bonds open. Let $P_\infty^d(p)$, $d = 2,3$, denote the probability that the origin belongs to an infinite open cluster; clearly $P_\infty^d(0) = 0$, $P_\infty^d(1) = 1$, and $p \to P_\infty^d(p)$ is non-decreasing. Hence there is a threshold value p_H^d, called the critical probability:

(1.1) $p_H^d \equiv \sup\{p : P_\infty^d(p) = 0\}$.

Below the value p_H^d only local travel is possible, but above p_H^d "the system percolates".

Let $|C|$ denote the number of bonds in the open cluster containing the origin. Evidently $E_p^d(|C|) = \infty$ for $p > p_H^d$. Define

(1.2) $p_T^d \equiv \sup\{p : E_p^d(|C|) < \infty\}$.

Then $p_T^d \leq p_H^d$. It is known (see, e.g., [12]) that

(1.3) (i) $p_H^2 \geq \frac{1}{2}$

(ii) $p_T^2 + p_H^2 = 1$

and obviously,

(iii) $p_T^2 \geq p_T^3$, $p_H^2 \geq p_H^3$.

Simulation suggests that $p_H^2 = \frac{1}{2}$, $p_H^3 \approx .25$ ([11]).

Define a self-avoiding path in the obvious way and let $f_n^d \equiv \text{card}\{\text{all self-avoiding paths of } n \text{ steps from the origin}\}$. Then

(1.4) $\lim_n (f_n^d)^{\frac{1}{n}} = \lambda_d$,

where λ_d is the connectivity constant of the lattice. It is known that ([11], [12])

(1.5) $p_T^2 \geq \lambda_2^{-1} \approx .375$

$p_T^3 \geq \lambda_3^{-1} \approx .22.$

The model I shall discuss is a generalization of the one above, and was introduced by Hammersley and Welsh (1965). Let $\{\ell_i\}$ denote the collection of bonds in the lattice and let $\{U_i\}$ be a collection of nonnegative, i.i.d. random variables; for the present we assume the U_i have finite mean. In biological terms, one may think of U_i as the time needed to transmit an infection between the two neighbors joined by the bond ℓ_i. For $x,y \in Z^d$ let

$R_{xy} \equiv \{$all self-avoiding paths from x to $y\}$,

and if $r \in R_{xy}$ let

(1.6) $t_r(\omega) \equiv \sum_{\{i:\ell_i \in r\}} U_i(\omega).$

Then

(1.7) $a_{xy}(\omega) \equiv \inf\{t_r(\omega) : r \in R_{xy}\}$

is called the <u>first-passage time</u> trom x to y. If $d = 2$, there always exists a path which achieves travel time a_{xy}; for $d = 3$, there is such a path provided that $U(0) < p_H^3$, where $U(0)$ is the atom at zero of the "time coordinate distribution" U. I will be particularly interested in

(1.8) $b_{on}(\omega) \equiv \inf_k\{a_{(o,o)(n,k)}(\omega)\}$ if $d = 2$

$\inf_{(j,k)} \{a_{(o,o,o)(n,j,k)}(\omega)\}$ if $d = 3$

$v_{on}(\omega) \equiv \inf\{a_{ox} : \|x\| = n\},$

where in the last definition o is the origin in either Z^2 or Z^3. Thus b_{on} is the point-to-line (or point-to-plane) first-passage time, and v_{on} is the first-passage time to the boundary of the diamond $\{\|x\| = n\}$. For biological applications it may be desirable to consider only distributions U for which $U(0) = 0$; this restriction eliminates much of the mathematical nastiness associated with the theory.

For simplicity let us assume that we start at $t = 0$ with a single infected site at the origin. Let

(1.9) $A_t(\omega) \equiv \{x \in Z^d : a_{ox}(\omega) \leq t\}$

be the collection of sites infected by time t; our problem, broadly speaking, is to describe the configuration A_t as completely as possible. Note that unless U is taken to be an exponential distribution, the evolution of the infection is non-Markovian.

For $t > 0$ define the "reach of the process at time t":

(1.10) $S_t^v(\omega) \equiv \sup\{\|x\| : x \in A_t(\omega)\} = \sup\{n : v_{on}(\omega) \le t\}$.

The random variable $\frac{S_t^v}{t}$ has been called the front velocity of the process. Define also

(1.11) $S_t^b(\omega) = \sup\{n : b_{on}(\omega) \le t\}$,

the "front at time t" in the x-direction.

I want to discuss in some detail two problems for these percolation models:

(1) the asymptotic behavior of the front velocities;

(2) the asymptotic shape of the infected region.

II. The Front Velocities.

Probably the original first-passage percolation model is that of Morgan and Welsh (1965), the Markovian model on the oriented square lattice (passage along bonds is permitted only in the north and east directions). (The unoriented Markovian model on Z^2 is a special case of a model of tumor growth put forth by Williams and Bjerknes (1972).) Morgan and Welsh conjectured that

(i) $E(\frac{S_t^v}{t})$ converges to a finite limit as $t \to \infty$;

(ii) $\frac{d}{dt} E(S_t^v)$ exists and equals twice the number of infected sites on the "front at time t" (i.e., the line $\|x\| = k$ if $S_t^v(\omega) = k$).

Both conjectures were later verified by Hammersley (1966). Consider the first conjecture for the general (unoriented) percolation model on Z^d. Recalling the definition of b_{on} and v_{on} in (1.8), we can deduce from the ergodic theory of subadditive processes ([2], [6], [12], [13]) that

(2.1) $\underline{d = 2}$: $\frac{b_{on}}{n} \xrightarrow{a.s.} \mu_2^b$ (a constant)

$\frac{v_{on}}{n} \xrightarrow{a.s.} \mu_2^v$ (a constant)

where $\mu_2^{b,v} > 0$ if $U(0) < p_T^2$

$= 0$ if $U(0) > p_T^2$ or if $U(0) = \frac{1}{2}$.

$\underline{d = 3}$: $\frac{b_{on}}{n} \xrightarrow{a.s.} \mu_3^b$ (a constant)

$\frac{v_{on}}{n} \xrightarrow{a.s.} \mu_3^v$ (a constant)

where $\mu_3^{b,v} > 0$ if $U(0) < p_T^2$

$= 0$ if $U(0) > p_H^3$.

It is easy to show, e.g., that $\frac{1}{2}\mu_2^b \le \mu_2^v \le \mu_2^b$, $\frac{1}{3}\mu_3^b \le \mu_3^v \le \mu_3^b$. There is reason to believe that $\mu^v < \mu^b$ except in the degenerate case but this has not yet been shown. (The results above for v_{on} do not appear in the references cited but can be deduced fairly easily from known results.)

Looking again at the definitions of S_t^v and S_t^b in (1.10) and (1.11), from the obvious analogy with standard renewal theory one has immediately that

$$(2.2)\quad \frac{S_t^v}{t} \xrightarrow{a.s.} \frac{1}{\mu_d^v}\ ,\quad \frac{S_t^b}{t} \xrightarrow{a.s.} \frac{1}{\mu_d^b} \quad \text{as } t \uparrow \infty.$$

It can further be shown that if $U(0) < p_T^d$, the convergence in (2.2) takes place in L^p as well for all $0 < p < \infty$. Thus the asymptotic value (and expected value) of the front velocities $\frac{S_t^v}{t}$ and $\frac{S_t^b}{t}$ is seen to be the reciprocal of the <u>time constants</u> μ_d^b and μ_d^v. For the oriented model we can say more; with an exponential time distribution (the Morgan-Welsh model), $E(\frac{S_t^v}{t}) \le \frac{1}{\mu^v}$ for <u>all</u> $t > 0$, since the function $t \to E(S_t^v)$ is then superadditive, while for a general distribution which is bounded above (by L, say), it is easily shown that (cf. Lemma 6.8 of [12]) $\sup_t E(\frac{S_{t-L}^v}{t}) \le \frac{1}{\mu^v}$.

Unfortunately, subadditive ergodic theory does not help in determining μ_d^b and μ_d^v. Obviously $\mu_3^{b,v} \le \mu_2^{b,v}$, and it can be shown that (except in the degenerate case $\mu_2^b < \underline{u}$, the mean of the distribution U. For certain distributions, upper bounds for $\mu^{b,v}$ can be found by various schemes ([5], [12], [13]; those for μ_d^v are new):

	exp(1)	$\mu_2^v \le .458$	$\mu_2^b \le .597$	$\mu_3^v \le .296$	$\mu_3^b \le .45$
(2.3)	U(0,1)		$\mu_2^b \le .425$		$\mu_3^b \le .377$
	Bernoulli (.25)	$\mu_2^v \le .485$	$\mu_2^b \le .64$	$\mu_3^v \le .34$	$\mu_3^b \le .54.$

Because of the convergence in (2.2), these upper bounds on $\mu^{b,v}$ give lower bounds on the asymptotic front velocities.

Upper bounds for the asymptotic front velocity are harder to obtain. Let $\xi_d^{v,b} = \frac{1}{\mu_d^{v,b}}$; in the case when $U(0) < \lambda_d^{-1}$, Hammersley (1966) got upper bounds on $\xi_d^v (= \lim_t \frac{S_t^v}{t})$:

(2.4)	exp(1)	$\xi_2^v \leq 6.16$	$\xi_3^v \leq 11.7$
	U(0,1)	$\xi_2^v \leq 7.22$	$\xi_3^v \leq 12.8$
	Bernoulli (.25)	$\xi_2^v \leq 10.0$	DNA
	Bernoulli (.10)	$\xi_2^v \leq 3.2$	$\xi_3^v \leq 6.6$.

A method for producing upper bounds on the front velocities was recently developed by Mollison (1978). This applies to the exponential distribution or to any distribution stochastically larger than an exponential distribution, i.e., any U such that $U(t) \leq 1 - e^{\alpha t}$ for some $\alpha > 0$ and all t. The method observes that the growth of a percolation process satisfying this condition is stochastically dominated by that of a contact birth process with density 1 at each site and nearest-neighbor contact distribution. The values ξ_d^b and ξ_d^v can be explicitly obtained for the CBP, giving an upper bound for the corresponding values in the percolation model. With an exp(1) time coordinate distribution, this gives

$$(2.5) \quad \xi_2^v \leq 6.03 \qquad \xi_3^v \leq 9.19$$
$$\xi_2^b \leq 4.47 \qquad \xi_3^b \leq 5.67.$$

Kesten (personal communication), using his results in [6], has developed a general method for finding lower bounds for μ_2^b, applicable to distributions U with $U(0) < p_T^2$. The resulting calculations are somewhat involved; preliminary computations for the exp(1) distribution give

$$(2.6) \quad \xi_2^b \leq 3.35,$$

an improvement over (2.5). The methods of Mollison and Kesten give the best lower bounds to date on the time constant for the exponential distribution, e.g.,

$$(2.7) \quad \mu_2^b \geq .2983, \ \mu_3^b \geq .176.$$

Again in the exponential case we may assert that $E(\frac{S_t^v}{t})$ and $E(\frac{S_t^b}{t})$ are bounded, for all $t > 0$, by the limiting value $\xi^{v,b}$.

Concerning conjecture (ii) of Morgan and Welsh for the general unoriented percolation model, the same work of Mollison shows that for time distributions which are exponential or stochastically larger than an exponential, $E(S_t^v)$ (and the higher moments of S_t^v as well) have finite time derivatives of all orders. For other distributions U it seems that no results of this type are known.

One would expect that a small change in the time coordinate distribution should product a correspondingly small change in the front velocity. Recent evidence for this is provided by a theorem of Cox ([1]):

(2.8) Let $U_n \xrightarrow{w} U$ (weak convergence), $U_n \geq V$ for all n, where V has finite mean.

If $d = 2$ and $U(0) \neq p_T^2$, $\mu_2^{v,b}(U_n) \to \mu_2^{v,b}(U)$. (If $p_T^2 = \frac{1}{2}$ no qualification is necessary.)

If $d = 3$ and $U(0) \notin [\lambda_3^{-1}, p_H^3]$, $\mu_3^{b,v}(U_n) \to \mu_3^{b,v}(U)$.

III. The asymptotic shape of the infected region.

The first substantial results in this direction were those of Richardson (1973), for $d = 2$. Richardson's results again applied (among other cases) to percolation models with time coordinate distributions stochastically larger than an exponential distribution. Partition the plane into unit squares, centered at the points of Z^2; for $x \in R^2$, let a_{ox} be the first-passage time (recall (2.7)) from (0,0) to the center of the square in which x lies. Using subadditivity, Richardson shows that there is a norm $N(x)$ on R^2 such that for any $\varepsilon > 0$,

$$(3.1) \quad P[\{x : N(x) \leq 1 - \varepsilon\} \subset t^{-1}A_t \subset \{x : N(x) \leq 1 + \varepsilon\}] \xrightarrow{t \uparrow \infty} 1.$$

Recent work on this problem has brought into focus the role of the assumption of moment conditions on the time coordinate distribution (recall that we have been operating under the assumption that U has finite mean, a most reasonable assumption from a biological point of view). Schürger (1979) showed that for $d \geq 2$, with any time coordinate distribution satisfying

$$(3.2) \quad U(0) < \lambda_d^{-1}, \int_0^\infty x^{1+\delta} dU(x) < \infty \quad \text{for some} \quad \delta > 0,$$

the strong version of Richardson's theorem holds, i.e., with probability one the inclusions in (3.1) hold for all t sufficiently large. (Vahidi ([13]) has noted that Kesten's recent result ([6]) permits the extension of Schürger's theorem to the case when $U(0) < p_T^d$, when $d = 2$ or 3.) Schürger also derives as an easy corollary that, a.s. for sufficiently large t, the convex hull of $t^{-1}A_t$ is included between $\{x : N(x) \leq 1 - \varepsilon\}$ and $\{x : N(x) \leq 1 + \varepsilon\}$.

Still more recently Cox and Durrett have extended Schürger's result significantly ([2]). Let Y be the minimum of the time coordinates of the $2d$ bounds with the origin as one vertex. Suppose first that $\mu_d > 0$ (which is always the case if $U(0) < p_T^d$). Then a necessary and sufficient condition for the strong version of Richardson's theorem (3.1) is that Y have a moment of order d (it is easy to see that this is implied by the assumption that U has a moment of order one-half). If $\mu_d = 0$ (cf. (2.1)) then $t^{-1}A_t$ will never contain $\{x : N(x) \leq 1 - \epsilon\}$, but in this case (again assuming Y has a moment of order d), given any compact set $K \subset R^d$,

(3.3) $P\{K \subset t^{-1}A_t$ for all t sufficiently large$\} = 1$.

If Y does not have a moment of order d, one cannot hope to get the first inclusion of (3.1), but the other inclusion is still valid: for any time coordinate distribution U,

(3.4) $P\{t^{-1}A_t \subset \{x : N(x) \leq 1 + \epsilon\}$ for all t sufficiently large$\} = 1$.

(Vahidi has shown that $\mu_3 = 0$ if $U(0) > p_H^3$ and U has a finite first moment; probably this is true without the moment assumption.)

Cox and Durrett also show that for d = 2, the a.s. convergence of $\frac{b_{on}}{n}$ and $\frac{v_{on}}{n}$ to constants as in (2.1) holds for any time coordinate distribution U (this improves a result of Wierman (1979), who showed it for the case when U has a moment of some order $\alpha > 0$). Thus the front velocity $\frac{S_t^{b,v}}{t}$ still converges a.s. to $\frac{1}{\mu_d^{v,b}}$ without any moment restriction, and all results of §2 continue to hold, at least for d = 2. For d = 3 it follows from the Cox-Durrett result that the front velocities converge when U has a moment of order one-half; if $U(0) < \lambda_3^{-1}$ and U has a moment of any positive order, Vahidi has extended Wierman's argument to show a.s. convergence still holds.

Almost all results on asymptotic shape given so far leave open the question of the true geometric shape determined by the norm N(x). Evidently the asymptotic shape is a diamond if the time coordinate distribution is degenerate; simulations suggest that the shape becomes rounder for more realistic distributions, but not quite circular. Recent work of Cox, Durrett and Liggett (1979) for the geometric distribution proves that, if d = 2 and p exceeds a certain critical probability, the limiting shape has a.s. a linear segment in its boundary.

References

1. Cox, J. Theodore (1979), The time constant of first-passage percolation on the square lattice, Adv. Appl. Prob. (to appear).

2. Cox, J. Theodore, and Durrett, Richard (1979), Some limit theorems for percolation processes with necessary and sufficient conditions, submitted to Ann. Prob.

3. Cox, J. Theodore, Durrett, Richard, and Liggett, Thomas (1979), personal communication.

4. Hammersley, J. M., (1966), First passage percolation, J. Roy. Stat. Soc. Ser. B 28, 491-496.

5. Hammersley, J. M., and Welsh, D.J.A. (1965), First passage percolation, subadditive processes, stochastic networks, and generalized renewal theory, Bernoulli-Bayes-Laplace Anniversary Volume, J. Neyman and L. M. LeCam, eds., Springer-Verlag.

6. Kesten, H. (1979), On the time constant and path length of first-passage percolation, Adv. Appl. Prob. (to appear).

7. Mollison, D. (1978), Markovian contact processes, Adv. Appl. Prob. 10, 85-108.

8. Morgan, R. W., and Welsh, D.J.A. (1965), A two-dimensional Poisson growth process, J. Roy. Stat. Soc. Ser. B 27, 497-504.

9. Richardson, D. (1973), Random growth in a tessellation, Proc. Camb. Phil. Soc. 74, 515-528.

10. Schürger, K. (1979), On the asymptotic geometrical behavior of percolation processes, J. Appl. Prob. (to appear).

11. Shante, V.K.S., and Kirkpatrick, S. (1971), An introduction to percolation theory, Adv. Phys. 20, 325-357.

12. Smythe, R. T., and Wierman, John C. (1978), First-passage percolation on the square lattice, Springer-Verlag Lecture Notes in Mathematics, Vol. 671.

13. Vahidi, M. (1979), First-passage percolation on the cubic lattice, Doctoral thesis, University of Oregon.

14. Wierman, John C. (1979), Weak moment conditions for time coordinates in first-passage percolation models (to appear).

15. Williams, T., and Bjerknes, R. (1972), A stochastic model for abnormal clone spread through epithelial basal layer, Nature 236, 19-21.

Lecture Notes in Biomathematics

Vol. 26: M. B. Katz, Questions of Uniqueness and Resolution in Reconstruction from Projections. IX, 175 pages. 1978.

Vol. 27: N. MacDonald, Time Lags in Biological Models. VII, 112 pages. 1978.

Vol. 28: P. C. Fife, Mathematical Aspects of Reacting and Diffusing Systems. IV, 185 pages. 1979.

Vol. 29: Kinetic Logic – A Boolean Approach to the Analysis of Complex Regulatory Systems. Proceedings, 1977. Edited by R. Thomas. XIII, 507 pages. 1979.

Vol. 30: M. Eisen, Mathematical Models in Cell Biology and Cancer Chemotherapy. IX, 431 pages. 1979.

Vol. 31: E. Akin, The Geometry of Population Genetics. IV, 205 pages. 1979.

Vol. 32: Systems Theory in Immunology. Proceedings, 1978. Edited by G. Bruni et al. XI, 273 pages. 1979.

Vol. 33: Mathematical Modelling in Biology and Ecology. Proceedings, 1979. Edited by W. M. Getz. VIII, 355 pages. 1980.

Vol. 34: R. Collins, T. J. van der Werff, Mathematical Models of the Dynamics of the Human Eye. VII, 99 pages. 1980.

Vol. 35: U. an der Heiden, Analysis of Neural Networks. X, 159 pages. 1980.

Vol. 36: A. Wörz-Busekros, Algebras in Genetics. VI, 237 pages. 1980.

Vol. 37: T. Ohta, Evolution and Variation of Multigene Families. VIII, 131 pages. 1980.

Vol. 38: Biological Growth and Spread: Mathematical Theories and Applications. Proceedings, 1979. Edited by W. Jäger, H. Rost and P. Tautu. XI, 511 pages. 1980.

This series reports new developments in biomathematics research and teaching – quickly, informally and at a high level. The type of material considered for publication includes:

1. Preliminary drafts of original papers and monographs
2. Lectures on a new field or presentations of new angles in a classical field
3. Seminar work-outs
4. Reports of meetings, provided they are
   a) of exceptional interest and
   b) devoted to a single topic.

Texts which are out of print but still in demand may also be considered if they fall within these categories.

The timeliness of a manuscript is more important than its form, which may be unfinished or tentative. Thus, in some instances, proofs may be merely outlined and results presented which have been or will later be published elsewhere. If possible, a subject index should be included. Publication of Lecture Notes is intended as a service to the international scientific community, in that a commercial publisher, Springer-Verlag, can offer a wide distribution of documents which would otherwise have a restricted readership. Once published and copyrighted, they can be documented in the scientific literature.

Manuscripts

Manuscripts should be no less than 100 and preferably no more than 500 pages in length.
They are reproduced by a photographic process and therefore must be typed with extreme care. Symbols not on the typewriter should be inserted by hand in indelible black ink. Corrections to the typescript should be made by pasting in the new text or painting out errors with white correction fluid. Authors receive 75 free copies and are free to use the material in other publications. The typescript is reduced slightly in size during reproduction; best results will not be obtained unless the text on any one page is kept within the overall limit of 18 x 26.5 cm (7 x 10½ inches). On request, the publisher will supply special paper with the typing area outlined.

Manuscripts in English, German or French should be sent to Dr. Simon Levin, 235 Langmuir, Cornell University, Ithaca, NY 14850/USA or directly to Springer-Verlag Heidelberg.

Springer-Verlag, Heidelberger Platz 3, D-1000 Berlin 33
Springer-Verlag, Neuenheimer Landstraße 28–30, D-6900 Heidelberg 1
Springer-Verlag, 175 Fifth Avenue, New York, NY 10010/USA

ISBN 3-540-10257-4
ISBN 0-387-10257-4